ADVANCES IN CHEMICAL PHYSICS

VOLUME LXI

EDITORIAL BOARD

Advances in
CHEMICAL PHYSICS

EDITED BY

I. PRIGOGINE

University of Brussels
Brussels, Belgium
and
University of Texas
Austin, Texas

AND

STUART A. RICE

Department of Chemistry
and
The James Franck Institute
The University of Chicago
Chicago, Illinois

VOLUME LXI

AN INTERSCIENCE® PUBLICATION
JOHN WILEY & SONS
NEW YORK · CHICHESTER · BRISBANE · TORONTO · SINGAPORE

An Interscience® Publication

Library of Congress Cataloging Number: 58-9935

ISBN 0-471-82055-5

Printed in the United States of America

10 9 8 7 6 5 4 3 2 1

CONTRIBUTORS TO VOLUME LXI

YEHUDA BAND, Ben Gurion University of the Negev, Department of Chemistry, Beer-Sheva, Israel

A. I. BOLDYREV, Institute of Chemical Physics, USSR Academy of Sciences, Moscow, USSR

JOEL BOWMAN, Department of Chemistry, Illinois Institute of Technology, Lews College of Science and Letters, Chicago, Illinois

KARL F. FREED, The James Franck Institute, Department of Chemistry, The University of Chicago, Chicago, Illinois

C. GEORGE, Service de Chimie-Physique II, Universite Libre de Bruxelles, Brussels, Belgium

G. L. GUTSEV, Institute of Chemical Physics, USSR Academy of Sciences, Chernogolovka, Moscow Region, USSR

F. MAYNE, Service de Chimie-Physique II, Universite Libre de Bruxelles, Brussels, Belgium

Y. OONO, Department of Physics, University of Illinois at Champaign-Urbana, Urbana, Illinois

I. PRIGOGINE, Service de Chimie-Physique II, Universite Libre de Bruxelles, Brussels, Belgium

SHERWIN J. SINGER, University of Pennsylvania, Department of Chemistry, Philadelphia, Pennsylvania

INTRODUCTION

Few of us can any longer keep up with the flood of scientific literature, even in specialized subfields. Any attempt to do more and be broadly educated with respect to a large domain of science has the appearance of tilting at windmills. Yet the synthesis of ideas drawn from different subjects into new, powerful, general concepts is as valuable as ever, and the desire to remain educated persists in all scientists. This series, *Advances in Chemical Physics*, is devoted to helping the reader obtain general information about a wide variety of topics in chemical physics, which field we interpret very broadly. Our intent is to have experts present comprehensive analyses of subjects of interest and to encourage the expression of individual points of view. We hope that this approach to the presentation of an overview of a subject will both stimulate new research and serve as a personalized learning text for beginners in a field.

ILYA PRIGOGINE

STUART A. RICE

CONTENTS

ADVANCES IN CHEMICAL PHYSICS

VOLUME LXI

PHOTODISSOCIATION OF DIATOMIC MOLECULES TO OPEN SHELL ATOMS

SHERWIN J. SINGER AND KARL F. FREED

The Department of Chemistry and The James Franck Institute
The University of Chicago, Chicago, Illinois 60637

and

YEHUDA B. BAND

The Department of Chemistry
Ben Gurion University of the Negev, Beer-Sheva, Israel

CONTENTS

I. INTRODUCTION

New insight into the nature of intermolecular forces can be gained from the study of photodissociation in which at least one of the fragments has nonvanishing electronic angular momentum. The molecule then dissociates to a multiplet of degenerate or nearly degenerate fragment electronic states arising from the different projections of fragment electronic angular momentum. The interpretation of populations and coherences among the fragment electronic states reveals details about the potential energy surfaces between open shell atoms or molecules, information that is difficult to obtain by other methods.

The prediction of state-to-state photodissociation cross sections to individual fragment electronic states involves complications that are not encountered when the photofragments are electronically structureless. Band, Freed, and Kouri[1,2] observe that the Born-Oppenheimer approximation must *always* fail upon dissociation to fragments with nonvanishing electronic angular momentum. In the case of diatomic molecule photodissociation, the mechanism of Born-Oppenheimer breakdown can be understood by considering a molecule whose constituent atoms are light enough to be described by Russell-Saunders coupling (see Fig. 1). The fragment electronic states belonging to a particular atomic term limit are slightly split in energy due to fine (spin-orbit) or hyperfine interactions. The atomic fine structure states are labeled by quantum numbers $j_a m_a$ and $j_b m_b$, total electronic angular momentum and projection on a space-fixed axis for atomic fragments a and b, respectively. A different molecular potential energy surface correlates with each magnetic sublevel $m_a m_b$. At large interfragment separation, these molecular electronic states lie close to each other in energy. Nonadiabatic interactions, although often weak, become increasingly important as molecular states approach a common atomic term limit. The most crucial region along

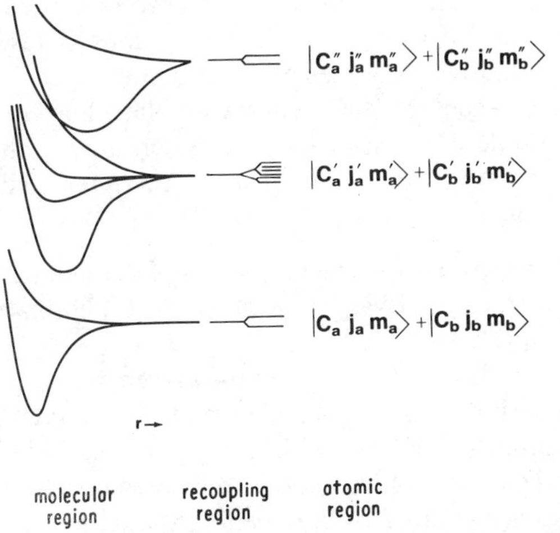

Figure 1. Schematic representation of molecular electronic potential energy curves for a diatomic molecule. The levels of each atomic limit are split by spin-orbit interaction (greatly exaggerated in the figures). A possible curve crossing in the molecular region is shown. Three regions along the internuclear coordinate r are differentiated according to the intramolecular coupling in each region.

the dissociation coordinate for the nonadiabatic processes described here is at large enough interfragment separation so that the several molecular states correlating with an atomic multiplet are near each other in energy, yet the interfragment distance is not so large that the fragments can be treated as separated atoms.

A. The Importance of Nonadiabatic Processes in Photodissociation

The bulk of experimental and theoretical work on photodissociation to date is concerned with the nuclear dynamics of the dissociation fragments, that is, state-to-state rotational and vibrational cross sections. The goal has been to understand state-to-state cross sections in terms of the potential energy surfaces for nuclear motion. (For reviews, see Refs. 3–7.) Use of a single energy surface for the nuclear dynamics as the fragments recede from one another rests on the assumption that the dissociation is electronically adiabatic. The Born-Oppenheimer approximation is often a very good one and the simplicity gained by following nuclear motion on just one potential energy surface is evident. However, there are several important motivations for extending our study of photodissociation to include the nonadiabatic transitions that occur when the fragments possess nonzero electronic angular momentum.

The first reason for treating nonadiabatic processes upon photodissociation is simply that neglect of these interactions leads to theoretical predictions in qualitative disagreement with experiment. This situation is most likely found when the total energy of the photofragments slightly exceeds the threshold for dissociation to an atomic limit with nonzero electronic angular momentum. Among the effects predicted[1,2,8] and illustrated by some model calculations[9-11] are:

1. Branching ratios to fine structure states of the same atomic term limit are nonstatistical and vary as a function of photon energy near the dissociation threshold.
2. The angular distribution of photofragments changes dramatically across the threshold region. Different fine structure states have different angular distributions. Interesting variation with photon energy and atomic multiplet state is also predicted for the polarization of fluorescence emitted by fragments produced in excited electronic states.
3. Photodissociation cross sections, angular distributions, and polarized fluorescence from excited atomic fragments reflect a wealth of resonance phenomena.

The resonance behavior arises from formation of quasibound levels due to the presence of a centrifugal barrier through which the fragments must tunnel to dissociate (shape resonance), or from the trapping of flux on electronic surfaces which are energetically closed to dissociation at near-threshold photon energies, but are coupled to unbound states by nonadiabatic interactions (Feshbach or Fano-type resonance). Nonstatistical branching ratios to fragment fine structure levels, as well as polarized fluorescence from excited fragments, have been observed experimentally in diatomic molecule photodissociation. To our knowledge there is not yet any experimentally confirmed observation of low-energy resonances in photodissociation to fragments with fine structure, although some may have been observed in CH^+.

Close-coupled calculations for photodissociation of NaH to the first excited atomic limit, $Na(3\,^2P) + H(1\,^2S)$, by Singer et al.[9,10] predict resonance peaks associated with dissociation along the $A\,^1\Sigma^+$ and $b\,^3\Pi$ potential energy surfaces of NaH. Besides demonstrating the importance of nonadiabatic interactions in calculating near-threshold cross sections, these calculations illustrate how electronic states that otherwise have no effect on the dissociation process can be observed as a consequence of nonadiabatic interactions. In the case of NaH, resonance peaks associated with the $b\,^3\Pi$ state appear in cross sections for dissociation from the singlet ground state by virtue of spin-orbit coupling among the molecular states that correlate with the Na(3 ^2P) + H(1 ^2S) atomic limit. Thus near-threshold dissociation offers the possibility of studying molecular potential energy surfaces that cannot be probed in ordinary bound-state spectroscopic techniques.

A coherent superposition of fragment atomic levels (e.g., the various m_j levels of an atomic fragment with total electronic angular momentum j) is produced in the photodissociation process. Inclusion of nonadiabatic effects is essential in predicting the distribution of fragment electronic states at all photon energies. The population and coherences among fragment electronic levels determine important observable properties such as fine structure state branching ratios, intensity and polarization of fluorescence emitted by excited photofragments, and, closely related to the latter, the response of the fragments to polarized light in a Laser Induced Fluorescence (LIF) or Multi-Photon Ionization (MPI) experiment. Therefore, another reason for studying nonadiabatic effects on dissociation in the presence of fragment electronic angular momentum is to establish methods by which the coherent superposition of fragment states can be characterized and calculated.

At photon energies near a threshold for dissociation a full numerical calculation is required to predict the energy variation of fine structure state branching ratios, polarization of fluorescence from photofragments, and other properties which depend on the superposition of electronic states of the dissociation products. At photon energies that are large in comparison with the magnitude of nonadiabatic couplings (see Section V for a more precise statement of this condition), the branching ratio, fluorescence polarization, and so on, does not vary with photon energy if the molecule is optically excited to only one dissociative molecular surface. In the large excess dissociation energy regime, a dynamical approximation which we call the diabatic or recoil limit allows calculation of branching ratios and fluorescence polarization. The recoil approximation is similar to the theory of Fano and Macek[12] for polarized emission from collisionally excited atoms. However, photodissociation requires a more complicated treatment than the collision problem because of the greater initial state selectivity possible with optical excitation.

To a degree, the traditional picture of photodissociation along a single electronic surface for nuclear motion is recovered in the recoil limit. However, populations and coherences among fragment electronic levels can be calculated using the recoil limit expressions, whereas these quantities are ignored in the traditional picture. This advantage is crucial because the electronic state distribution determines the fine structure state branching ratio and the polarization of fluorescence from each fragment level. If a single molecular surface is involved in the dissociative process, fragment angular distributions, fine structure state branching ratios, and polarization of fluorescence from electronically excited fragments can be determined with absolutely no dynamical calculation even for two-photon dissociation. These observable parameters are determined solely by the rotational quantum numbers and electronic symmetry of the molecular states involved in the dissociation. Explicit formulas by which these parameters are calculated, and some tabulated results are presented in Section IV. Depending on the initial state selectivity (as is possible in resonant two-photon dissociation), fine structure state branching ratios can be nonstatistical even at photon energies where the recoil approximation is valid.

Among the reasons provided in this introduction for studying photodissociation to fragments with nonvanishing angular momentum, we must not omit the great advantage of photodissociation as an experimental technique, namely, the state selectivity possible with optical excitation. With initial rotational state selection (achievable, for instance, by resonant two-photon dissociation, discussed later), the total angular momentum of the photofragments is fixed within one unit. Crossed molecular beam experiments are more problemmatic, especially when the atomic fragments are electronically excited. Crossed beam data is an average over all partial waves, or, in classical terms, impact parameters. This average is difficult to deconvolute for comparison of experimental data with theoretical calculations based on model molecular potential energy surfaces. In contrast, only a few values of total angular momentum need to be considered in a photodissociation experiment with initial state selectivity. The enhanced selectivity of the photodissociation process is equivalent to having the ability to examine a small subset of the partial waves involved in the collision problem.

Dissociation by resonant two-photon excitation of the molecule is emphasized in this review as a technique for enhancing initial state selectivity. Resonant two-photon, two-color dissociation enables greater experimental initial state selectivity relative to single-photon dissociation because the first transition between two bound states isolates a particular photo-excited sublevel which in turn is dissociated by the second photon. This overcomes the loss of information in one-photon dissociation due to the necessity of averaging over a thermal distribution of initial states. In addition, the intermediate state is prepared with a polarization equal to that of the first photon. The polarization of the intermediate state can thereby be varied in a precise manner. This enables the extraction of added information which is inaccessible in single-photon dissociation from an unpolarized initial ensemble, even if only a single initial rovibronic state is populated.

B. Basic Physical Ideas

The general scheme for dissociation of a diatom into light atom (*L-S* coupling) fragments is shown in Fig. 1. The various atomic term limits for each of the fragments are denoted by c_a and c_b. (Throughout this chapter, *a* and b refer to the two atomic fragments.) In Fig. 1 the first three atomic term limits (dissociation thresholds) are shown. We identify three regions along the dissociation coordinate, the *molecular region* at small internuclear separation r, the *atomic region* at large r, and the *recoupling region* which connects the molecular and atomic regions.

At small internuclear separations, the molecular region, the interaction between the two atoms is dominated by forces arising from the electronic interactions among all the electrons and nuclei, including exchange interactions. Because of the dominance of these forces, which are symmetric with respect to rotation about the nuclear axis, projections of angular momentum along the internuclear axis are nearly conserved quantum numbers in the molecular region. The system in the molecular region is best described by use of the adiabatic Born-Oppenheimer (ABO) approximation in which electronic motion is separated from nuclear motion and projections of angular momentum along the internuclear axis are conserved to lowest order. In general, the nonadiabatic coupling, which arises from nonseparability of electronic and nuclear motion plus the splittings due to relativistic interactions, is small and may be neglected or taken into account in the molecular region by perturbation theory. Exceptions occur when potential energy surfaces cross or nearly cross and these cases must be treated as multichannel problems.

When the fragments possess nonvanishing electronic angular momentum, each different projection of electronic angular momentum in the separated atom limit correlates with a different molecular electronic state. Nonadiabatic coupling may not be neglected between molecular states that approach the same atomic limit because these ABO molecular electronic states become energetically degenerate in the atomic limit, so the nonadiabatic coupling here eventually exceeds the separations between adiabatic curves. Molecular states having definite projections of angular momentum along the internuclear axis are not a proper basis for description of the fragments in large r atomic region. Coriolis interactions, which slowly vanish as r^{-2} as the fragments separate, couple molecular states with projections of angular momentum on the diatomic axis differing by one unit. Spin-orbit interaction, which persists at all interfragment separation, destroys total spin S as a good quantum number of the molecular states when these states lie very close in energy at large r.

Small nonadiabatic interactions like Coriolis or spin-orbit cause a recoupling of electronic angular momentum when the energy splitting between molecular states falls to the magnitude of the nonadiabatic coupling. Eventually projections of angular momentum along the internuclear axis are lost as good quantum numbers. In the large r atomic region they are replaced by the atomic fine structure state quantum numbers, total electronic angular momentum of each photofragment and its projection on a space-fixed axis. The range of r in which neither the molecular nor atomic description applies to the fragments, and in which the nonadiabatic processes are localized, is denoted as the recoupling region.

The theory presented in this chapter can treat all nonadiabatic interactions, including those that arise in curve crossings or avoided crossings in the molecular region. However, we focus on the nonadiabatic transitions that occur in the recoupling region because the effect is found in a wide class of molecules, and because these processes are amenable to systematic approximation in which the exact form of the molecular potential surfaces is not crucial. In contrast, the dynamics of crossings or avoided crossings at small r must be approached on a case-by-case basis with knowledge of the molecular potential energy surfaces.

C. Outline of this Review

In Section II we review experiments in which a diatomic molecule is photodissociated to fragments with nonzero electronic angular momentum. A considerable body of data about fine structure state branching ratios and polarizations already exists and is yet to be fully analyzed theoretically. The examples discussed in Section II are seen to be variants of the basic physical model just described.

The following three sections contain a theoretical analysis of photodissociation in the presence of nonadiabatic interactions. Section III is devoted to the various sets of basis functions which are used to expand the full Hamiltonian in the molecular and atomic regions. We review Hund's coupling cases (a), (b), and (c) with particular emphasis on the explicit form of the basis functions in each coupling scheme. We also present the orthogonal transformation coefficients by which the various molecular and atomic basis representations are interrelated.

The derivation of photodissociation cross sections in Section IV relies on a perturbative (Golden Rule) treatment of the matter-radiation field interaction. Equations presented in that section enable the calculation of cross sections for observable processes starting from transition amplitudes, which are matrix elements of the matter-radiation field interaction between bound and dissociative molecular states in the Golden Rule approximation. We derive cross sections for an extremely detailed experiment, namely angularly resolved detection of photofragments and fluorescence from excited fragments produced upon resonant two-photon dissociation of a diatomic molecule. These general expressions are conveniently reduced to describe simpler experiments, as for example, angular resolution of either photofragments or fluorescence but not both, integral cross sections, or single-photon dissociation. Detailed cross sections are given by very complex equations which we break into simpler subunits by use of a density matrix formalism. The density matrix formalism also allows our equations to be extended to treat photodissociation involving resonant absorption of an arbitrary number of photons.

We have already mentioned that dynamical information enters the cross section equations derived in Section IV through photodissociation transition amplitudes, dipole matrix elements between bound and dissociative molecular states. Methods by which the dynamical problem is solved in order to calculate transition amplitudes are reviewed in Section V. The discussion of diabatic or recoil approximations for transition amplitudes is extensive because these approximations are applicable in light atom diatomics away from the near-threshold energy region. Therefore a large portion of the photon energy spectrum can be treated in the diabatic or recoil approximation. The intramolecular coupling often changes as fragments recede from each other, for instance, from Hund's case (b) to (a) in light atom diatomics, or from case (b) to (c) in heavier diatomics. In this work we introduce new recoil limit approximations for molecules in which the coupling scheme changes within the molecular region (see Fig. 1).

II. REVIEW OF EXPERIMENTAL MEASUREMENTS

Many experimental investigations of photodissociation phenomena have been performed over the past quarter of a century. These studies have been directed toward the determination of thresholds for dissociation into particular fragment states, predissociation lifetimes, population distributions among possible final states, fragment angular and translational energy distributions, and fluorescence angular and polarization distributions arising from product emission. Some recent motivation has come from the prospect of discovering new lasers, improving laser isotope separation processes, and advancing our understanding of elementary processes in combustion and atmospheric physics.

A number of excellent reviews exist which discuss experimental studies of photodissociation of small molecules. Simons[13] presents an account of photofragment processes that contains a detailed description of experimental findings. Another outstanding review is in the book by Okabe,[5] and Ashford et al.[14] review vacuum ultraviolet photochemistry of small polyatomic molecules. Leone[6] presents a complete account of the photofragment dynamics of diatomic and simple polyatomic species. Dunbar[15] reviews ion photodissociation phenomena. Moseley and Durup[16] describe ion photofragment spectroscopy performed using fast ion beams, and Moseley[17] reviews a number of aspects of the photodissociation dynamics of ions. Here we discuss only those experiments relating to the influence of the existence of nonzero fragment angular momenta on diatomic photodissociation.

Powerful new light sources have hastened progress in the experimental probing of photodissociation dynamics. Tunable and frequency-doubled dye lasers, as well as rare-gas halide excimer lasers are now readily available, and vacuum ultraviolet synchrotron sources are greatly improved. These new lasers and new light detectors enable laser induced fluorescence studies in the infrared and visible to be performed on photoproducts at low enough pressures that collison processes can be neglected. Narrow-band lasers are available to probe the energy regions very near threshold for dissociation and to excite specific molecular rotational-vibrational states as the initial state prior to dissociation.

Several other technical advances have aided in probing the dynamics of molecular photodissociation. Time-of-flight spectrometers provide improved high-resolution translational and angular distributions, and extremely short excited state lifetimes can now be detected. Supersonic nozzle beams are available to cool the initial molecular states and thereby to characterize better the initial molecular states before photodissociation. Improved drift tube mass spectrometers and ion cyclotron resonance traps have aided the study of ion photofragment dynamics.

Perhaps the simplest type of experiment to study photodissociation involves using a light source with sufficient energy (typically UV or VUV radiation) to dissociate molecules in a dilute gas. Then the fluorescence from excited fragments is monitored, or molecules in the ground electronic state are probed with laser induced fluorescence. The gas pressure should be low enough to insure that a negligible fraction of excited states encounter a collision before radiation is emitted. The fluorescence angular distribution and polarization of excited fragments can be studied for varying polarizations of the incident light. These distributions provide information regarding the orientation of the transition dipole moment with respect to the molecular axis and the lifetime of the excited state.

A more detailed probe of photodissociation dynamics emerges from experiments that cross a light beam with a molecular beam and that measure angular distributions and fragment time of flight. Even more detailed information is obtained by resolving the fragment emission or laser induced fluorescence in a particular direction (and polarization) to determine the angular differential cross sections for production of specific atomic fragment states as a function of incident photon energy. Double differential cross sections (particle and fluorescence angular distributions) can be obtained from coincidence experiments. Analysis of this experimental data provides information about the ground and excited potential energy surfaces and the dipole transition moments. If nonadiabatic processes occur when the fragments recede from each other, these experiments can probe the properties of optically inaccessible states. Hence, molecular photodissociation experiments appear to be ideally suited for studying the interactions and collision dynamics of open shell atoms. Such information is not readily accessible by other experimental probes. In addition, initial rotational state selection is equivalent in the semiclassical approximation to the selection of impact parameters, a simplifying feature of analyzing photodissociation experiments that is not readily afforded by other experimental methods.

We now illustrate some experiments on diatomic molecule photodissociation. Emphasis is placed on experiments that have observed fine structure distributions or that appear to afford an excellent opportunity for their study. This discussion is certainly not exhaustive, and the reviews mentioned previously provide a more thorough account of specific systems that have been studied.

The best-understood molecule is H_2. Potential surfaces and transition dipole moments as a function of internuclear distance have been calculated theoretically and measured experimentally. Figure 2 shows a simplified potential energy diagram for H_2.[18] The threshold for photodissociation

$$H_2 + h\nu \rightarrow H(1s) + H(nl)$$

with $n = 2$ occurs at 14.9 eV (83.9 nm) and with $n = 3$ at 16.8 eV (74.4 nm). A

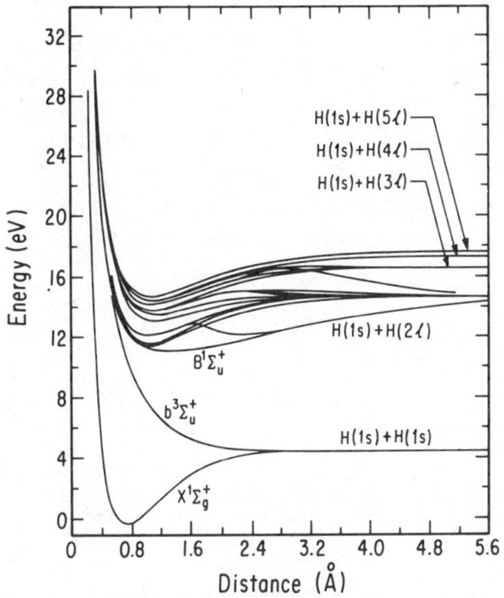

Figure 2. Potential energy curves for some of the lower electronic energy surfaces of H_2 taken from the data of T. E. Sharp.[18]

number of experiments have been performed to study the final state distributions of the excited hydrogen atom fragments. Mentall and Guyon[19] monitor the dissociation of H_2 near the threshold for $n = 2$. The contribution from the direct photodissociation on several electronic surfaces ($B\,^1\Sigma_u^+$, $C\,^1\Pi_u$, $B'\,^1\Sigma_u^+$) and a series of bands due to predissociation, produced in the P and R branch excitation of the $X\,^1\Sigma_g^+ \rightarrow D\,^1\Pi_u$ transition, are followed by observing the peak of the fluorescence cross section [$H(2p) \rightarrow H(1s) + h\nu$] of Lyman-$\alpha$ radiation near 83.9 nm. The $D\,^1\Pi_u$ states, produced by P and R branch excitation, are known to predissociate by nonadiabatic coupling through the $B'\,^1\Sigma_u^+$ state. The $D\,^1\Pi_u$ levels, that are produced by Q branch excitation, have opposite parity from the B' state and remain unperturbed. Thus the states excited in the Q-branch excitation decay by fluorescing back down to $X\,^1\Sigma_g^+$ state. Mentall and Guyon[19] have measured the hydrogen $(2\,^2P)/(2\,^2S)$ branching ratio produced in the photodissociation of H_2 at 83.9 nm using synchrotron radiation. The $2\,^2P_{1/2,\,3/2}$ states emit Lyman-α radiation, decaying to the $1\,^2S_{1/2}$ state, whereas the $2\,^2S_{1/2}$ state is metastable with a lifetime of 0.12 sec. The $2\,^2P_{3/2}$ state lies 0.365 cm^{-1} above the $2\,^2P_{1/2}$ state, but the $2\,^2S_{1/2}$ differs in energy from the $2P_{1/2}$ by only the Lamb shift (0.036 cm^{-1}). When an external electric field is applied, the ^2P and ^2S states are mixed. By monitoring the fluorescence as a function of the electric field strength, the percentage of atoms excited into $^2P_{1/2,\,3/2}$ and $^2S_{1/2}$ can be determined. Mentall and Guyon find that 57% of the excited atoms are in the ^2S state. If all the products result from adiabatic dissociation along the $B'\,^1\Sigma_u^+$, no ^2P atoms would be produced since the B' state correlates with $2\,^2S_{1/2} + 1\,^2S_{1/2}$ limit as discussed by Mulliken.[20] The substantial fraction of ^2P state fragments indicates that nonadiabatic interactions admix other electronic states. Moreover, the measured branching ratio is not at all statistical, so the dynamics leading to this branching ratio is nontrivial. The measured branching ratio is presented in Fig. 3 as a function of incident wavelength. Considerable structure is present in Fig. 3.

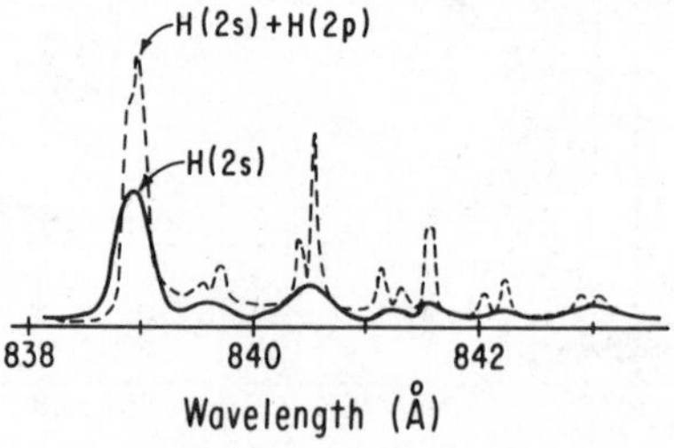

Figure 3. Variation of total photofragment fluorescence (dashed line) and production of H(2s) fragments (solid line) with excitation wavelength. The 2p/2s branching ratio is inferred from experimental data by Mentall and Guyon.[19]

A more complete analysis of the dependence of the Lyman-α intensity on the electric field strength can be made following the methods discussed by Gabrielse and Band[21] in order to determine the relative phase of the amplitudes for producing $2P$ and $2S$ atomic fragments. This relative phase is a very sensitive function of the dynamics.[21] The variation of the relative phase with incident laser frequency, as well as the angular distributions of atomic fragments and fluorescence polarization distributions, are quantities that would be of interest to compare with calculations in order to maximize the test of theory for this crucial example where *ab initio* results can be accurately computed.

The Balmer series emission has also been studied. Measurement have been made by Borrell et al.[22] and by Lee and Judge[23] of the branching ratios for producing H atoms with $n = 3, 4$, and 5 from the dissociation of H_2.

Oxygen photodissociation has been studied by Stone et al.[24] with 120 and 124 nm light, which is above the threshold at 7.047 eV (175.9 nm) for the dissociation to produce an excited oxygen atom

$$O_2 + h\nu \rightarrow O(^3P) + O(^1D)$$

They determine the oxygen atom kinetic energies using an atomic beam time-of-flight technique. The oxygen atom angular distributions are measured by varying the angle between the direction of the photon beam and the atomic beam flight path. Figure 4 shows the potential curves of O_2.[25] The O_2

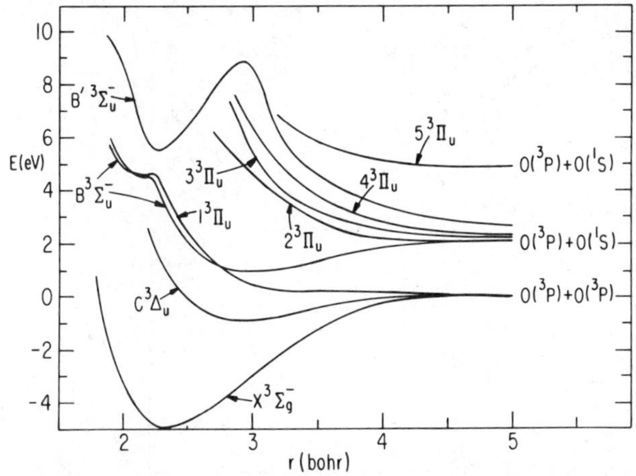

Figure 4. Potential energy curves for some triplet states of O_2 which may be important in photodissociation from the $X\ ^3\Sigma_g^-$ ground state. Only excited levels of u symmetry are shown. A large number of singlet and quintet electronic states, as well as additional triplet levels, which correlate with the $O(^3P) + O(^1S, ^1D, ^3P)$ atomic limits are omitted from this diagram. The potential curves shown here were sketched based on *ab initio* data from several calculations.[25]

angular distribution is found to be consistent with the assumption that predissociation of the $B\,^3\Sigma_u^-$ states is important, and an average lifetime of 2×10^{-13} sec is attributed to the predissociated levels. Lee et al.[26] observe the wavelength dependence of the quantum yield of $O(^1D)$ production in the photon wavelength region between 116 and 177 nm. Below 139 nm the $O(^1D)$ quantum yield is very strongly dependent on wavelength, whereas above 139 nm it is unity until a sharp cut-off of $O(^1D)$ production appears at 175 nm, the point at which the probability of producing $O(^3P) + O(^3P)$ becomes large. This system is very rich in complexity, because of the large number of fine structure states of the atomic fragments present, and should be a fertile hunting area for interesting phenomena. But work with easier systems may be desirable at first.

Alkali dimers present theoretically simpler systems in some respects. Figure 5 shows the Born-Oppenheimer adiabatic potential surfaces of Na_2 taken from the *ab initio* calculations of Konowalow et al.[27] The threshold for photodissociation into $Na(^2S) + Na(^2P)$ is at 22,952 cm^{-1}. Rothe et al.[28] measure the polarization of the $^2P_{3/2}$ fluorescence. Na_2 fragmentation above

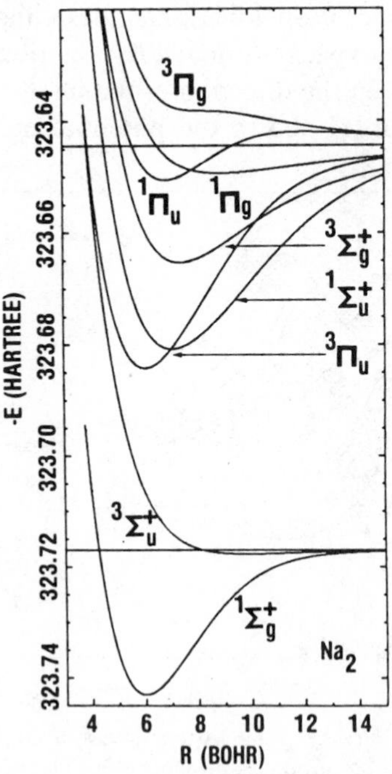

Figure 5. Potential energy curves for the electronic states of Na_2 which correlate with the $Na(^2S) + Na(^2S)$ and $Na(^2P) + Na(^2S)$ atomic limits. These are results of the *ab initio* calculations of Konowalow et al.[27]

this threshold produces excited atoms predominantly in the $^2P_{3/2}$ state. The same trend is also found in K_2, Cs_2, Rb_2 dissociation, as observed by Kraulings and Yanson.[29] Also supersonic nozzle experiments of Feldman and Zare[30] find a preferential production of the $^2P_{3/2}$ component of Rb from predissociation of Rb_2 at 476.5 nm. Bredford and Engelke[31] excited Rb_2 to the $C\,^1\Pi_u$ and $D\,^1\Pi_u$ bands and find Rb*(5 $^2P_{3/2}$) from the C state, and predominantly Rb*(4 $^2D_{3/2}$) from the D state, followed by Rb ($^2P_{1/2}$ and $^2P_{3/2}$) radiation. This suggests different predissociation processes for the C and D states.

The homonuclear alkali dimers have an r^{-3} long range resonance interaction for their states dissociating to $^2S + {}^2P$ limits which makes them qualitatively different in certain respects from the heteronuclear alkali dimers which have been studied by Bredford and Engelke.[31] Laser Induced Fluorescence studies have determined accurate values for the $X\,^1\Sigma^+$, $D\,^1\Pi$, $C\,^1\Sigma^+$, and $a\,^3\Sigma^+$ potential curves of NaK (see Bredford and Engelke,[32] McCormack and McCaffery,[33] Eisel et al.[34]). Also accurate potential curves for NaH have been calculated by Sachs et al.[35] Calculations on Na_2 photodissociation are discussed in Section VI along with ones for NaH. No resonance interactions arise for NaH photodissociation, and the reduced mass is higher, leading to some qualitative differences in its photodissociation dynamics.

Photodissociation processes in diatomic halogen molecules have been intensely studied. The lowest lying potential curves of IBr and I_2 are presented in Figs. 6 and 7 and show the general trends present in many of the

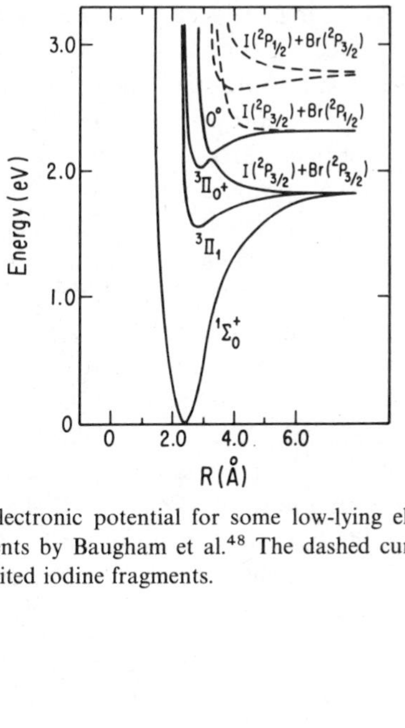

Figure 6. Approximate electronic potential for some low-lying electronic states of IBr, as constructed from experiments by Baugham et al.[48] The dashed curves represent unobserved states of IBr leading to excited iodine fragments.

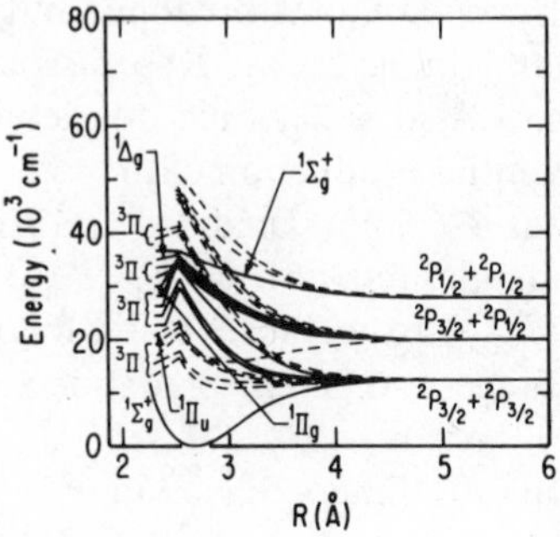

Figure 7. Potential energy curves for the valence states of I_2 estimated by Mulliken.[40]

dihalogens. Child and Bernstein[36] discuss the lowest-lying potential curves of the dihalogens and their observed dissociation products. Many time-of-flight and angular distribution experiments have been reported. Wilson[37] has studied the two-photon dissociation of I_2 and Sander and Wilson[38] use a double-laser pulsed technique to dissociate the *B* state of I_2. Burde et al.[39] study I_2 in the energy region near threshold for dissociation to $I^*(^2P_{1/2}) + I(^2P_{3/2})$ using a tunable laser source. A number of photodissociation experiments have been performed for the Br_2 molecule[41–47] as well as IBr[42,48,49] Furthermore, Clyne and McDermid[50] have studied the predissociation of the *B* state of Cl_2. Krasnoperov and Paniflov[51] note the preferential production of the excited $Cl(^2P_{1/2})$ over the Cl $(^2P_{3/2})$ state in the photodissociation of IC1 using the second harmonic of the Nd:Yag laser at 532 nm. In all of these molecules the spin-orbit interaction is sufficiently large that it is unreasonable to treat spin-orbit as a perturbation. Hund's case (c) coupling must be used as the appropriate zero-order basis. As a general rule, the electronic potential surfaces in I_2, Br_2, and IBr are far enough apart at large internuclear distances that nonadiabatic effects due to the nuclear kinetic energy coupling are unimportant at these large distances.

The study of metal halide molecules has been spurred by the potential for developing atomic and Raman lasers. Bersohn[52] reviews much of the work on the photodissociation of metal halides in a systematic and comprehensive manner. Figure 8 shows the lowest-lying potential curves of HI as a function of internuclear distance. Similar trends appear for many metal halides. Clear et al.[53] photodissociate HI and DI using the fourth harmonic of the Nd:Yag laser (at 226 nm). They find 36% of the iodine in the excited $I^*(^2P_{1/2})$ state from the dissociation of HI, but only 26% from DI. The effect of reduced mass on the importance of nonadiabatic transitions is discussed in Section V. Clear et al.[53] analyze the photofragment angular distributions to determine that the

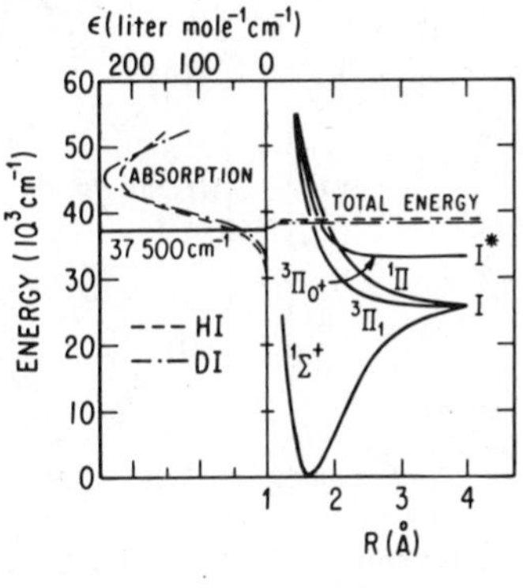

Figure 8. Absorption spectra and calculated potential energy surfaces for HI (dashed curves) and DI (dotted curves) from Clear et al.[53]

transition to H + I* is a parallel one, whereas the transition to H + I is perpendicular. This can be understood based on the knowledge that the transition $^3\Pi_0^+ \leftarrow {}^1\Sigma_{0^+}$ is parallel and the transitions $^3\Pi_1 \leftarrow {}^1\Sigma_{0^+}$, $^1\Pi \leftarrow {}^1\Sigma_{0^+}$ are perpendicular. Su and Riley[54] examine the photodissociation of LiI, NaI, KI, RbI, and CsI at 266 nm. They find that all except LiI yield $I^*(^2P_{1/2})$ exclusively, whereas LiI produces both ground and excited state iodine atoms. This result reinforces the view that the reduced mass scales the importance of nonadiabatic interactions and can be viewed, in part, as a nonadiabaticity parameter. Su and Riley[55] observed branching ratios Br*/Br of 3.0, 2.5, 5.7, and 0.8 from NaBr, KBr, RbBr, and CsBr photodissociation, respectively, and only ground-state bromine from LiBr. The angular distributions indicate a 50% mixture of parallel and perpendicular transitions for the heavier bromides and a 29% parallel character for the transition to the ground-state Br atom from LiBr.

Considerable interest exists in producing Raman lasers based on stimulated anti-Stokes scattering from excited atoms to up-convert the frequency of a pump laser. The development of this type of laser is difficult because of the need for creating a large metastable population inversion (see the I_2 case, Fig. 7). The first coherent anti-Stokes emission has been produced by Sorokin et al.[56] in K vapor, where a transient population inversion between $K(4p\,^2P^0_{3/2})$ and $K(^2P_{1/2})$ is created by optical pumping. White and Henderson[57] have made a tunable VUV laser using a metastable $I^*(^2P_{1/2})$ inversion resulting from the photodissociation NaI + $h\nu$(248 nm) → Na + $I^*(^2P_{1/2})$. The pump laser at 206 nm is used to drive the Raman process laser emission resulting from the anti-Stokes process. NaI is chosen as the iodine atom donor because the Na liberated atoms do not strongly absorb

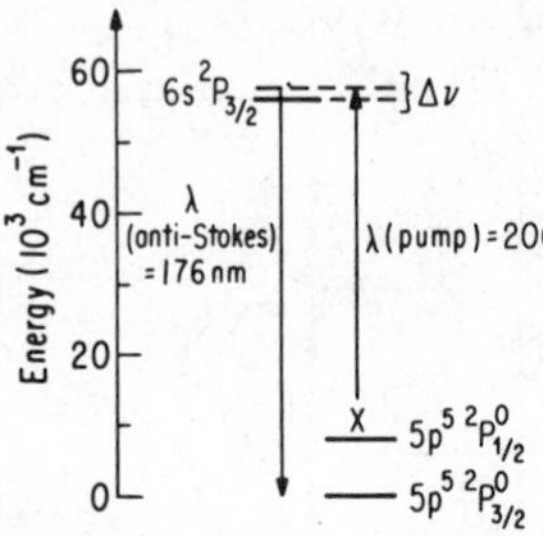

Figure 9. Anti-Stokes Raman lasing scheme of White and Henderson[57] using inverted I($5p^2\ ^2P_{1/2}$) atoms. The pump laser is detuned by 5–6 cm^{-1} from the $5p^2\ ^2P_{1/2} \rightarrow 6p^2\ ^2P_{3/2}$ resonance line, resulting in a tunable anti-Stokes Raman laser at 178 nm.

206 or 178 nm radiation since these wavelengths fall within a Cooper minimum of Na. Another such anti-Stokes laser has been developed by White and Henderson[58] using an inverted population of Tl*($6p^2\ ^2P_{3/2}$) produced by the photodissociation process TlCl + λ(248 nm) → Tl*($^2P_{3/2}$) + Cl driven by a KrF* excimer laser. Raman upconversion is carried out with tunable dye laser operating near the Tl($6p^2\ ^2P_{3/2}$) → Tl($7s^2\ S_{1/2}$) at 532 nm. The Raman scattering by the Tl*($6p^2\ ^2P_{3/2}$) photoproduct produces UV light near the Tl($7s^2\ S_{1/2}$) → Tl($^2P_{1/2}$) transition at 376 nm. Rather wide tunability is insured because the frequency mismatch $\Delta\omega$ (see Fig. 9) can be quite large and yet provide sufficient gain. Lüthy et al.[59] have studied the possibility of a very efficient two-photon dissociation of TlI to produce a thallium laser emitting at 535 nm.

Photofragment spectroscopy of ions has yielded information on the photodissociation dynamics of H_2^+, CH^+, O_2^+, Ar_2^+, NH^+, PH^+, SH^+, NO^+, and N_2^{2+}. The reader is referred to the comprehensive reviews by Moseley and Durup,[16] and Moseley[17] for details. In Section VI, special attention is given to CH^+ photodissociation, partly because of the important astrophysical implications of this process and partly because the system potential curve and transition moments are well enough known to allow comparison of theory and experiment. Some of the photodissociation experimenta on CH^+ are described in conjunction with the theoretical calculations in Section VI.

III. MOLECULAR AND ATOMIC BASIS FUNCTIONS

In this section we begin to develop the theoretical apparatus necessary to predict nonadiabatic processes that occur during photodissociation. One requirement of the theory is to provide an exact description of the photodissociation dynamics given the relevant nonadiabatic interactions.

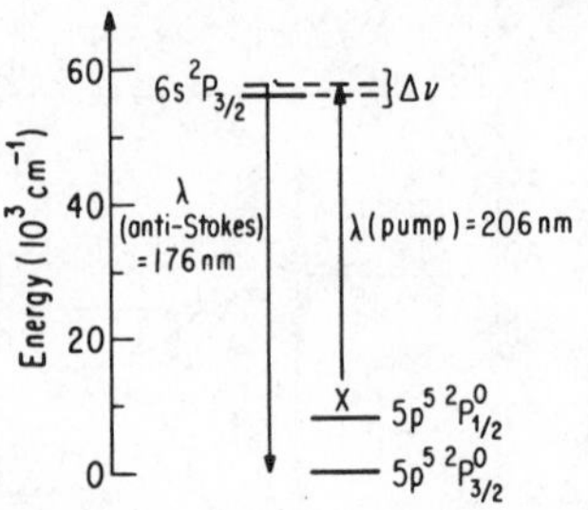

More importantly, we seek a theory that provides a vocabulary by which the qualitative physical ideas introduced in the introduction may be analyzed and used as a basis for dynamical approximations. Specifically, these qualitative ideas are based on the separation of the dissociation coordinate into several regions (Fig. 1): (1) a *molecular region* at small internuclear separation r in which the photofragments can still be usefully described in the same manner as the stable molecule, (2) an *atomic region* in which the fragments behave essentially as separated atoms, and (3) a *recoupling region* between the molecular and atomic regions in which neither the molecular nor atomic description is very successful and in which nonadiabatic effects are important.

We present several basis sets for expansion of the electronic and nuclear rotational coordinates in the full wave function. The various representations are related by orthogonal transformations (given later), and any one of them can be used to provide an exact description of photodissociation dynamics. The basis sets are grouped into two categories. The first group, the molecular basis functions, are likely to provide a good zeroth order description in the molecular region. In contrast, the atomic basis functions are eigenstates of the separated fragments at large r, while at small r the atomic basis functions are complicated linear combinations of molecular basis states. Nonadiabatic processes are treated in Section V, where the utility of the molecular and atomic representations described here becomes fully apparent.

Each of the molecular basis sets we introduce corresponds to a differing Hund's coupling case for the diatomic molecule. Both bound and dissociative levels are expressed as linear combinations of primitive Hund's case (a), (b) or (c) molecular basis functions. These basis sets are related by orthogonal transformations which are explicitly given below. Use of a particular primitive Hund's case basis has no effect on predicted photodissociation cross sections so long as all states of a particular set are included in the expansion of the total wave function. Matrix elements of the total diatomic molecule Hamiltonian between Hund's case basis functions are easily evaluated because several quantum numbers, which describe the angular motion of electrons and nuclei, are well defined for each Hund's case.

We exhibit the theory in terms of several different primitive molecular bases because each Hund's case basis set forms a convenient point of departure for generating approximate bound and continuum wave functions. In limiting situations, determined by the relative magnitude of the splitting between adiabatic Born-Oppenheimer (ABO) energy levels, and the Coriolis or spin-orbit coupling between ABO levels, the total molecular wave function is approximated by a single Hund's case basis function instead of a linear combination.

This section provides a detailed description of the molecular basis functions that correspond to each of the Hund's coupling cases. Symmetry properties of these basis states are given, as well as the transformations that interrelate each of the Hund's case bases. We also exhibit the r-independent transformation by which eigenstates for the separated atoms are constructed from each of the molecular basis functions at large r. Although the various Hund's cases have been discussed by many authors,[60-66] it is usually with the goal of evaluating multiplet splittings in the molecule as opposed to our interest in describing nonadiabatic effects in molecular photodissociation. The symmetry properties of the matrix elements and the molecular states determine the secular determinant which must be solved to obtain multiplet energy levels. Hence, the individual basis functions themselves are not always explicitly analyzed in the bound state work as we do here. Our treatment of the coupling of molecular basis functions to form atomic eigenstates relies heavily on the work of Mies,[67] Weisheit and Lane,[68] and Launay and Roueff.[69]

We use the notation $|JM\eta\rangle$ to designate a molecular eigenstate. Total angular momentum and its projection on a space-fixed axis (quantum numbers J, M) are always conserved in each of the Hund's cases as well as in the wave function that describes the separated atomic fragments. The symbol η stands for all other quantum numbers for the molecule, including inversion symmetry. In the lowest order of approximation, η is a set that contains separate electronic, vibrational, and nuclear rotational quantum numbers. Usually the true molecular eigenstates can still be well approximated by a zero-order state and thus η can be designated by the electronic, vibrational, and nuclear rotational quantum numbers of the zero-order state. When coupling between these degrees of freedom is so strong that the true molecular eigenstates can no longer be identified by separate quantum numbers for each degree of freedom, the symbol η simply labels the eigenstates $|JM\eta\rangle$ obtained by diagonalizing the full Hamiltonian for the diatom.

A. Hund's Case (a) Representation

The molecular eigenstate $|JM\eta\rangle$ is expanded in a Hund's case (a) basis for rotational and electronic degrees of freedom.

$$|JM\eta\rangle = \sum_{\Lambda\Sigma} a(\eta|\Lambda\Sigma)\frac{1}{r}\chi_{Jv\eta}(r)|JMc\Lambda S\Sigma\rangle. \tag{1}$$

$\chi_{Jv\eta}(r)$ is a vibrational wave function, or when combined with the expansion coefficient $a(\eta|\Lambda\Sigma)$, forms the radial wave function matrix for a continuum molecular wave function. The case (a) basis states for electronic and rotational coordinates have the explicit form,

$$|JMc\Lambda S\Sigma\rangle = \left(\frac{2J+1}{4\pi}\right)^{1/2} D^{J*}_{M\Omega}(\alpha\beta 0)|c\Lambda S\Sigma\rangle. \tag{2}$$

The internuclear vector **r** has magnitude r and orientation $\hat{\mathbf{r}}$, described by the Euler angles (α, β, 0), with respect to a space-fixed (SF) coordinate system. $D^J_{M\Omega}(\alpha\beta 0)$ is a Wigner rotation matrix,[70,71] and $|c\Lambda S\Sigma\rangle$ is an adiabatic Born-Oppenheimer (ABO) electronic state whose parametric dependence on r is not explicitly shown. The index c serves to distinguish molecular electronic states with the same values of $|\Lambda|$ and S. (For instance, if the atomic term limits are ^1P for two nonidentical atoms, the channel index c differentiates the two $^1\Pi$ electronic states that correlate with this limit.)

The $a(\eta|\Lambda\Sigma)$ are eigenvector coefficients of the primitive case (a) multiplet states obtained by diagonalization of the multiplet Hamiltonian in the case (a) bases. When constructed in the case (a) basis, the off-diagonal matrix elements of this Hamiltonian include spin-orbit coupling between states of different S but identical $\Omega = \Lambda + \Sigma$, Coriolis interaction between states with the same S but Ω differing by ± 1, and radial derivative coupling between states with the same, Λ, S and Σ. The derivative coupling arises from action of the kinetic energy operator on the ABO wave function $|c\Lambda S\Sigma\rangle$. At small r, the largest contribution to the diagonal matrix elements are the Born-Oppenheimer electronic energies $\varepsilon_{c|\Lambda|S}(r)$. Coriolis and spin-orbit interactions also contribute to the diagonal matrix elements. The nonadiabatic couplings discussed above are diagonalized by forming the linear combination specified by the $a(\eta|\Lambda\Sigma)$. In principle, the sum in Eq. (1) must extend over all electronic states labeled by c and all vibrational levels labeled by v in order to exactly treat the coupling between electronic, vibrational, and rotational motion. Often this coupling is weak, so only intramultiplet coupling among the $(2S+1)(2-\delta_\Lambda 0)$ states with $\Lambda = \pm|\Lambda|, \Sigma = -S, -S+1, \ldots, S$ are indicated here, with small off-resonance intermultiplet contributions assumed to be incorporated into the $\varepsilon_{c|\Lambda|S}$.[63] Our treatment can be trivially generalized to include all types of coupling by writing the expansion coefficients in Eq. (1) as $a(\eta|c\Lambda S\Sigma v)$; we use the form of Eq. (1) for notational convenience.

There is a certain redundancy in Eq. (1). On one hand, the $a(\eta|\Lambda\Sigma)$ combined with the radial wave $\chi_{Jv\eta}(r)$ function may be regarded as r-dependent coefficients of basis functions $|JMc\Lambda S\Sigma\rangle$ [Eq. (2)] for electronic and rotational coordinates. Continuum wave functions are generally written in this form. The r-dependent coefficients obtained by combining $\chi_{Jv\eta}(r)$ and $a(\eta|\Lambda\Sigma)$ are organized as the rows of a matrix which multiplies a vector whose elements are the basis functions $|JMc\Lambda S\Sigma\rangle$. Each row represents an independent wave function. The label η for continuum states differentiates the rows of the wave function matrix. On the other hand, $\chi_{Jv\eta}(r)$ and $a(\eta|\Lambda\Sigma)$ are usually written separately in a bound-state wave function. This is because the $a(\eta|\Lambda\Sigma)$ are first determined by solving a secular determinant among electronic-rotational basis states $|JMc\Lambda S\Sigma\rangle$ to obtain potential energy surfaces for vibrational motion. The eigenvalues of the secular determinant are the energies of the multiplet levels $|JM\eta\rangle$, and a different potential energy surface is found for each η. Therefore, vibrational wave functions $\chi_{Jv\eta}(r)$ must be designated by a multiplet index η as well as the vibrational quantum number v. We retain the ambiguity in expressions like Eq. (1) to avoid separate discussions for bound and dissociative molecular eigenstates.

The state $|JM\eta\rangle$ is an eigenstate with respect to inversion of all coordinates in the space-fixed (SF) frame with parity eigenvalue $(-)^p$. States with definite inversion symmetry are formed by taking linear combinations of basis states with $\pm\Lambda$, $\pm\Sigma$.[61,72,73] Therefore the relative phase of the expansion coefficients $a(\eta|\Lambda\Sigma)$ and $a(\eta|-\Lambda-\Sigma)$ determine the parity of the molecular state. Molecular eigenstates $|JM\eta\rangle$ with parity eigenvalue $(-)^p$ are formed when the expansion coefficients have the following phase relation,[8,72,73]

$$a(\eta|\Lambda\Sigma) = (-)^{p+J-S+\sigma}\, a(\eta|-\Lambda-\Sigma) \tag{3}$$

The symbol σ is equal to 1 if the electronic state has Σ^- electronic state symmetry and is zero otherwise.

Relation (3) is demonstrated[8] by noting that inversion of the coordinates of all electrons and nuclei in the diatomic molecule, as referred to a space-fixed (SF) coordinate system, is equivalent to the following two operations:

1. Rotation of the body-fixed (BF) coordinate frame by an angle π about the BF y-axis. This operation is denoted by $C_2(y_{\mathrm{BF}})$, and its effect on the case (a) basis functions in Eq. (2) is to replace the Euler angles (α, β, γ) by $(\pi+\alpha,\ \pi-\beta,\ \pi-\gamma)$. Under this transformation, the Wigner rotation becomes (Ref. 70, Appendix V)

$$C_2(y_{\mathrm{BF}})D^{J*}_{M\Omega}(\alpha\beta\gamma) = (-)^{J-\Omega}D^{J*}_{M-\Omega}(\alpha\beta\gamma) \tag{4}$$

2. Reflection of the BF coordinate system through the BF xz-plane. This operation, designated by $\sigma(xz_{\mathrm{BF}})$, acts on those parts of the wave function [Eq. (1)] that are functions of coordinates referred to the BF axes, namely, the vibrational and electronic basis functions. The vibrational wave functions are symmetric under this transformation because they only depend on r. Reflection of the electronic coordinates (Kronig symmetry) has the effect,

$$\sigma(xz_{\mathrm{BF}})|c\Lambda S\Sigma\rangle = (-)^{S-\Lambda-\Sigma+\sigma}|c-\Lambda S-\Sigma\rangle \tag{5}$$

Equation (5) implies a definite phase convention for the electronic wave functions when $\Lambda \neq 0$: We assume the state $|c\Lambda S\Sigma\rangle$ is, in principle, built from combinations of atomic orbitals whose spacial parts have the same phase conventions as the spherical harmonics of Condon and Shortley. (This point is discussed by Y.-N. Chiu[72, 73] and Singer et al.[8]) It is easy to follow this phase convention consistently through our calculations; consistency is all that is ultimately relevant about the choice of phase convention.

In order to further clarify our notation, we exhibit the expansion coefficients for a pure Hund's case (a) molecular eigenstate with parity $(-)^p$ and multiplet quantum number Ω,

$$a(\eta|\Lambda\Sigma) = (-)^{p+J-S+\sigma} a(\eta|-\Lambda-\Sigma) \tag{6a}$$

$$|a(\eta|\Lambda\Sigma)| = \delta_{\Omega,|\Lambda+\Sigma|}(2 - \delta_{\Lambda 0}\,\delta_{\Sigma 0})^{-1/2} \tag{6b}$$

The square root in Eq. (6b) is the normalization factor which must be included when the wave function is a sum of terms with $\pm\Lambda$, $\pm\Sigma$. Pure Hund's case (a) states with different Ω are split by the spin-orbit interaction, which is approximately given as[60]

$$H_{SO} = A\Lambda\Sigma \tag{7}$$

In general, states with Ω differing by at most ± 1 are coupled by Coriolis forces, which are neglected in the pure case (a) limit [Eq. (6)] where state of different Ω are not mixed. Otherwise the full sum over Ω, as in [Eq. (1)], must be retained.

B. The Asymptotic Large Distance Limit

The relation between the case (a) coupling scheme basis functions [Eq. (2)] at large r and eigenstates for the separated atomic fragments must be known to set up the theory for molecular dissociation. This is because Coriolis and spin-orbit couplings between case (a) basis states can persist even at r large enough so that a separated atom description of the fragments is valid. This crucial point is demonstrated explicitly below when we exhibit matrix elements of the full Hamiltonian.

At large r the spatial part of the electronic wave function in Eq. (2) factors into a sum of products of wave functions for each atomic fragment. The spin can be taken to remain coupled as in the molecule,

$$|c\Lambda S\Sigma\rangle = \sum_{\Lambda_a\Lambda_b(\Lambda_a+\Lambda_b=\Lambda)} |c_a l_a \Lambda_a\rangle |c_b l_b \Lambda_b\rangle \langle \Lambda_a \Lambda_b | c\Lambda\rangle |S\Sigma\rangle, \; r \to \infty \tag{8}$$

The subscripts a and b in Eq. (8) refer to the two atomic fragments. Here we assume the atomic term limit can be described by Russell-Saunders coupling in both fragments. In this case, the term limit is specified by total electronic orbital (l_a and l_b) and spin (s_a and s_b) quantum numbers for each fragment. In Eq. (8) Λ_a and Λ_b are the projections of orbital electronic angular momentum for each atom on the internuclear axis. The indices c_a and c_b distinguish atomic term limits with the same values of $l_a s_a$ or $l_b s_b$.

The orbital part of the molecular electronic wave function $|c\Lambda S\Sigma\rangle$ becomes a sum of degenerate product states $|c_a l_a \Lambda_a\rangle|c_b l_b \Lambda_b\rangle$ in Eq. (8) with coefficients $\langle\Lambda_a\Lambda_b|c\Lambda\rangle$. If one of the atoms is in a S-state (l_a or l_b is zero), then the sum in Eq. (8) consists of a single term. For instance, if $l_b = 0$, then $\langle\Lambda_a\Lambda_b|c\Lambda\rangle = \delta_{\Lambda\Lambda_a}$. Otherwise the coefficients $\langle\Lambda_a\Lambda_b|c\Lambda\rangle$ are determined by noting that the ABO wave function $|c\Lambda S\Sigma\rangle$ is an eigenstate of the *clamped nuclei* electronic Hamiltonian $H_{\text{elec}}(r)$ at all r, and the $\langle\Lambda_a\Lambda_b|c\Lambda\rangle$ are the eigenvector coefficients of $H_{\text{elec}}(r)$ in the product basis $|c_a l_a \Lambda_a\rangle|c_b l_b \Lambda_b\rangle|S\Sigma\rangle$ as $r \to \infty$. $H_{\text{elec}}(r)$ consists of kinetic energy operators for all electrons plus Coulomb interactions between all electrons and nuclei. At large r, $H_{\text{elec}}(r)$ can be developed in a multipolar expansion in inverse powers of r, and matrix elements of the Hamiltonian are taken in the product basis $|c_a l_a \Lambda_a\rangle|c_b l_b \Lambda_b\rangle|S\Sigma\rangle$. The leading nonvanishing term in the multipolar expansion for neutral fragments is proportional to r^{-5} if neither l_a nor l_b is zero. (Otherwise the $\langle c\Lambda|\Lambda_a\Lambda_b\rangle$ are trivial, as explained previously.)

As $r \to \infty$, coupling between atomic states with different term limits $c_a c_b$ and electron exchange interactions may be safely neglected or included afterwards as a secondary effect. The determination of the coefficients $\langle\Lambda_a\Lambda_b|c\Lambda\rangle$ thereby reduces to a small configuration interaction calculation among product states with the same atomic term limit $c_a c_b$ and with $\Lambda_a + \Lambda_b = \Lambda$. All elements of the matrix, which has to be diagonalized in the $r \to \infty$ limit, are proportional to the same atomic radial integral. Therefore, the coefficients $\langle\Lambda_a\Lambda_b|c\Lambda\rangle$ can be evaluated without knowledge of the atomic integral. The configuration interaction calculation we describe has been carried out by Knipp,[74] Fontana,[75,76] and Chang[77] to determine asymptotic splittings of ABO energy levels. To our knowledge, the eigenvector coefficients $\langle\Lambda_a\Lambda_b|c\Lambda\rangle$, which depend only on the symmetry of the atomic term limits, have not been tabulated in general.

Since the atomic fragments are described by Russell-Saunders coupling, the asymptotic form of the electronic state in Eq. (8) is easily expressed in terms of a basis of atomic fine structure states using standard angular momentum techniques. We first expand the product state $|c_a l_a \Lambda_a\rangle|c_b l_b \Lambda_b\rangle|S\Sigma\rangle$ in terms of atomic fine structure states $|c_a j_a \Omega_a\rangle|c_b j_b \Omega_b\rangle$ that are quantized along the internuclear axis,

$$|c_a l_a \Lambda_a\rangle|c_b l_b \Lambda_b\rangle|S\Sigma\rangle = \sum_{jj_aj_b} |j\Omega c_a j_a c_b j_b\rangle\langle j\Omega j_a j_b|\Lambda_a\Lambda_b S\Sigma\rangle \tag{9}$$

where

$$|j\Omega c_a j_a c_b j_b\rangle = \sum_{\Omega_a\Omega_b} |c_a j_a \Omega_a\rangle |c_b j_b \Omega_b\rangle \langle j\Omega | j_a j_b \Omega_a \Omega_b\rangle \tag{10}$$

The fine structure (total electronic angular momentum) quantum numbers of each of the fragments are j_a and j_b. The quantum number j is for total electronic angular momentum of both fragments $\mathbf{j} = \mathbf{j}_a + \mathbf{j}_b$. The recoupling coefficients in Eq. (9) are given as

$$\begin{aligned}\langle j\Omega j_a j_b | \Lambda_a \Lambda_b S\Sigma\rangle &= \sum_{\Sigma_a\Sigma_b\Omega_a\Omega_b} \langle s_a s_b \Sigma_a \Sigma_b | S\Sigma\rangle \langle j_a \Omega_a | l_a s_a \Lambda_a \Sigma_a\rangle \\ &\quad \times \langle j_b \Omega_b | l_b s_b \Lambda_b \Sigma_b\rangle \langle j\Omega | j_a j_b \Omega_a \Omega_b\rangle \qquad (11) \\ &= \sum_L [(L)(S)(j_a)(j_b)]^{1/2} \langle L\Lambda | l_a l_b \Lambda_a \Lambda_b\rangle \\ &\quad \times \langle j\Omega | LS\Lambda\Sigma\rangle \begin{Bmatrix} l_a & s_a & j_a \\ l_b & s_b & j_b \\ L & S & j \end{Bmatrix}\end{aligned}$$

where Eq. (24) of Ref. 70 is used in the last step to introduce the simpler expression in terms of 9-j symbols. The symbol (L) is used to denote the quantity $2L + 1$.

Combining results in Eqs. (8) and (9), we can write the primitive Hund case (a) electronic-rotation basis state of Eq. (2) as

$$\begin{aligned}|JMc\Lambda S\Sigma\rangle &= \left(\frac{2J+1}{4\pi}\right)^{1/2} D^{J^*}_{M\Omega}(\alpha\beta 0) \sum_{jj_a j_b \Lambda_a \Lambda_b} |j\Omega c_a j_a c_b j_b\rangle \qquad (12) \\ &\quad \times \langle j\Omega j_a j_b | \Lambda_a \Lambda_b S\Sigma\rangle \langle \Lambda_a \Lambda_b | c\Lambda\rangle, \; r \to \infty\end{aligned}$$

The atomic electronic state $|j\Omega c_a j_a c_b j_b\rangle$, with definite projection Ω of electronic angular momentum along the internuclear axis, is related to the state $|jmc_a j_a c_b j_b\rangle$, having a definite value for the projection of $\mathbf{j}$ along a spaced-fixed (SF) z-axis, by Wigner rotation matrices,

$$|j\Omega c_a j_a c_b j_b\rangle = \sum_m |jmc_a j_a c_b j_b\rangle D^j_{m\Omega}(\alpha\beta 0) \tag{13}$$

Substitution of Eq. (13) into Eq. (12) and use of a standard contraction of the product of two Wigner rotation matrices leads to the expression for $|JMc\Lambda S\Sigma\rangle$,

$$|JMc\Lambda S\Sigma\rangle = \sum_{jlj_aj_b} |JMjlc_aj_ac_bj_b\rangle\langle jlc_aj_ac_bj_b|c\Lambda S\Sigma\rangle_J \tag{14}$$

The atomic eigenstates $|JMjlc_aj_ac_bj_b\rangle$ in Eq. (14) have the limiting form

$$|JMjlc_aj_ac_bj_b\rangle = \sum_{m\mu} |jmc_aj_ac_bj_b\rangle Y_{l\mu}(\hat{\mathbf{r}})\langle JM|jlm\mu\rangle, \; r \to \infty \tag{15}$$

The quantum number l is for orbital angular momentum of the nuclei about their center of mass, and $Y_{l\mu}(\hat{\mathbf{r}})$ is a spherical harmonic. Using Eq. (11), the transformation coefficients in (14) are given by

$$\begin{aligned}\langle jlc_aj_ac_bj_b|c\Lambda S\Sigma\rangle_J &= [(S)(j_a)(j_b)]^{1/2}(-)^{l-\Omega-J}\langle l0|jJ-\Omega\Omega\rangle \\ &\quad\times \sum_{L\Lambda_a\Lambda_b} (L)^{1/2}\langle L\Lambda|l_al_b\Lambda_a\Lambda_b\rangle\langle j\Omega|LS\Lambda\Sigma) \\ &\quad\times \begin{Bmatrix} l_a & s_a & j_a \\ l_b & s_b & j_b \\ L & S & j \end{Bmatrix}\langle\Lambda_a\Lambda_b|c\Lambda\rangle\end{aligned} \tag{16}$$

We pause to examine the significance of the transformation in Eq. (14). The coefficients $\langle jlc_aj_ac_bj_b|c\Lambda S\Sigma\rangle_J$ [Eq. (16)] are independent of r since they are derived from states for $r \to \infty$. The case (a) basis states, appearing in Eqs. (2) and (12), may be subjected to the transformation Eq. (14) at any r. This defines a new basis which we call the *atomic states* $|JMjlc_aj_ac_bj_b\rangle$ as specified linear combinations of case (a) basis functions *for all r*. It is only at large interfragment separations that $|JMjlc_aj_ac_bj_b\rangle$ becomes the simple eigenstate for the separated fragments given in Eq. (15). The atomic basis functions defined in Eq. (14) may appear to have limited utility for small r. The significance of these basis states is made clear later in deriving approximate expressions for the nonadiabatic interactions that couple electronic states as the molecule dissociates.

C. Hund's Case (b) Representation

The electron spin is quantized along the space-fixed (SF) z-axis for the case (b) basis functions. The latter are obtained from the case (a) wave functions by relating electronic wave functions with spin quantized along the internuclear axis to those with definite projection of spin along the space-fixed z-axis,

$$|c\Lambda S\Sigma\rangle = \sum_{M_S} |c\Lambda SM_S\rangle D^S_{M_S\Sigma}(\alpha\beta 0) \tag{17}$$

Substituting Eq. (17) into Eq. (2), we obtain

$$|JMc\Lambda S\Sigma\rangle = \sum_N \frac{(N)^{1/2}}{(J)^{1/2}} \langle J\Omega|NS\Lambda\Sigma\rangle \times \left[\sum_{M_S} \langle JM|NSM_N M_S\rangle \left(\frac{2N+1}{4\pi}\right)^{1/2} D^{N*}_{M_N\Lambda}(\alpha\beta 0)|c\Lambda SM_S\rangle\right] \tag{18}$$

The quantity in square brackets is a primitive case (b) wave function for which N, the quantum number for the vector difference $\mathbf{J} - \mathbf{S}$, is conserved. From Eq. (18) we easily identify the coefficients that relate the case (a) and (b) basis functions,

$$\frac{(N)^{1/2}}{(J)^{1/2}} \langle J\Omega|NS\Lambda\Sigma\rangle \tag{19}$$

The molecular eigenstate $|JM\eta\rangle$ is alternatively expanded in primitive case (a) electronic-rotation basis functions $|JMc\Lambda S\Sigma\rangle$, as defined in Eq. (2), or the analogous case (b) basis states, $|JMc\Lambda SN\rangle$ given as,

$$|JMc\Lambda SN\rangle = \sum_{M_S} \langle JM|NSM_N M_S\rangle \left(\frac{2N+1}{4\pi}\right)^{1/2} D^{N*}_{M_N\Lambda}(\alpha\beta 0)|c\Lambda SM_S\rangle \tag{20}$$

In terms of each of these primitive bases the full wave function is written

$$|JM\eta\rangle = \sum_{\Lambda\Sigma} a(\eta|\Lambda\Sigma)\frac{1}{r}\chi_{Jv\eta}(r)|JMc\Lambda S\Sigma\rangle \qquad \text{case (a)} \tag{21}$$

$$= \sum_{\Lambda N} a(\eta|\Lambda N)\frac{1}{r}\chi_{Jv\eta}(r)|JMc\Lambda SN\rangle \qquad \text{case (b)} \tag{22}$$

The expansion coefficients and basis states are related by the transformation

$$a(\eta|\Lambda N) = \sum_\Sigma \frac{(N)^{1/2}}{(J)^{1/2}} \langle J\Omega|NS\Lambda\Sigma\rangle a(\eta|\Lambda\Sigma) \tag{23}$$

$$|JMc\Lambda SN\rangle = \sum_\Sigma \frac{(N)^{1/2}}{(J)^{1/2}} \langle J\Omega|NS\Lambda\Sigma)|JMc\Lambda S\Sigma\rangle \tag{24}$$

Molecular eigenstates with definite inversion symmetry are constructed with case (b) expansion coefficients having the following relative phase,

$$a(\eta|\Lambda N) = (-)^{p+N+\sigma}\, a(\eta|-\Lambda N) \tag{25}$$

As examples of the use of these coefficients, we give $a(\eta|\Lambda N)$ for a pure case (b) molecular state with nuclear rotation quantum number $N = N_1$,

$$|a(\eta|\Lambda N)| = \delta_{NN_1}(2 - \delta_{\Lambda o})^{-1/2} \tag{26}$$

or for a pure case (a) molecular state expanded in the primitive case (b) basis,

$$|a(\eta|\Lambda N)| = \frac{(N)^{1/2}}{(J)^{1/2}} \langle J\Omega|NS\Lambda\Sigma\rangle \tag{27}$$

In addition to Eqs. (26) or (27), the phase convention specified in Eq. (25) is followed to produce molecular states with definite parity.

The orthogonal transformation, which relates the case (b) electronic-rotation states $|JMc\Lambda SN\rangle$ of Eq. (20) to the atomic basis states of Eqs. (14) and (15), is easily found using results previously derived for the case (a) basis [see Eqs. (8)–(16) and accompanying discussion]. The coefficients $\langle jlc_aj_ac_bj_b|c\Lambda S\Sigma\rangle_J$ of Eqs. (14) and (16) are subjected to the transformation in Eq. (18) which relates the case (a) and (b) bases, yielding new coefficients $\langle jlc_aj_ac_bj_b|c\Lambda SN\rangle_J$,

$$\begin{aligned}\langle jlc_aj_ac_bj_b|c\Lambda SN\rangle_J &= \sum_{\Sigma} \langle jlc_aj_ac_bj_b|c\Lambda S\Sigma\rangle_J \frac{(N)^{1/2}}{(J)^{1/2}} \langle J\Omega|NS\Lambda\Sigma\rangle \\ &= [(S)(j_a)(j_b)(N)(j)]^{1/2}(-)^{\Lambda-j-S}\sum_L (L)^{1/2} \\ &\quad \times \langle l0|NL-\Lambda\Lambda\rangle \begin{Bmatrix} N & L & l \\ j & J & S \end{Bmatrix} \begin{Bmatrix} l_a & s_a & j_a \\ l_b & s_b & j_b \\ L & S & j \end{Bmatrix} \\ &\quad \times \sum_{\Lambda_a\Lambda_b} \langle L\Lambda|l_al_b\Lambda_a\Lambda_b\rangle\langle\Lambda_a\Lambda_b|c\Lambda\rangle\end{aligned} \tag{28}$$

At large r, the transformation in Eq. (28) turns the case (b) basis states into the atomic eigenstates in Eq. (15). When r is not large, this basis transformation is still defined, although the significance of the basis vectors so obtained is not clear unless certain approximations are invoked.

D. Hund's Case (c) Representation

In Hund's case (c) the projections of electron orbital and spin angular momentum are no longer separately conserved because the spin-orbit splitting is large compared to the splitting of ABO levels. Mulliken distinguishes two types of case (c) coupling.[78, 83] The first kind is typified by very strong spin-orbit coupling between two ABO levels with different spins S so that Λ, S, and Σ fail to have any meaning in the resulting molecular eigenstates. (All three of these quantum numbers may not always lose their meaning, as when the strong coupling occurs between two states with the same Λ but different S.) Spin-orbit coupling only mixes states with the same value of $\Omega = |\Lambda + \Sigma|$, and this, in general, is the only good electronic angular momentum quantum number in case (c). The characterization of case (c) states of the first kind requires knowledge of the ordering of ABO levels and their splittings, and so can only be approached on a molecule-by-molecule basis by examination of the electronic potential energy surfaces.

The second kind of case (c) coupling, referred to as the "far-nuclei" case (c), results when molecular states can be identified by the fine structure quantum numbers of the atomic levels with which they correlate at large r. The spacing between molecular levels is principally determined by the atomic fine structure splitting, and the electronic part of the full Hamiltonian acts as a perturbation that splits levels identified by the same values of atomic quantum numbers $c_a j_a c_b j_b$. We confine our discussion of Hund's case (c) to the far-nuclei type because it can be treated systematically and is very relevant to the dynamics of photodissociation.

To obtain far-nuclei case (c) basis functions we subject the large r form of the case (a) states to a recoupling transformation so that $c_a j_a c_b j_b$ and Ω are good quantum numbers in the new basis. We have already performed this task in the discussion of the Hund's case (a) basis, and the desired transformation is given in Eq. (11). According to Eq. (12), the case (c) electronic-rotation basis states are generated as a linear combination of the case (a) states of Eq. (2) through

$$|JMj\Omega c_a j_a c_b j_b\rangle = \sum_{\Lambda\Sigma} |JMc\Lambda S\Sigma\rangle\langle j\Omega c_a j_a c_b j_b|c\Lambda S\Sigma\rangle \tag{29}$$

where

$$\langle j\Omega c_a j_a c_b j_b|c\Lambda S\Sigma\rangle = \sum_{\Lambda_a\Lambda_b} \langle j\Omega j_a j_b|\Lambda_a\Lambda_b S\Sigma\rangle\langle\Lambda_a\Lambda_b|c\Lambda\rangle \tag{30}$$

The coefficients in Eq. (30) are defined in terms of angular momentum recoupling coefficients in Eq. (11), and the $\langle\Lambda_a\Lambda_b|c\Lambda\rangle$ are explained in our discussion of the Hund's case (a) basis functions.

For all r, the expansion of the molecular eigenstate $|JM\eta\rangle$ in the case (c) basis states [Eq. (29)] is well defined,

$$|JM\eta\rangle = \sum_{j\Omega j_a j_b} a(\eta|j\Omega j_a j_b)\frac{1}{r}\chi_{Jv\eta}(r)|JMj\Omega c_a j_a c_b j_b\rangle \tag{31}$$

This expression is analogous to Eq. (1) for case (a) and Eq. (22) for case (b). At *large r*, the basis states $|JMj\Omega c_a j_a c_b j_b\rangle$, produced by the transformation in Eq. (29), resemble atomic fine structure states, as written in Eqs. (9)–(11). When the spin-orbit splitting exceeds the difference of ABO energies $\varepsilon_{c|\Lambda|S}(r)$ in a substantial region of internuclear separations, as often occurs for diatomics composed of heavy atoms, the expansion given in Eq. (31) can be employed with great advantage. Use of the case (c) basis brings the spin-orbit energies to the diagonal of the full Hamiltonian matrix, whereas the electronic Hamiltonian contributes matrix elements that couple primitive case (c) states with different $jj_a j_b$ but the same Ω. Coriolis interactions mix levels with Ω differing by ± 1, but this coupling is generally quite small for heavier diatomics. The large spin-orbit splitting and weak Coriolis coupling in typical case (c) molecules inhibits coupling between states $|JMj\Omega c_a j_a c_b j_b\rangle$ with different $c_a j_a c_b j_b$ or Ω. The residual electronic coupling between case (c) states with identical $\Omega c_a j_a c_b j_b$ is treated by forming linear combinations of case (c) states with different j. Knipp,[74] Fontana,[75,76] and Chang[77] calculate multiplet splittings caused by the action of $H_{\text{elec}}(r)$ among far-nuclei case (c) levels in this manner, that is, spin-orbit splittings so much larger than differences of ABO levels that coupling between states with different $c_a j_a c_b j_b$ is neglected.

The transformation that relates the case (c) electronic-rotation basis [Eq. (29)] to the atomic basis [Eqs. (14) and (15)] is obtained by comparing Eq. (6) and Eq. (30) as

$$|JMjlc_a j_a c_b j_b\rangle = \sum_{\Omega} |JMj\Omega c_a j_a c_b j_b\rangle\langle\Omega|l\rangle_{Jj} \tag{32a}$$

$$\langle\Omega|l\rangle_{Jj} = (-)^{l-\Omega-J}\langle l0|jJ - \Omega\Omega\rangle \tag{32b}$$

Even though the basis functions in Eq. (32a) are composed of separated atomic fragment eigenstates for large r only, the transformation in Eq. (32b) is valid for all r. When expressed in the case (c) basis, molecular states of definite parity $(-)^p$ have expansion coefficients with the symmetry property,

$$a(\eta|j\Omega j_a j_b) = (-)^{p+l_a+l_b+J-j} a(\eta|j-\Omega j_a j_b) \tag{33}$$

E. Evaluation of Matrix Elements of the Hamiltonian

The full Hamiltonian for the diatomic molecule may be written as a sum of an electronic Hamiltonian $H_{\text{elec}}(r)$, a nuclear kinetic energy term $T_{\text{nuc}}(r)$, plus the relativistic interaction $H_{\text{rel}}(r)$ which contains spin-orbit and hyperfine couplings.

$$H(r) = H_{\text{elec}}(r) + T_{\text{nuc}}(r) + H_{\text{rel}}(r) \tag{34}$$

As described above, $H_{\text{elec}}(r)$, the "clamped-nuclei" electronic Hamiltonian, is a sum of kinetic energy operators for all the electrons plus Coulomb interactions between all electrons and nuclei. The nuclear kinetic energy operator is given as

$$T_{\text{nuc}}(r) = -\frac{\hbar^2}{2\mu}\frac{1}{r}\frac{\partial^2}{\partial r^2}r + \frac{\hbar^2}{2\mu r^2}\mathbf{l}^2 \tag{35}$$

where $\mathbf{l}$ is the orbital angular momentum operator for the nuclei about their center of mass. Only spin-orbit coupling is included in $H_{\text{rel}}(r)$ here.

$$H_{\text{rel}}(r) \approx H_{SO}(r)$$

Hyperfine interactions can be handled similarly, but at the expense of more complicated angular momentum algebra.

The electronic Hamiltonian is diagonal in any basis for which Λ and S are good quantum numbers. Among the basis sets we have considered, $H_{\text{elec}}(r)$ is diagonal in the Hund's case (a) and (b) bases with eigenvalues $\varepsilon_{c|\Lambda|S}(r)$, the ABO potential energy surfaces. Estimates for $\varepsilon_{c|\Lambda|S}(r)$ are obtained from *ab initio* calculations or from experiment. The matrix elements of $H_{\text{elec}}(r)$, when transformed to the case (c) or atomic basis, are given in Table I.

To calculate photodissociation cross sections, we require matrix elements of $T_{\text{nuc}}(r)$ and $H_{\text{rel}}(r)$ as a function of internuclear distance from the Franck-Condon (small r) to the atomic (large r) region. These matrix elements are difficult to estimate theoretically, and it is hoped that inferences about these quantities will be made by interpreting experimental photodissociation cross sections to open shell fragments. It is well known that the action of the radial derivative operator in $T_{\text{nuc}}(r)$ on the ABO electronic states gives rise to nonadiabatic interactions. These off-diagonal couplings have been calculated as a function of bond distance for simple one- or two-electron molecules.[84-86]

Nonadiabatic interactions generated by the second term in Eq. (34) are known as Coriolis coupling. Matrix elements of $\mathbf{l}^2$ in the molecular basis sets are evaluated by writing $\mathbf{l}$ as the difference of total angular momentum and electronic angular momentum. Specifically, $\mathbf{l} = \mathbf{J} - \mathbf{L} - \mathbf{S}$ for case (a), where

TABLE I

Matrix Elements of $H_{\mathrm{elec}}(r)$, $T_{\mathrm{nuc}}(r)$ and $H_{SO}(r)$ in the Hund's Case (a), (b), (c), and the Atomic Bases[a]

	Case (a): $\langle JMc'\Lambda'S'\Sigma'\|H\|JMc\Lambda S\Sigma\rangle$	Case (b): $\langle JMc'\Lambda'S'N'\|H\|JMc\Lambda SN\rangle$
$H_{\mathrm{elec}}(r)$	$\delta_{c'c}\delta_{\Lambda'\Lambda}\delta_{S'S}\delta_{\Sigma'\Sigma}\varepsilon_{c\|\Lambda\|S}(r)$	$\delta_{c'c}\delta_{\Lambda'\Lambda}\delta_{S'S}\delta_{N'N}\varepsilon_{c\|\Lambda\|S}(r)$
$T_{\mathrm{nuc}}(r)^b$	$-\frac{\hbar^2}{2\mu}\langle c'\Lambda S\|\frac{\partial^2}{\partial r^2}\|c\Lambda S\rangle\delta_{\Lambda'\Lambda}\delta_{S'S}\delta_{\Sigma'\Sigma}+\frac{\hbar^2}{2\mu r^2}$	$-\frac{\hbar^2}{2\mu}\langle c'\Lambda S\|\frac{\partial^2}{\partial r^2}\|c\Lambda S\rangle\delta_{\Lambda'\Lambda}\delta_{S'S}\delta_{N'N}+\frac{\hbar^2}{2\mu r^2}$
	$\times\{\delta_{\Lambda'\Lambda}\delta_{\Sigma'\Sigma}[J(J+1)+S(S+1)-2\Omega^2+2\Lambda\Sigma+\langle c'\Lambda S\|L^2\|c\Lambda S\rangle$	$\times\{\delta_{\Lambda'\Lambda}[N(N+1)-2\Lambda^2+\langle c'\Lambda S\|L^2\|c\Lambda S\rangle]+\delta_{\Lambda',\Lambda\pm 1}$
	$-\delta_{\Lambda'\Lambda}\delta_{\Sigma',\Sigma\pm 1}[(J\mp\Omega)(J\pm\Omega+1)(S\mp\Sigma)(S\pm\Sigma+1)]^{1/2}$	$\times[(N\pm\Lambda)(N\mp\Lambda+1)]^{1/2}\langle c'\Lambda'S\|L_{\pm}\|c\Lambda S\rangle\}\delta_{S'S}$
	$+\delta_{\Lambda',\Lambda\pm 1}\delta_{\Sigma'\Sigma}[(J\mp\Omega)(J\pm\Omega+1)]^{1/2}\langle c'\Lambda'S\|L_{\pm}\|c\Lambda S\rangle$	
	$+\delta_{\Lambda',\Lambda\pm 1}\delta_{\Sigma',\Sigma\mp 1}[(S\mp\Sigma)(S\pm\Sigma+1)]^{1/2}\langle c'\Lambda'S\|L_{\pm}\|c\Lambda S\rangle\}\delta_{S'S}$	
$H_{SO}(r)^c$	$\delta_{\Omega'\Omega}\sum_{\substack{jj_aj_b\\ L'LF}}(j)(j_a)(j_b)[(L')(L)(S')(S)]^{1/2}(-)^{S'-S+j+\Omega-\phi}$	$\sum_{\substack{jj_aj_b\\ L'LF}}(j)(j_a)(j_b)[(L')(L)(S')(S)(N')(N)]^{1/2}(-)^{J-j}$
	$\times\begin{Bmatrix} l_a & s_a & j_a\\ l_b & s_b & j_b\\ L' & S' & j\end{Bmatrix}\begin{Bmatrix} l_a & s_a & j_a\\ l_b & s_b & j_b\\ L & S & j\end{Bmatrix}\begin{Bmatrix} L & L' & F\\ S' & S & j\end{Bmatrix}\langle F\phi\|SS'\Sigma-\Sigma'\rangle$	$\times\begin{Bmatrix} l_a & s_a & j_a\\ l_b & s_b & j_b\\ L' & S' & j\end{Bmatrix}\begin{Bmatrix} l_a & s_a & j_a\\ l_b & s_b & j_b\\ L & S & j\end{Bmatrix}\begin{Bmatrix} L & L' & F\\ S' & S & j\end{Bmatrix}\begin{Bmatrix} N & N' & F\\ S' & S & J\end{Bmatrix}$
	$\times\langle F-\phi\|LL'\Lambda-\Lambda'\rangle\langle c'\Lambda'\|L'\Lambda'\rangle\langle L\Lambda\|c\Lambda\rangle(\varepsilon_{c_aj_a}+\varepsilon_{c_bj_b})$	$\times\langle F-\phi\|NN'\Lambda-\Lambda'\rangle\langle F-\phi\|LL'\Lambda-\Lambda'\rangle$
		$\times\langle c'\Lambda'\|L'\Lambda'\rangle\langle L\Lambda\|c\Lambda\rangle(\varepsilon_{c_aj_a}+\varepsilon_{c_bj_b})$

	Case (c): $\langle JMj'\Omega' c'_a j'_a c'_b j'_b \| H \| JMj\Omega c_a j_a c_b j_b \rangle$	atomic: $\langle JMjl c_a j_a c_b j_b \| H \| JMjl c_a j_a c_b j_b \rangle$
$H_{elec}(r)$	$\delta_{\Omega'\Omega} \sum_{c\Lambda SL'LF} (S)[(L')(L)(j'_a)(j_a)(j'_b)(j_b)(j')(j)]^{1/2}(-)^{F-S+\Lambda-\Omega+L'+L}$ $\times \begin{Bmatrix} l_a & s_a & j'_a \\ l_b & s_b & j'_b \\ L' & S & j' \end{Bmatrix} \begin{Bmatrix} l_a & s_a & j_a \\ l_b & s_b & j_b \\ L & S & j \end{Bmatrix} \begin{Bmatrix} L' & L & F \\ j & j' & S \end{Bmatrix} \langle F0 \| LL'\Lambda - \Lambda \rangle$ $\times \langle F0 \| jj'\Omega - \Omega \rangle \langle c'\Lambda' \| L'\Lambda' \rangle \langle L\Lambda \| c\Lambda \rangle \varepsilon_{c\|\Lambda\|S}(r)$	$\sum_{c\Lambda SL'LF} (S)[(L')(L)(j'_a)(j_a)(j'_b)(j_b)(j')(j)(l')(l)]^{1/2}$ $\times (-)^{F+\Lambda+L'+L+J-S}$ $\times \begin{Bmatrix} l_a & s_a & j'_a \\ l_b & s_b & j'_b \\ L' & S & j' \end{Bmatrix} \begin{Bmatrix} l_a & s_a & j_a \\ l_b & s_b & j_b \\ L & S & j \end{Bmatrix} \begin{Bmatrix} L' & L & F \\ j & j' & S \end{Bmatrix} \begin{Bmatrix} l' & l & F \\ j & j' & J \end{Bmatrix}$ $\times \langle F0 \| LL'\Lambda - \Lambda \rangle \langle F0 \| ll'00 \rangle \langle c'\Lambda' \| L'\Lambda' \rangle \langle L\Lambda \| c\Lambda \rangle \varepsilon_{c\|\Lambda\|S}(r)$
$T_{nuc}(r)$[a]	$-\frac{\hbar^2}{2\mu} \langle j'\Omega c'_a j'_a c'_b j'_b \| \frac{\partial^2}{\partial r^2} \| j\Omega c_a j_a c_b j_b \rangle \delta_{\Omega\Omega'} + \frac{\hbar^2}{2\mu r^2}$ $\times \{\delta_{\Omega'\Omega}[J(J+1) - 2\Omega^2 + j(j+1)] - \delta_{\Omega',\Omega\pm 1}$ $\times [(J \mp \Omega)(J \pm \Omega + 1)(j \mp \Omega)(j \pm \Omega + 1)]^{1/2}\}$	$-\frac{\hbar^2}{2\mu} \langle J'jl'c'_a j'_a c'_b j'_b \| \frac{\partial^2}{\partial r^2} \| Jjlc_a j_a c_b j_b \rangle$ $+ \delta_{j'j}\delta_{l'l}\delta_{c'_a j'_a c'_b j'_b; c_a j_a c_b j_b} \frac{\hbar^2 l(l+1)}{2\mu r^2}$
$H_{SO}(r)$[c]	$\delta_{j'j}\delta_{\Omega'\Omega}\delta_{c'_a j'_a c'_b j'_b; c_a j_a c_b j_b}(\varepsilon_{c_a j_a} + \varepsilon_{c_b j_b})$	$\delta_{j'j}\delta_{l'l}\delta_{c'_a j'_a c'_b j'_b; c_a j_a c_b j_b}(\varepsilon_{c_a j_a} + \varepsilon_{c_b j_b})$

[a] See Eqs. (16), (28), and (32b) for definitions of these basis sets. $\varepsilon_{c|\Lambda|S}(r)$ denotes an adiabatic Born-Oppenheimer (ABO) potential energy surface. The symbol $\langle L\Lambda | c\Lambda \rangle$ designates the quantity $\Sigma_{\Lambda_a\Lambda_b} \langle L\Lambda | l_a l_b \Lambda_a \Lambda_b \rangle \langle \Lambda_a \Lambda_b | c\Lambda \rangle$ which is unity if one of the atomic fragments is in an S state [l_a or l_b is zero; see Eq. (8)].

[b] Matrix elements of $\mathbf{j}^2$ and $j_\pm$ which appear in the Coriolis coupling term are given in the pure precession approximation for case (c) and the atomic representation [see Eq. (39)]. Similar matrix elements of $\mathbf{L}^2$ and $L_\pm$ for cases (a) and (b) are left unspecified in the table. The pure precession approximation for these quantities is given in Eqs. (36)–(38).

[c] The spin-orbit interaction is approximated by its asymptotic ($r \to \infty$) value, which is diagonal in the case (c) and atomic representations.

L is total electronic angular momentum; **l** = **N** − **L** for case (b); **l** = **J** − **j** for case (c), where **j** = **L** + **S**. The Reversed Angular Momentum (RAM) method of Van Vleck[87] furnishes a convenient technique for evaluation $\mathbf{l}^2$ in the various molecular bases. (Also see Freed[62,63] for the generalization of RAM to spherical tensor form.) Matrix elements of $\mathbf{l}^2$ are exhibited in Table I. They are given in terms of matrix elements of **L** (cases a, b) or **j** (case c) between ABO electronic states.

Unfortunately, matrix elements of **L** are not evaluated as part of *ab initio* electronic structure calculations and little is known about the r-dependence of these quantities. The standard approximation in the theory of fine structure transitions upon atomic collisions is to approximate $\mathbf{l}^2$ by its asymptotic ($r \to \infty$) value. Then $\mathbf{l}^2$ is diagonal in the atomic representation in Eq. (15) at all r:

$$\langle JMj'l'c_a'j_a'c_b'j_b'|\mathbf{l}^2|JMjlc_aj_ac_bj_b\rangle \approx l(l+1)\delta_{j'j}\delta_{l'l}\delta_{c_a'j_a'c_b'j_b';c_aj_ac_bj_b} \tag{36}$$

The Coriolis interaction matrix in the case (a), (b), or (c) molecular bases is generated by similarity transformation of the diagonal matrix of Eq. (36) using the r-independent transformations in Eqs. (16), (28), and (32b), respectively. Another way to state the approximation in Eq. (36) is that the asymptotic product form of the ABO electronic states [Eq. (8)] are used in place of the true r-dependent states $|c\Lambda S\Sigma\rangle$ in the electronic matrix elements of **L** or **j**. For case (a) or (b) [see Eq. (8)], consistency with Eq. (36) implies that the matrix elements of **L** in Table I are approximated as

$$\begin{aligned}\langle c\Lambda'S|\mathbf{L}|c\Lambda S\rangle = &\sum_{\Lambda_a'\Lambda_b'\Lambda_a\Lambda_b} \langle c\Lambda'|\Lambda_a'\Lambda_b'\rangle \\ &\times \langle c_al_a\Lambda_a'c_bl_b\Lambda_b'|\mathbf{L}|c_al_a\Lambda_ac_bl_b\Lambda_b\rangle\langle\Lambda_a\Lambda_b|c\Lambda\rangle\end{aligned} \tag{37}$$

for all r, where $|c_al_a\Lambda_ac_bl_b\Lambda_b\rangle$ is the product of separated atoms state $|c_al_a\Lambda_a\rangle|c_bl_b\Lambda_b\rangle$, and the matrix elements of **L** in the product basis are explicitly given as

$$\begin{aligned}\langle c_al_a\Lambda_ac_bl_b\Lambda_b|L_z|c_al_a\Lambda_ac_bl_b\Lambda_b\rangle &= \Lambda_a + \Lambda_b \\ \langle c_al_a\Lambda_a'c_bl_b\Lambda_b'|L_\pm|c_al_a\Lambda_ac_bl_b\Lambda_b\rangle &= -\delta_{\Lambda_a',\Lambda_a\pm1}[(l_a \mp \Lambda_a)(l_a \pm \Lambda_a + 1)]^{1/2} \\ &\quad - \delta_{\Lambda_b',\Lambda_b\pm1}[(l_b \mp \Lambda_b)(l_b \pm \Lambda_b + 1)]^{1/2} \\ \langle c_al_a\Lambda_a'c_bl_b\Lambda_b'|\mathbf{L}^2|c_al_a\Lambda_ac_bl_b\Lambda_b\rangle &= \delta_{\Lambda_a'\Lambda_a}\delta_{\Lambda_b'\Lambda_b} \\ &\quad \times [l_a(l_a+1) + l_b(l_b+1) + 2\Lambda_a\Lambda_b] \\ &\quad + \delta_{\Lambda_a'\Lambda_a\pm1}\delta_{\Lambda_b'\Lambda_b\mp1}[(l_a \mp \Lambda_a)(l_a \pm \Lambda_a + 1)(l_b \pm \Lambda_b)(l_b \mp \Lambda_b + 1)]^{1/2}\end{aligned} \tag{38}$$

Approximation (36) for case (c) implies that

$$\langle j\Omega c_a j_a c_b j_b | j_z | j\Omega c_a j_a c_b j_b \rangle = \Omega$$
$$\langle j\Omega' c_a j_a c_b j_b | j_\pm | j\Omega c_a j_a c_b j_b \rangle = -\delta_{\Omega', \Omega\pm 1}[(j \mp \Omega)(j \pm \Omega + 1)]^{1/2}$$
$$\langle j\Omega c_a j_a c_b j_b | \mathbf{j}^2 | j\Omega c_a j_a c_b j_b \rangle = j(j+1). \tag{39}$$

Approximation equations (36)–(39) are equivalent to the pure precession hypothesis of Van Vleck.[88]

The evaluation of the spin-orbit Hamiltonian $H_{SO}(r)$, even in the approximate form

$$H_{SO}(r) = \sum_i a_i(r)\mathbf{l}_i \cdot \mathbf{s}_i \tag{40}$$

requires complicated matrix elements between ABO electronic states. Some attempts at realistic calculation of $H_{SO}(r)$ have been made,[89,90] but, like the Coriolis operator, little is known about the bond length dependence of $H_{SO}(r)$ and it is generally approximated by its asymptotic ($r \to \infty$) value. Using the asymptotic approximation, matrix elements of $H_{SO}(r)$ in the atomic basis are given as

$$\langle JMj'l'c'_a j'_a c'_b j'_b | H_{SO} | JMjlc_a j_a c_b j_b \rangle \approx (\varepsilon_{c_a j_a} + \varepsilon_{c_b j_b})\delta_{j'j}\delta_{l'l}\delta_{c'_a j'_a c'_b j'_b; c_a j_a c_b j_b} \tag{41}$$

for all r. The spin-orbit Hamiltonian in the approximate form of Eq. (41) is exhibited in the Hund's case (a), (b), and (c) representations in Table I.

IV. DIFFERENTIAL PHOTODISSOCIATION CROSS SECTIONS

A unified treatment is presented in this section for the angular distributions of photofragments and fluorescence, emitted by atomic fragments produced in excited states by the photodissociation. We exhibit differential cross sections for one-photon and resonant two-photon dissociation. The two-photon case is developed for experiments that use the first resonance photon to select molecules in a particular rotation-vibration level. In some instances more than one photon may be required to select a desired initial level, so we show how to use our theory to obtain results for dissociation involving absorption or emission of an arbitrary number of resonance photons. Our expressions are presented in terms of a Liouville or dual-space formalism which has been previously shown to simplify calculation of photofragmentation and photoionization cross sections.[91-96]

An essential difference between single-photon and two-photon dissociation is that in the former case the molecule is dissociated from an isotropic initial state, whereas in the latter case it is dissociated from an intermediate state that is anisotropic due to its preparation by the first photon. This can be illustrated with a simple example. All states with projections M_0 of initial total angular momentum J_0 are equally populated if the molecular ensemble is isotropic. When the molecule is promoted to an intermediate level with total angular momentum $J_1 = J_0 + 1$ by radiation that is linearly polarized along the z-axis, then states with projections M_1 of J_1 are not equally populated. Levels with $M_1 = \pm J_1$ have zero probability while the probability of finding the molecule in other levels is weighted by the factor $\langle J_1 M_1 | 1 J_0 0 M_0 \rangle^2$ (Wigner-Eckart theorem). This demonstrates that the intermediate state is not isotropic. This anisotropy is reflected in the angular distribution of photofragments which, in contrast to single-photon dissociation, can have multipolar components with angular momentum index K_D greater than two, and hence, can contain additional dynamical information concerning photodissociation.

The precise form of two-photon cross sections depends on experimental conditions. If the transition between ground and intermediate levels is weakly pumped by the radiation field, then perturbation theory is appropriate to describe the multiphoton process. Second-order perturbation theory is needed to derive the photofragment cross section for the two-photon process, whereas emission of fluorescence by photofragments produced in excited states requires a third-order description. When the first transition is strongly pumped and begins to saturate, the appropriate procedure is to obtain a steady-state solution of the Liouville equation to all orders in the field strength for the intermediate density matrix.[94,97,98]

The angular distribution of photofragments produced in resonant two-photon dissociation has been treated semiclassically by Zare[99] and Sander and Wilson,[38] and fully quantum mechanically by Chen and Yeung.[95] The angular distribution of polarized fluorescence, emitted by excited photofragments following single-photon molecular photodissociation, has been described by van Brunt and Zare,[100] Chamberlain and Simons,[101] MacPherson et al.,[102] and Loge and Zare.[103,104] Vigué et al.[105,106] have shown that quantum interference effects between the two Λ-doubling components that form multiplet states of definite inversion symmetry can affect the polarization of fluorescence from excited photofragments. Some earlier predictions must be modified in light of their work. Lambropoulos and Berry,[92] Hansen,[93] Chien,[96] and Chien et al.[107] consider the closely related subject of the angular distribution of electrons produced by two-photon ionization of atoms and diatomic molecules. Singer et al.[8,108]

generalized previous studies to properly include nonadiabatic interactions upon dissociation, and provided a unified treatment of angular distributions of both fragments and fragment fluorescence produced by one-photon and resonant two-photon dissociation.

The strategy, employed here in the construction of the differential cross section, is to consider the theoretical description of a very detailed measurement, and then to reduce it in a stepwise fashion to obtain expressions for less detailed cross sections. Accordingly, we first ask for the differential cross section for dissociation from an initial molecular eigenstate $|J_0 M_0 \eta_0\rangle$ via a transition to intermediate molecular state $|J_1 M_1 \eta_1\rangle$ to a set of dissociative molecular surfaces. The various atomic fine structure states open to each of the fragments are labeled $|c_a j_a m_a\rangle$ and $|c_b j_b m_b\rangle$, the indices c_a and c_b serving to distinguish term limits that might have the same quantum numbers $j_a m_a$ or $j_b m_b$. The molecular eigenstates $|JM\eta\rangle$ are discussed in detail in section III above.

To obtain photofragment differential cross sections, we ask for the probability of detecting fragments receding from the molecule's center of mass with momentum vector **k**. Furthermore, we consider the angular distribution of polarized fluorescence emitted if an excited fragment, chosen with no loss of generality to be the "a" fragment, decays from state $|c_a j_a m_a\rangle$ to $|c'_a j'_a m'_a\rangle$. (Our expressions are equally applicable to a laser induced fluorescence (LIF) experiment in which a fragment produced in state $|c_a j_a m_a\rangle$ is excited to state $|c'_a j'_a m'_a\rangle$ by a probe laser and the total population of the latter state is measured.)

To illustrate how less detailed cross sections may be obtained, we write the differential cross section for the two-photon process described above as

$$\frac{d^2\sigma}{d\Omega_\varepsilon\, d\Omega_k}(\hat{\varepsilon}_S, \mathbf{k}) = \sum_{K_S Q_S K_D Q_D} \frac{\sigma_{K_S Q_S; K_D Q_D}}{32\pi^2} C^*_{K_S Q_S}(\hat{\varepsilon}_S) C_{K_D Q_D}(\hat{\mathbf{k}}) \tag{42}$$

where the $C_{KQ}(\theta\phi)$ are renormalized spherical harmonics,

$$C_{KQ}(\theta\phi) = \left(\frac{4\pi}{2K+1}\right)^{1/2} Y_{KQ}(\theta\phi) \tag{43}$$

and $\hat{\varepsilon}_S$ is the polarization vector of the spontaneously emitted light in the $c_a \to c'_a$ transition. Y_{KQ} is the usual spherical harmonic. Both $\hat{\mathbf{k}}$ and $\hat{\varepsilon}_S$ are defined with reference to the z-axis of the space-fixed coordinate system. (A slightly more general form for the differential cross section is derived later.) Equation (42) may be viewed as an expansion of the differential cross section, viewed as a function of angles $\hat{\varepsilon}_S$ and $\hat{\mathbf{k}}$ in an orthogonal basis set. The use of the $C_{KQ}(\theta\phi)$ are traditional; they reduce to Legendre polynomials if $Q = 0$,

$$C_{KQ}(\theta\phi) = P_K(\cos\theta),\ Q = 0 \tag{44}$$

Often less-detailed experiments are performed than the photon-photofragment coincidence detection which Eq. (42) describes. For example, the photofragment angular distribution may be measured without resolving the angular distribution or polarization of spontaneously emitted light. The differential cross section for this experiment, which is a function of only one angle $\hat{\mathbf{k}}$, is obtained from Eq. (42) by integration over all possible directions in which the fluorescence photon may propagate and by summing over two polarizations for each direction of propagation,

$$\frac{d\sigma}{d\Omega_k}(\mathbf{k}) = \sum_{K_DQ_D} \frac{\sigma_{00;K_DQ_D}}{4\pi} C_{K_DQ_D}(\hat{\mathbf{k}}) \tag{45}$$

From Eq. (45) it is seen that the photofragment angular distribution is determined if the partial cross sections $\sigma_{K_SQ_S;K_DQ_D}$ for the photon-photofragment coincidence detection measurement are known. Similarly, the angular distribution of polarized fluorescence is obtained by integration of Eq. (42) over all angles $\hat{\mathbf{k}}$,

$$\frac{d\sigma}{d\Omega_\varepsilon}(\hat{\boldsymbol{\varepsilon}}_S) = \sum_{K_SQ_S} \frac{\sigma_{K_SQ_S;00}}{8\pi} C^*_{K_SQ_S}(\hat{\boldsymbol{\varepsilon}}_S) \tag{46}$$

The quantity $\sigma_{00;00}$ is proportional to the integral photofragment cross section, which describes an experiment without angular detection of either fluorescence or fragments.

Using perturbation theory to describe the interaction between the molecule and radiation, the differential cross section for the detailed two-photon, photon-photofragment coincidence experiment described is

$$\begin{aligned}\frac{d^2\sigma}{d\Omega_\varepsilon\, d\Omega_k}(\hat{\boldsymbol{\varepsilon}}_S, \mathbf{k}) = \Gamma(J_0)^{-1} \sum_{M_0 m'_a m_b \eta_0} \Bigg| \sum_{m_a \eta_1} &\langle c'_a j'_a m'_a | \hat{\boldsymbol{\varepsilon}}_S \cdot \mathbf{x}_a | c_a j_a m_a \rangle \\ &\times \langle \mathbf{k}, c_a j_a m_a c_b j_b m_b^{(-)} | \hat{\boldsymbol{\varepsilon}}_2 \cdot \mathbf{x} | J_1 M_1 \eta_1 \rangle \\ &\times \langle J_1 M_1 \eta_1 | \hat{\boldsymbol{\varepsilon}}_1 \cdot \mathbf{x} | J_0 M_0 \eta_0 \rangle \Bigg|^2\end{aligned} \tag{47}$$

The constant Γ contains factors which depend on the frequencies of the two absorbed photons as well as the spontaneously emitted ones that are necessary to obtain the proper units for a cross section. Γ is specified in Ref. 8 for the special case of one-photon dissociation. Without loss of generality, we assume fragment a is the one that undergoes fluorescence decay following photodissociation. The polarization vectors of the photon that induces the bound–bound transition, the photon that dissociates the molecule, and the spontaneously emitted photon are $\hat{\boldsymbol{\varepsilon}}_1$, $\hat{\boldsymbol{\varepsilon}}_2$, and $\hat{\boldsymbol{\varepsilon}}_S$, respectively; $\mathbf{x}$ is a vector of all electron coordinates, $\mathbf{x}_a$ those of fragment a. $|\mathbf{k}, c_a j_a m_a c_b j_b m_b^{(-)}\rangle$ is a wave function for motion on the dissociative molecular surfaces. The asymptotic form of this wavefunction has incoming spherical waves in all channels and plane wave in direction $\mathbf{k}$ and in the fine structure state $|c_a j_a m_a\rangle |c_b j_b m_b\rangle$,

$$|\mathbf{k}, c_a j_a m_a c_b j_b m_b^{(-)}\rangle = \sum_{c'_a j'_a m'_a c'_b j'_b m'_b} \left\{ \exp(i\mathbf{k}\cdot\mathbf{r}) \delta_{c_a j_a m_a c_b j_b m_b; c'_a j'_a m'_a c'_b j'_b m'_b} + \frac{\exp(-ikr)}{r} g(\hat{\mathbf{r}})_{c_a j_a m_a c_b j_b m_b; c'_a j'_a m'_a c'_b j'_b m'_b} \right\} \times |c'_a j'_a m'_a\rangle |c'_b j'_b m'_b\rangle, \quad r \to \infty \tag{48}$$

The wave vector $\mathbf{k}$ depends on the internal energy of the fragments and, therefore, on the quantum numbers c_a, j_a, c_b, and j_b. These labels are omitted from $\mathbf{k}$ and k for notational convenience. Since the first photon is absorbed in a resonant transition between bound levels, energy conservation implies

$$\frac{\hbar^2 k^2}{2\mu} = E(\eta_1) + \hbar\omega_2 - \varepsilon_{c_a j_a} - \varepsilon_{c_b j_b} \tag{49}$$

where $E(\eta_1)$ is the energy of the intermediate bound level, ω_2 is the frequency of the photon which dissociates the molecule, and μ is the reduced mass of the two fragments. The internal energies of fragments a and b in fine structure levels $|c_a j_a m_a\rangle |c_b j_b m_b\rangle$ are $\varepsilon_{c_a j_a}$ and $\varepsilon_{c_b j_b}$, respectively. The sums over η_0 and η_1 in Eq. (47) are over all molecular levels whose energy difference falls within the bandwidth of the radiation inducing the first transition,

$$|E(\eta_1) - E(\eta_0)| \lesssim \hbar\, \Delta\omega_1$$

The continuum wave function of Eq. (48) which appears in Eq. (47) may be expanded in energy normalized scattering states $|EJ_2 M_2 j l c_a j_a c_b j_b^{(-)}\rangle$ of definite total angular momentum J_2, which are defined in Eqs. (14) and (15) through

$$|\mathbf{k}, c_a j_a m_a c_b j_b m_b^{(-)}\rangle = \sum_{J_2 j l m \mu} |EJ_2 M_2 j l c_a j_a c_b j_b^{(-)}\rangle \langle J_2 M_2 | j l m \mu\rangle \times \langle j m | j_a j_b m_a m_b\rangle i^l Y^*_{l\mu}(\hat{\mathbf{k}}) \tag{50}$$

In Eq. (50) j is the angular momentum quantum number for the vector sum $\mathbf{j}_a + \mathbf{j}_b$, and l is the quantum number for the difference $\mathbf{J}_2 - \mathbf{j}$ which is the nuclear orbital angular momentum. Equation (50) is easily derived using the plane wave expansion in Eq. (49) and Clebsch-Gordan coefficients to successively couple $\mathbf{j}_a + \mathbf{j}_b = \mathbf{j}$ and $\mathbf{j} + \mathbf{l} = \mathbf{J}$. The scattering state $|EJ_2M_2jlc_aj_ac_bj_b^{(-)}\rangle$ is subject to the asymptotic boundary conditions.

$$|EJ_2M_2jlc_aj_ac_bj_b^{(-)}\rangle = \left(\frac{2\mu}{\pi\hbar^2}\right)^{1/2} \sum_{j'l'c'_aj'_ac'_bj'_b} \left\{\frac{\hat{j}_l(kr)}{k^{1/2}}\,\delta_{jlc_aj_ac_bj_b;j'l'c'_aj'_ac'_bj'_b} + \frac{\hat{h}_{l'}^{(-)}(kr)}{k^{1/2}}\mathscr{T}^{J_2}_{jlc_aj_ac_bj_b;j'l'c'_aj'_ac'_bj'_b}\right\}|J_2M_2j'l'c'_aj'_ac'_bj'_b\rangle,$$

$$r \to \infty \tag{51}$$

where $\mathscr{T}^{J_2}_{jlc_aj_ac_bj_b;j'l'c'_aj'_ac'_bj'_b}$ is a scattering T-matrix for outgoing plane wave and incoming spherical wave boundary conditions; $\hat{h}_l(kr)$ and $\hat{j}_l^{(-)}(kr)$ are Riccati-Bessel and Riccati-Hankel functions, respectively. As above, we omit subscripts that indicate the dependence of the channel wave vector k on the quantum numbers $c_aj_ac_bj_b$ for notational convenience. Substitution of Eq. (50) into Eq. (47), and remembering that the wave function of Eq. (50) appears as a bra, not a ket, in (47), yields

$$\frac{d^2\sigma}{d\Omega_\varepsilon\, d\Omega_k}(\hat{\boldsymbol{\varepsilon}}_S, \mathbf{k}) = \Gamma(J_0)^{-1} \sum_{M_0m'_am_b\eta_0} \Bigg| \sum_{m_a\eta_1J_2jlm\mu} \langle c'_aj'_am'_a|\hat{\boldsymbol{\varepsilon}}_S\cdot\mathbf{x}_a|c_aj_am_a\rangle \times \langle J_2M_2|jlm\mu\rangle\langle jm|j_aj_bm_am_b\rangle i^{-l}Y_{l\mu}(\hat{\mathbf{k}}) \times \langle EJ_2M_2jlc_aj_ac_bj_b^{(-)}|\hat{\boldsymbol{\varepsilon}}_2\cdot\mathbf{x}|J_1M_1\eta_1\rangle \times \langle J_1M_1\eta_1|\hat{\boldsymbol{\varepsilon}}_1\cdot\mathbf{x}|J_0M_0\eta_0\rangle\Bigg|^2 \tag{52}$$

As discussed in Ref. 8, the differential cross section in Eq. (52) is restricted to cases where spontaneous emission occurs principally after the fragments stop interacting with each other.

We now proceed to reexpress the differential cross section in Eq. (52) as a product of four Liouville space operators or vectors. When exhibited in irreducible form, the physical significance of each component of the cross section is made transparent.

A. The Differential Cross Section in Terms of Irreducible Liouville Space Operators

The differential cross section in Eq. (52) is decomposed into a product of five matrices,

$$\frac{d^2\sigma}{d\Omega_\varepsilon\, d\Omega_k}(\hat{\boldsymbol{\varepsilon}}_S, \mathbf{k}) = \Gamma \sum_{\substack{\bar{m}_a J_2 \bar{M}_2 \bar{j}\bar{l} \bar{J}_1 \bar{M}_1 \bar{\eta}_1 J_0 \bar{M}_0 \bar{\eta}_0 \\ m_a J_2 M_2 j l J_1 M_1 \eta_1 J_0 M_0 \eta_0}} \mathscr{D}^\dagger(j_a m_a; j_a \bar{m}_a)$$
$$\times\, \mathscr{U}^\dagger(j_a m_a; j_a \bar{m}_a \,|\, J_2 M_2 j l; \bar{J}_2 \bar{M}_2 \bar{j}\bar{l})$$
$$\times\, \mathscr{L}(J_2 M_2 j l; \bar{J}_2 \bar{M}_2 \bar{j}\bar{l} \,|\, J_1 M_1 \eta_1; \bar{J}_1 \bar{M}_1 \bar{\eta}_1)$$
$$\times\, \mathscr{A}(J_1 M_1 \eta_1; \bar{J}_1 \bar{M}_1 \bar{\eta}_1 \,|\, J_0 M_0 \eta_0; J_0 M_0 \eta_0)$$
$$\times\, \rho_0(J_0 M_0 \eta_0 \,|\, \bar{J}_0 \bar{M}_0 \bar{\eta}_0) \tag{53}$$

where we continue the notation of the previous section in which $(J_0) = 2J_0 + 1$ for J_0 a total angular momentum quantum number and where

$$\rho_0(J_0 M_0 \eta_0 \,|\, \bar{J}_0 \bar{M}_0 \bar{\eta}_0) = (J_0)^{-1} \delta_{J_0 \bar{J}_0} \delta_{M_0 \bar{M}_0} \delta_{\eta_0 \bar{\eta}_0} \tag{54}$$

$$\mathscr{A}(J_1 M_1 \eta_1; \bar{J}_1 \bar{M}_1 \bar{\eta}_1 \,|\, J_0 M_0 \eta_0; \bar{J}_0 \bar{M}_0 \bar{\eta}_0) = \langle J_1 M_1 \eta_1 | \hat{\boldsymbol{\varepsilon}}_1 \cdot \mathbf{x} | J_0 M_0 \eta_0 \rangle$$
$$\times\, \langle J_0 M_0 \eta_0 | \hat{\boldsymbol{\varepsilon}}_1 \cdot \mathbf{x} | \bar{J}_1 \bar{M}_1 \bar{\eta}_1 \rangle \tag{55}$$

$$\mathscr{L}(J_2 M_2 j l; J_2 M_2 j l \,|\, J_1 M_1 \eta_1; \bar{J}_1 \bar{M}_1 \bar{\eta}_1) = \langle E J_2 M_2 j l c_{aJ} j_a c_{bJ} j_b^{(-)} | \hat{\boldsymbol{\varepsilon}}_2 \cdot \mathbf{x} | J_1 M_1 \eta_1 \rangle$$
$$\times\, \langle \bar{J}_1 \bar{M}_1 \bar{\eta}_1 | \hat{\boldsymbol{\varepsilon}}_2 \cdot \mathbf{x} | E \bar{J}_2 \bar{M}_2 \bar{j}\bar{l} c_{aJ} j_a c_{bJ} j_b^{(-)} \rangle \tag{56}$$

$$\mathscr{U}^\dagger(j_a m_a; j_a \bar{m}_a \,|\, J_2 M_2 j l; \bar{J}_2 \bar{M}_2 \bar{j}\bar{l}) = \sum_{m_b m \mu \bar{m} \bar{\mu}} \langle J_2 M_2 | j l m \mu \rangle \langle j m | j_a j_b m_a m_b \rangle$$
$$\times\, i^{-l} Y_{l\mu}(\hat{\mathbf{k}}) \langle \bar{J}_2 \bar{M}_2 | \bar{j}\bar{l}\bar{m}\bar{\mu} \rangle \langle \bar{j}\bar{m} | \bar{j}_a \bar{j}_b \bar{m}_a \bar{m}_b \rangle$$
$$\times\, i^{\bar{l}} Y^*_{\bar{l}\bar{\mu}}(\hat{\mathbf{k}}) \tag{57}$$

and

$$\mathscr{D}^\dagger(j_a m_a; j_a \bar{m}_a) = \sum_{m'_a} \langle j'_a m'_a | \hat{\boldsymbol{\varepsilon}}_S \cdot \mathbf{x}_a | j_a m_a \rangle \langle j_a \bar{m}_a | \hat{\boldsymbol{\varepsilon}}_S . \mathbf{x}_a | j'_a m'_a \rangle \tag{58}$$

The first matrix $\rho_0(J_0 M_0 \eta_0; \bar{J}_0 \bar{M}_0 \bar{\eta}_0)$ of Eq. (54) is the density matrix for the molecule prior to absorption of photons. It is diagonal because we assume no coherences among molecular levels are present initially. Otherwise, a different expression for ρ_0 is substituted in Eq. (53), but the rest of that expression remains unchanged. The density matrix ρ_0 can alternatively be considered a vector in the dual space, or Liouville space, spanned by the quantum numbers $J_0 M_0 \eta_0; \bar{J}_0 \bar{M}_0 \bar{\eta}_0$.

The operator $\mathscr{A}(J_1 M_1 \eta_1; \bar{J}_1 \bar{M}_1 \bar{\eta}_1 | J_0 M_0 \eta_0; \bar{J}_0 \bar{M}_0 \bar{\eta}_0)$ contains dynamical information about the initial transition between bound levels $|J_0 M_0 \eta_0\rangle$ and $|J_1 M_1 \eta_1\rangle$, such as molecular dipole moment integrals and polarization of the incident photon. The operator $\mathscr{A}$ transforms the Liouville space vector $\rho_0(J_0 M_0 \eta_0; \bar{J}_0 \bar{M}_0 \bar{\eta}_0)$ into another vector describing the coherent superposition of molecular states produced by absorption of photon. $\mathscr{L}(J_2 M_2 jl; \bar{J}_2 \bar{M}_2 \bar{j}\bar{l} | J_1 M_1 \eta_1; \bar{J}_1 \bar{M}_1 \bar{\eta}_1)$ is the half-collision operator accounting for all photodissociation dynamics.

The remaining adjoint matrices $\mathscr{U}^\dagger$ of Eq. (57) and $\mathscr{D}^\dagger$ of Eq. (58), that make up the differential cross section, are detection operators. They probe the coherent superposition of fragment states specified by the product $\mathscr{L}$, $\mathscr{A}$ and ρ_0 to extract the observables of interest to us. The operator $\mathscr{U}^\dagger(j_a m_a; j_a \bar{m}_a | J_2 m_2 jl; \bar{J}_2 \bar{M}_2 \bar{j}\bar{l})$ contains geometrical, not dynamical, information. It transforms a dual-space vector from the partial wave basis with quantum numbers $J_2 M_2 jl; \bar{J}_2 \bar{M}_2 \bar{j}\bar{l}$ to the plane wave basis, labeled by $\mathbf{k} j_a m_a j_b m_b$; $\mathbf{k} j_a \bar{m}_a j_b m_b$. The former is the basis most convenient for dynamical calculations, while the latter corresponds to experimental measurement of fragments receding from each other with wave vector $\mathbf{k}$. The magnetic quantum number of the nonfluorescing fragment is not determined in the experiment, so $\mathscr{U}^\dagger$ is traced over m_b. (For greatest generality, the transformation operator $\mathscr{U}^\dagger$ is not traced over m_b. Then our expressions can be recovered by introduction of a detection operator for atom b that is proportional to $\delta_{m_b \bar{m}_b}$.) The probability of detecting fluorescence with polarization vector $\hat{\boldsymbol{\varepsilon}}_S$ is given by contraction of the dual-space bra vector $\mathscr{D}^\dagger(j_a m_a; j_a \bar{m}_a)$ of Eq. (58) with the product of $\mathscr{U}^\dagger$, $\mathscr{L}$, $\mathscr{A}$, and ρ_0, a dual space ket vector, that describes the superposition of magnetic sublevels m_a produced in the photodissociation.

Sums are independently performed over J_1 and $\bar{J}_1$ if the rotational spacing of the intermediate bound levels is smaller than the frequency bandwidth of the radiation which excites the bound–bound transition. However, if the bandwidth is sufficiently narrow to resolve P, Q, and R branch transitions, we fix $J_1 = \bar{J}_1 = J_0 - 1$, J_0, and $J_0 + 1$ separately to obtain the cross sections for each of these transitions, respectively. In the same manner, the multiplet quantum numbers, signified by η_1 and $\bar{\eta}_1$, are summed independently if intermediate multiplet levels are not spectroscopically resolved. The subset of the quantum numbers η_1 and $\bar{\eta}_1$ that are distinguished by sufficiently narrow bandwidth excitation are equal to each other and are not summed.

Each of the five components [Eqs. (54)–(58)] of the differential cross section becomes more concise and physically transparent if a change of Liouville space basis is made to one in which the elements transform as irreducible representations of the rotation group. We use the density matrix [Eq. (54)] for the initial bound state as an example. It transforms in the following manner under a rotation of the coordinate system through Euler angles α, β, γ:

$$\rho_0(J_0M_0\eta_0;\, \bar{J}_0\bar{M}_0\bar{\eta}_0) = \sum_{M_0'\bar{M}_0'} \rho_0(J_0M'_0\eta_0;\, \bar{J}_0\bar{M}'_0\bar{\eta}_0)D^{J_0}_{M_0'M_0}(\alpha\beta\gamma) \times D^{\bar{J}_0^*}{}_{\bar{M}_0'\bar{M}_0}(\alpha\beta\gamma) \tag{59}$$

while the linear combination of density matrix elements,

$$\rho_0[B_0\beta_0(J_0\eta_0;\, \bar{J}_0\bar{\eta}_0)] = \sum_{M_0\bar{M}_0} (-)^{\bar{J}_0-\bar{M}_0}\langle B_0\beta_0|\, J_0\bar{J}_0M_0 - \bar{M}_0\rangle \times \rho_0(J_0M_0\eta_0;\, \bar{J}_0\bar{M}_0\bar{\eta}_0) \tag{60}$$

transforms irreducibly as,

$$\rho_0[B_0\beta_0(J_0\eta_0;\, \bar{J}_0\bar{\eta}_0)] = \sum_{\beta_0'} \rho_0[B_0\beta'_0(J_0\eta_0;\, \bar{J}_0\bar{\eta}_0)]D^{B_0}_{\beta_0'\beta_0}(\alpha\beta\gamma) \tag{61}$$

On a more pedestrian level, the change of Liouville space basis can be viewed as the insertion of the identity operator,

$$\delta_{M_0M_0'}\delta_{\bar{M}_0\bar{M}_0'} = \sum_{B_0\beta_0} (-)^{\bar{J}_0-\bar{M}_0}\langle B_0\beta_0|\, J_0\bar{J}_0M_0 - \bar{M}_0\rangle(-)^{\bar{J}_0-\bar{M}_0} \times \langle B_0\beta_0|J_0\bar{J}_0M_0 - \bar{M}_0\rangle$$

between $\mathscr{L}(J_1M_1\eta_1;\, \bar{J}_1\bar{M}_1\bar{\eta}_1|\, J_0M_0\eta_0;\, \bar{J}_0\bar{M}_0\bar{\eta}_0)$ and $\rho_0(J_0M_0\eta_0;\, \bar{J}_0\bar{M}_0\bar{\eta}_0)$ in Eq. (53). By similar insertions we transform the five dual-space vectors and operators [Eqs. (54)–(58)] to an irreducible basis. The starting expressions for the irreducible components of ρ_0 [Eq. (54)] is given in Eq. (60). The other four basic quantities are

$$\begin{aligned}\mathscr{A}&[B_1\beta_1(J_1\eta_1;\, \bar{J}_1\bar{\eta}_1)|\, B_0\beta_0(J_0\eta_0;\, \bar{J}_0\bar{\eta}_0)] \\ &= \sum_{M_0\bar{M}_0M_1\bar{M}_1} (-)^{\bar{J}_1-\bar{M}_1}\langle B_1\beta_1|\, J_1\bar{J}_1M_1 - \bar{M}_1\rangle \\ &\quad\times \mathscr{A}(J_1M_1\eta_1;\, \bar{J}_1\bar{M}_1\bar{\eta}_1|\, J_0M_0\eta_0;\, \bar{J}_0\bar{M}_0\bar{\eta}_0)(-)^{\bar{J}_0-\bar{M}_0} \\ &\quad\times \langle B_0\beta_0|\, J_0\bar{J}_0M_0 - \bar{M}_0\rangle\end{aligned} \tag{62}$$

$$\begin{aligned}\mathscr{L}[B_2\beta_2(J_2jl;\ \bar{J}_2\bar{j}\bar{l})|B_1\beta_1(J_1\eta_1;\ \bar{J}_1\bar{\eta}_1)] &= \sum_{M_1\bar{M}_1M_2\bar{M}_2} (-)^{\bar{J}_2-\bar{M}_2}\langle B_2\beta_2|J_2\bar{J}_2M_2-\bar{M}_2\rangle \\ &\times \mathscr{L}(J_2M_2jl;\ \bar{J}_2\bar{M}_2\bar{j}\bar{l}|J_1M_1\eta_1;\ \bar{J}_1\bar{M}_1\bar{\eta}_1)(-)^{\bar{J}_1-\bar{M}_1} \\ &\times \langle B_1\beta_1|J_1\bar{J}_1M_1-\bar{M}_1\rangle \end{aligned} \tag{63}$$

$$\begin{aligned}\mathscr{U}^{\dagger}[B_3\beta_3(j_a\colon j_a)|B_2\beta_2(J_2jl;\ \bar{J}_2\bar{j}\bar{l})] = \sum_{M_2\bar{M}_2m_a\bar{m}_a} & (-)^{j_a-\bar{m}_a}\langle B_3\beta_3|j_aj_am_a-\bar{m}_a\rangle \\ &\times \mathscr{U}^{\dagger}(j_am_a;\ j_a\bar{m}_a|J_2M_2jl;\ \bar{J}_2\bar{M}_2\bar{j}\bar{l})(-)^{\bar{J}_2-\bar{M}_2} \\ &\times \langle B_2\beta_2|J_2\bar{J}_2M_2-\bar{M}_2\rangle \end{aligned} \tag{64}$$

and

$$\mathscr{D}^{\dagger}[B_3\beta_3(j_a;\ j_a)] = \sum_{m_a\bar{m}_a} \mathscr{D}^{\dagger}(j_am_a;\ j_am_a)(-)^{j_a-m_a}\langle B_3\beta_3|j_aj_am_a-\bar{m}_a\rangle \tag{65}$$

We obtain the differential cross section in terms in irreducible components by substitution of Eqs. (60) and (63)–(66) into Eq. (53),

$$\begin{aligned}\frac{d^2\sigma}{d\Omega_\varepsilon\, d\Omega_k}(\hat{\varepsilon}_S, \mathbf{k}) = \Gamma \sum_{\substack{B_1B_2B_3B_4\\ \beta_1\beta_2\beta_3\beta_4}} \sum_{\substack{\bar{\eta}_1\bar{J}_2\bar{j}\bar{l}\\ \eta_1J_2jl}} & \mathscr{D}^{\dagger}[B_3\beta_3(j_a;\ j_a)] \\ &\times \mathscr{U}^{\dagger}[B_3\beta_3(j_a;\ j_a)|B_2\beta_2(J_2jl;\ \bar{J}_2\bar{j}\bar{l})] \\ &\times \mathscr{L}[B_2\beta_2(J_2jl;\ \bar{J}_2\bar{j}\bar{l})|B_1\beta_1(J_1\eta_1;\ \bar{J}_1\bar{\eta}_1)] \\ &\times \mathscr{A}[B_1\beta_1(J_1\eta_1;\ \bar{J}_1\bar{\eta}_1)|B_0\beta_0(J_0\eta_0;\ \bar{J}_0\bar{\eta}_0)] \\ &\times \rho_0[B_0\beta_0(J_0\eta_0;\ \bar{J}_0\bar{\eta}_0)] \end{aligned} \tag{66}$$

The detailed angular momentum algebra necessary to perform the sums in Eqs. (60) and (63)–(66) can be found in Ref. 108. We now exhibit and discuss the results.

B. Analysis of Intermediate-State Density Matrix

Using standard angular momentum recoupling techniques, Eq. (60) simplifies to the following expression,

$$\rho_0[B_0\beta_0(J_0\eta_0;\ \bar{J}_0\bar{\eta}_0)] = \delta_{B_00}\delta_{\beta_00}\delta_{J_0\bar{J}_0}\delta_{\eta_0\bar{\eta}_0}(J_0)^{-1/2} \tag{67}$$

As expected for the isotropic initial distribution that ρ_0 describes, only the scalar component of ρ_0 does not vanish when transformed to the irreducible representation.

Upon performance of the sums over M_0, $\bar{M}_0$, M_1 and $\bar{M}_1$, the irreducible components of $\mathscr{A}$ of Eq. (62) are found to be

$$\mathscr{A}[B_1\beta_1(J_1\eta_1;\ \bar{J}_1\bar{\eta}_1)|B_0\beta_0(J_0\eta_0;\ \bar{J}_0\bar{\eta}_0)]$$

$$= \sum_{K_1Q_1} \langle B_1\beta_1|K_1B_0Q_1\beta_0\rangle[(B_0)(K_1)(J_1)(\bar{J}_1)]^{1/2}(-)^{J_0-\bar{J}_0} \quad (68)$$

$$\times \begin{Bmatrix} B_1 & K_1 & B_0 \\ J_1 & 1 & J_0 \\ \bar{J}_1 & 1 & \bar{J}_0 \end{Bmatrix} \phi^{K_1}_{Q_1}(\hat{\boldsymbol{\varepsilon}}_1)\langle J_1\eta_1||\mathbf{x}||J_0\eta_0\rangle\langle \bar{J}_0\bar{\eta}_0||\mathbf{x}||\bar{J}_1\bar{\eta}_1\rangle$$

the symbol $\phi^{K_1}_{Q_1}(\hat{\boldsymbol{\varepsilon}}_1)$ denotes a photon polarization matrix as defined by Omont,[94]

$$\phi^{K_1}_{Q_1}(\hat{\boldsymbol{\varepsilon}}_1) = \sum_{q_1\bar{q}_1} (-)^{1-\bar{q}_1}\langle K_1Q_1|11q_1-\bar{q}_1\rangle\varepsilon^*_{q_1}\varepsilon_{\bar{q}_1} \quad (69)$$

The elements of the polarization matrix [Eq. (69)] are the irreducible components of the (adjoint) density matrix for the radiation field. They are related to the Stokes parameters for radiation with complex polarization vector $\hat{\boldsymbol{\varepsilon}}_1$ as given by Omont.[94] When the Cartesian components of $\hat{\boldsymbol{\varepsilon}}_1$ are real, it is useful to express the elements of the polarization matrix in terms of the renormalized spherical harmonics $C_{K_1Q_1}(\hat{\boldsymbol{\varepsilon}}_1)$, of Eq. (43),

$$\phi^{K_1}_{Q_1}(\hat{\boldsymbol{\varepsilon}}_1) = \sum_{q_1\bar{q}_1} (-)^{1-\bar{q}_1}\langle K_1Q_1|11q_1-\bar{q}_1\rangle C^*_{1q_1}(\hat{\boldsymbol{\varepsilon}}_1)C_{1\bar{q}_1}(\hat{\boldsymbol{\varepsilon}}_1) \quad (70)$$

$$= -\langle K_1 0|1100\rangle C^*_{K_1Q_1}(\hat{\boldsymbol{\varepsilon}}_1)$$

The reduced molecular transition amplitudes are defined as

$$\langle J_1M_1\eta_1|x_q|J_0M_0\eta_0\rangle = \langle J_1M_1|1J_0qM_0\rangle\langle J_1\eta_1||\mathbf{x}||J_0\eta_0\rangle, \quad (71)$$

where x_q is a component of the electric dipole vector in the space-fixed coordinate system. This x_q is related to the components $x_{q'}$ in the body-fixed system by a rotation,[70,71]

$$x_q = \sum_{q'} D^{1*}_{qq'}(\alpha\beta 0)x_{q'} \quad (72)$$

Reduced molecular transition amplitudes between bound levels expanded in the Hund's case (a) and (b) bases, discussed in Section III, are given in Table II. The symbol $\langle J_1 v_1 c_1 \Lambda_1 S_1 | x_{q'} | J_0 v_0 c_0 \Lambda_0 S_0 \rangle$ in Table II denotes the usual integral over electronic and vibrational coordinates,

$$\langle J_1 v_1 c_1 \Lambda_1 S_1 | x_{q'} | J_0 v_0 c_0 \Lambda_0 S_0 \rangle = \int_0^\infty dr \chi^*_{J_1 v_1 \eta_1}(r) \mu(r) \chi_{J_0 v_0 \eta_0}(r) \tag{73}$$

in which $\mu(r)$ is the electronic dipole moment function,

$$\mu(r) = \langle c_1 \Lambda_1 S_1 | x_{q'} | c_0 \Lambda_0 S_0 \rangle \tag{74}$$

Equation (74) vanishes unless $S_1 = S_0$ and $\Lambda_1 = \Lambda_0 + q'$.

The absorption operator [Eq. (63)] does not change, even when the initial density matrix is no longer a scalar as it is in Eq. (67). For the simple form of ρ_0 presented in Eq. (67), we can evaluate the product of $\mathscr{A}$ and ρ_0 to obtain $\rho_1[B_1\beta_1(J_1\eta_1; \bar{J}_1\bar{\eta}_1)]$, the density matrix for the coherent superposition of intermediate molecular bound states produced upon absorption of a photon by an initially isotropic ensemble, as

$$\begin{aligned}\rho_1[B_1\beta_1(J_1\eta_1; \bar{J}_1\bar{\eta}_1)] &= \sum_{B_0\beta_0\eta_0\bar{\eta}_0} \mathscr{A}[B_1\beta_1(J_1\eta_1; \bar{J}_1\bar{\eta}_1)|B_0\beta_0(J_0\eta_0; \bar{J}_0\bar{\eta}_0)] \\ &\quad \times \rho_0[B_0\beta_0(J_0\eta_0; \bar{J}_0\bar{\eta}_0)] \\ &= \delta_{B_1K_1}\delta_{\beta_1Q_1}(J_0)^{-1}[(J_1)(\bar{J}_1)]^{1/2} \\ &\quad \times W(1K_1J_0\bar{J}_1; 1J_1)\phi^{K_1}_{Q_1}(\hat{\varepsilon}_1)\langle J_1\eta_1||\mathbf{x}||\bar{J}_0\bar{\eta}_0\rangle \\ &\quad \times \langle J_0\eta_0||\mathbf{x}||\bar{J}_1\bar{\eta}_1\rangle\end{aligned} \tag{75}$$

TABLE II

Reduced Molecular Transition Amplitudes for Dipole Transitions between Bound Levels $|J_0M_0\eta_0\rangle$ and $|J_1M_1\eta_1\rangle$[a]

Case (a):

$$\langle J_1\eta_1||\mathbf{x}||J_0\eta_0\rangle = \sum_{\Lambda_0\Sigma_0\Lambda_1\Sigma_1} a^*(\eta_1|\Lambda_1\Sigma_1)a(\eta_0|\Lambda_0\Sigma_0)(J_0)^{1/2}(J_1)^{-1/2}\langle J_1\Omega_1|1J_0q_1'\Omega_0\rangle \times \langle J_1v_1c_1\Lambda_1S_1|x_{q'}|J_0v_0c_0\Lambda_0S_0\rangle\delta_{S_1S_0}\delta_{\Sigma_1\Sigma_0}$$

Case (b):

$$\langle J_1\eta_1||\mathbf{x}||J_0\eta_0\rangle = \sum_{\Lambda_0N_0\Lambda_1N_1} a^*(\eta_1|\Lambda_1N_1)a(\eta_0|\Lambda_0\Sigma_0)[(N_0)(J_0)]^{1/2}\langle N_1\Lambda_1|1N_0q'\Lambda_0\rangle \times W(1J_1N_0S_0; J_0N_1)\langle J_1v_1c_1S_1|x_{q'}|J_0v_0c_0\Lambda_0S_0\rangle\delta_{S_1S_0}$$

[a] The symbol $\langle J_1v_1c_1\Lambda_1S_1|x_{q'}|J_0v_0c_0\Lambda S_0\rangle$ is the usual transition overlap integral which is defined by Eqs. (73) and (74).

The irreducible representation of ρ_1 in Eq. (75) is equal to that of the polarization matrix ($B_1 = K_1$). Initially there is no preferred molecular orientation, so we take an equal weighted average of the projections M_0 of J_0. The initial molecular density matrix transforms as a scalar and, therefore, when coupled to the photon density matrix $\phi^{K_1}_{Q_1}(\boldsymbol{\varepsilon}_1)$, leads to an intermediate density matrix with $B_1 = K_1$ and $\beta_1 = Q_1$. Single photon cross sections are recovered (within a constant) if the anisotropy due to the first photon is neglected ($K_1 = 0$) and all coherences within the intermediate multiplet levels due to preparation by optical excitation are destroyed ($\eta_1 = \bar{\eta}_1$). In this case the intermediate density matrix [Eq. (75)] reduces to

$$\rho_1[B_1\beta_1(J_1\eta_1;\, \bar{J}_1\bar{\eta}_1)] = \delta_{B_1 0}\delta_{\beta_1 0}\delta_{J_1 \bar{J}_1}\delta_{\eta_1\bar{\eta}_1}(J_0)^{-1/2} \tag{76}$$

Therefore, we can quickly reduce our expressions for two-photon cross sections to cross sections for single-photon dissociation by setting $K_1 = 0$, $\langle J_1\eta_1 \| \mathbf{x} \| J_0\eta_0 \rangle = 1$, and $\eta = \bar{\eta}$.

The methods of optical pumping theory [94,97,98] can be used to extend our results to describe particular features of the excitation process or the effect of relaxation pathways available to the molecule. Examples of the special situations that may arise are strong optical pumping which saturates a transition, loss of coherence among the intermediate levels through spontaneous emission, radiationless decay, or collisions. For these problems, the intermediate density matrix ρ_1 is described by the stationary solution of the master equation obtained by adding relaxation terms to the quantum Liouville equation. Alternatively, the explicit time evolution of the intermediate molecular levels can be studied using pulsed excitation from ground to intermediate state, and then the molecule is dissociated with a second pulsed excitation after a variable time delay. This scheme has been used to study the time evolution of coherently prepared atomic levels. Chien[96] has developed an elegant density matrix formalism that emphasizes the graphical techniques of Yutsis et al.[109] to describe the results of these experiments.

Another direction for extension of our results is the description of resonant dissociation processes involving more than two photons. Multiple resonant absorptions are described by inserting a sequence of absorption operators [Eq. (62)] between $\mathscr{L}$ and ρ_0 in the differential cross section [Eq. (66)]. The general form of the density operator which describes the molecule after absorption of n resonant photons, but prior to dissociation of the molecule is

$$\begin{aligned}\rho_n[B_n\beta_n(J_n\eta_n;\, \bar{J}_n\bar{\eta}_n)] = \prod_{\alpha=0}^{n-1} \sum_{B_\alpha\beta_\alpha\eta_\alpha} &\Gamma_n[B_{\alpha+1}\beta_{\alpha+1}(J_{\alpha+1}\eta_{\alpha+1};\, \bar{J}_{a+1}\bar{\eta}_{a+1})] \\ &\times \mathscr{A}[B_{\alpha+1}\beta_{\alpha+1}(J_{\alpha+1}\eta_{\alpha+1};\, \bar{J}_{\alpha+1}\bar{\eta}_{\alpha+1}) | B_\alpha\beta_\alpha(J_\alpha\eta_\alpha;\, \bar{J}_\alpha\bar{\eta}_\alpha)] \\ &\times \rho_0[B_0\beta_0(J_0\eta_0;\, \bar{J}_0\bar{\eta}_0)]\end{aligned} \tag{77}$$

The relaxation matrices $\Gamma_n[B_{\alpha+1}\beta_{\alpha+1}(J_{\alpha+1}\eta_{\alpha+1};\ \bar{J}_{\alpha+1}\bar{\eta}_{\alpha+1})]$ are included in Eq. (77) to account for process which destroy coherence among the intermediate levels, as discussed earlier. If such processes do not occur, the relaxation matrices may be omitted.

C. Analysis of Half-Collision Operator

The half collision operator of Eq. (63) contains photodissociation amplitudes $\langle EJ_2M_2jlc_aj_ac_bj_b^{(-)}|\hat{\varepsilon}_2\cdot\mathbf{x}|J_1M_1\eta_1\rangle$ that describe transitions between the intermediate and dissociative molecular states in addition to the effects of nonadiabatic interactions between electronic levels as the molecule dissociates. The same amplitudes are used to construct the differential cross section for single-photon dissociation. Using the Wigner-Eckart theorem, we factor out a Clebsch-Gordan coefficient, containing all the M_1 and M_2 dependence of the probability amplitude, and define a reduced transition amplitude,

$$\langle EJ_2M_2jlc_aj_ac_bj_b^{(-)}|\varepsilon_2\cdot\mathbf{x}|J_1M_1\eta_1\rangle \equiv \sum_{q_2}\langle J_2M_2|1J_1q_2M_1\rangle \times \varepsilon_{q_2}^{*}\tau(J_2jlc_aj_ac_bj_b|J_1\eta_1) \qquad (78)$$

Equations (69) and (78) are substituted into Eq. (63), and the sums over M_1, $\bar{M}_1$, $\bar{M}_2$, and $\bar{M}_2$ are carried out[108] resulting in the following expression for the irreducible components of $\mathscr{L}$:

$$\mathscr{L}[B_2\beta_2(J_2jl;\ \bar{J}_2\bar{j}\bar{l})|B_1\beta_1(J_1\eta_1;\ \bar{J}_1\bar{\eta}_1)]$$
$$= \sum_{K_2Q_2}(-)^{J_2-\bar{J}_2}(-)^{J_1-\bar{J}_1}[(J_2)(\bar{J}_2)(K_1)(K_2)]^{1/2}\langle B_2\beta_2|B_1K_1\beta_1Q_1\rangle \times \begin{Bmatrix} B_1 & K_2 & B_2 \\ J_1 & 1 & J_2 \\ \bar{J}_1 & 1 & \bar{J}_2 \end{Bmatrix} \phi_{Q_2}^{K_2}(\hat{\varepsilon}_2)\tau(J_2jlc_aj_ac_bj_b|J_1\eta_1) \times \tau^{*}(\bar{J}_2\bar{j}\bar{l}c_aj_ac_bj_b|\bar{J}_1\bar{\eta}_1) \qquad (79)$$

The irreducible representations B_2 of the dissociating system after absorption of the second photon represent the combined anisotropies of the two radiation fields, K_1 ($= B_1$, if the initial ensemble is isotropic [see Eq. (75)]) and K_2. If $B_1 = 0$, we recover the half collision operator for single photon dissociation,

$$\mathscr{L}[B_2\beta_2(J_2jl;\ \bar{J}_2\bar{j}\bar{l})|B_1 = 0\,(J_1\eta_1;\ \bar{J}_1\bar{\eta}_1)]$$
$$= \delta_{B_2K_2}\delta_{\beta_2Q_2}\delta_{J_1\bar{J}_1}[(J_2)(\bar{J}_2)]^{1/2}(J_1)^{-1/2}W(1K_2J_1\bar{J}_2;\ 1J_2) \times \phi_{Q_2}^{K_2}(\hat{\varepsilon}_2)\,\tau(J_2jlc_aj_ac_bj_b|J_1\eta_1)\tau^{*}(\bar{J}_2\bar{j}\bar{l}c_aj_ac_bj_b|\bar{J}_1\bar{\eta}_1) \qquad (80)$$

while the two-photon case permits $B_1 \neq 0$ and thereby gives new information about the molecular amplitudes $\tau(J_2 jlc_a j_a c_b j_b | J_1 \eta_1)$.

D. Analysis of Geometrical Matrix $\mathscr{U}^\dagger$

The next component of the differential cross section to be considered is the geometrical factor $\mathscr{U}^\dagger$ of Eqs. (57) and (64) which transforms the Liouville space basis and traces the result over m_b. The angular momentum algebra necessary to simplify Eq. (64) is performed in Ref. 108, and we quote the final result,

$$\begin{aligned}
&\mathscr{U}^\dagger[B_3\beta_3(j_a;\, j_a)|B_2\beta_2(J_2jl;\, \bar{J}_2\bar{j}\bar{l})] \\
&\quad = \sum_{K_D Q_D} [(K_D)(K_S)(J_2)(\bar{J}_2)(j)(\bar{j})(l)(\bar{l})]^{1/2} \frac{i^{\bar{l}-l}}{4\pi} (-)^{\bar{j}-j-\bar{l}} C_{K_D Q_D}(\hat{\mathbf{k}}) \\
&\qquad \times \langle K_D 0 | l\bar{l}00\rangle \langle B_2\beta_2 | B_3 K_D \beta_3 Q_D\rangle W(jB_3 j_b j_a;\, \bar{j} j_a) \\
&\qquad \times \begin{Bmatrix} K_D & B_2 & B_3 \\ l & J_2 & j \\ \bar{l} & \bar{J}_2 & \bar{j} \end{Bmatrix}
\end{aligned} \tag{81}$$

The operator $\mathscr{U}^\dagger$ resolves the total anisotropy B_2 into two components, namely K_D, the anisotropy of the photofragment angular distribution, and B_3, the representation of the density matrix that describes the polarization of the magnetic sublevels of fragment a. Actually, there is a third component to B_2 representing the fragment b density matrix, but this is zero because $\mathscr{U}^\dagger$ is traced over m_b. It is shown later that B_3 and β_3 must equal K_S and Q_S, respectively, which are the components of the polarization matrix for the spontaneously emitted light.

Integration over all photofragment detection angles $\hat{\mathbf{k}}$ to obtain total photofragment cross sections forces K_D to vanish in Eq. (81). Likewise, we demonstrate below that collection of all fragment fluorescence forces $B_3(=K_S) = 0$. Accordingly we give the $K_D = 0$ limit of $\mathscr{U}^\dagger$ for integral photofragment and differential fluorescence cross sections, the $K_S = 0$ limit of $\mathscr{U}^\dagger$ for integral fluorescence and differential photofragment cross sections, and the $K_S = K_D = 0$ limit for integral fragment and fluorescence cross sections as

$$\begin{aligned}
&\mathscr{U}^\dagger[B_3\beta_3(j_a;\, j_a)|B_2\beta_2(J_2jl;\, \bar{J}_2\bar{j}\bar{l})] \\
&\quad \overset{(K_D=0)}{=} [(J_2)(\bar{J}_2)(j)(\bar{j})]^{1/2} \frac{(-)^{\bar{j}-j}}{4\pi} \delta_{\bar{l}l}\delta_{B_3B_2}\delta_{\beta_3\beta_2} W(jlB_3\bar{J}_2;\, J_2\bar{j}) \\
&\qquad \times W(jB_3 j_b j_a;\, \bar{j} j_a)
\end{aligned} \tag{82a}$$

$$\overset{(B_3=K_S=0)}{=} [(J_2)(\bar{J}_2)(l)(\bar{l})]^{1/2}(j_a)^{-1/2}\frac{i^{\bar{l}-l}}{4\pi}(-)^{l}\delta_{j\bar{j}}\delta_{K_DB_2}\delta_{Q_D\beta_2}C_{K_DQ_D}(\hat{\mathbf{k}})$$

$$\times \langle K_D0|l\bar{l}00\rangle W(J_2jK_D\bar{l};\, l\bar{J}_2) \tag{82b}$$

and

$$\overset{(K_D=0,\,B_3=K_S=0)}{=} (J_2)^{1/2}(j_a)^{-1/2}\frac{1}{4\pi}\,\delta_{B_20}\delta_{\bar{j}\bar{j}}\delta_{l\bar{l}}\delta_{J_2\bar{J}_2} \tag{82c}$$

respectively.

E. Analysis of Fluorescence Detection Operator $\mathscr{D}^\dagger$

The final component of the differential cross section we must consider is $\mathscr{D}^\dagger$ the Liouville space vector which accounts for spontaneous emission by excited photofragments. After performing the sum in Eq. (65), the irreducible components of $\mathscr{D}^\dagger$ are found to be

$$\mathscr{D}^\dagger[B_3\beta_3(j_a;\,j_a)] = \delta_{B_3K_S}\delta_{\beta_3Q_S}(j_a')(-)^{Q_S}\,\phi^{K_S}_{-Q_S}(\hat{\boldsymbol{\varepsilon}}_S)$$
$$\times\, W(1K_Sj_a'j_a;\,1j_a)|\langle c_a'j_a'||\mathbf{x}_a||c_aj_a\rangle|^2 \tag{83}$$

Using Eq. (70), we also exhibit $\mathscr{D}^\dagger$ in a form that is appropriate for detection of spontaneously emitted light with polarization vector $\hat{\boldsymbol{\varepsilon}}_S$ having real Cartesian components,

$$\mathscr{D}^\dagger[B_3\beta_3(j_a;\,j_a)] = \delta_{B_3K_S}\delta_{\beta_3Q_S}(j_a')(-)^{Q_S+1}\langle K_S0|1100\rangle C^*_{K_SQ_S}(\hat{\boldsymbol{\varepsilon}}_S)$$
$$\times\, W(1K_Sj_a'j_a;\,1j_a)|\langle c_a'j_a'||\mathbf{x}_a||c_aj_a\rangle|^2 \tag{84}$$

Integration over all propagation vectors for spontaneously emitted light with two orthogonal polarizations for each direction of propagation leads to a factor of $8\pi\delta_{K_S0}$ in Eqs. (83) and (84).

Equation (84) for $\mathscr{D}^+$, though not quite as general as Eq. (83), allows us to write the differential cross in the form of Eq. (42). The familiar parameters, characterizing the angular distribution of photofragments and fluorescence, can be expressed as simple algebraic functions of the partial cross sections $\sigma_{K_SQ_S;\,K_DQ_D}$ defined in Eq. (42). For example, the anisotropy parameter β_D for the photofragment angular distribution is defined by writing the photofragment cross section [Eq. (45)] in the form,

$$\frac{d\sigma}{d\Omega_k}(\mathbf{k}) = \frac{\sigma_0}{4\pi}[1 + \beta_D P_2(\hat{\mathbf{k}}\cdot\hat{\mathbf{z}})] \tag{85}$$

where P_2 is the second Legendre polynomial and σ_0 is the integral photofragment cross section. The molecule is assumed to be dissociated by radiation which is linearly polarized along the z-axis. In terms of the partial cross sections $\sigma_{K_SQ_SK_DQ_D}$ of Eq. (42), β_D is

$$\beta_D = \frac{\sigma_{00;20}}{\sigma_{00;00}} \tag{86}$$

We construct a similar anisotropy parameter β_S for the angular distribution of fluorescence,

$$\beta_S = \frac{\sigma_{20;00}}{\sigma_{00;00}} \tag{87}$$

The familiar polarization ratio,

$$\mathscr{P} = \frac{I_\parallel - I_\perp}{I_\parallel + I_\perp} \tag{88}$$

where $I_\parallel$ and $I_\perp$ are the intensities of spontaneously emitted light polarized parallel and perpendicular to incident linearly polarized radiation, is simply related to β_S,

$$\mathscr{P} = \frac{3\beta_S}{4 + \beta_S} \tag{89}$$

In $I_\parallel$ and $I_\perp$ are obtained from Eq. (46) by setting $\hat{\boldsymbol{\varepsilon}}_S = \hat{\mathbf{z}}$ and $\hat{\boldsymbol{\varepsilon}}_S = \hat{\mathbf{y}}$, respectively, and by insertion of these results into Eq. (88).

If the initial molecular ensemble is isotropic, that is ρ_0 given by Eq. (67), the indices K_S and K_D assume the following values for single-photon dissociation,

$$\begin{aligned} &K_S \leqslant 2 \\ &K_D \leqslant K_S + 2 \end{aligned} \tag{90}$$

while for two-photon dissociation, they satisfy

$$\begin{aligned} &K_S \leqslant 2 \\ &K_D \leqslant K_S + 4 \end{aligned} \tag{91}$$

In principle, the number of partial cross sections $\sigma_{K_S Q_S; K_D Q_D}$, needed to characterize the angular distribution of photofragments and fluorescence, is 182 for one-photon and 378 for two-photon dissociation. In most experimental situations the number of independent partial cross sections is far less than that allowed by Eq. (90) or Eq. (91) because of symmetry in the experimental arrangement. Illuminating discussions of the relation of the number of independent parameters to the symmetry of the experimental arrangement are provided by Hertel and Stoll[110] and Greene and Zare.[11]

V. THE DYNAMICAL PROBLEM

Electronically nonadiabatic transitions occur as the photofragments separate and they modify the nascent distribution of electronic states produced by optical excitation of the molecule. Nonadiabatic processes affect observable properties of the fragments such as total cross sections, branching ratios, and angular distributions. The dependence of observable properties on the photodissociation dynamics is manifested in the equations of Section IV for calculating cross sections. These theoretical expressions contain the transition amplitudes $\tau(J_2 jlc_a j_a c_b j_b | J_1 \eta_1)$ [Eq. (78)], which are the quantities that carry information about the photodissociation dynamics in the cross section equations. The efficiency with which particle flux is driven from the bound level $|J_1 M_1 \eta_1\rangle$ to dissociative molecular surfaces and the amplitude for undergoing nonadiabatic transitions are two examples of dynamical effects that are described by the amplitudes $\tau(J_2 jlc_a j_a c_b j_b | J_1 \eta_1)$. Specifically, these amplitudes appear in the dual space operator $\mathscr{L}[B_2\beta_2(J_2 jl; \bar{J}_2 \bar{j}\bar{l}) | B_1\beta_1(J_1\eta_1; \bar{J}_1\bar{\eta}_1)]$ [Eq. (79)] which describes the half-collision dynamics.

We consider photodissociation induced by weak light fields, so the electric dipole approximation and perturbation theory are employed to treat the matter-radiation field interaction that induces photodissociation. Consequently, $\tau(J_2 jlc_a j_a c_b j_b | J_1 \eta_1)$ is a matrix element of the electric dipole operator between bound and dissociative states both of which are unperturbed by the radiation field. If many fragment electronic states are accessible, the continuum state $|EJ_2 M_2 jlc_a j_a c_b j_b^{(-)}\rangle$, that appears in $\tau(J_2 jlc_a j_a c_b j_b | J_1 \eta_1)$ [see Eqs. (14), (15), (50), and (78)], is a complicated multichannel wave function. In some cases numerical methods are the only reliable way to evaluate the dipole matrix element between $|EJ_2 M_2 jlc_a j_a c_b j_b^{(-)}\rangle$ and $|J_1 M_1 \eta_1\rangle$. Available numerical methods are listed in Section V.A where the reader is referred to more extensive treatments of this subject.

The remainder of this section is devoted to a description of dynamical approximations which take advantage of the fact that nonadiabatic transitions are usually insignificant in regions along the dissociation coordinate where the electronic interaction between the fragments is strong, so long as there are no curve crossings or near crossings there. We designate this interval the molecular region, and it is shown schematically in Fig. 1. The adiabatic Born-Oppenheimer (ABO) energy levels are well separated in the molecular region because of the strength of the electronic interaction. Members of a multiplet group, that is, those $(2 - \delta_{\Lambda 0})$ $(2S + 1)$ states associated with the ABO level with quantum numbers $|\Lambda|$ and S, lie close to one another in energy. However, the direct coupling between multiplet levels can be taken to vary slowly with internuclear separation in the molecular region.

The large splitting between ABO levels and the weak dependence of intramultiplet coupling on r, taken together, indicate that motion along uncoupled adiabatic potential energy surfaces provides an accurate picture of dissociation dynamics in the molecular region. The same scheme usually provides an excellent description of bound molecular levels, and we are merely applying it to describe dissociative motion. Of course, crossings or avoided crossings of ABO levels can occur in the molecular region, and they necessitate a more complicated treatment of those levels that lie close to each other in energy at these crossings. We discuss the adiabatic description of dissociative motion in the molecular region in Section V.B.

Even when the adiabatic molecular state description is valid for most of the interaction region, we must separately assess its applicability in the crucial recoupling region which lies beyond the molecular region, but before the atomic region in which the fragments behave as separated atoms (see Fig. 1). The recoupling region is so designated because the angular momentum quantum numbers that differentiate eigenstates of the r-dependent Hamiltonian matrix $\mathsf{H}(r)$ [Eq. (34)] change from projections along the internuclear axis, like Λ or Ω in the molecular region, to atomic fine or hyperfine quantum numbers.

The adiabatic picture is valid for all r when nonadiabatic transitions do not occur in the recoupling region. Otherwise the wave function can be propagated through the recoupling region using numerical techniques, or if the fragment kinetic energy is large, by employing a diabatic or recoil approximation as described below. We obtain the following rough estimate of the likelihood of nonadiabatic transitions in the recoupling region from the Landau-Zener model:[112] The nonadiabatic transition probability is expressed in terms of the parameter P.

$$P = \frac{\Delta r \, \Delta E}{v \hbar} \tag{92}$$

where Δr is the length of the recoupling region, ΔE the splitting between adiabatic surfaces and v the relative velocity of the fragments there. This parameter P is the time to traverse the recoupling region, $v^{-1}\,\Delta r$, divided by the internal period of the system, one inverse Bohr frequency $\hbar\,\Delta E^{-1}$. Because we emphasize separate regions along the dissociation coordinates, it is also useful to think of the nonadiabaticity parameter [Eq. (92)] as the length of the recoupling region measured in a natural unit of distance, $v\hbar\,\Delta E^{-1}$, the distance the fragments travel in a time period equal to an inverse Bohr frequency.

We introduce three approximation schemes that are useful when nonadiabatic transitions are localized to the recoupling region. In each of these schemes we assume that nonadiabatic transitions do not take place in the molecular region, or, at most that they occur at isolated crossing or avoided crossing points. The dynamics of the recoupling region may be treated by (1) direct numerical solution, (2) a diabatic or recoil approximation if $P \ll 1$ (Sections V.C–V.F), or (3) an adiabatic approximation if $P \gg 1$ (Section V.G). The greatest simplification for the photodissociation differential cross sections occurs in the diabatic limit. Because of this simplification and of the relevance of this limit to high energy processes, much discussion is devoted to this topic.

A. Numerical Methods

Numerical generation of the transition amplitudes $\tau(J_2jlc_aj_ac_bj_b|\,J_1\eta_1)$ is sometimes the only way to reliably predict cross sections from a set of model potential surfaces and nonadiabatic interactions. This is most likely to occur at low excess dissociation energy (unless the fragment kinetic energy is so small that the adiabatic approximation is valid). Numerical solution of the full multichannel dissociation problem is necessary to accurately predict the positions and shapes of resonances peaks in the low-energy region. These peaks are usually very sensitive to the shape of molecular potential energy surfaces and the nonadiabatic coupling between dissociation channels.

Numerical solution for diatomic dissociation is a relatively painless alternative because the number of dissociation channels (energetically available fragment electronic states) is relatively small. The dimensionality of the problem, and, consequently, the difficulty of numerical solution, is much greater for larger molecules in which energy is partitioned among fragment rotational and vibrational states, as well as electronic levels. A full discussion of numerical techniques for generating photodissociation amplitudes is beyond the scope of this chapter. Instead, we briefly describe several methods to guide the interested reader to more extensive discussions. Shapiro and Bersohn[7] also survey numerical techniques for obtaining photodissociation cross sections.

Techniques for numerical solution of the photodissociation problem can be divided according to whether the wave function for the initial bound state is required explicitly as input for the calculation, or is implicitly determined with photodissociation amplitudes in the same calculation that includes bound and dissociative channels. The wave functions of bound and dissociative states are most easily expressed using different coordinate systems or basis functions, or both. If the explicit form of the initial level can be compactly represented with natural bound-state coordinates and basis functions, and can be transformed to the natural representation of the continuum wave function, the most efficient numerical methods are those that employ an explicitly form of the initial states. Morse, Band, and Freed[113-120] have shown how a bound-state wave function for a triatomic molecule can be explicitly represented in the natural coordinates and basis functions of continuum states. If, on the other hand, the bound wave function is complicated, or is transformed into the natural representation of continuum states with difficulty, the methods of choice are those in which the bound state is implicitly determined along with the dissociative wave function. Even when many molecular electronic states need to be considered, diatomic photodissociation is not a particularly difficult numerical problem. Either of the two groups of methods here can be used with relative ease. However, the large number of channels encountered in polyatomic dissociation forces careful selection of the method best tailored to handle the individual problem of interest.

Kulander and Light[121] were the first to obtain transition amplitudes as the overlap of a multichannel continuum wave function with a bound state which is brought into the calculation explicitly. The continuum wave function matrix is generated using the stable and efficient R-matrix method, and the normalization of the stabilized wave function and transition amplitudes determined at the end of the calculation. Alternatively, Johnson's renormalized Numerov method[122] can be used to generate a stabilized continuum wave function.[11,123,124]

Band et al.[125] have shown that transition amplitudes can be obtained as the asymptotic coefficients of outgoing spherical waves in the solution of the driven Schrödinger equation,

$$[E\mathsf{1} - \mathsf{H}]|\psi^{(+)}_{\text{driven}}\rangle = \mathsf{V}|\chi\rangle$$

In the above equation H is the Hamiltonian matrix for the dissociative channels, and V is the coupling matrix (i.e., of dipole matrix elements) that induces dissociation of the molecule from the bound state $|\chi\rangle$. Boundary conditions of regularity at the origin and only outgoing spherical waves at large r are imposed on $|\psi^{(+)}_{\text{driven}}\rangle$. Using the method of Sams and Kouri,[126,127] the driven Schrödinger equation can be transformed into a Volterra integral equation. Band and Aizenbud[128,129] obtained $|\psi^{(+)}_{\text{driven}}\rangle$ numerically in this manner. The driven (inhomogeneous) Schrödinger equation cannot be stabilized in the same manner as the usual (homogeneous) Schrödinger equation. Singer et al.[130] have used invariant embedding concepts to derive nonlinear differential equations by which the asymptotic form of $|\psi^{(+)}_{\text{driven}}\rangle$ can be obtained in a stable manner.

Shapiro[131,132] first developed a practical method to calculate photodissociation transition amplitudes in which the bound-state wave function is implicitly determined in the same calculation as the dissociative state. Sharpiro constructs an augmented non-Hermitian Hamiltonian matrix which spans the space of bound and continuum channels, and, in addition, an "artificial" channel. The artificial channel introduces flux onto the dissociative channels while keeping the convenient homogeneous form of the Schrödinger equation. Photodissociation transition amplitudes are extracted from the poles of certain scattering T-matrix elements calculated from the artificial channel Hamiltonian. The procedure requires iterative convergence to a bound state energy after which photodissociation amplitudes at various energies may be calculated without further iteration. Shapiro and Balint-Kurti[133] have illustrated how the artificial channel method may be used to determine bound-state energies of a triatomic molecule.

The group of techniques known as complex scaling or complex coordinate rotation (reviewed by Reinhardt[134]) is another example of a method in which an explicit representation of the bound state wave function is not required prior to the calculation of photodissociation amplitudes. The common feature of these methods is the imposition of homogeneous boundary conditions on the eigenvectors of the molecular Hamiltonian, above the dissociation threshold and the interpretation of the resulting complex eigenvalues as the positions and widths of quasibound levels. Using Siegert boundary conditions,[135] Atabek and Lefebvre[136–139] have determined complex energy eigenvalues for a triatomic molecule which dissociates collinearly. While complex coordinate methods may not be suited for rapid calculation of photodissociation amplitudes to individual fragment states, they may prove useful when extremely narrow resonances arise in near threshold dissociation. These narrow resonances may be exceedingly difficult to characterize using conventional numerical techniques.

B. Adiabatic Treatment of the Molecular Region

For many cases electronically nonadiabatic transitions do not occur in the dissociating molecule when the fragments are in the molecular region (Fig. 1). This section shows how an adiabatic approximation can be used for the molecular region, and how this adiabatic solution can be joined to either an exact or approximate solution for the recoupling region in order to calculate photodissociation transition amplitudes to atomic fine structure states.

The adiabatic approximation is implemented by the following procedure: (1) Expand the Hamiltonian [Eq. (34)] for the dissociating fragments in any of the basis sets discussed in Section III. The dimensionality of $\mathsf{H}(r)$ is the total multiplicity of the atomic limit (or limits when curve crossings are present). (2) At each internuclear separation of the atomic limit find the eigenvectors of an adiabatic Hamiltonian $\mathsf{H}_{ad}(r)$, which is the full Hamiltonian minus the radial nuclear kinetic energy operator,

$$H_{ad}(r) = H(r) - \left(\frac{-\hbar^2}{2\mu} \frac{1}{r} \frac{\partial^2}{\partial r^2} r \right). \tag{93}$$

These eigenvectors are obtained through

$$\mathsf{O}^t(r)\mathsf{H}_{ad}(r)\mathsf{O}(r) = \boldsymbol{\varepsilon}(r) \tag{94}$$

where $\mathsf{O}(r)$ is an r-dependent orthogonal matrix by which $\mathsf{H}_{ad}(r)$ is transformed to the diagonal representation $\boldsymbol{\varepsilon}(r)$ with r-dependent eigenvalues $\varepsilon_{\alpha_2}(r)$. The label α_2 designates the adiabatic molecular states. (The subscript "2" is used along with quantum numers for the dissociative surfaces when they may be confused with similar quantum numbers for intermediate bound levels.)

The eigenvectors $|J_2M_2\alpha_2\rangle$ of $\mathsf{H}_{ad}(r)$ depend parametrically on r by virtue of the coefficients contained in the matrix $\mathsf{O}(r)$,

$$|J_2M_2\alpha_2\rangle = \sum_{\eta_2} |J_2M_2\eta_2\rangle O^{J_2}_{\eta_2;\alpha_2} \tag{95}$$

This r-dependence is not explicitly represented for notational convenience. The basis functions $|J_2M_2\eta_2\rangle$, used to expand the Hamiltonian in Eq. (95), are left unspecified for the moment. They may be any one of the Hund's case representations discussed in Section III or even the atomic basis [Eq. (14)]. (As explained in Section III, the atomic basis is not a basis of separated atom functions at finite r, but is a full basis for the photodissociation.) The same unique eigenvalues $\boldsymbol{\varepsilon}(r)$ are obtained no matter which of the alternative representations just mentioned are used to evaluate matrix elements of $\mathsf{H}_{ad}(r)$.

At large r the eigenvectors $|J_2M_2\alpha_2\rangle$ of $\mathsf{H}_{ad}(r)$ become the atomic states $|J_2M_2jlc_aj_ac_bj_b\rangle$ [Eq. (15)], by definition of the latter. If all relativistic and Coriolis coupling are included in $\mathsf{H}_{ad}(r)$, none of the eigenvalues $\varepsilon_{\alpha_2}(r)$ with identical good quantum numbers cross. The lowest energy adiabatic state $|J_2M_2\alpha_2\rangle'$ correlates adiabatically with the lowest energy atomic level $|J_2M_2jlc_aj_ac_bj_b\rangle'$, the next lowest energy adiabatic state $|J_2M_2\alpha_2\rangle''$ with the next lowest atomic level $|J_2M_2jlc_aj_ac_bj_b\rangle''$, and so on until all adiabatic states are paired with an atomic limit. The noncrossing rule guarantees that the pairing of adiabatic levels with their atomic limits does not depend on the internuclear separation at which we rank the eigenvalues $\varepsilon_{J_2\alpha_2}(r)$. Therefore, each adiabatic level can just as well be identified by the quantum numbers of its atomic limit,

$$|J_2M_2\alpha_2\rangle = |J_2M_2\alpha_2(jlc_aj_ac_bj_b)\rangle \tag{96}$$

The adiabatic approximation consists of replacing the multichannel continuum wave function $|EJ_2M_2jlc_aj_ac_bj_b^{(-)}\rangle$ of Eqs. (14), (15), and (50) by a linear combination of continuum states between which there are no couplings in the molecular region,

$$|EJ_2M_2jlc_aj_ac_bj_b^{(-)}\rangle = \sum_{\alpha_2} \frac{1}{r} \psi^{(-)}_{EJ_2\alpha_2}(r)|J_2M_2\alpha_2\rangle C^{J_2}_{\alpha_2;jlc_aj_ac_bj_b} \tag{97}$$

The adiabatic radial wave functions satisfy a Schrödinger equation for motion on the potential energy surface $\varepsilon_{J_2\alpha_2}(r)$,

$$\left[E + \frac{\hbar^2}{2\mu}\frac{d^2}{dr^2} - \varepsilon_{J_2\alpha_2}(r)\right]\psi^{(-)}_{EJ_2\alpha_2}(r) = 0 \tag{98}$$

As is well known, the adiabatic approximation involves neglect of terms that arise from action of the radial derivatives in the kinetic energy operator on the basis states $|J_2M_2\alpha_2\rangle$.

The adiabatic wave function of Eqs. (97) and (98) is regular at the origin,

$$\psi^{(-)}_{EJ_2\alpha_2}(r) = 0, \ r \to 0 \tag{99}$$

In addition, the wave function of Eq. (97) must match smoothly onto the solution used to describe the recoupling region. The value and first derivative of $\psi^{(-)}_{EJ_2\alpha_2}(r)$ at the boundary joining the molecular and recoupling regions can be used to begin numerical propagation of a multichannel wave function out to large r at which point the asymptotic boundary conditions stated in Eq. (51) are imposed to determine the overall normalization of the wave function. A diabatic or adiabatic approximation may be used in certain limiting circumstances instead of full-scale numerical propagation to continue the wave function through the recoupling region. Imposition of boundary conditions under the diabatic or adiabatic approximations is explained below in Sections III.C and III.G, respectively. In the discussion that follows, it is assumed that we know the $C^{J_2}_{\alpha_2;jlc_aj_ac_bj_b}$ which guarantee that asymptotic boundary conditions [Eq. (51)] are obeyed. In practice, these coefficients are determined at the end of the calculation, after the rest of the wave function is obtained at larger r and the asymptotic analysis is performed.

Let r_1 be the boundary between the molecular and recoupling regions. The precise value of r_1 is never important in any of the approximation schemes we describe. In almost all circumstances, r_1 falls far outside the Franck-Condon region. Therefore the only significant contribution to the transition amplitude comes from the molecular region although, to be sure, the adiabatic wave function is strongly affected by the dynamics of the recoupling region.

Under the weakly restrictive assumption that r_1 lies outside the Franck-Condon region, the photodissociation amplitudes are given as,

$$\tau(J_2jlc_aj_ac_bj_b|J_1\eta_1) = \sum_{\alpha_2} \tau(J_2\alpha_2|J_1\eta_1)C^{J_2*}_{\alpha_2;jlc_aj_ac_bj_b} \tag{100}$$

where $\tau(J_2\alpha_2|J_1\eta_1)$ is the adiabatic approximation transition amplitude,

$$\tau(J_2\alpha_2|J_1\eta_1) = \int_0^{r_1} dr\psi^{(-)*}_{EJ_2\alpha_2}(r)\mu_{J_2\alpha_2;J_1\eta_1}(r)\chi_{J_1v_1}(r) \tag{101}$$

The transition dipole moment function in Eq. (101) depends on the elements of the matrix $\mathsf{O}(r)$. For concreteness, $\mu_{J_2\alpha_2;J_1\eta_1}(r)$ is presented for $\mathsf{H}(r)$ expanded in the Hund's case (a) representation prior to diagonalization,

$$\begin{aligned}\mu_{J_2\alpha_2;J_1\eta_1}(r) = \sum_{\substack{c_2\Lambda_2S_2\Sigma_2q_2' \\ \Lambda_1\Sigma_1}} & O^{J_2}_{\alpha_2;c_2\Lambda_2S_2\Sigma_2}(r)\langle c_2\Lambda_2S_2| \\ & \times x_{q_2'}|c_1\Lambda_1S_1\rangle\delta_{S_1S_2}\delta_{\Sigma_1\Sigma_2}(J_1)^{1/2}(J_2)^{-1/2} \\ & \times \langle J_2\Omega_2|1J_1q_2'\Omega_1\rangle a(\eta_1|\Lambda_1\Sigma_1)\end{aligned} \tag{102a}$$

or the Hund's case (b) representation,

$$\begin{aligned}\mu_{J_2\alpha_2;J_1\eta_1}(r) = \sum_{\substack{c_2\Lambda_2S_2N_2q_2' \\ \Lambda_1N_1}} & O^{J_2}_{\alpha_2;c_2\Lambda_2S_2N_2}(r)\langle c_2\Lambda_2S_2|x_{q_2'}|c_1\Lambda_1S_1\rangle\delta_{S_1S_2} \\ & \times (N_1)^{1/2}(J_1)^{1/2}\langle N_2\Lambda_2|1N_1q_2'\Lambda_1\rangle \\ & \times W(1J_2N_1S_1;\, J_1N_2)a(\eta_1|\Lambda_1N_1)\end{aligned} \tag{102b}$$

Equation (100) reflects the physical assumptions inherent in treating the photodissociation dynamics separately in the molecular and recoupling regions. The amplitudes $\tau(J_2\alpha_2|J_1\eta_1)$ describe selection rules for the optical transitions and effects arising from the efficiency with which the bound and continuum wave functions in Eq. (101) overlap in the Franck-Condon region. Equations (100) and (101) may be easily generalized to treat crossings or near crossings of ABO levels in the molecular region. This generalization requires substitution of a multichannel wave function in Eq. (101) for those particular sets of levels that do cross.

Electronically nonadiabatic effects in the recoupling region are included through the coefficients $C^{J}_{\alpha_2;\,jlc_aj_ac_bj_b}$. As mentioned before, these coefficients can be obtained by matching at the point r_1 a numerically generated wave function for $r > r_1$ to the adiabatic approximation for $r < r_1$. In addition, we introduce diabatic and adiabatic approximations by which the $C^{J_2}_{\alpha_2;\,jlc_aj_ac_bj_b}$ can be found in the limit of very small or large values of the parameter P in Eq. (92).

C. Diabatic or Recoil Approximation

The dissociating molecule behaves according to the diabatic or recoil approximation when the width Δr of the recoupling region is small compared to the length, $v\hbar\,\Delta E^{-1}$. ΔE is the splitting between adiabatic levels in the recoupling region and v is the relative velocity of the fragments. [See Eq. (92) and accompanying discussion.] The following conditions tend to make the dissociation diabatic: (1) The asymptotic electronic interactions between the fragments are short-ranged dispersion forces which quickly become smaller than the magnitude of nonadiabatic interactions implying a short range Δr of the recoupling region; (2) A small reduced mass or a large kinetic energy of the fragments. Either of these makes v large; (3) Small splittings between adiabatic levels in the recoupling region.

The recoil limit approximation photodissociation amplitudes $\tau(J_2jlc_aj_ac_bj_b|\,J_1\eta_1)$ are derived by matching an adiabatic solution for the molecular region ($r < r_1$) of the form of Eq. (97) to a solution for $r > r_1$ in the limit of vanishing or negligible Δr and ΔE. Because the interval Δr is small, we match the adiabatic solution [Eq. (97)] at r_1 to the asymptotic form of the wave function for the atomic region [Eq. (51)]. For simplicity, suppose that the adiabatic Hamiltonian [Eq. (93)] is expanded in the atomic basis set [Eq. (14)] prior to the diagonalization of Eq. (94). (This is *not* an expansion in wave functions for noninteracting atomic fragments, as explained in Section III.) The orthogonal matrix, which effects the diagonalization in Eq. (94) at r_- slightly less than r_1, is designated by $\mathsf{O}(r_-)$. In other words, $\mathsf{O}(r_-)$ specifies the transformation to the representation which best describes the dissociating molecule immediately before the fragments enter the recoupling region.

Under the recoil or diabatic approximation, we assume that the width of the recoupling region is negligible in size. Therefore we match the adiabatic solution from the molecular region [Eq. (97)] directly with the asymptotic atomic solution [Eq. (51)] at the point r_1. We first express the atomic basis functions in Eq. (51) in terms of the adiabatic basis functions [Eq. (95)] as

$$|J_2M_2jlc_aj_ac_bj_b\rangle = \sum_{\alpha_2} O^{J_2}_{jlc_aj_ac_bj_b;\alpha_2}(r_-)|J_2M_2\alpha_2\rangle$$

This equation is valid only for r very close, but slightly less than r_1. The matching conditions obtained by requiring the value and first derivative of the wave function be continuous at r_1 are given as,

$$\psi^{(-)}_{EJ_2\alpha_2}(r_1)C^{J_2}_{\alpha_2;jlc_aj_ac_bj_b} - \sum_{j'l'c'_aj'_ac'_bj'_b}\left(\frac{2\mu}{\pi\hbar^2}\right)^{1/2}\left\{\frac{\hat{j}_l(kr_1)}{k^{1/2}} \times \delta_{jlc_aj_ac_bj_b;j'l'c'_aj'_ac'_bj'_b} + \frac{\hat{h}^{(-)}_{l'}(kr_1)}{k^{1/2}}\mathcal{T}_{jlc_aj_ac_bj_b;j'l'c'_aj'_ac'_bj'_b}\right\} \times O^{J_2}_{j'l'c'_aj'_ac'_bj'_b;\alpha_2}(r_-) = 0 \qquad (103a)$$

and

$$\frac{d}{dr}\left[\psi^{(-)}_{EJ_2\alpha_2}(r)C^{J_2}_{\alpha_2;jlc_aj_ac_bj_b} - \sum_{j'l'c'_aj'_ac'_bj'_b}\left(\frac{2\mu}{\pi\hbar^2}\right)^{1/2}\left\{\frac{\hat{j}_l(kr)}{k^{1/2}} \times \delta_{jlc_aj_ac_bj_b;j'l'c'_aj'_ac'_bj'_b} + \frac{\hat{h}^{(-)}_{l'}(kr)}{k^{1/2}}\mathcal{T}_{jlc_aj_ac_bj_b;j'l'c'_aj'_ac'_bj'_b}\right\} \times O^{J_2}_{j'l'c'_aj'_ac'_bj'_b;\alpha_2}(r_-)\right]_{r=r_1} = 0 \qquad (103b)$$

An algebraic equation is obtained for the scattering T-matrix elements in terms of the logarithmic derivatives of the adiabatic wave functions $\psi^{(-)}_{EJ_2\alpha_2}(r)$ at r_1 by eliminating the coefficients $C^{J_2}_{\alpha_2;jlc_aj_ac_bj_b}$ between Eqs. (103a) and (103b). Similarly, the $C^{J_2}_{\alpha_2;jlc_aj_ac_cj_b}$ can be found by elimination of the T-matrix between Eqs. (103a) and (103b).

The solution to Equations (103a) and (103b) is found in closed form when r_1 is sufficiently large so that the Bessel functions in these expressions can be replaced by their asymptotic limits,

$$\hat{j}_l(kr) = \sin(kr - \tfrac{1}{2}l\pi),\ r \to \infty \qquad (104a)$$

$$\hat{h}^{(-)}_l(kr) = -\exp[-i(kr - \tfrac{1}{2}l\pi)],\ r \to \infty \qquad (104b)$$

and $\psi^{(-)}_{EJ_2\alpha_2}(r_1)$ can likewise be represented in its asymptotic form which is written as if the outgoing plane wave, incoming spherical wave boundary conditions were imposed in the adiabatic molecular basis,

$$\psi^{(-)}_{EJ_2\alpha_2}(r) = \left(\frac{2\mu}{\pi\hbar^2 k}\right)^{1/2} \mathrm{e}^{-i\delta_{J_2\alpha_2}} \sin(kr + \delta_{J_2\alpha_2}) \tag{105}$$

Because of the assumption of small ΔE underlies the recoil limit approximation, the wave vectors for all channels are taken to be equal at r_1, and Eqs. (103a) and (103b) are solved to yield

$$\mathscr{T}^{J_2}_{jlc_aj_ac_bj_b;j'l'c'_aj'_ac'_bj'_b} = i^{l-l'} \sum_{\alpha_2} O^{J_2}_{jlc_aj_ac_bj_b;\alpha_2}(r_-)\, \mathrm{e}^{-i\delta_{J_2\alpha_2}} \times \sin(\delta_{J_2\alpha_2}) O^{J_2}_{j'l'c'_aj'_ac'_bj'_b;\alpha_2}(r_-) \tag{106}$$

$$C^{J_2}_{\alpha_2;jlc_aj_ac_bj_b} = i^{-l} O^{J_2}_{jlc_aj_ac_bj_b;\alpha_2} \tag{107}$$

According to Eq. (106), the recoil limit T-matrix is constructed by an orthogonal similarity transform of the diagonal molecular region adiabatic basis T-matrix with elements $\exp(-i\delta_{J_2\alpha_2})\sin(\delta_{J_2\alpha_2})$. The T-matrix of Eq. (106) is obtained by enforcing the proper boundary conditions for photodissociation, an outgoing plane wave in atomic state $c_aj_ac_bj_b$ and incoming spherical waves in all other fine structure states $c'_aj'_ac'_bj'_b$. The usual T-matrix for inelastic scattering is the complex conjugate of Eq. (106).

Substituting Eq. (107) in Eq. (100), we obtain the recoil limit approximation for photodissociation transition amplitudes,

$$\tau(J_2jlc_aj_ac_bj_b|J_1\eta_1) = \sum_{\alpha_2} \tau(J_2\alpha_2|J_1\eta_1) i^l O^{J_2}_{jlc_aj_ac_bj_b;\alpha_2}(r_-) \tag{108}$$

The amplitudes $\tau(J_2\alpha_2|J_1\eta_1)$ depend on the behavior of the matrix $\mathsf{O}(r)$ in the Franck-Condon region [see Eqs. (101) and (102)], while the value of $\mathsf{O}(r_-)$ scrambles the flux in the recoupling region.

Many molecules have $\mathsf{O}(r)$ in the Franck-Condon region differ considerably from $\mathsf{O}(r_-)$. For instance, coupling in diatoms composed of two light atoms tends to be closer to Hund's case (b) at small internuclear separations where large Coriolis interactions arise due to their proportionality to r^{-2}. As r increases, the Coriolis interaction decreases relative to the spin-orbit splitting, and the molecular coupling scheme becomes closer to Hund's case (a). Another common example is found in heavier diatomics, such as halogen dimers, where spin-orbit coupling overtakes the ABO level splitting as the latter decreases with increasing r. These heavier diatomics evolve from case (a) to case (c) with increasing internuclear separation. In the latter example, the amplitude $\tau(J_2\alpha_2|J_1\eta_1)$ is evaluated using case (a) basis functions while the matrix $\mathsf{O}(r_-)$ in Eq. (108) is given by the transformation from the case (c) to the atomic basis.

Equation (108) describes the dissociation of molecules in the approximation where intramolecular coupling evolves adiabatically from the Franck-Condon to the recoupling region, and then the molecule undergoes a sudden change from molecular to atomic coupling at large r_1. The approximation given in Eq. (108) is called the J-dependent recoil limit because the amplitudes $\tau(J_2\alpha_2|J_1\eta_1)$ depend on J_2 though both dynamical and geometrical factors. We neglect the dynamical effect of molecular rotation in a further approximation called the J-independent recoil limit. In this J-independent approximation $\tau(J_2\alpha_2|J_1\eta_1)$ varies with J_2 only through geometrical factors which can be removed from the rest of the amplitude.

The power of Eq. (108) to simplify formulas for cross sections is shown dramatically when the dissociating molecule can be described by one of the Hund's coupling cases before the fragments enter the recoupling region. Then the matrix elements $O^{J_2}_{jlc_aj_ac_bj_b;\alpha_2}(r_-)$ are simply given analytically in terms of angular momentum recoupling coefficients, as previously derived in Section III for Hund's coupling cases (a), (b), and (c). In the discussion of the J-dependent recoil limit which follows immediately, we explore the consequences of replacing the $O^{J_2}_{jlc_aj_ac_bj_b;\alpha_2}(r_-)$ in Eq. (108) by the transformation coefficients that relate the atomic basis functions to either the Hund's case (a), (b) or (c) basis states. Subsequently, we discuss the J-independent recoil limit, in which further simplification is made possible by neglecting the variation of $\tau(J_2\alpha_2|J_1\eta_1)$ with J_2 which arises from the dynamical effects of molecular rotation.

D. J-Dependent Recoil Limit

According to Eq. (108), photodissociation transition amplitudes to atomic fine structure states can be written as sums of products of two factors in the diabatic or large kinetic energy limit. The first factor $\tau(J_2\alpha_2|J_1\eta_1)$ is a transition amplitude to an adiabatic molecular level. The second factor is a unitary transformation coefficient $i^lO^{J_2}_{jlc_aj_ac_bj_b;\alpha_2}(r_-)$ that relates basis functions of the adiabatic molecular representation to the atomic basis. Taken in isolation, Eq. (108) is merely an expression of the transition amplitudes $\tau(J_2jlc_aj_ac_bj_b|J_1\eta_1)$ in an alternative basis; the amplitudes $\tau(J_2\alpha_2|J_1\eta_1)$ could, in principle, be calculated exactly, from a close coupled calculation at all r, and then Eq. (108) would not represent an approximation. Equation (108) does represent a physical approximation when nonadiabatic interactions in the molecular region are not used to calculate the $\tau(J_2\alpha_2|J_1\eta_1)$ [Eq. (101)] and when the diabatic approximation [Eq. (107)] is used to treat nonadiabatic processes in the recoupling region. Since the form of Eq. (108) is exact, we can retain some nonadiabatic coupling in the molecule, which, for instance, may arise from a curve crossing, yet still avoid a full-scale numerical calculation. In the curve crossing example, a multichannel wave function is required in Eq. (101) for just those adiabatic molecular levels that cross. The matrix $\mathsf{O}(r_-)$ describes intramolecular coupling at r_- near but slightly less than the internuclear separation where the angular momenta begin to recouple to atomic levels. If intramolecular coupling at r_- falls under one of the Hund's coupling cases discussed in Section III, $\mathsf{O}(r_-)$ is given analytically in terms of angular momentum recoupling coefficients. Since Eq. (108) is in the form of a unitary transformation of the transition amplitudes, this transformation can be absorbed in the Liouville space operator $\mathscr{U}^\dagger[B_3\beta_3(j_a;\, j_a)$ $|B_2\beta_2(J_2jl;\, \bar{J}_2\bar{j}\bar{l})]$ [Eq. (81)] when the transformation is given in terms of angular momentum recoupling coefficients. This procedure generates a new half-collision operator $\tilde{\mathscr{L}}[B_2\beta_2(J_2\eta_2;\, \bar{J}_2\bar{\eta}_2)|B_1\beta_1(J_1\eta_1;\, \bar{J}_1\bar{\eta}_1)]$ and geometrical matrix $\tilde{\mathscr{U}}^\dagger[B_3\beta_3(j_a;\, j_a)|B_2\beta_2(J_2\eta_2;\, \bar{J}_2\bar{\eta}_2)]$ for each Hund's case representation.

In the approximation $O^{J_2}_{jlc_aj_ac_bj_b;\alpha_2}(r_-)$ in Eq. (108) can be replaced by $\langle\eta_2|jlc_aj_ac_bj_b\rangle_{J_2}$, a transformation coefficient which projects a Hund's case basis function onto an atomic basis state (i.e. $\alpha_2 \rightarrow \eta_2$), the photodissociation amplitude is given by

$$\tau(J_2jlc_aj_ac_bj_b|J_1\eta_1) = \sum_{\eta_2} \tau(J_2\eta_2|J_1\eta_1)i^l \times \langle\eta_2|jlc_aj_ac_bj_b\rangle_{J_2} \qquad (109)$$

where $\langle\eta_2|jlc_aj_ac_bj_b\rangle_{J_2}$ is $\langle c\Lambda_2S_2\Sigma_2|jlc_aj_ac_bj_b\rangle_{J_2}$ [Eq. (16)] for Hund's case (a), $\langle c\Lambda_2S_2N_2|jlc_aj_ac_bj_b\rangle_{J_2}$ [Eq. (28)] for Hund's case (b), or $\langle\Omega|l\rangle_{J_2j}$ [Eq. (32b)] for Hund's case (c). Residual nonadiabatic couplings, which are not described by the transformation coefficient $\langle\eta_2|jlc_aj_ac_bj_b\rangle_{J_2}$ at large fragment kinetic energy, may still be included in the calculation of the amplitudes $\tau(J_2\eta_2|J_1\eta_1)$.

Substitution of J-dependent recoil limit amplitudes [Eq. (109)] into the half-collision operation $\mathscr{L}$ [Eq. (79)] is found to lead to a product of two dual-space operators: One is a half-collision operator, which describes dissociation to molecular states and which we label $\tilde{\mathscr{L}}[B_2\beta_2(J_2\eta_2;$ $\bar{J}_2\bar{\eta}_2|B_1\beta_1;\bar{J}_1\bar{\eta}_1)]$, while the remaining operator effects a unitary transformation in Liouville space from the molecular to the atomic basis,

$$\mathscr{L}[B_2\beta_2(J_2jl;\ \bar{J}_2\bar{j}\bar{l})|\mathrm{B}_1\beta_1(\mathrm{J}_1\eta_1;\ \bar{J}_1\bar{\eta}_1)]$$
$$\equiv \sum_{\eta_2\bar{\eta}_2} i^{l-\bar{l}}\langle jlc_aj_ac_bj_b|\eta_2\rangle_{J_2}\langle\eta_2|\bar{j}\bar{l}c_aj_ac_bj_b\rangle_{\bar{J}_2}$$
$$\times\ \tilde{\mathscr{L}}[B_2\beta_2(J_2\eta_2;\ \bar{J}_2\bar{\eta}_2)|B_1\beta_1(J_1\eta_1;\ \bar{J}_1\bar{\eta}_1)] \tag{110}$$

where

$$\tilde{\mathscr{L}}[B_2\beta_2(J_2\eta_2;\ \bar{J}_2\bar{\eta}_2)|B_1\beta_1(J_1\eta_1;\ \bar{J}_1\bar{\eta}_1)]$$
$$= \sum_{K_2Q_2}(-)^{J_2-\bar{J}_2}(-)^{J_1-\bar{J}_1}[(J_2)(\bar{J}_2)(K_1)(K_2)]^{1/2}\langle B_2\beta_2|B_1K_2\beta_1Q_2\rangle \tag{111}$$
$$\times\begin{Bmatrix} B_1 & K_2 & B_2\\ J_1 & 1 & J_2\\ \bar{J}_1 & 1 & \bar{J}_2\end{Bmatrix}\phi^{K_2}_{Q_2}(\hat{\varepsilon}_2)\tau(J_2\eta_2|J_1\eta_1)\tau^*(\bar{J}_2\bar{\eta}_2|\bar{J}_1\bar{\eta}_1)$$

The transformation operator, that is factored from $\mathscr{L}$ in Eq. (110) by invoking the diabatic approximation, is now multipled into $\mathscr{U}^\dagger$ [Eq. (81)]. This gives rise to a J-dependent recoil limit transformation operator, labeled $\tilde{\mathscr{U}}^\dagger[B_3\beta_3(j_a;\ j_a)|B_2\beta_2(J_2\eta_2;\ \bar{J}_2\bar{\eta}_2)]$, which directly transforms from the molecular to the plane wave atomic dual-space basis. We carry out these steps explicitly as follows:

$$\sum_{\substack{B_2J_2\bar{j}\bar{l}\\ \beta_2J_2jl}}\mathscr{U}^\dagger[B_3\beta_3(j_a;\ j_a)|B_2\beta_2(J_2jl;\ \bar{J}_2\bar{j}\bar{l})]\mathscr{L}[B_2\beta_2(J_2jl;\ \bar{J}_2\bar{j}\bar{l})$$
$$\times\ |B_1\beta_1(J_1\eta_1;\ \bar{J}_1\bar{\eta}_1)] = \sum_{\substack{B_2\bar{J}_2\bar{j}\bar{l}\\ \beta_2J_2jl}}\mathscr{U}^\dagger[B_3\beta_3(j_a;\ j_a)|B_2\beta_2(J_2jl;\ \bar{J}_2\bar{j}\bar{l})$$
$$\times\sum_{\eta_2\bar{\eta}_2} i^{l-\bar{l}}\langle jlc_aj_ac_bj_b|\eta_2\rangle_{J_2}\langle\bar{\eta}_2|\bar{j}\bar{l}c_aj_ac_bj_b\rangle_{\bar{J}_2}\tilde{\mathscr{L}}[B_2\beta_2(J_2\eta_2;\ \bar{J}_2\bar{\eta}_2)$$
$$\times\ |B_1\beta_1(J_1\eta_1;\ \bar{J}_1\bar{\eta}_1)] \equiv \sum_{\substack{B_2\bar{J}_2\bar{\eta}_2\\ \beta_2J_2\eta_2}}\tilde{\mathscr{U}}^\dagger[B_3\beta_3(j_a;\ j_a)|B_2\beta_2(J_2\eta_2;\ \bar{J}_2\bar{\eta}_2)]$$
$$\times\ \tilde{\mathscr{L}}[B_2\beta_2(J_2\eta_2;\ \bar{J}_2\bar{\eta}_2)|B_1\beta_1(J_1\eta_1;\ \bar{J}_1\bar{\eta}_1)] \tag{112}$$

The sums over j, l, $\bar{j}$, and $\bar{l}$ necessary to obtain $\tilde{\mathscr{U}}^\dagger$ in Eq. (112) are performed in Ref. 108. When the dissociative molecular states are expanded in a Hund's case (a) basis, the angular momentum algebra yields

$$\begin{aligned}
&\tilde{\mathscr{U}}^{\dagger}[B_3\beta_3(j_a;\, j_a)|B_2\beta_2(J_2\eta_2;\, \bar{J}_2\bar{\eta}_2)] \\
&\quad = \sum_{K_DQ_D} [(K_D)(B_3)(J_2)(\bar{J}_2)]^{1/2}\, \frac{1}{4\pi}\, C_{K_DQ_D}(\hat{\mathbf{k}}) \langle B_2\beta_2|B_3K_D\beta_3Q_D\rangle \\
&\quad \times \langle \Lambda_a\Lambda_b|c_2\Lambda_2\rangle\langle \bar{c}_2\bar{\Lambda}_2|\bar{\Lambda}_a\bar{\Lambda}_b\rangle \sum_{L\bar{L}XZ} (L)(\bar{L})(X)(Z)(S_2)^{1/2}(\bar{S}_2)^{1/2} \\
&\quad \times (j_a)(j_b)\begin{pmatrix} L & \bar{L} & X \\ \Lambda_2 & -\bar{\Lambda}_2 & -\lambda_2 \end{pmatrix}\begin{pmatrix} L & l_a & l_b \\ -\Lambda_2 & \Lambda_a & \Lambda_b \end{pmatrix}\begin{pmatrix} \bar{L} & l_a & l_b \\ -\bar{\Lambda}_2 & \bar{\Lambda}_a & \bar{\Lambda}_b \end{pmatrix} \\
&\quad \times \left\{\begin{array}{cccccc} Z & \bar{S}_2 & s_b & l_b & L & X \\ & & & j_b & & \\ & S_2 & s_a & & s_a & \bar{L} \\ & & & K_s & & \\ s_b & s_a & j_a & j_a & l_a & l_b \end{array}\right\} \\
&\quad \times \left[(-)^{J_2+\Omega_2}(-)^{\bar{S}_2-\Sigma_2}(-)^{\bar{L}-\bar{\Lambda}_2}\begin{pmatrix} K_D & B_2 & B_3 \\ 0 & \omega_2 & -\omega_2 \end{pmatrix} \right. \\
&\quad \times \left. \begin{pmatrix} B_2 & J_2 & \bar{J}_2 \\ \omega_2 & -\Omega_2 & \bar{\Omega}_2 \end{pmatrix}\begin{pmatrix} B_3 & Z & X \\ \omega_2 & -\sigma_2 & -\lambda_2 \end{pmatrix}\begin{pmatrix} Z & S_2 & \bar{S}_2 \\ -\sigma_2 & \Sigma_2 & -\bar{\Sigma}_2 \end{pmatrix} \right]
\end{aligned} \tag{113}$$

In the Hund's case (b) basis we find

$$\begin{aligned}
&\tilde{\mathscr{U}}^{\dagger}[B_3\beta_3(j_a;\, j_a)|B_2\beta_2(J_2\eta_2;\, \bar{J}_2\bar{\eta}_2)] \\
&\quad = \sum_{K_DQ_D} [(K_D)(B_3)(J_2)(\bar{J}_2)]^{1/2}\, \frac{1}{4\pi}\, C_{K_DQ_D}(\hat{\mathbf{k}}) \langle B_2\beta_2|B_3K_D\beta_3Q_D\rangle \\
&\quad \times \langle \Lambda_a\Lambda_b|c_2\Lambda_2\rangle\langle \bar{c}_2\bar{\Lambda}_2|\bar{\Lambda}_a\bar{\Lambda}_b\rangle \sum_{L\bar{L}XZ} (L)(\bar{L})(X)(Z)(S_2)^{1/2}(\bar{S}_2)^{1/2} \\
&\quad \times (j_a)(j_b)\begin{pmatrix} L & \bar{L} & X \\ \Lambda_2 & -\bar{\Lambda}_2 & -\lambda_2 \end{pmatrix}\begin{pmatrix} L & l_a & l_b \\ -\Lambda_2 & \Lambda_a & \Lambda_b \end{pmatrix}\begin{pmatrix} \bar{L} & l_a & l_b \\ -\bar{\Lambda}_2 & \bar{\Lambda}_a & \bar{\Lambda}_b \end{pmatrix} \\
&\quad \times \left\{\begin{array}{cccccc} Z & \bar{S}_2 & s_b & l_b & L & X \\ & & & j_b & & \\ & S_2 & s_a & & s_a & \bar{L} \\ & & & K_s & & \\ s_b & s_a & j_a & j_a & l_a & l_b \end{array}\right\}
\end{aligned}$$

$$\times \Bigg[(-)^{\bar{L}-N_2+R_2+B_3+S_2+\bar{S}_2}(R_2)(N_2)^{1/2}(\bar{N}_2)^{1/2}$$

$$\times \begin{pmatrix} N_2 & \bar{N}_2 & R_2 \\ \Lambda_2 & -\bar{\Lambda}_2 & -\lambda_2 \end{pmatrix} \begin{pmatrix} X & R_2 & K_D \\ -\lambda_2 & \lambda_2 & 0 \end{pmatrix} \begin{Bmatrix} B_2 & B_3 & K_D \\ X & R_2 & Z \end{Bmatrix}$$

$$\times \left\{ \begin{matrix} B_2 & R_2 & Z \\ J_2 & N_2 & S_2 \\ \bar{J}_2 & \bar{N}_2 & \bar{S}_2 \end{matrix} \right\} \Bigg] \tag{114}$$

while in the Hund's case (c) basis we have

$$\tilde{\mathscr{U}}^{\dagger}[B_3\beta_3(j_a; j_a)|B_2\beta_2(J_2\eta_2; \bar{J}_2\bar{\eta}_2)]$$

$$= \sum_{K_D Q_D} (K_D)[(B_3)(J_2)(\bar{J}_2)(j)(\bar{j})]^{1/2} \frac{1}{4\pi} C_{K_D Q_D}(\hat{\mathbf{k}})(-)^{\omega_2+\bar{j}+J_2}$$

$$\times \langle B_2\beta_2|B_3K_D\beta_3Q_D\rangle W(jB_3j_bj_a; \bar{j}\bar{j}_a) \begin{pmatrix} K_D & B_2 & B_3 \\ 0 & \omega_2 & -\omega_2 \end{pmatrix}$$

$$\times \begin{pmatrix} B_2 & J_2 & \bar{J}_2 \\ \omega_2 & -\Omega_2 & \bar{\Omega}_2 \end{pmatrix} \begin{pmatrix} B_3 & j & \bar{j} \\ -\omega_2 & \Omega_2 & -\bar{\Omega}_2 \end{pmatrix} \tag{115}$$

The symbol $\{\ldots\}$ in Eqs. (113) and (114) is a particular type of 18-j symbol which is explicitly given as

$$\sum_{F\bar{F}} (-)^{j_a+j_b+F}(F)(\bar{F}) \begin{Bmatrix} \bar{F} & F & B_3 \\ j_a & j_a & j_b \end{Bmatrix} \left\{ \begin{matrix} S & F & L \\ s_a & j_a & l_a \\ s_b & j_b & l_b \end{matrix} \right\}$$

$$\times \left\{ \begin{matrix} \bar{S} & \bar{F} & \bar{L} \\ s_a & j_a & l_a \\ s_b & j_b & l_b \end{matrix} \right\} \left\{ \begin{matrix} X & B_3 & Z \\ L & F & S \\ \bar{L} & \bar{F} & \bar{S} \end{matrix} \right\} \tag{116}$$

A thorough discussion of the 18-j symbols can be found in Appendix 4 of Ref. 109. Equations (113) and (114) are arranged to highlight the similarities between the case (a) and case (b) transformation operators. The portions of $\tilde{\mathscr{U}}^{\dagger}$ which are different in the case (a) and case (b) versions are put into square brackets at the end of Eqs. (113) and (114). These factors are equal when total spin vanishes ($S_2 = \bar{S}_2 = 0$) and therefore case (a) and (b) bases are identical.

The results of this section are now summarized by stating the steps necessary to calculate J-dependent recoil limit differential cross sections:

1. The density matrix for intermediate bound levels [Eq. (75)] is retained exactly within the recoil limit approximation scheme. This density matrix is evaluated by insertion of expansion coefficients for the ground and intermediate multiplet levels and the Franck-Condon factors $\langle J_1 v_1 c_1 \Lambda_1 S | x_{q_1'} | J_0 v_0 c_0 \Lambda_0 S \rangle$ for the bound–bound transition.
2. The labor involved in evaluating the J-dependent recoil limit collision operator $\tilde{\mathscr{L}}$ [Eq. (111)] mostly lies in calculating the amplitudes $\tau(J_2\eta_2|J_1\eta_1)$. These are dipole matrix elements [see Eq. (101)] between the intermediate bound level $|J_1 M_1 \eta_1\rangle$ and a continuum wave function which is generated by using adiabatic molecular states at small r. This continuum molecular wave function [Eq. (105)] has a plane wave $\exp(i\mathbf{k}\cdot\mathbf{r})$ in the channel labeled by η_2 and possible incoming spherical waves in all other channels. If there is no coupling between adiabatic molecular states at small r, an uncoupled radial wave function $\psi^{(-)}_{EJ_2\eta_2}(r)$ [Eq. (98)] is generated for each molecular channel, and $\tau(J_2\eta_2|J_1\eta_1)$ is the usual bound-free Franck-Condon factor for these adiabatic functions. On the other hand, if nonadiabatic couplings due to, say, a curve crossing at small r, is assumed to be significant, a multichannel wave function which reflects the nonadiabaticity caused by this interaction is required.
3. Once the dynamical factors, contained in $\tilde{\mathscr{L}}$ [Eq. (111)] and ρ_1 [Eq. (75)], are in hand, the rest of the calculation of J-dependent recoil limit cross sections consists of manipulation of the pure geometrical factors $\mathscr{U}^\dagger$ [Eqs. (113)–(115)] and $\mathscr{D}^\dagger$ [Eq. (83)]. The detailed J-dependent recoil limit cross section, can then be evaluated as

$$\frac{d^2\sigma}{d\Omega_\varepsilon\, d\Omega_k}(\hat{\varepsilon}_S, \mathbf{k}) = \Gamma \sum_{\substack{B_1B_2B_3B_4\\ \beta_1\beta_2\beta_3\beta_4}} \sum_{\substack{\bar{\eta}_1\bar{\eta}_2\bar{J}_2\\ \eta_1\eta_2J_2}} \mathscr{D}^\dagger[B_3\beta_3(j_a; j_a)]$$
$$\times \tilde{\mathscr{U}}^\dagger[B_3\beta_3(j_a; j_a)|B_2\beta_2(J_2\eta_2; \bar{J}_2\bar{\eta}_2)]$$
$$\times \tilde{\mathscr{L}}[B_2\beta_2(J_2\eta_2; \bar{J}_2\bar{\eta}_2)|B_1\beta_1(J_1\eta_1; \bar{J}_1\bar{\eta}_1)]$$
$$\times \rho_1[B_1\beta_1(J_1\eta_1; \bar{J}_1\bar{\eta}_1))] \tag{117}$$

The cross sections corresponding to less detailed measurements are generated by taking the same limits discussed in Section IV. For instance, single-photon dissociation ($K_1 = B_1 = 0$), integral photofragment cross sections ($K_D = 0$), integral fluorescence cross sections ($K_S = B_3 = 0$).

E. J-Independent Recoil Limit

The J-independent recoil limit approximation consists of neglecting those dependences of the amplitudes $\tau(J_2\alpha_2|J_1\eta_1)$ [Eq. (101)] on J_2 which arises from dynamical factors. Geometrical factors are present in $\tau(J_2\alpha_2|J_1\eta_1)$ through the dipole moment function $\mu_{J_2\alpha_2;J_1\eta_1}(r)$ [Eq. (102)] which depend on J_2, but these geometrical factors may be separated from the amplitude and treated exactly. The J-independent recoil limit neglects the variation of the adiabatic radial wave function $\psi_{EJ_2\alpha_2}^{(-)}(r)$ on J_2 over the range $|J_2 - J_1| < 1$, not for all J_2. Conditions favoring this slow dependence on J_2 are large reduced mass of the fragments or large fragment kinetic energy.

A further condition for validity of the J-independent recoil approximation is that there be no transitions between electronic states in the molecular region. Otherwise dissociative flux, weighted by the J_2-dependent geometric factors appropriate for the channel (or channels) carrying oscillator strength from the intermediate bound level, can be diverted to other molecular channels during the dissociation. The flux appearing in dissociative channels by virtue of nonadiabatic transitions in the molecular region is then incorrectly weighted if the J-independent recoil approximation is employed. In these cases, such as may arise from curve crossing, use of the J-dependent formalism presented in the previous section is necessary. The J-independent recoil limit does, however, properly include interference between molecular channels that simultaneously carry oscillator strength from a bound level.

Again a great simplification of the cross-section expressions is achieved and physical insight is gained if the intramolecular coupling can be described by one of the Hund's cases in the Franck-Condon region. This allows us to express elements of the matrix $\mathsf{O}(r)$, which appear in Eqs. (102a) and (102b), in terms of angular momentum recoupling coefficients. If $\psi_{EJ_2\alpha_2}^{(-)}(r)$ in these equations approximately does not vary over the range $|J_2 - J_1| \leqslant 1$ for fixed J_1, the J_2 dependence of $\tau(J_2\alpha_2|J_1\eta_1)$ is then known analytically, and the sums over J_2 and $\bar{J}_2$ in Eq. (117) can be performed in closed form.

The procedure described in the preceding paragraph is now carried through explicitly. We write the amplitudes $\tau(J_2\alpha_2|J_1\eta_1)$, where the molecular adiabatic level $|J_2M_2\alpha_2\rangle$ coincides with one of the Hund's case basis functions in the Franck-Condon region, as a product of geometrical factors times a J_2 independent amplitude. The adiabatic level α_2 may evolve into a different Hund's case or to an intermediate case outside the Franck-Condon region without affecting the validity of our results. In the Hund's case (a) representation, this procedure yields

$$\tau(J_2\alpha_2|J_1\eta_1) = (J_1)^{1/2}(J_2)^{-1/2}\langle J_2\Omega_2|1J_1q_2'\Omega_1\rangle\tau(c_2\Lambda_2S_2\Sigma_2|J_1\eta_1) \quad (118)$$

where

$$\tau(c_2\Lambda_2S_2\Sigma_2|J_1\eta_1) = \sum_{\Lambda_1\Sigma_1}\int_0^{r_1} dr\psi^{(-)}_{EJ_2c_2\Lambda_2S_2\Sigma_2}(r)\langle c_2\Lambda_2S_2|x_{q_2'}|c_1\Lambda_1S_1\rangle$$

$$\times\, \chi_{J_1v_1}(r)\,\delta_{S_2S_1}\delta_{\Sigma_2\Sigma_1}a(\eta_1|\Lambda_1\Sigma_1) \tag{119}$$

is independent of J_2 under the assumptions of the J-independent recoil approximation. The analogous results for Hund's case (b) are given as,

$$\tau(J_2\alpha_2|J_1\eta_1) = (N_1)^{1/2}(J_1)^{1/2}\langle N_2\Lambda_2|1N_1q_2'\Lambda_1\rangle W(1J_2N_1S_1;\, J_1N_2)$$

$$\times\, \tau(c_2\Lambda_2S_2N_2|J_1\eta_1) \tag{120}$$

where

$$\tau(c_2\Lambda_2S_2N_2|J_1\eta_1) = \sum_{\Lambda_1N_1}\int_0^{r_1} dr\psi^{(-)}_{EJ_2c_2\Lambda_2S_2N_2}(r)\langle c_2\Lambda_2S_2|x_{q_2'}|c_1\Lambda_1S_1\rangle$$

$$\times\, \chi_{J_1v_1}(r)\,\delta_{S_2S_1}a(\eta_1|\Lambda_1N_1) \tag{121}$$

In Eqs. (118) and (119) above we have allowed for the possibility that the case (a) amplitudes $\tau(c_2\Lambda_2S_2\Sigma_2|J_1\eta_1)$ may depend on Σ_2 (besides the obvious selection rule $\Sigma_1 = \Sigma_2$) due to, say, large spin-orbit splitting in the molecule or selective interaction of one of the $(2S_2 + 1)$ multiplet states with another close lying state. Similarly we allow for non-negligible dependence of the case (b) amplitudes on N_2 in Eqs. (120) and (121). The Hund's case (a) [Eqs. (118) and (119)] and case (b) [Eqs. (120) and (121)] amplitudes are related by the transformation [Eq. (19)] and have the same physical content unless the radial wave functions in Eqs. (119) and (121) are affected by the quantum numbers Σ_2 or N_2, respectively. When the case (a) and (b) representations are physically equivalent, Eq. (23) may be used to show

$$\tau(c_2\Lambda_2S_2\Sigma_2|J_1\eta_1) = (N_2)^{1/2}(J_2)^{-1/2}\langle J_2\Omega_2|N_2S_2\Lambda_2\Sigma_2\rangle\tau(c_2\Lambda_2S_2N_2|J_1\eta_1) \tag{122}$$

Upon insertion of the J-independent recoil limit transition amplitudes for case (a) [Eq. (118)] into the half-collision operator [Eq. (111)], we obtain

$$\tilde{\mathcal{L}}[B_2\beta_2(J_2\eta_2;\bar{J}_2\bar{\eta}_2)|B_1\beta_1(J_1\eta_1;\bar{J}_1\bar{\eta}_1)]$$

$$= \sum_{K_2Q_2\bar{\Lambda}_1\bar{\Sigma}_1\Lambda_1\Sigma_1} (-)^{J_2-\bar{J}_2}(-)^{J_1-\bar{J}_1}[(J_1)(\bar{J}_1)(K_2)(K_1)]^{1/2}$$

$$\times \langle B_2\beta_2|B_1K_2\beta_1Q_2\rangle \begin{Bmatrix} B_1 & K_2 & B_2 \\ J_1 & 1 & J_2 \\ \bar{J}_1 & 1 & \bar{J}_2 \end{Bmatrix} \phi^{K_2}_{Q_2}(\hat{\varepsilon}_2)$$

$$\times \langle J_2\Omega_2|1J_1q'_2\Omega_1\rangle\langle \bar{J}_2\bar{\Omega}_2|1\bar{J}_1\bar{q}'_2\bar{\Omega}_1\rangle \delta_{\bar{S}_2\bar{S}_1}\delta_{S_2S_1}\delta_{\bar{\Sigma}_2\bar{\Sigma}_1}\delta_{\Sigma_2\Sigma_1}$$

$$\times \tau(c_2\Lambda_2S_2\Sigma_2|J_1\eta_1)\tau^*(\bar{c}_2\bar{\Lambda}_2\bar{S}_2\bar{\Sigma}_2|\bar{J}_1\bar{\eta}) \tag{123}$$

and for the case (b) amplitudes [Eq. (120)],

$$\tilde{\mathcal{L}}[B_2\beta_2(J_2\eta_2;\bar{J}_2\bar{\eta}_2)|B_1\beta_1(J_1\eta_1;\bar{J}_1\bar{\eta}_1)]$$

$$= \sum_{K_2Q_2\bar{\Lambda}_1\bar{N}_1\Lambda_1N_1} (-)^{J_2-\bar{J}_2}(-)^{J_1-\bar{J}_1}[(J_1)(\bar{J}_1)(N_1)(\bar{N}_1)(K_2)(K_1)]^{1/2}$$

$$\times \langle B_2\beta_2|B_1K_1\beta_1Q_1\rangle \begin{Bmatrix} B_1 & K_2 & B_2 \\ J_1 & 1 & J_2 \\ \bar{J}_1 & 1 & \bar{J}_2 \end{Bmatrix} \phi^{K_2}_{Q_2}(\hat{\varepsilon}_2)\langle N_2\Lambda_2|1N_1q'_2\Lambda_1\rangle$$

$$\times \langle \bar{N}_2\bar{\Lambda}_2|1\bar{N}_1\bar{q}'_2\bar{\Lambda}_1\rangle\delta_{S_2\bar{S}_1}\delta_{S_2S_1}W(1J_2N_1S_1;J_1N_2)$$

$$\times W(1\bar{J}_2\bar{N}_1\bar{S}_1;\bar{J}_1\bar{N}_2)\tau(c_2\Lambda_2S_2N_2|J_1\eta_1)\tau(c_2\bar{\Lambda}_2\bar{S}_2\bar{N}_2|\bar{J}_1\bar{\eta}_1) \tag{124}$$

The product of the matrices $\tilde{\mathcal{L}}$ and $\tilde{\mathcal{U}}^\dagger$ in Eq. (117) involves a sum over J_2 and $\bar{J}_2$. The J-independent recoil limit enables the factorization of $\tilde{\mathcal{L}}$ into the product of an analytic term, containing all J_2 and $\bar{J}_2$ dependence, and a dynamical portion. This factorization allows the J_2 and $\bar{J}_2$ sums in Eq. (117) to be performed in closed form. Before carrying out these sums below for some special cases, we review the several versions of $\tilde{\mathcal{L}}$ and $\tilde{\mathcal{U}}^\dagger$ which we have generated.

The half-collision operator $\tilde{\mathcal{L}}$ of Eq. (123) describes promotion of dissociative flux to adiabatic molecular levels that are characterized by the Hund's case (a) coupling scheme in the Franck-Condon region, and Eq. (124) is for radiative coupling to molecular states that are characterized by Hund's case (b) coupling. The $\tilde{\mathcal{U}}^\dagger$ matrices [Eqs. (113), (114), or (115)] describe the influence of asymptotic (large r) nonadiabatic processes when the intramolecular coupling, just prior to entry of the fragments into the recoupling region, is characterized by Hund's case (a), (b), or (c), respectively.

To evaluate cross sections in the J-independent recoil limit, we choose the $\tilde{\mathscr{L}}$ matrix [Eq. (123) or (124)] appropriate for the type of intramolecular coupling found in the Franck-Condon region. This $\tilde{\mathscr{L}}$ is then contracted (by summing over J_2 and $\bar{J}_2$) with the $\tilde{\mathscr{U}}^\dagger$ matrix [Eqs. (113), (114), or (115)] that describes nonadiabatic transitions at large r. For instance, promotion of flux to a Hund's case (a) adiabatic molecular level followed by recoupling of angular moments from case (a) to atomic states at larger r is described by use of the half-collision operator Eq. (123) and the geometrical matrix Eq. (113). The analogous process with case (b) adiabatic molecular states is treated with the half-collision operator Eq. (124) and geometrical matrix Eq. (113).

Our formalism is also capable of describing propagation of flux on dissociative molecular surfaces that are characterized by one Hund's coupling scheme in the Franck-Condon region and a different coupling scheme near the recoupling region. This "mix-and-match" scheme is useful because intramolecular coupling often changes with internuclear distance: Levels described by Hund's case (b) can evolve into case (a) states at larger r, or case (a) type coupling can be replaced by case (c). In the former example, optical excitation occurs to a dissociative case (b) level with quantum numbers $\Lambda_2 N_2$ that evolves into a case (a) level with quantum numbers $\Lambda_2\Sigma_2$ before recoupling to atomic states. This situation is treated with the case (b) $\tilde{\mathscr{L}}$ matrix Eq. (124) and the case (a) $\mathscr{U}^\dagger$ matrix Eq. (113). It is necessary to determine the case (a) state which at large r adiabatically correlates with the optically excited case (b) level.

The J_2 and $\bar{J}_2$ sums in Eq. (117) are carried out for two examples in Ref. 108: Hund's case (a) coupling in both Franck-Condon and recoupling region, and Hund's case (b) in both regions. The expressions obtained after the sums are performed are somewhat arbitrarily broken into the product of a new half-collision matrix $\tilde{\tilde{\mathscr{L}}}$ and geometrical matrix $\tilde{\tilde{\mathscr{U}}}^\dagger$ in order to mimic the form of Eq. (117). The final results for dissociation involving Hund's case (a) coupling in the molecular region are

$$\begin{aligned}
&\tilde{\tilde{\mathscr{L}}}[B_2\beta_2(\eta_2;\,\bar{\eta}_2)|B_1\beta_1(J_1\eta_1;\,\bar{J}_1\bar{\eta}_1)] \\
&\quad= \sum_{K_2Q_2} (-)^{1-J_1+\Omega_2}\langle B_2\beta_2|B_1K_2Q_1Q_2\rangle[(J_1)(\bar{J}_1)(K_2)]^{1/2} \\
&\quad\times \begin{pmatrix} B_2 & B_1 & K_2 \\ \omega_2 & -\omega_1 & -\delta_2 \end{pmatrix}\begin{pmatrix} J_1 & \bar{J}_1 & B_1 \\ \Omega_1 & -\bar{\Omega}_1 & -\omega_1 \end{pmatrix}\begin{pmatrix} 1 & 1 & K_2 \\ q_2' & -\bar{q}_2' & -\delta_2 \end{pmatrix} \\
&\quad\times \phi_{Q_2}^{K_2}(\hat{\varepsilon}_2)\delta_{S_2S_1}\delta_{\bar{S}_2\bar{S}_1}\delta_{\Sigma_2\Sigma_1}\delta_{\bar{\Sigma}_2\bar{\Sigma}_1}\tau(c_2\Lambda_2S_2\Sigma_2|J_1\eta_1) \\
&\quad\times \tau^*(\bar{c}_2\bar{\Lambda}_2\bar{S}_2\bar{\Sigma}_2|\bar{J}_1\bar{\eta}_1)
\end{aligned} \tag{125}$$

$$\tilde{\tilde{\mathscr{U}}}^{\dagger}[B_3\beta_3(j_a;\, j_a)|B_2\beta_2(\eta_2;\, \bar{\eta}_2)]$$

$$= \sum_{\substack{K_D L \bar{L} \\ Q_D X Z}} (K_D)(B_3)^{1/2}\frac{1}{4\pi} C_{K_D Q_D}(\hat{\mathbf{k}})\langle B_2\beta_2|B_3K_D\beta_3Q_D\rangle\langle\Lambda_a\Lambda_b|c_2\Lambda_2\rangle$$

$$\times \langle\bar{c}_2\bar{\Lambda}_2|\bar{\Lambda}_a\bar{\Lambda}_b\rangle\left[(-)^{\bar{S}-\Sigma_2}\begin{pmatrix} K_D & B_2 & B_3 \\ 0 & \omega_2 & -\omega_2\end{pmatrix}\begin{pmatrix} B_3 & Z & X \\ \omega_2 & -\sigma & -\lambda_2\end{pmatrix}\right.$$

$$\left.\times\begin{pmatrix} Z & S_2 & \bar{S}_2 \\ -\sigma & \Sigma_2 & -\bar{\Sigma}_2\end{pmatrix}\right](-)^{\bar{L}-\bar{\Lambda}_2}(L)(\bar{L})(X)(Z)(S_2)^{1/2}(\bar{S}_2)^{1/2}(j_a)(j_b)$$

$$\times\begin{pmatrix} L & \bar{L} & X \\ \Lambda_2 & -\bar{\Lambda}_2 & -\lambda_2\end{pmatrix}\begin{pmatrix} L & l_a & l_b \\ -\Lambda_2 & \Lambda_a & \Lambda_b\end{pmatrix}\begin{pmatrix} \bar{L} & l_a & l_b \\ -\bar{\Lambda}_2 & \bar{\Lambda}_a & \bar{\Lambda}_b\end{pmatrix}$$

$$\times\left\{\begin{array}{cccccc} Z & \bar{S}_2 & s_b & l_b & L & X \\ & & j_b & & & \\ S_2 & s_a & & & s_a & \bar{L} \\ & & K_s & & & \\ s_b & s_a & j_a & j_a & l_a & l_b\end{array}\right\} \tag{126}$$

and for Hund's case (b) coupling in the molecular region,

$$\tilde{\tilde{\mathscr{L}}}[B_2\beta_2(\eta_2;\, \bar{\eta}_2)|B_1\beta_1(J_1\eta_1;\, \bar{J}_1\bar{\eta}_1)]$$

$$= \sum_{R_1}(-)^{R_2+K_1+Z+1+K_2+N_1+\bar{\Lambda}_2}\langle B_2\beta_2|B_1K_2Q_1Q_2\rangle(R_1)$$

$$\times[(R_2)(N_1)(\bar{N}_1)(K_2)(J_1)(\bar{J}_1)]^{1/2}\begin{pmatrix} R_2 & R_1 & K_2 \\ \lambda_2 & -\lambda_1 & -\delta_2\end{pmatrix}$$

$$\times\begin{pmatrix} N_1 & \bar{N}_1 & R_1 \\ \Lambda_1 & -\bar{\Lambda}_1 & -\lambda_1\end{pmatrix}\begin{pmatrix} 1 & 1 & K_2 \\ q_2' & -\bar{q}_2' & -\delta_2\end{pmatrix}$$

$$\times\begin{Bmatrix} R_1 & R_2 & K_2 \\ B_2 & B_1 & Z\end{Bmatrix}\begin{Bmatrix} B_1 & R_1 & Z \\ J_1 & N_1 & S_1 \\ \bar{J}_1 & \bar{N}_1 & \bar{S}_1\end{Bmatrix}\delta_{\bar{S}_2\bar{S}_1}\delta_{S_2S_1}$$

$$\times\,\phi_{Q_2}^{K_2}(\hat{\boldsymbol{\varepsilon}}_2)\tau(c_2\Lambda_2S_2N_2|J_1\eta_1)\,\tau^*(\bar{c}_2\bar{\Lambda}_2\bar{S}_2\bar{N}_2|\bar{J}_1\bar{\eta}_1) \tag{127}$$

$$\tilde{\tilde{\mathscr{U}}}^{\dagger}[B_3\beta_3(j_a;\, j_a)|B_2\beta_2(\eta_2;\, \bar{\eta}_2)]$$

$$= \sum_{\substack{K_D L \bar{L} \\ Q_D X}} (K_D)(B_2)^{1/2} \frac{1}{4\pi} C_{K_D Q_D}(\hat{\mathbf{k}}) \langle B_2\beta_2 | B_3 K_D \beta_3 Q_D \rangle \langle \Lambda_a \Lambda_b | c_2 \Lambda_2 \rangle$$

$$\times \langle \bar{c}_2 \bar{\Lambda}_2 | \bar{\Lambda}_a \bar{\Lambda}_b \rangle \Bigg[(-)^{B_3 + \lambda_2} (-)^{K_D + X + R_2} (R_2)^{1/2} \begin{pmatrix} K_D & R_2 & X \\ 0 & \lambda_2 & -\lambda_2 \end{pmatrix}$$

$$\times \begin{Bmatrix} B_2 & B_3 & K_D \\ X & R_2 & Z \end{Bmatrix} \Bigg] (-)^{\bar{L} - \bar{\Lambda}_2} (L)(\bar{L})(X)(Z)(S_2)^{1/2}(\bar{S}_2)^{1/2}(j_a)(j_b)$$

$$\times \begin{pmatrix} L & \bar{L} & X \\ \Lambda_2 & -\bar{\Lambda}_2 & -\lambda_2 \end{pmatrix} \begin{pmatrix} L & l_a & l_b \\ -\Lambda_2 & \Lambda_a & \Lambda_b \end{pmatrix} \begin{pmatrix} \bar{L} & l_a & l_b \\ -\bar{\Lambda}_2 & \bar{\Lambda}_a & \bar{\Lambda}_b \end{pmatrix}$$

$$\times \left\{ \begin{array}{ccccccccccc} Z & & \bar{S}_2 & & s_b & & l_b & & L & & X \\ & & & & & j_b & & & & & \\ & S_2 & & s_a & & & & s_a & & \bar{L} & \\ & & & & & K_s & & & & & \\ s_b & & s_a & & j_a & & j_a & & l_a & & l_b \end{array} \right\}. \tag{128}$$

Multiplication of the matrices $\tilde{\tilde{\mathscr{U}}}^{\dagger}$ and $\tilde{\tilde{\mathscr{L}}}$ involves summation over B_2, β_2, ω_2, Σ_2, and $\bar{\Sigma}_2$ for the case (a) expressions, Eqs. (125) and (126), or summation over B_2, β_2, R_2, λ_2, and Z for the case (b) expressions, Eqs. (127) and (128). The two sets of operators give identical cross sections if the transition amplitudes in Eqs. (125) and (127) do not depend on Σ_2 or N_2 and if the multiplet expansions coefficients $a(\eta_1|\Lambda_1\Sigma_1)$ and $a(\eta_1|\Lambda_1 N_1)$ are related as in Eq. (23). The case (a) and (b) expressions are also trivially equivalent if $S_1 = S_2 = 0$. However, they yield different results if dissociative multiplet levels are significantly split or different coupling schemes are used for the intermediate bound levels.

Several aspects of the J-independent recoil liimits for $\tilde{\tilde{\mathscr{L}}}$ and $\tilde{\tilde{\mathscr{U}}}^{\dagger}$ are noteworthy. Because of the J-dependences of amplitudes for the initial resonant transition between bound levels, the effect of molecular rotation for two-photon dissociation is not entirely eliminated in this limit. This is in contrast to the J-independent limit of single-photon cross sections which do not depend on molecular rotation. Equations (125)–(128) are cast in a form that emphasizes the similarities between the case (a) and (b) versions of $\tilde{\tilde{\mathscr{L}}}$ and $\tilde{\tilde{\mathscr{U}}}^{\dagger}$. The portions of the transformation operators which change on going from case (a) to case (b) are set off in square brackets in Eqs. (126) and (128). Although nonadiabatic transitions in the molecular region must be absent if the J-independent recoil approximation is to be valid, interferences are properly treated in this approximation when the interferences are between bound or dissociative electronic states that simultaneously carry oscillator strength from the ground state. For instance, if the molecule can dissociate along two different dissociative surfaces, $c_2\Lambda_2S_2\Sigma_2$ and $c'_2\Lambda'_2S_2\Sigma_2$ (since nonadiabatic transitions are not allowed, total spin S is conserved), cross terms between amplitudes $\tau(c_2\Lambda_2S_2\Sigma_2|J_1\eta_1)$ and $\tau^*(c'_2\Lambda'_2S_2\Sigma_2|J_1\eta_1)$ arise in Eq. (125).

The greatest simplification occurs if dissociation proceeds along a single electronic energy surface. In this case, the coefficients $\sigma_{K_SQ_S;K_DQ_D}$ of $C_{K_DQ_D}(\hat{\mathbf{k}})C^*_{K_SQ_S}(\hat{\boldsymbol{\varepsilon}}_S)$ in the differential cross section (42), are all proportional to an overall dynamical factor $|\tau(c_2\Lambda_2S_2\Sigma_2|J_1\eta_1)|^2$. The anisotropy parameters that specify the angular distribution of photofragments and fluorescence and ratios of the coefficients $\sigma_{K_SQ_S;K_DQ_D}$ are therefore determined by geometrical factors without any dynamical calculation. Examples are the anisotropy parameters for the angular distribution of fragments, β_D [Eq. (86)], and of fluorescence, β_S [Eq. (87)]. In addition to β_D, we also define a second anisotropy parameter γ_D for the photofragment angular distribution,

$$\gamma_D = \frac{\sigma_{00;40}}{\sigma_{00;00}} \tag{129}$$

The fragment angular distribution is completely specified by β_D and γ_D if both photons that dissociate the molecule are linearly polarized along the z-axis. Branching ratios to atomic fine structure states, which are the relative magnitude of $\sigma_{00;00}$ as a function of $c_aj_ac_bj_b$, are also completely determined without dynamical calculation in the J-independent recoil limit when dissociation occurs on a single electronic surface.

For the many cases in which the requirements for validity of the J-independent recoil approximation are met, a small number of experimental quantities, the asymmetry parameters and branching ratios, yields considerable information about the molecular electronic structure. This is illustrated in Tables III through X where we present asymmetry parameters and branching ratios for some common molecular electronic state symmetries and atomic term limits. For simplicity, results for pure case (a) [Eqs. (125) and (126)] or case (b) [Eqs. (127) and (128)] intramultiplet coupling in the molecular region, are given, although recoil limit anisotropy parameters and branching ratios are readily evaluated for more complicated examples. The strong variation of branching ratios and asymmetry parameters with the symmetry of those electronic states that are involved in the dissociation demonstrates the information content of these quantities.

The fluorescence anisotropy parameters and polarization ratio predicted for one-photon dissociation of a Hund's case (a) or (b) diatomic are presented in Tables III through VI. For illustration, anisotropy parameters for

TABLE III
Fluorescence anisotropy parameters for one-photon dissociation: Hund's case (a)

Atomic limit	Molecular transition	Atomic j-state	β_S	P
$^1P + ^1S$	$^1\Sigma_0 \rightarrow {}^1\Sigma$	1	4/5	1/2
	$^1\Sigma_0 \rightarrow {}^1\Pi$	1	7/5	7/9
	$^1\Pi_1 \rightarrow {}^1\Sigma$	1	−2/5	−1/3
	$^1\Pi_1 \rightarrow {}^1\Pi$	1	−2/5	−1/3
$^2P + ^2S$	$^1\Sigma_0 \rightarrow {}^1\Sigma$	3/2	2/5	3/11
	$^1\Sigma_0 \rightarrow {}^1\Pi$	3/2	7/10	21/47
	$^1\Pi_1 \rightarrow {}^1\Sigma$	3/2	−1/5	−3/19
	$^1\Pi_1 \rightarrow {}^1\Pi$	3/2	−1/5	−3/19
	$^3\Sigma_{0,1} \rightarrow {}^3\Sigma$	3/2	2/5	3/11
	$^3\Sigma_{0,1} \rightarrow {}^3\Pi$	3/2	7/10	21/47
	$^3\Pi_{0,1,2} \rightarrow {}^3\Sigma$	3/2	−1/5	−3/19
	$^3\Pi_0 \rightarrow {}^3\Pi$	3/2	2/5	3/11
	$^3\Pi_1 \rightarrow {}^3\Pi$	3/2	−1/5	−3/19
	$^3\Pi_2 \rightarrow {}^3\Pi$	3/2	−2/5	−1/3
$^3P + ^2S$	$^2\Sigma_{1/2} \rightarrow {}^2\Sigma$	1	1/5	1/7
		2	7/25	21/107
	$^2\Sigma_{1/2} \rightarrow {}^2\Pi$	1	7/20	7/29
		2	49/100	147/449
	$^2\Pi_{1/2,3/2} \rightarrow {}^2\Sigma$	1	−1/10	−1/13
		2	−7/50	−21/193
	$^2\Pi_{1/2} \rightarrow {}^2\Pi$	1	−1/5	−3/19
		2	7/25	21/107
	$^2\Pi_{3/2} \rightarrow {}^2\Pi$	1	1/5	1/7
		2	−7/25	−21/93
$^3P + ^2S$	$^4\Sigma_{1/2,3/2} \rightarrow {}^4\Sigma$	1	1/5	1/7
	$^4\Sigma_{1/2} \rightarrow {}^4\Sigma$	2	19/55	57/239
	$^4\Sigma_{3/2} \rightarrow {}^4\Sigma$	2	1/5	1/7
	$^4\Sigma_{1/2} \rightarrow {}^4\Pi$	1	11/25	11/37
	$^4\Sigma_{3/2} \rightarrow {}^4\Pi$	1	1/5	1/7
	$^4\Sigma_{1/2} \rightarrow {}^4\Pi$	2	47/95	141/427
	$^4\Sigma_{3/2} \rightarrow {}^4\Pi$	2	17/35	51/157
	$^4\Pi_{1/2(\Sigma=1/2)} \rightarrow {}^4\Sigma$	1	−1/10	−1/13
		2	−19/110	−57/421
	$^4\Pi_{1/2(\Sigma=3/2)} \rightarrow {}^4\Sigma$	1	−1/10	−1/13
		2	−1/10	−1/13
	$^4\Pi_{5/2(\Sigma=3/2)} \rightarrow {}^4\Sigma$	1	−1/10	−1/13
		2	−1/10	−1/13
	$^4\Pi_{1/2(\Sigma=1/2)} \rightarrow {}^4\Pi$	1	0	0
		2	8/35	6/37
	$^4\Pi_{3/2(\Sigma=1/2)} \rightarrow {}^4\Pi$	1	1/5	1/7
		2	−1/10	−1/13
	$^4\Pi_{1/2(\Sigma=3/2)} \rightarrow {}^4\Pi$	1	−2/5	−1/3
		2	2/5	3/11
	$^4\Pi_{5/2(\Sigma=3/2)} \rightarrow {}^4\Pi$	1	—	—
		2	−2/5	−1/3

TABLE IV
Branching ratios for one-photon dissociation: Hund's case (a)

Atomic limit	Molecular transition	Atomic j-state branching ratio		
$^2P + {}^2S$		$j = \frac{1}{2}$	$\frac{3}{2}$	
	$^3\Pi_0 \rightarrow {}^3\Pi$	1	$\frac{1}{2}$	
	$^3\Pi_1 \rightarrow {}^3\Pi$	1	2	
	$^3\Pi_2 \rightarrow {}^3\Pi$	0	$\frac{3}{2}$	
$^3P + {}^2S$		$j = 0$	1	2
	$^2\Pi_{1/2} \rightarrow {}^2\Pi$	1	9/4	5/4
	$^2\Pi_{3/2} \rightarrow {}^2\Pi$	0	3/4	15/4
	$^4\Sigma_{1/2} \rightarrow {}^4\Sigma$	1	3/4	11/4
	$^4\Sigma_{3/2} \rightarrow {}^4\Sigma$	0	9/4	9/4
	$^4\Sigma_{1/2} \rightarrow {}^4\Pi$	1	15/2	19/2
	$^4\Sigma_{3/2} \rightarrow {}^4\Pi$	3	9/2	21/2
	$^4\Pi_{1/2(\Sigma=1/2)} \rightarrow {}^4\Sigma$	1	3/4	11/4
	$^4\Pi_{3/2} \rightarrow {}^4\Sigma$	1	3/4	11/4
	$^4\Pi_{1/2(\Sigma=3/2)} \rightarrow {}^4\Sigma$	0	9/4	9/4
	$^4\Pi_{5/2} \rightarrow {}^4\Sigma$	0	9/4	9/4
	$^4\Pi_{1/2(\Sigma=1/2)} \rightarrow {}^4\Pi$	1	9/2	7/2
	$^4\Pi_{3/2} \rightarrow {}^4\Pi$	0	3	6
	$^4\Pi_{1/2(\Sigma=3/2)} \rightarrow {}^4\Pi$	3	9/2	3/2
	$^4\Pi_{5/2} \rightarrow {}^4\Pi$	0	0	9

TABLE V

Fluorescence anisotropy parameters for one-photon dissociation: Hund's case (b)[a]

Atomic limit	Molecular transition	Atomic j-state	$\beta_S\left(=\dfrac{4P}{3-P}\right)$	$P\left(=\dfrac{3\beta_S}{4+\beta_S}\right)$
$^1P + {}^1S$	$^1\Sigma \rightarrow {}^1\Sigma$	1	4/5	1/2
	$^1\Sigma \rightarrow {}^1\Pi$	1	7/5	7/9
	$^1\Pi \rightarrow {}^1\Sigma$	1	−2/5	−1/3
	$^1\Pi \rightarrow {}^1\Pi$	1	−2/5	−1/3
$^2P + {}^2S$	$^1\Sigma \rightarrow {}^1\Sigma$	3/2	2/5	3/11
	$^1\Sigma \rightarrow {}^1\Pi$	3/2	7/10	21/47
	$^1\Pi \rightarrow {}^1\Sigma$	3/2	−1/5	−3/19
	$^1\Pi \rightarrow {}^1\Pi$	3/2	−1/5	−3/19
	$^3\Sigma_{N=2,3,4} \rightarrow {}^3\Sigma$	3/2	2/5	3/11
	$^3\Sigma_{N=2,3,4} \rightarrow {}^3\Pi$	3/2	7/10	21/47
	$^3\Pi_{N=2,3,4} \rightarrow {}^3\Sigma$	3/2	−1/5	−3/19
	$^3\Pi_{N=2} \rightarrow {}^3\Pi$	3/2	−2/7	−3/13
	$^3\Pi_{N=3} \rightarrow {}^3\Pi$	3/2	−4/23	−3/22
	$^3\Pi_{N=4} \rightarrow {}^3\Pi$	3/2	−4/35	−3/34
$^3P + {}^2S$	$^2\Sigma_{N=3,4} \rightarrow {}^2\Sigma$	1	1/5	1/7
		2	7/25	21/107
	$^2\Sigma_{N=3,4} \rightarrow {}^2\Pi$	1	7/20	7/29
		2	49/100	147/449
	$^2\Pi_{N=3,4} \rightarrow {}^2\Sigma$	1	−1/10	−1/13
		2	−7/50	−21/193
	$^2\Pi_{N=3} \rightarrow {}^2\Pi$	1	−2/35	−1/23
		2	−14/75	−21/143
	$^2\Pi_{N=4} \rightarrow {}^2\Pi$	1	−2/15	−3/29
		2	−2/25	−3/49
	$^4\Sigma_{N=2} \rightarrow {}^4\Sigma$	1	1/5	1/7
		2	3/10	9/43
	$^4\Sigma_{N=3} \rightarrow {}^4\Sigma$	1	1/5	1/7
		2	8/35	6/37
	$^4\Sigma_{N=4} \rightarrow {}^4\Sigma$	1	1/5	1/7
		2	22/85	33/203
	$^4\Sigma_{N=5} \rightarrow {}^4\Sigma$	1	1/5	1/7
		2	13/40	39/173

[a] Integral total spin: $J_0 = 2$, $J_1 = 3$; half integral total spin: $J_0 = 5/2$, $J_1 = 7/2$.

TABLE VI
Branching ratios for one-photon dissociation: Hund's case (b)[a]

Atomic limit	Molecular transition		Atomic j-state / Branching ratio		
$^2P + {}^2S$		$j =$	$\frac{1}{2}$	$\frac{3}{2}$	
	$^3\Pi_{N=2} \rightarrow {}^3\Pi$		1	7/2	
	$^3\Pi_{N=3} \rightarrow {}^3\Pi$		13/8	23/8	
	$^3\Pi_{N=4} \rightarrow {}^3\Pi$		15/8	21/8	
$^3P + {}^2S$		$j =$	0	1	2
	$^2\Pi_{N=3} \rightarrow {}^2\Pi$		1	7/2	15/2
	$^2\Pi_{N=4} \rightarrow {}^2\Pi$		5/3	9/2	35/6
	$^4\Sigma_{N=2} \rightarrow {}^4\Sigma$		1	2	4
	$^4\Sigma_{N=3} \rightarrow {}^4\Sigma$		7/27	28/9	98/27
	$^4\Sigma_{N=4} \rightarrow {}^4\Sigma$		5/9	8/3	34/9
	$^4\Sigma_{N=5} \rightarrow {}^4\Sigma$		35/27	14/9	112/27

[a] Integral total spin: $J_0 = 2$, $J_1 = 3$; half integral total spin: $J_0 = 5/2$, $J_1 = 7/2$.

TABLE VII
Fluorescence Anisotropy Parameters for Two-photon Dissociation: Hund's Case (a)[a]

Atomic limit	Molecular transition	Atomic j-state	Anisotropy parameters β_D	γ_D	β_S
$^1P + {}^1S$	$^1\Sigma \rightarrow {}^1\Sigma \rightarrow {}^1\Sigma$	1	2.46753	0.62338	0.98701
	$^1\Sigma \rightarrow {}^1\Sigma \rightarrow {}^1\Sigma$	1	−0.51020	−0.48980	1.20408
	$^1\Sigma \rightarrow {}^1\Pi \rightarrow {}^1\Sigma$	1	−0.92437	−0.075630	−0.36975
	$^1\Sigma \rightarrow {}^1\Pi \rightarrow {}^1\Pi$	1	2.37327	0.49770	−0.308756
$^2P + {}^2S$	$^1\Sigma \rightarrow {}^1\Sigma \rightarrow {}^1\Sigma$	1/2	2.46753	0.62338	0.0
		3/2	2.46753	0.62338	0.49351
	$^1\Sigma \rightarrow {}^1\Sigma \rightarrow {}^1\Pi$	1/2	−0.51020	−0.48980	0.0
		3/2	−0.51020	−0.48980	0.60204
	$^1\Sigma \rightarrow {}^1\Pi \rightarrow {}^1\Sigma$	1/2	−0.92437	−0.075630	0.0
		3/2	−0.92437	−0.075630	−0.18487
	$^1\Sigma \rightarrow {}^1\Pi \rightarrow {}^1\Pi$	1/2	2.37327	0.49770	0.0
		3/2	2.37327	0.49770	−0.15438
	$^3\Sigma_0 \rightarrow {}^3\Sigma_0 \rightarrow {}^3\Sigma$	1/2	2.46753	0.62338	0.0
	$^3\Sigma_1 \rightarrow {}^3\Sigma_1 \rightarrow {}^3\Sigma$	1/2	2.37327	0.49770	0.0
	Unresolved multiplet[b]	1/2	2.40858	0.54477	0.0

TABLE VII
(Continued)

Atomic limit	Molecular transition	Atomic j-state	Anisotropy parameters β_D	γ_D	β_S
$^2P + ^2S$	$^3\Sigma_0 \rightarrow {}^3\Sigma_0 \rightarrow {}^3\Sigma$	3/2	2.46753	0.62338	0.493506
	$^3\Sigma_1 \rightarrow {}^3\Sigma_1 \rightarrow {}^3\Sigma$	3/2	2.37327	0.49770	0.47465
	Unresolved multiplet[b]	3/2	2.40858	0.54477	0.481715
	$^3\Sigma_0 \rightarrow {}^3\Sigma_0 \rightarrow {}^3\Pi$	1/2	−0.51020	−0.48980	0.0
	$^3\Sigma_1 \rightarrow {}^3\Sigma_1 \rightarrow {}^3\Pi$	1/2	−0.64935	−0.35065	0.0
	Unresolved multiplet[b]	1/2	−0.60074	−0.39926	0.0
	$^3\Sigma_0 \rightarrow {}^3\Sigma_0 \rightarrow {}^3\Pi$	3/2	−0.51020	−0.48980	0.60204
	$^3\Sigma_1 \rightarrow {}^3\Sigma_1 \rightarrow {}^3\Pi$	3/2	−0.64935	−0.35065	0.62987
	Unresolved multiplet[b]	3/2	−0.60074	−0.39926	0.62015
	$^3\Sigma_0 \rightarrow {}^3\Pi_1 \rightarrow {}^3\Sigma$	1/2	−0.92437	−0.07563	0.0
	$^3\Sigma_1 \rightarrow {}^3\Pi_0 \rightarrow {}^3\Sigma$	1/2	−0.51020	−0.48980	0.0
	$^3\Sigma_1 \rightarrow {}^3\Pi_2 \rightarrow {}^3\Sigma$	1/2	−1.00000	0.0	0.0
	Unresolved multiplet[b]	1/2	−1.00807	0.00080736	0.0
	$^3\Sigma_0 \rightarrow {}^3\Pi_1 \rightarrow {}^3\Sigma$	3/2	−0.92437	−0.07563	−0.18487
	$^3\Sigma_1 \rightarrow {}^3\Pi_0 \rightarrow {}^3\Sigma$	3/2	−0.51020	−0.48980	−0.10204
	$^3\Sigma_1 \rightarrow {}^3\Pi_2 \rightarrow {}^3\Sigma$	3/2	−1.00000	0.0	−0.20000
	Unresolved multiplet[b]	3/2	−1.00807	−0.0080736	−0.20161
	$^3\Sigma_0 \rightarrow {}^3\Pi_1 \rightarrow {}^3\Pi$	1/2	2.37327	0.49770	0.0
	$^3\Sigma_1 \rightarrow {}^3\Pi_0 \rightarrow {}^3\Pi$	1/2	2.46753	0.62338	0.0
	$^3\Sigma_1 \rightarrow {}^3\Pi_2 \rightarrow {}^3\Pi$	1/2	—	—	—
	Unresolved multiplet[b]	1/2	2.42187	0.56250	0.0
	$^3\Sigma_0 \rightarrow {}^3\Pi_1 \rightarrow {}^3\Pi$	3/2	2.37327	0.49770	0.154378
	$^3\Sigma_1 \rightarrow {}^3\Pi_0 \rightarrow {}^3\Pi$	3/2	2.46753	0.62338	0.493506
	$^3\Sigma_1 \rightarrow {}^3\Pi_2 \rightarrow {}^3\Pi$	3/2	2.00000	0.00000	−0.40000
	Unresolved multiplet[b]	3/2	2.15163	0.20218	−0.210600

[a] Integral total spin: $J_0 = 2$, $J_1 = 3$; half integral total spin: $J_0 = 5/2$, $J_1 = 7/2$.

[b] "Unresolved multiplet" indicates average over initial multiplet levels and coherent sum over intermediate multiplet levels.

TABLE VIII
Branching Ratios for Two-photon Dissociation: Hund's case (a)

Atomic limit	Molecular transition	Atomic *j*-state Relative intensities		
$^2P + {}^2S$	$(J_0 = 2, J_1 = 3)$		$j = 1/2$	$3/2$
	$^3\Sigma_0 \rightarrow {}^3\Pi_1 \rightarrow {}^3\Pi$		1.0	2.0
	$^3\Sigma_1 \rightarrow {}^3\Pi_0 \rightarrow {}^3\Pi$		0.53226	0.26613
	$^3\Sigma_1 \rightarrow {}^3\Pi_2 \rightarrow {}^3\Pi$		0.0	2.01613
	Unresolved multiplet[a]		1.0	3.1797
$^3P + {}^2S$	$(J_0 = 5/2, J_1 = 7/2)$	$j = 0$	1	2
	$^2\Sigma_{1/2} \rightarrow {}^2\Pi_{1/2} \rightarrow {}^2\Pi$	1.0	2.25	1.25
	$^2\Sigma_{1/2} \rightarrow {}^2\Pi_{3/2} \rightarrow {}^2\Pi$	0.0	1.13888	5.69444
	Unresolved multiplet[a]	1.0	3.38888	6.94444

[a] "Unresolved multiplet" indicates an average over initial multiplet levels and coherent sum over intermediate multiplet levels. *It is not the sum of the intensities for resolved fine structure states.*

TABLE IX

Fluorescence Anisotropy Parameters for Two-photon Dissociation: Hund's case (b)[a]

Atomic limit	Molecular transition	Atomic j-state	Anisotropy parameters β_D	γ_D	β_S
$^1P + {^1S}$	$^1\Sigma \rightarrow {^1\Sigma} \rightarrow {^1\Sigma}$	1	2.46753	0.62338	0.98701
	$^1\Sigma \rightarrow {^1\Sigma} \rightarrow {^1\Pi}$	1	−0.51020	−0.48980	1.20408
	$^1\Sigma \rightarrow {^1\Pi} \rightarrow {^1\Sigma}$	1	−0.92437	−0.075630	−0.36975
	$^1\Sigma \rightarrow {^1\Pi} \rightarrow {^1\Pi}$	1	2.37327	0.49770	−0.308756
$^2P + {^2S}$	$^1\Sigma \rightarrow {^1\Sigma} \rightarrow {^1\Sigma}$	1/2	2.46753	0.62338	0.0
		3/2	2.46753	0.62338	0.49351
	$^1\Sigma \rightarrow {^1\Sigma} \rightarrow {^1\Pi}$	1/2	−0.51020	−0.48980	0.0
		3/2	−0.51020	−0.48980	0.60204
	$^1\Sigma \rightarrow {^1\Pi} \rightarrow {^1\Sigma}$	1/2	−0.92437	−0.075630	0.0
		3/2	−0.92437	−0.075630	−0.18487
	$^1\Sigma \rightarrow {^1\Pi} \rightarrow {^1\Pi}$	1/2	2.37327	0.49770	0.0
		3/2	2.37327	0.49770	−0.15438
	$^1\Sigma_1 \rightarrow {^3\Sigma_2} \rightarrow {^3\Sigma}$	1/2	2.41512	0.55349	0.0
	$^3\Sigma_2 \rightarrow {^3\Sigma_3} \rightarrow {^3\Sigma}$	1/2	2.37327	0.49770	0.0
	$^3\Sigma_3 \rightarrow {^3\Sigma_2} \rightarrow {^3\Sigma}$	1/2	2.43382	0.57843	0.0
	$^3\Sigma_3 \rightarrow {^3\Sigma_4} \rightarrow {^3\Sigma}$	1/2	2.42857	0.571428	0.0
	Unresolved multiplet[b]	1/2	2.40857	0.54477	0.0
	$^3\Sigma_1 \rightarrow {^3\Sigma_2} \rightarrow {^3\Sigma}$	3/2	2.41512	0.55349	0.483024
	$^3\Sigma_2 \rightarrow {^3\Sigma_3} \rightarrow {^3\Sigma}$	3/2	2.37327	0.49770	0.474654
	$^3\Sigma_3 \rightarrow {^3\Sigma_2} \rightarrow {^3\Sigma}$	3/2	2.43382	0.57843	0.483024
	$^3\Sigma_3 \rightarrow {^3\Sigma_4} \rightarrow {^3\Sigma}$	3/2	2.42857	0.571428	0.485714
	Unresolved multiplet[b]	3/2	2.40857	0.54477	0.481715
	$^3\Sigma_1 \rightarrow {^3\Sigma_2} \rightarrow {^3\Pi}$	1/2	−0.59130	−0.408704	0.0
	$^3\Sigma_2 \rightarrow {^3\Sigma_3} \rightarrow {^3\Pi}$	1/2	−0.64935	−0.35065	0.0
	$^3\Sigma_3 \rightarrow {^3\Sigma_2} \rightarrow {^3\Pi}$	1/2	−0.591296	−0.408704	0.0
	$^3\Sigma_3 \rightarrow {^3\Sigma_4} \rightarrow {^3\Pi}$	1/2	−0.571429	−0.42857	0.0
	Unresolved multiplet[b]	1/2	−0.600739	−0.39926	0.0
	$^3\Sigma_1 \rightarrow {^3\Sigma_2} \rightarrow {^3\Pi}$	3/2	−0.59130	−0.408704	0.618259
	$^3\Sigma_2 \rightarrow {^3\Sigma_3} \rightarrow {^3\Pi}$	3/2	−0.64935	−0.350649	0.62987
	$^3\Sigma_3 \rightarrow {^3\Sigma_2} \rightarrow {^3\Pi}$	3/2	−0.59130	−0.408704	0.618259
	$^3\Sigma_3 \rightarrow {^3\Sigma_4} \rightarrow {^3\Pi}$	3/2	−0.571429	−0.42857	0.614286
	Unresolved multiplet[b]	3/2	−0.600739	−0.39926	0.62015

TABLE IX
(Continued)

Atomic limit	Molecular transition	Atomic j-state	Anisotropy parameters		
			β_D	γ_D	β_S
$^2P + ^2S$	$^3\Sigma_1 \to {}^3\Pi_2 \to {}^3\Sigma$	1/2	−1.00000	0.0	0.0
	$^3\Sigma_2 \to {}^3\Pi_2 \to {}^3\Sigma$	1/2	−0.321821	−0.67818	0.0
	$^3\Sigma_2 \to {}^3\Pi_3 \to {}^3\Sigma$	1/2	−0.93926	−0.06074	0.0
	$^3\Sigma_3 \to {}^3\Pi_2 \to {}^3\Sigma$	1/2	−1.00000	0.0	0.0
	$^3\Sigma_3 \to {}^3\Pi_3 \to {}^3\Sigma$	1/2	−0.298701	−0.70130	0.0
	$^3\Sigma_3 \to {}^3\Pi_4 \to {}^3\Sigma$	1/2	−0.90164	−0.098361	0.0
	Unresolved multiplet[b]	1/2	−1.00807	−0.0080736	0.0
	$^3\Sigma_1 \to {}^3\Pi_2 \to {}^3\Sigma$	3/2	−1.00000	0.0	−0.2
	$^3\Sigma_2 \to {}^3\Pi_2 \to {}^3\Sigma$	3/2	−0.321821	−0.67818	−0.064364
	$^3\Sigma_2 \to {}^3\Pi_3 \to {}^3\Sigma$	3/2	−0.93926	−0.06074	−0.18785
	$^3\Sigma_2 \to {}^3\Pi_2 \to {}^3\Sigma$	3/2	−1.00000	0.0	−0.2
	$^3\Sigma_3 \to {}^3\Pi_3 \to {}^3\Sigma$	3/2	−0.298701	−0.70130	−0.0597403
	$^3\Sigma_3 \to {}^3\Pi_4 \to {}^3\Sigma$	3/2	−0.90164	−0.098361	−0.18033
	Unresolved multiplet[b]	3/2	−1.00807	0.0080736	−0.201615
	$^3\Sigma_1 \to {}^2\Pi_2 \to {}^3\Pi$	1/2	2.41512	0.55349	0.0
	$^3\Sigma_2 \to {}^3\Pi_2 \to {}^3\Pi$	1/2	2.41512	0.55349	0.0
	$^3\Sigma_2 \to {}^3\Pi_3 \to {}^3\Pi$	1/2	2.46070	0.61425	0.0
	$^3\Sigma_3 \to {}^3\Pi_2 \to {}^3\Pi$	1/2	2.41512	0.55349	0.0
	$^3\Sigma_3 \to {}^3\Pi_3 \to {}^3\Pi$	1/2	2.46070	0.61425	0.0
	$^3\Sigma_3 \to {}^3\Pi_4 \to {}^3\Pi$	1/2	2.42857	0.57143	0.0
	Unresolved multiplet[b]	1/2	2.42187	0.56250	0.0
	$^3\Sigma_1 \to {}^3\Pi_2 \to {}^3\Pi$	3/2	2.17207	0.22942	−0.067764
	$^3\Sigma_2 \to {}^3\Pi_2 \to {}^3\Pi$	3/2	2.17207	0.22942	−0.20542
	$^3\Sigma_2 \to {}^3\Pi_3 \to {}^3\Pi$	3/2	2.18223	0.24297	−0.069066
	$^3\Sigma_2 \to {}^3\Pi_3 \to {}^3\Pi$	3/2	2.17207	0.229421	−0.067764
	$^3\Sigma_3 \to {}^3\Pi_3 \to {}^3\Pi$	3/2	2.18223	0.24297	−0.214848
	$^3\Sigma_3 \to {}^3\Pi_4 \to {}^3\Pi$	3/2	2.33766	0.450216	−0.142857
	Unresolved multiplet[b]	3/2	2.15163	0.202176	−0.210600

[a] Integral total spin: $J_0 = 2$, $J_1 = 3$; half integral total spin: $J_0 = 5/2$, $J_1 = 7/2$.
[b] "Unresolved multiplet" indicates average over initial multiplet levels and coherent sum over intermediate multiplet levels.

TABLE X
Branching Ratios for two-photon dissociation: Hund's case (b)

Atomic limit	Molecular Transition	Atomic j-state Relative intensities	
$^2P + {}^2S$	$(J_0 = 2, J_1 = 3)$	$j = 1/2$	3/2
	$^3\Sigma_1 \to {}^3\Pi_2 \to {}^3\Pi$	1.0	3.01570
	$^3\Sigma_2 \to {}^3\Pi_2 \to {}^3\Pi$	0.185185	0.558461
	$^3\Sigma_2 \to {}^3\Pi_3 \to {}^3\Pi$	1.41836	2.109284
	$^3\Sigma_3 \to {}^3\Pi_2 \to {}^3\Pi$	0.0015117	0.0045589
	$^3\Sigma_3 \to {}^3\Pi_3 \to {}^3\Pi$	0.221620	0.329576
	$^3\Sigma_3 \to {}^3\Pi_4 \to {}^3\Pi$	1.608687	2.123467
	Unresolved multiplet[a]	1.0	3.1797

[a] "Unresolved multiplet" indicates an average over initial multiplet levels and coherent sum over intermediate multiplet levels. *It is not the sum of the intensities for resolved fine structure states.*

molecules prepared in particular multiplet states are given, although this would be difficult to achieve experimentally. We have shown for the case when atom b is in an S-state that multiplet-averaged branching ratios are always statistical.[8] This no longer holds true if fragments obtained by dissociating molecules prepared in a particular initial multiplet level can be isolated. Also, note that all the multiplet-averaged parameters β_S for a given atomic limit are related to β_S for transitions to the $^1P + {}^1S$ limit with the same electronic orbital symmetry ($\Sigma \to \Pi$, etc.) by the same ratio. (In the following section we discuss the origin of this behavior.) The particle anisotropy parameter β_D is always 2 and -1 for a parallel and perpendicular one-photon dissociative transition, respectively.

F. Reduction of J-Independent Recoil Limit to Some Well-Known Results

Cross sections calculated under the J-independent recoil approximation reduce to results that can be obtained by other methods when neither rotational nor fine structure is resolved in the two-photon process. Without resolution of individual rotational levels in the initial transition between bound levels, the photofragment angular distribution agrees with that produced by classical interaction of an ensemble of molecular dipoles with an electric field. When neither rotational or fine structure of the molecule affects the photodissociation dynamics, the effect of fragment electronic angular momentum can be conveniently summarized in a depolarization coefficient. The notion of a depolarization coefficient was introduced by Fano[140] and its use illustrated by Fano and Macek[12] in their treatment of fluorescence from collisionally excited atoms.

Additional physical assumptions beyond those of the J-independent recoil limit are necessary to extract classical photofragment angular distributions or a depolarization coefficient. The assumptions inherent in the J-independent recoil approximation pertain to dissociation dynamics, and involve the photodissociation amplitudes $\tau(J_2 jlc_a j_a c_b j_b | J_1 \eta_1)$ exclusively. The validity of the J-independent recoil approximation is a consequence of the nature of the molecular potential energy surfaces and intramolecular coupling. The further approximations necessary to derive classical angular distributions or a depolarization coefficient from the J-independent recoil limit involve the process by which the molecule is excited. These further approximations depend on the properties of the radiation used to dissociate the molecule. Either the classical or depolarization coefficient description fails when the radiation used to dissociate the molecule has sufficiently narrow bandwidth to resolve molecular rotation and fine structure. However, the recoil approximations just described are based on intrinsic molecular properties and their validity is not dependent on excitation process. The classical and depolarization coefficient descriptions are explained in detail below.

Angular distributions produced by two-photon dissociation have been derived previously based on classical considerations.[38,99] To illustrate the classical approach, let us consider the case in which the two steps of the two-photon dissociation are both parallel transitions induced by the absorption of radiation that is linearly polarized in the z-direction. According to classical arguments, the probability of finding a molecule with dipole moment oriented with angle θ with respect to the z-axis after absorption of one photon in $\cos^2 \theta$. The probability that a molecule so oriented will absorb a second linearly polarized photon is also proportional to $\cos^2 \theta$. Regardless of the rotational state of the molecule, the angular distribution has the functional form $\cos^4 \theta$, or in terms of the fragment anisotropy parameters defined in the previous chapter, $\beta_D = 20/7$, $\gamma_D = 8/7$. This is clearly at odds with the J-independent recoil limit anisotropies in Tables VII through X, which depend on the initial and intermediate rotational quantum numbers. The classical limit must be recovered for heavy molecules, for which intermediate rotational levels cannot be spectroscopically resolved and, as we demonstrate below, a coherent sum of J_1 and $\bar{J}_1$ in the product of [Eq. (125) or (127)] and ρ_1 [Eq. (78)] leads to the classical result.

Another interesting question is the relation of our expressions for the polarization of fluorescence from excited photofragments to Fano and Macek's celebrated treatment of the effect of fine or hyperfine interactions on polarized emission from collisionally excited atoms.[12] Fano and Macek consider the case in which nonvanishing electron spin has no effect on the actual collision dynamics, except for the recoupling of atomic angular momenta into fine structure levels as the collision partners separate. This situation is analogous to the rapid recoupling of fragment angular momentum in photodissociation described in the diabatic or recoil approximation. In Fano and Macek's treatment, the effect of nonvanishing electron spin on anisotropy parameters of the atomic fragments is contained in a depolarization coefficient. Anisotropy parameters for the system with fine structure are expressed as a depolarization coefficient times an anisotropy parameter for a similar system in which the atomic electronic spin is zero. In this way, geometrical factors contained in the depolarization coefficient completely describe the recoupling of angular momenta to fine structure states and are separate from dynamical information.

An analogous separation of geometrical and dynamical factors in the photodissociation problem can be achieved if we can identify a depolarization coefficient that relates the partial cross sections $\sigma_{K_S Q_S; K_D Q_D}$ [Eq. (42)] to partial cross sections $\sigma^{(0)}_{K_S Q_S; K_D Q_D}$ for a similar system in which total spin of each fragment vanishes ($s_a = s_b = 0$). Judging by the complexity of our expressions [Eqs. (125)–(128)], the coefficient that relates $\sigma_{K_S Q_S; K_D Q_D}$ [Eq. (88)] to its zero-spin ($s_a = s_b = 0$) value, $\sigma^{(0)}_{K_S Q_S; K_D Q_D}$ would not be especially revealing. The problem we treat is also more complex because a greater number of angular momenta are involved, but this is not the essential difference between our work and Fano-Macek. Rather, the essential difference lies in the fact that our expressions [Eqs. (125)–(128)] allow for resolution of fine structure in the intermediate bound electronic state and, therefore, for coherence in the final distribution of atomic spins. As appropriate for collisional excitation, Fano and Macek may assume that the distribution of spins, nuclear or electronic for hyperfine or spin-orbit interactions, respectively, to be statistical. This suggests that the relationship between $\sigma_{K_S Q_S; K_D Q_D}$ and $\sigma^{(0)}_{K_S Q_S; K_D Q_D}$ can be conveniently summarized in a depolarization coefficient when no dynamical effect in the dissociation due to electron spin is observed. This occurs when the molecular fine structure levels are not resolved in the bound–bound transition of the two-photon process and, as is almost always the case, the initial distribution of multiplet states is statistical.

Geometrical and dynamical factors are distinguished in the J-independent recoil approximation, although these cross sections do not factor into a geometrical and dynamical part, as in the Fano-Macek theory, and fragment angular distributions vary with intermediate rotational state, in contrast with classically derived results. Having identified the physical conditions which distinguish our expressions from results of either the classical or Fano-Macek approach, we now explicitly demonstrate that neglect of rotational and fine structure in our expressions yields classical fragment angular distributions and allows us to extract a depolarization coefficient for photodissociation.

Failure to spectroscopically resolve intermediate rotational levels is modeled in our theoretical expressions by summing coherently over J_1 and $\bar{J}_1$ in the product of $\bar{\bar{\mathscr{L}}}$ [Eq. (125) or (127)] and ρ_1 [Eq. (75)]. Averaging over initial molecular fine structure levels with neglect of spectroscopic resolution of fine structure levels is most conveniently performed using the Liouville space operator for Hund's case (a) molecular states. The transition amplitudes $\tau(c_2\Lambda_2S_2\Sigma_2|J_1\eta_1)$ in Eq. (125) is assumed not to depend on Σ_2, and the full expression for the differential cross section is averaged over initial multiplet levels Σ_0. Identical results are obtained by neglecting the dependence of $\tau(c_2\Lambda_2S_2N_2|J_1\eta_1)$ on N_2 in the case (b) expression Eq. (127), and averaging cross sections over initial case (b) levels labeled by N_0.

The Liouville space operators that describe unresolved rotational structure and multiplet averaging are denoted here by angle brackets $\langle\ldots\rangle$. Explicitly, these operators are given as,

$$\langle\rho_1\rangle[B_1\beta_1(\eta_1;\bar{\eta}_1)] = (-)^{1+q_1'}\begin{pmatrix} 1 & 1 & K_1 \\ q_1' & -\bar{q}_1' & \omega_1 \end{pmatrix}\phi^{K_1}_{Q_1}(\hat{\boldsymbol{\varepsilon}}_1)$$

$$\times\langle v_1c_1\Lambda_1S_0|x_{q_1'}|J_0v_0c_0\Lambda_0S_0\rangle$$

$$\times\langle J_0v_0c_0\Lambda_0S_0|x_{\bar{q}_1'}|\bar{v}_1\bar{c}_1\bar{\Lambda}_1S_0\rangle\delta_{\bar{S}_1S_0}\delta_{S_1S_0}\delta_{B_1K_1}\delta_{\beta_1Q_1} \tag{130}$$

$$\langle\bar{\bar{\mathscr{L}}}\rangle[B_2\beta_2(\eta_2;\bar{\eta}_2)|B_1\beta_1(\eta_1;\bar{\eta}_1)]$$

$$= \sum_{K_2Q_2}(-)^{\omega_2+q_2'+1}[(K_1)(K_2)]^{1/2}\langle B_2\beta_2|K_1K_2Q_1Q_2\rangle$$

$$\times\begin{pmatrix} B_2 & K_1 & K_2 \\ \omega_2 & -\omega_1 & -\delta_2 \end{pmatrix}\begin{pmatrix} 1 & 1 & K_2 \\ q_2' & -\bar{q}_2' & -\delta_2 \end{pmatrix}\phi^{K_2}_{Q_2}(\hat{\boldsymbol{\varepsilon}}_2)$$

$$\times\,\delta_{\bar{S}_2\bar{S}_1}\delta_{S_2S_1}\tau(c_2\Lambda_2S_2|\eta_1)\tau^*(\bar{c}_2\bar{\Lambda}_2\bar{S}_2|\bar{\eta}_1) \tag{131}$$

$$
\begin{aligned}
&\langle\tilde{\tilde{\mathcal{U}}}^{\dagger}\rangle[B_3\beta_3(j_a;j_a)|B_2\beta_2(\eta_2;\bar{\eta}_2)] \\
&\quad = \sum_{K_D Q_D L\bar{L}} \frac{(B_3)^{1/2}(K_D)}{(S_2)^{1/2}} \frac{1}{4\pi} C_{K_D Q_D}(\hat{\mathbf{k}})\langle B_2\beta_2|B_3K_D\beta_3Q_D\rangle\langle c_2\Lambda_2|\Lambda_a\Lambda_b\rangle \\
&\quad \times \langle\bar{\Lambda}_a\bar{\Lambda}_b|\bar{c}_2\bar{\Lambda}_2\rangle\begin{pmatrix} K_D & B_2 & B_3 \\ 0 & \omega_2 & -\omega_2 \end{pmatrix}(L)(\bar{L})(j_a)(j_b)\begin{pmatrix} L & \bar{L} & B_3 \\ \Lambda_2 & -\bar{\Lambda}_2 & -\lambda_2 \end{pmatrix} \\
&\quad \times \begin{pmatrix} L & l_a & l_b \\ -\Lambda_2 & \Lambda_a & \Lambda_b \end{pmatrix}\begin{pmatrix} \bar{L} & l_a & l_b \\ -\bar{\Lambda}_2 & \bar{\Lambda}_a & \bar{\Lambda}_b \end{pmatrix}(-)^{j_a+j_b+\lambda_2+\bar{\Lambda}_2-S_2} \\
&\quad \times \begin{bmatrix} j_b & & l_b & & \bar{L} & & L & & l_b \\ & s_b & & l_a & & B_3 & & s_b & \\ S_2 & & s_a & & j_a & & j_a & & s_a \end{bmatrix}
\end{aligned}
\tag{132}
$$

where the notation [...] in Eq. (132) indicates the 15-j symbol,

$$
\begin{aligned}
&\begin{bmatrix} j_b & & l_b & & \bar{L} & & L & & l_b \\ & s_b & & l_a & & B_3 & & s_b & \\ S_a & & s_a & & j_a & & j_a & & s_a \end{bmatrix} \\
&\quad \equiv \sum_{F\bar{F}} (F)(\bar{F})\begin{Bmatrix} \bar{F} & F & B_3 \\ j_a & j_a & j_b \end{Bmatrix}\begin{Bmatrix} S & F & L \\ s_a & j_a & l_a \\ s_b & j_b & l_b \end{Bmatrix}\begin{Bmatrix} S & \bar{F} & \bar{L} \\ s_a & j_a & l_a \\ s_b & j_b & l_b \end{Bmatrix} \\
&\quad \times \begin{Bmatrix} L & F & S \\ \bar{F} & \bar{L} & B_3 \end{Bmatrix}
\end{aligned}
\tag{133}
$$

The fluorescence detection operation [Eqs. (83), (84)] only serves to probe the multipole moments of the electronic density matrix and is Eq. (133) unaffected by averaging over rotational and spin quantum numbers.

The transition amplitudes in Eqs. (125) and (127) are assumed to be independent of J_1 in the range $|J_1 - J_0| \leqslant 1$. When only one intermediate and one dissociative molecular state participate in the dissociation process, these transition amplitudes determine the overall normalization of the differential cross section and angular anisotropy parameters are completely determined, just as for the J-independent recoil cross sections. In this limit, anisotropy parameters β_D and γ_D for the angular distribution of photofragments, given in Table XI, agree with classical results.

TABLE XI
Classical Limit of Anisotropy Parameters for Photofragment Angular Distribution

Symmetry of molecular transition			β_D	γ_D
$\lvert q_1'\rvert = 0$	$\lvert q_2'\rvert = 0$	$(\parallel, \parallel)$	20/7	8/7
$\lvert q_1'\rvert = 0$	$\lvert q_2'\rvert = 1$	$(\parallel, \perp)$	5/7	−12/7
$\lvert q_1'\rvert = 1$	$\lvert q_2'\rvert = 0$	$(\perp, \parallel)$	5/7	−12/7
$\lvert q_1'\rvert = 1$	$\lvert q_2'\rvert = 1$	$(\perp, \perp)$	−10/7	3/7

Depolarization coefficients that depend only on quantum numbers for the dissociative molecular state and fragment fine structure level can be extracted from Eq. (132) when quantum interference due to dissociation through more than one intermediate or dissociative electronic state does not occur. Except for the transition amplitudes which provide an overall normalization, the operators $\langle \rho_1 \rangle$ and $\langle \tilde{\tilde{\mathscr{L}}} \rangle$ are completely independent of spin quantum numbers. Letting $K_D = 0$ to determine the angular distribution of fluorescence, it is seen from Eq. (132) that the operator $\langle \tilde{\tilde{\mathscr{U}}}^\dagger \rangle$ is reduced by the factor,

$$g(B_3) = \left\{ \sum_{\substack{\bar{\Lambda}_a\bar{\Lambda}_b \\ \Lambda_a\Lambda_b}} \sum_{L\bar{L}} (-)^{j_a + j_b - S} \frac{(\bar{L})(L)}{(S)^{1/2}} \right.$$
$$\times \begin{pmatrix} L & \bar{L} & B_3 \\ \Lambda_2 & -\bar{\Lambda}_2 & -\lambda_2 \end{pmatrix} \begin{pmatrix} L & l_a & l_b \\ -\Lambda_2 & \Lambda_a & \Lambda_b \end{pmatrix} \begin{pmatrix} \bar{L} & l_a & l_b \\ -\bar{\Lambda}_2 & \bar{\Lambda}_a & \bar{\Lambda}_b \end{pmatrix}$$
$$\times \left\{ \begin{matrix} j_b & & l_b & & \bar{L} & & L & & l_b & \\ & s_b & & l_a & & B_3 & & l_a & & s_b \\ S & & s_a & & j_a & & j_a & & s_a & \end{matrix} \right\}$$
$$\left. \times \langle c_2\Lambda_2 | \Lambda_a\Lambda_b \rangle \langle \bar{\Lambda}_a\bar{\Lambda}_b | c_2\Lambda_2 \rangle \right\}$$
$$\times \left\{ \sum_{\substack{\bar{\Lambda}_a\bar{\Lambda}_b \\ \Lambda_a\Lambda_b}} \delta_{\Lambda_a\bar{\Lambda}_a} \delta_{\Lambda_b\bar{\Lambda}_b} (-)^{l_a + \Lambda_a} \begin{pmatrix} B_3 & l_a & l_a \\ 0 & \Lambda_a & -\Lambda_a \end{pmatrix} \right.$$
$$\left. \times \langle c_2\Lambda_2 | \Lambda_a\Lambda_b \rangle \langle c_2\Lambda_2 | \Lambda_a\Lambda_b \rangle \right\}^{-1} \tag{134}$$

for nonzero spin. This is the depolarization coefficient that relates multipole moments of the electronic density matrix to their zero-spin values, analogous to the Fano-Macek treatment. Taking the dependence of $\mathscr{D}^\dagger$ on fine structure quantum numbers into account, we can relate the fluorescence anisotropy parameter β_S to its zero-spin value $\beta_S^{(0)}$,

$$\beta_S = g(K_S)\frac{(j_a')(j_a)W(1K_Sj_a'j_a;\ 1j_a)[W(j_a1s_al_a';\ j_a'l_a)]^2}{W(1K_Sl_a'l_a;\ 1l_a)}\beta_S^{(0)} \tag{135}$$

When atom b, from which fluorescence is not detected, is produced in an S-state, as is the case for all the examples in Tables III–X,

$$\beta_S = (j_a)(-)^{l_a'+s_a-j_a'}\frac{\begin{Bmatrix} j_a & j_a & 2 \\ 1 & 1 & j_a' \end{Bmatrix}}{\begin{Bmatrix} l_a & l_a & 2 \\ 1 & 1 & l_a \end{Bmatrix}}\begin{Bmatrix} j_a & j_a & 2 \\ l_a & l_a & s_a \end{Bmatrix}\beta_S^{(0)} \tag{136}$$

G. Completely Adiabatic Dissociation

In this section we consider photodissociation dynamics in which the adiabaticity parameter $P = (\Delta r\,\Delta E)/(v\hbar)$ Eq. (92) is very large. This indicates that the distance $(v\hbar)/(\Delta E)$ the fragments travel in one Bohr frequency period is much smaller than the length of the recoupling region. Therefore, the intramolecular coupling changes very slowly on a time scale based on the spacing between molecular levels, and the electronic states respond adiabatically to the slow perturbation. Physical conditions that lead to large P are long-range forces between the fragments, which implies that Δr is large, large splitting ΔE between molecular levels in the recoupling region, and large reduced mass.

We have already described how the adiabatic approximation is implemented in the molecular region [see Eqs. (93)–(102) and accompanying discussion], and it is straightforward to extend the adiabatic treatment to the whole length of the dissociation coordinate. We recall that adiabatic basis functions $|J_2M_2\alpha_2\rangle$ [Eq. (95)] are defined previously as the r-dependent eigenvectors of the full dissociating molecule Hamiltonian minus the radial kinetic energy term [Eq. (93)]. Coupling between adiabatic states arises from the action of the radial kinetic energy operator on these levels.

Each adiabatic level $|J_2M_2\alpha_2\rangle$ is uniquely associated with an eigenstate $|J_2M_2jlc_aj_ac_bj_b\rangle$ of the atomic fragments with which it correlates adiabatically [see Eq. (96)]. This simply means that at very large r the level $|J_2M_2\alpha_2\rangle$ evolves into the fragment eigenstate $|J_2M_2jlc_aj_ac_bj_b\rangle$. Neglecting coupling between the basis states $|J_2M_2\alpha_2\rangle$ in the adiabatic limit, the continuum wave function $|EJ_2M_2jlc_aj_ac_bj_b^{(-)}\rangle$ and Eq. (50) can be replaced by the much simpler expression,

$$|EJ_2M_2jlc_aj_ac_bj_b^{(-)}\rangle = \frac{1}{r}\psi^{(-)}_{EJ_2\alpha_2}(r)|J_2M_2\alpha_2\rangle \tag{137}$$

The radial wave function $\psi^{(-)}_{EJ_2\alpha_2}(r)$ satisfies the radial Schrödinger equation [Eq. (98)] with boundary conditions Eqs. (99) and (105).

The wave function Eq. (137) is a special case of Eq. (97) which is obtained by the following choice of coëfficients $C^{J_2}_{\alpha\,;jlc_aj_ac_bj_b}$ in Eq. (97),

$$C^{J_2}_{\alpha_2;jlc_aj_ac_bj_b} = \begin{cases} i^{-l} & \text{if } |J_2M_2\alpha_2\rangle \text{ correlates} \\ & \text{adiabatically with } |J_2M_2jlc_aj_ac_bj_b\rangle \\ 0 & \text{otherwise} \end{cases} \tag{138}$$

The continuum wave function describes photodissociation in which non-adiabatic transitions do not take place in the molecular region. The coefficients $C^{J_2}_{\alpha_2;jlc_aj_ac_bj_b}$ account for possible nonadiabatic transitions outside the molecular region. The special choice of coefficients of Eq. (138) represents the physical assumption that nonadiabatic transitions do not occur in the recoupling region, and may be compared with the diabatic coefficients of Eq. (107) which describe the opposite dynamical limit.

Transition amplitudes $\tau(J_2jlc_aj_ac_bj_b|J_1\eta_1)$ in the adiabatic limit are given by Eq. (100) using the coefficients $C^{J_2}_{\alpha_2;jlc_aj_ac_bj_b}$ specified by Eq. (138). If the dissociation is completely adiabatic, fragments are produced only in those atomic states which correlate with the adiabatic levels that receive dissociative flux from bound states. Cross sections are obtained by substitution of the adiabatic transition amplitudes $[\tau(J_2\alpha_2|J_1\eta_1)C^{J_2}_{\alpha_2;jlc_aj_ac_bj_b}]$ into the various equations derived in Section IV. Just as in the diabatic limit, the expressions for cross sections simplify enormously under the following two conditions: (1) Only one adiabatic level receives dissociative flux from the bound levels which are initially populated in the photodissociation process. This condition often occurs due to a combination of selection rules and huge differences in Franck-Condon factors to different dissociative levels. (2) The amplitudes $\tau(J_2\alpha_2|J_1\eta_1)$ given in Eq. (101), which are dipole matrix elements between the continuum states [Eq. (137)] and the bound level $|J_1M_1\eta_1\rangle$, are approximately independent of J_2 over the range $|J_2 - J_1| \leqslant 1$. This occurs when the excess translational energy of the photofragments is much larger than the rotational constant of the molecule so that the radial wave function $\psi^{(-)}_{EJ_2\alpha_2}(r)$ in Eq. (137) is sensitive to changes in the centrifugal potential over the range $|J_2 - J_1| \leqslant |$. The adiabatic approximation is valid for molecules with very large reduced mass, and therefore this second condition is expected to be valid at moderate fragment kinetic energies.

Under the conditions specified, all transition amplitudes $\tau(J_2jlc_aj_ac_bj_b|J_1\eta_1)$ are proportional to $\tau(J_2\alpha_2|J_1\eta_1)$ for $|J_2 - J_1| \leqslant 1$. (Actually, the $\tau(J_2jlc_aj_aj_bj_b|J_1\eta_1)$ are zero or unity times $\tau(J_2\alpha_2|J_1\eta_1)$, as seen from Eqs. (100) and (138).) Therefore the partial cross sections [Eq. (42)], which are bilinear sums of transition amplitudes, are proportional to $|\tau(J_2\alpha_2|J_1\eta_1)|^2$. Branching ratio and anisotropy parameters are specified by the coefficients $C^{J_2}_{\alpha_2;\,jlc_aj_ac_bj_b}$ given in Eq. (138), and the details of photodissociation dynamics, which determine the $\tau(J_2\alpha_2|J_1\eta_1)$, do not affect these observable parameters. In contrast to the diabatic limit (in which the symmetry of the molecular electronic states involved in the dissociation is sufficient to determine branching ratios and anisotropy parameters), limited knowledge of the dissociative molecular potential energy surfaces is needed in the adiabatic limit. The precise shape of these curves is not important, but the energy ordering of molecular electronic levels must be known in order to link multiplet states in the molecular region to the atomic eigenstates with which they correlate adiabatically.

VI. NUMERICAL CALCULATIONS

The calculations of differential photodissociation cross sections described here illustrate how dissociation experiments can provide a sensitive probe of molecular potential surfaces. The effects of nonadiabatic interactions, which are predicted here, are especially suited to furnish understanding of the interaction between open shell fragments. We also demonstrate how nonadiabatic effects can be exploited to gain information about electronic states that are not directly reached from the ground molecular level by optical excitation.

Numerically calculated photodissociation cross sections for three diatomic molecules, NaH, CH^+ and Na_2, are presented in this section. Both NaH and CH^+ exhibit very strong electronically nonadiabatic effects, as may be expected for molecules with very small reduced mass and short ranged potentials. The reduced mass of Na_2 is about an order of magnitude greater than those of the hydrides and the resonance dipole interaction between $Na(3^2S)$ and $Na(3^2P)$ fragments give rise to a long range potential. Calculations confirm that dissociation of Na_2 is electronically adiabatic close to the dissociation threshold.

A. NaH + $h\nu \rightarrow$ Na(3 ^2P) + H(1 ^2S)

NaH can be dissociated to the first excited atomic limit by photoexcitation from the $X\,^1\Sigma^+$ ground state to either the $A\,^1\Sigma^+$ or $B\,^1\Pi$ excited electronic states (Fig. 10). Besides the $A\ ^1\Sigma^+$ and $B\ ^1\Pi$ level, the two triplet states with designations $b\ ^3\Pi$ and $c\ ^3\Sigma^+$ also correlate with the Na(^2P) + H(^2S) term limit. Calculations by Singer, Freed, and Band[9,10] predict that these triplet levels have an observable effect on the dissociation dynamics near the Na(^2P) + H(^2S) threshold. The presence of nonadiabatic coupling to the triplet states is manifested in the energy variation of photodissociation cross sections, fragment angular distributions and the polarization of the fragments. (Fragment polarization is conveniently probed in NaH dissociation by measuring the angular distribution and polarization of sodium *D*-line fluorescence.) Quasibound levels in both triplet and singlet states exhibit resonances in these cross sections. The quasibound levels arise from tunneling through a centrifugal barrier (shape resonance) or from a dissociative level that correlates with the upper sodium fine structure state when it is not energetically accessible (Feshbach resonance). This shows how so-called "dark" states, such as these triplet levels, are not merely spectators to the dissociation process, and can be probed in the energy region where nonadiabatic effects are significant.

Adiabatic Born-Oppenheimer potential energy surfaces and electronic transition dipole integrals for this calculation are taken from *ab initio* calculations of Sachs et al.[35] Spin-orbit and Coriolis interaction matrix elements are approximated by their value as $r \to \infty$, as described in Section

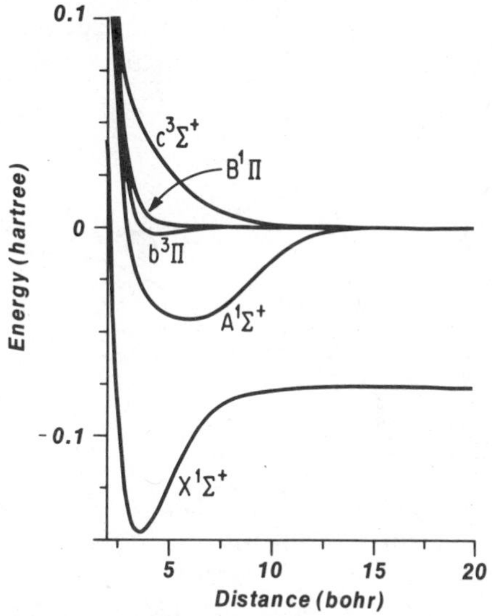

Figure 10. Molecular potential energy curves calculated by Sachs et al.[35] for the $X\ ^1\Sigma^+$ ground state of NaH, and the electronic levels correlating with the Na(3 ^2P) + H(1 ^2S) atomic limit.

III, but hyperfine interactions are neglected. Although hyperfine interactions do not need to be considered in the half-collision dynamics, they do affect the polarization of fluorescence emitted by the sodium atom. In order to compare with experiment, the anisotropy parameters presented here should be modified according to procedures given by Fano and Macek[12] and Greene and Zare.[111]

The branching ratio for dissociation to $Na(^2P_{3/2})$ fragments relative to $Na(^2P_{1/2})$ fragments is shown as a function of energy in Fig. 11. The branching ratio approaches 2:1 at high energies, as predicted in the J-independent recoil limit (Section V.E). Fragment kinetic energy of more than 10 times the fine structure state splitting is required before the recoil limit is observed. Dissociation to the upper fine structure level is suppressed when the excess dissociation energy is just barely above the $Na(^2P)$ threshold, as can be seen in the branching ratio curves for all the initial angular momenta in Fig. 11. It is also suppressed when the total angular momentum is large, making the $j_{Na} = 3/2$ channel energetically less favorable. This is illustrated by the $J_0 = 10$ branching ratio, which is less than 2 for all energies below $200\ cm^{-1}$, in contrast to the curves for other initial angular momenta. Branching ratios show considerable structure at very low energy (Figs. 12–14), but these features cannot be seen well on the scale of Fig. 11.

The very low translational energy cross section for single-photon dissociation from $J_0 = 3$ is shown in Fig. 12. A slightly asymmetric resonance peak is present at $-8.8\ cm^{-1}$, well below the $Na(^2P_{3/2})$ threshold. This peak can be associated with the $b\ ^3\Pi$ dissociative molecular surface and could not be calculated without properly treating nonadiabatic coupling. Because of the

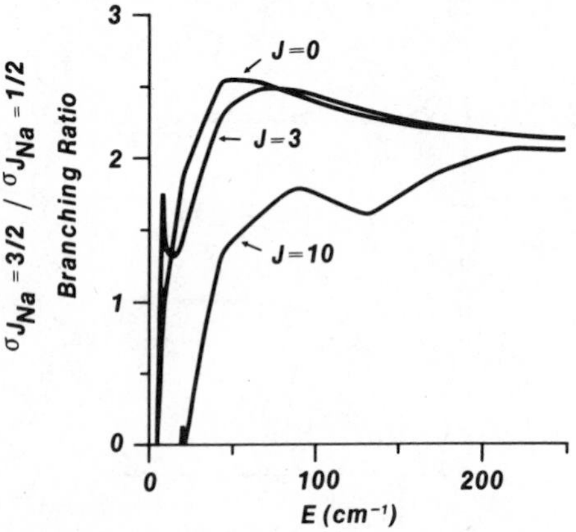

Figure 11. Calculated sodium $^2P_{3/2}/^2P_{1/2}$ branching ratio as a function of excess dissociation energy for one-photon dissociation of NaH from three different initial angular momentum levels of the $X\ ^1\Sigma^+$, $v_1 = 0$ vibronic state.[9,10]

slight asymmetry of the resonance peak, we attribute its formation to a quasibound level formed on the $b\,^3\Pi$ surfaces that correlates adiabatically with the $j_{\mathrm{Na}} = 3/2$ atomic limit, which is energetically closed at $-8.8\ \mathrm{cm}^{-1}$. The coupling of a bound state to the continuum leads to the presence of an asymmetric Feshbach or Fano type resonance peak. The fragment angular distribution exhibits dramatic energy dependence near the dissociation threshold, clearly reflecting the resonance at $-8.8\ \mathrm{cm}^{-1}$.

The cross sections and anisotropy parameters for $J_0 = 10$ (Fig. 13) are interesting because two resonances are observed in the calculated cross sections. The huge peak at $12.0\ \mathrm{cm}^{-1}$ in the $j_{\mathrm{Na}} = 1/2$ cross section is assigned as a shape resonance on the $A\,^1\Sigma^+$ surface. Its position and shape is predicted by single channel calculations in which the $A\,^1\Sigma^+$ state is the only dissociative surface, and, taken together with the absence of this feature from the $j_{\mathrm{Na}} = 3/2$ cross section, this suggests that dissociation dynamics are nearly adiabatic in this case. The resonance at $20.0\ \mathrm{cm}^{-1}$ in both the $j_{\mathrm{Na}} = 1/2$ and 3/2 cross sections is associated with the $b\,^3\Pi$ surface. The intensity of this second peak is diminished in relation to the first peak because of Franck-Condon factors. (However, compare Fig. 13 with Fig. 14, where the molecule is dissociated from a vibrational level of the $A\,^1\Pi^+$ surface in a two-photon process.) There Franck-Condon factors favor the $b\,^3\Pi$ resonance indicating the importance of interchannel coupling in this case.

Even though the intensity of the $b\,^3\Pi$ resonance at $20.0\ \mathrm{cm}^{-1}$ is quite low, its presence is clearly manifested in the anisotropy parameters for the fragment and fluorescence angular distributions. Since most of the dissociative flux remains on the lowest adiabatic surface during the half collision, that

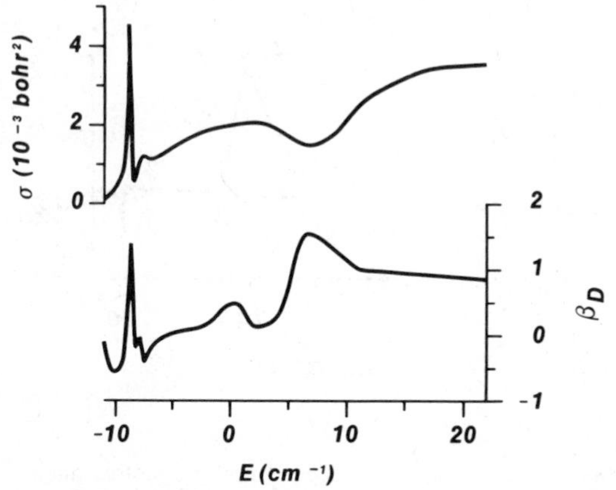

Figure 12. Variation of the sodium $^2P_{1/2}$ photofragment cross section and anisotropy parameters with excess dissociation energy for one-photon dissociation of NaH from the $X\,^1\Sigma^+$, $J_1 = 3$, $v_1 = 0$ vibronic level.[9,10]

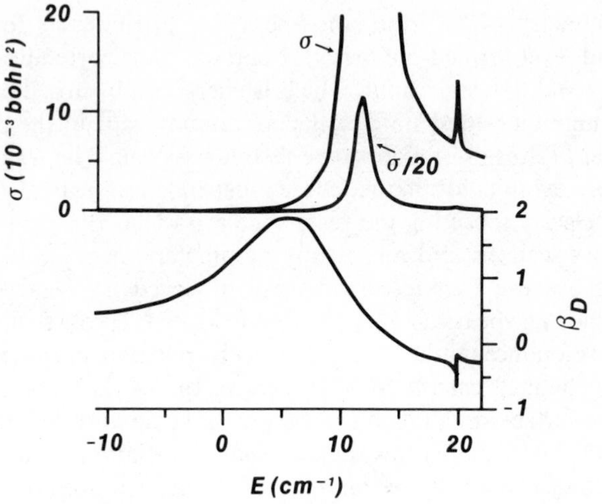

(a)

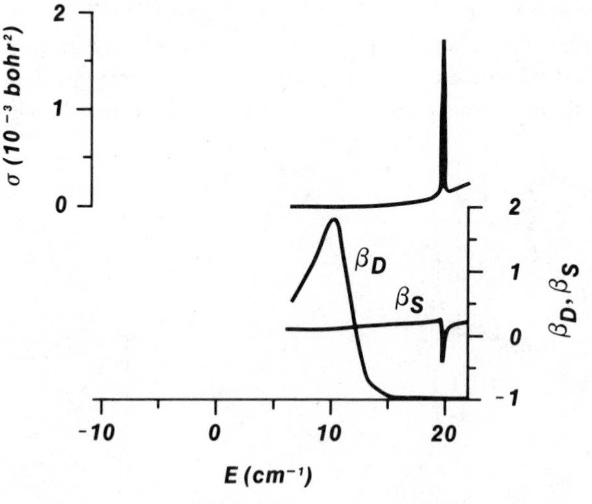

(b)

Figure 13. Excess dissociation energy dependence of photofragment cross section and anisotropy parameters for single-photon dissociation of NaH from the $X\,^1\Sigma^+$, $J_1 = 10$, $v_1 = 0$ level.[9,10] Parameters for fragments produced in the sodium $^2P_{1/2}$ and $^2P_{3/2}$ levels are shown in (*a*) and (*b*), respectively.

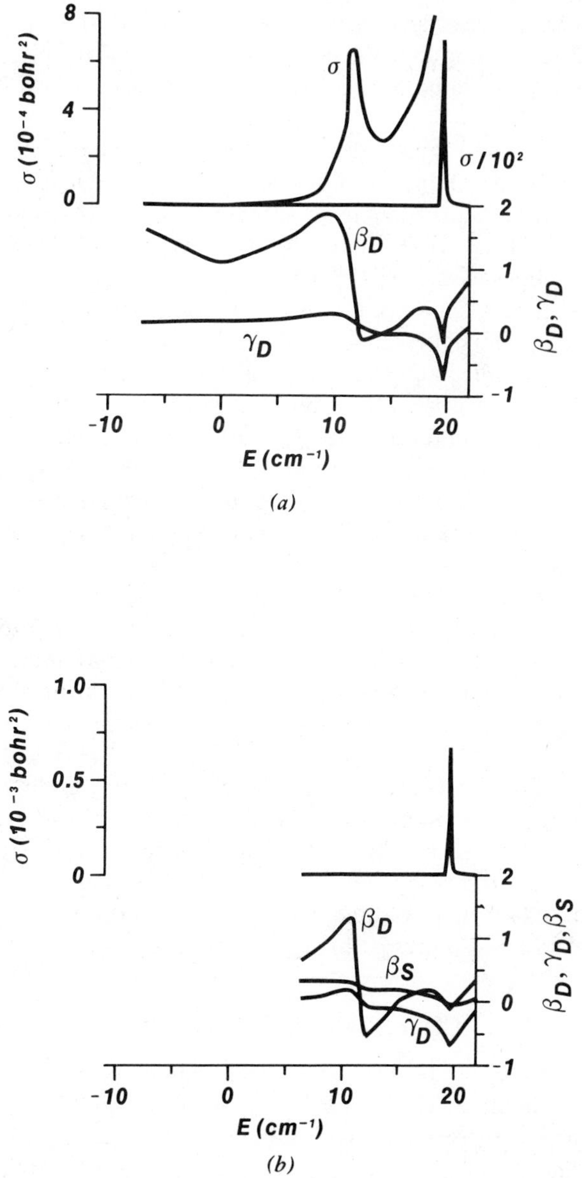

Figure 14. Photofragment cross section (arbitrary units) and anisotropy parameters for two-photon dissociation of NaH as a function of excess dissociation energy.[10] The molecule is assumed to be prepared in the $J_1 = 10$, $v_1 = 10$ level of the $A\ ^1\Sigma^+$ electronic state by a transition from a vibronic level of the $X\ ^1\Sigma^+$ surface with $J_0 = 9$. Parameters for fragments produced in the sodium $^1P_{1/2}$ and $^2P_{3/2}$ levels are shown in (*a*) and (*b*), respectively.

is, on the $A\,^1\Sigma^+$ curve at small r which correlates with the $j_{\mathrm{Na}} = 1/2$ atomic limit, it is not surprising that β_D varies smoothly through the region of the $A\,^1\Sigma^+$ resonance, while both β_D and β_S display sharp features for the $b\,^3\Pi$ resonance.

Cross sections and anisotropy parameters for resonant two-photon dissociation of NaH are displayed in Fig. 14. The molecule is excited from a bound state with total angular momentum $J_0 = 9$ to an intermediate rovibronic level of the $A\,^1\Sigma^+$ state with $J_1 = 10$, $v_1 = 10$. The molecular is subsequently photodissociated by absorption of a second photon. Both photons are linearly polarized along the same laboratory axis in the example shown in Fig. 14, although results of the dynamical calculations could be used to calculate cross sections and anisotropy parameters for arbitrary photon polarization. Two-photon transition amplitudes are formed using the same continuum wave functions as for the one-photon process, but with vibrational wave functions for the $A\,^1\Sigma^+$ state and electronic transition moment functions for the two possible bound–free transitions, $A\,^1\Sigma^+ \rightarrow A\,^1\Sigma^+$ (molecular dipole moment function for the excited state) and $A\,^1\Sigma^+ \rightarrow B\,^1\Pi$.

The most obvious difference between the one- and two-photon cross sections is the enhancement of the $b\,^3\Pi$ resonances compared to the $A\,^1\Sigma^+$ peak. Transitions from a bound $A\,^1\Sigma^+$ level to the continuum of the same electronic state would be forbidden if the dipole moment function were constant in the Franck-Condon region and nonadiabatic couplings not included. Even with both these conditions relaxed, the parallel transition is still strongly disfavored. Due to the additional anisotropy given to the system by absorption of two photons, the fragment angular distribution has a hexadecapolar component ($K_D = 4$) which is not present for single-photon dissociation. The anisotropy parameter for this additional component, γ_D [see Eqs. (42), (129)], is plotted in Fig. 14 along with the anisotropy parameters that are also present in single-photon dissociation, β_D [Eq. (86)] and β_S [Eq. (87)].

B. $\mathrm{CH}^+ + h\nu \rightarrow \mathrm{C}^+(2\,^2P) + \mathrm{H}(1\,^2S)$

Four molecular electronic states of the CH^+ ion correlate with the lowest energy atomic limit, $\mathrm{C}^+(2\,^2P) + \mathrm{H}(1\,^2S)$ (see Fig. 15).[141-143] In order of increasing energy, these molecular states are the $X\,^1\Sigma^+$, $a\,^3\Pi$, $A\,^1\Pi$, and $c\,^3\Sigma$ levels. The $X\,^1\Sigma^+$, $a\,^3\Pi$, and $A\,^1\Pi$ potential energy surfaces all have attractive wells. The CH^+ ion prepared in a bound level in any one of these states can be photoexcited to the continuum of another molecular electronic state, principally by perpendicular transitions, that is, $X\,^1\Sigma^+ \rightarrow A\,^1\Pi$, $a\,^3\Pi \rightarrow c\,^3\Sigma$, $A\,^1\Pi \rightarrow X\,^1\Sigma^+$, and so on.

The theory presented in this article and the experiments discussed below

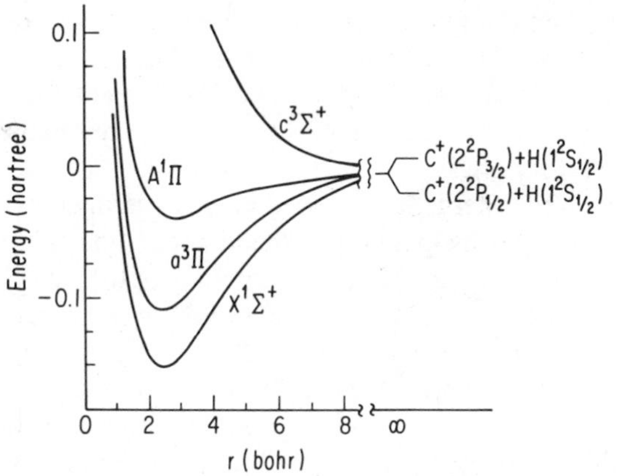

Figure 15. Potential energy curves of the CH^+ ion which correlate with the $C^+(2\,^2P) + H(1\,^2S)$ atomic limit. The spin-orbit splitting (64 cm^{-1}) is greatly exaggerated. These curves were constructed[11] using the *ab initio* data of Green et al. and singlet potential energy curves kindly supplied to us by Helm.[143]

lend credence to the belief that CH^+ is an excellent candidate for observation of numerous resonance peaks upon photodissociation. The spin-orbit splitting of the $^2P_{1/2}$ and $^2P_{3/2}$ states of the C^+ ion is 64 cm^{-1}, large enough to strongly mix the $X\,^1\Sigma^+$, $a\,^3\Pi$, $A\,^1\Pi$, and $c\,^3\Sigma$ ABO levels at large interfragment separation. The three attractive potential energy surfaces can each support quasibound levels (shape resonances) due to the formation of a centrifugal barrier. In addition, Feshbach type resonances are expected in the energy region between the $^2P_{1/2}$ and $^2P_{3/2}$ limits of the C^+ ion.

As mentioned in Section II, several groups have been working on the photofragment spectroscopy of CH^+. Cosby et al.[144] and Helm et al.[145] observe resonances in the optical and near UV region for CH^+ using a coaxial fast-ion-beam-laser photofragment spectrometer. Many resonances of strong and moderate intensity have been assigned by Helm et al. as shape resonances on the $A\,^1\Pi$ state of CH^+ based on single channel theoretical calculations and on the assumption that the transitions originate from high-lying rotational levels on the $X\,^1\Sigma^+$ ground state.[145,146] The wealth of other resonances in the data of Helm et al.[145] cannot be explained by the single channel calculations, but they are in qualitative agreement with multichannel theoretical calculations discussed later that include nonadiabatic couplings between molecular levels which correlate with the $C^+(2\,^2P) + H(1\,^2S)$ atomic limit.[11,147]

Carrington et al.[148] observe resonances for CH^+ in the infrared region of 970–1090 cm^{-1} using a CW carbon dioxide laser beam that is collinear with a CH^+ ion beam. The resonances are found to be as narrow as 10 MHz and many exhibit doublet splittings that are believed to be due to hyperfine interactions or Λ- and ρ-type doublings. The infrared resonances of Carrington et al. have not yet been assigned, but as mentioned previously and as predicted for NaH, it is believed that many of the resonances are the result of nonadiabatic couplings.

On the basis of the energetics involved in the experiments of Helm et al.,[145] it is believed that the principal perpendicular transition leading to photodissociation is the $X\,^1\Sigma^+ \rightarrow A\,^1\Pi$ transition, Carrington et al.[148] use infrared excitation to dissociate the CH^+ ions and therefore it is energetically possible that the $a\,^3\Pi \rightarrow c\,^3\Sigma$ and the $A\,^1\Pi \rightarrow X\,^1\Sigma^+$ transitions may also play a role in their experiments. The presence of an attractive r^{-4} term in the long range ion–atom interaction on the $c\,^3\Sigma$ potential surface produces a small attractive well on the $c\,^3\Sigma$ state centered at about 6.1 a.u. With the aid of a simple Franck-Condon picture, it is plausible that the coupling to higher vibrational and rotational levels of the $a\,^3\Pi$ state may lead to the formation of shape and Feshbach type resonances on the $c\,^3\Sigma$ surface.

The purpose of ongoing calculations described here is to confirm that proper treatment of nonadiabatic coupling in the photodissociation of CH^+ leads to the prediction of a very large number of resonance peaks in the photofragment spectrum (see Figs. 16–18), and to help assign the experiment-

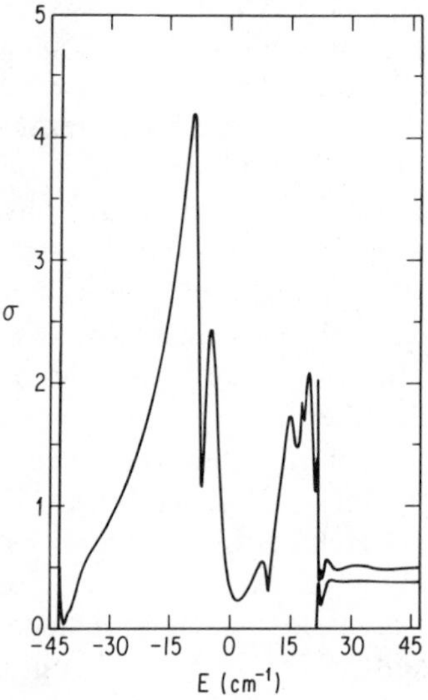

Figure 16. Calculated ion photofragment cross section (arbitrary units) as a function of excess dissociation energy for single-photon dissociation of CH^+. The ion is assumed initially to be in the $J_1 = 2$, $v_1 = 0$ level of the $X\,^1\Sigma^+$ state. Only the contribution from final angular momentum $J_2 = 3$ is included in the cross section.

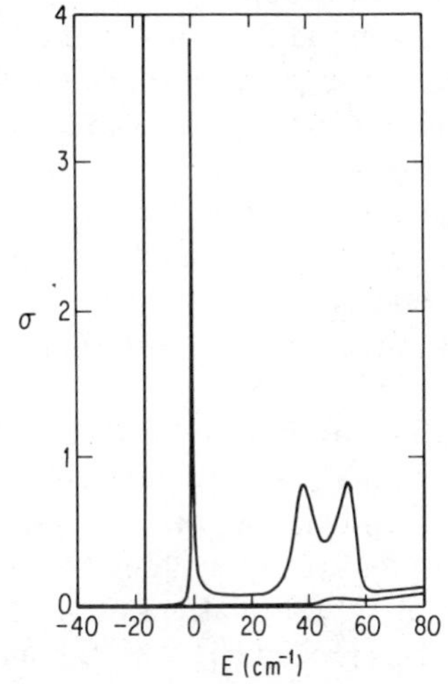

Figure 17. Partial photodissociation cross sections for initial $J_0 = 12$ and final $J = 12$. The CH^+ ion is assumed to be prepared in the $v_0 = 10$ level of the $X\ ^1\Sigma^+$ state. The partial cross section for production of $C^+(^2P_{3/2})$ fragments is negligibly small until 40 cm^{-1}, almost 20 cm^{-1} above the $C^+(^2P_{3/2})$ threshold. Also, the resonance near $-20\ cm^{-1}$ exceeds the scale of the graph by more than one order of magnitude.

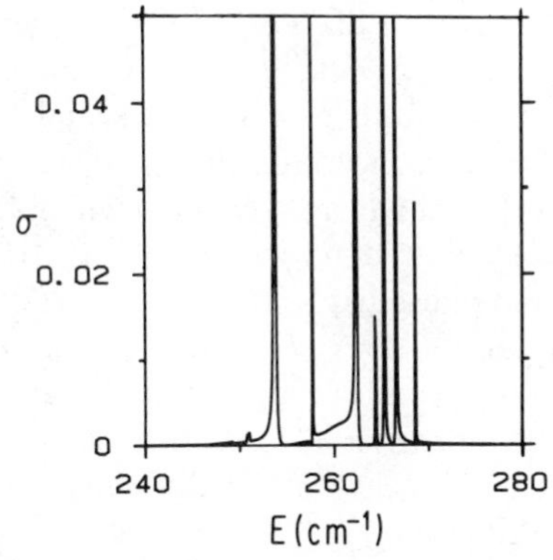

Figure 18. Partial photodissociation cross section for initial $J_0 = 19$ and final $J = 20$. Initially the CH^+ ion is in the $v_0 = 0$ level of the $X\ ^1\Sigma^+$ state. Only a small energy range of the $C^+(^2P_{1/2})$ photofragment cross section is shown. Other resonance peaks calculated for this combination of total angular momenta are much smaller. The maximum height of the resonance peaks is not converged, but many exceed the scale of the graph by several orders of magnitude.

ally resonance peaks. Direct comparison of the positions and shapes of individual resonance peaks in the calculated and observed spectra have been hampered by the large number of resonances, the fact that some are quite narrow, and by uncertainties in the potential energy surfaces. In particular, the threshold dissociation energy is not known exactly. This leads to problems in relating the energy scale of theoretical calculations (fragment kinetic energy) to the experimental scale (wavelength of optical excitation). Ongoing experimental efforts to achieve greater resolution of fragment kinetic energy and current calculations hopefully will help alleviate this difficulty.

In addition to the strong motivation for theoretical calculations of CH^+ provided by photodissociation experiments, there is a long standing interest in the rate of radiative association of $C^+(2\,^2P)$ with $H(1\,^2S)$ and the photodissociation rate of CH^+ to explain the observed abundance of CH and CH^+ in interstellar regions.[149] Calculations of radiative recombination rates ignoring all resonance effects predict CH–CH^+ abundances off by an order of magnitude, while those including $A\,^1\Pi$ shape resonances considerably revise earlier estimates but are insufficient to explain astrophysical data.[150-152] The calculations presented here predict a large number of resonances of nonadiabatic origin in addition to the $A\,^1\Pi$ shape resonances. More extensive calculations of the radiative association rate will be presented elsewhere.[147] Current calculations of the photodissociation rate of CH^+ in interstellar space are hampered by a lack of knowledge of the repulsive portion of the CH^+ potentials at higher energies.

In the calculations of cross sections displayed in Figs. 16–18, spin-orbit and Coriolis coupling between ABO levels of CH^+ are approximated by their $r \to \infty$ values, as described in Section III. The molecular potential energy curves and transition moment functions are estimated using experimental data which is only available for the singlet states, and using results of *ab initio* calculations.[141-143] Details concerning the construction of potential energy surfaces and numerical methods are given elsewhere.[11]

The quantity actually plotted in Figs. 16–18 is $\Sigma_{jl}\,|\tau(J_2 jlc_a j_a c_b j_b\,|\,J_1\eta_1)|^2$, as given in Eq. (78). (Total photodissociation cross sections are proportional to the sums of this quantity over $J_2 = J_1 - 1,\ J_1,\ J_1 + 1$.) Considerable structure is evident in the excess energy region between the two fine structure state thresholds of the $C^+(2\,^2P)$ ion (-42 and $+21\ \mathrm{cm}^{-1}$ on our energy scale in which the zero of energy is the statistically weighted average energy of the C^+ fine structure multiplet).

Simple theoretical models have been applied to CH^+ photodissociation in which motion is assumed to occur on a single adiabatic potential energy surface (Fig. 15).[145,146] Nonadiabatic effects are then included by the use of perturbation theory. These simple models predict only one or two resonances for each final angular momentum J_2. Full close-coupled quantum calculations in Fig. 16 for $J_1 = 2$, $J_2 = 3$ clearly exhibit a total of eight features occurring between the two fine structure state thresholds of C^+. The close-coupled calculations of Fig. 17 for $J_1 = 12$, $J_2 = 12$ have at least four features and in Fig. 18 for $J_1 = 19$, $J_2 = 20$ there is a large cluster of shape resonances in the energy region between 240 and 280 cm^{-1}. Additional resonances for $J_1 = 19$, $J_2 = 20$ exist outside of the region shown in Fig. 18. In addition, the resonances shown have half-widths as narrow as 0.01 MHz which is even narrower than the resonances observed by Carrington and co-workers.[148] Actual assignments of these resonances and a comparison with experimental data will be presented elsewhere.[147]

C. $Na_2(X^1\Sigma_g) + h\nu \rightarrow Na(3\,^2P) + Na(3\,^2S)$

The interaction between the excited and ground-state sodium fragments produced in photodissociation of Na_2 exhibits a number of interesting features. There is an extremely long-range resonance dipole interaction (RDI) between the photofragments of the homonuclear diatomic.[153-155] Also, the $B\,^1\Pi_u$ electronic potential energy surface (Fig. 5) has a substantial barrier because the short-range electronic forces are attractive whereas the RDI potential is repulsive for this state. The height of the $B\,^1\Pi_u$ barrier, still a matter of controversy, is believed to be ~400–500 cm^{-1},[156,157] and the consequences of the RDI potential on collisional and dissociation dynamics are yet to be fully explored, although some valuable studies exist.[153] The sodium ^2P–^2S interaction potentials have long been of interest because elastic and fine structure state changing collisions of $^2P_{1/2,3/2}$ and $^2S_{1/2}$ atoms are responsible for the broadening of the D_1 and D_2 resonance lines in sodium vapor.[158-160]

The RDI potential arises because the lowest excited electronic states of Na_2 correlate asymptotically with the degenerate atomic states $Na(3\,^2S_{1/2})$ + $Na(3\,^2P_{j_{Na}})$ and $Na(3\,^2P_{j_{Na}}) + Na(3\,^2S_{1/2})$ with $j_{Na} = 1/2, 3/2$. The degeneracy is lifted by the dipole term in the multipolar expansion of the electronic potential, resulting in long-range potential energy curves which behave as $(-)^S w(-)^\Lambda[2 - \Lambda](p^2)/R^3$, where w denotes $g(+)$ or $u(-)$ symmetry and p is the atomic transition moment $\langle 3\,^2S_{1/2}|e\mathbf{x}|3\,^2P_{j_{Na}}\rangle$; Λ and S have their usual meanings as ABO level quantum numbers. The short-range part of the ground ($X\,^1\Sigma_g^+$) and dissociative potential curves have been synthesized from the *ab initio* calculations of Konowalow et al.[27] (Fig. 5) and from experimental data,[156] and are smoothly joined to the RDI potentials given above at large R. Details about the construction of the potential energy curves and numerical method are given in Ref. 123.

Calculated photodissociation cross sections from initial level $v_1 = 0$, $J_1 = 10$ of the $X\,^1\Sigma_g^+$ state are exhibited in Fig. 19. Only the contribution from

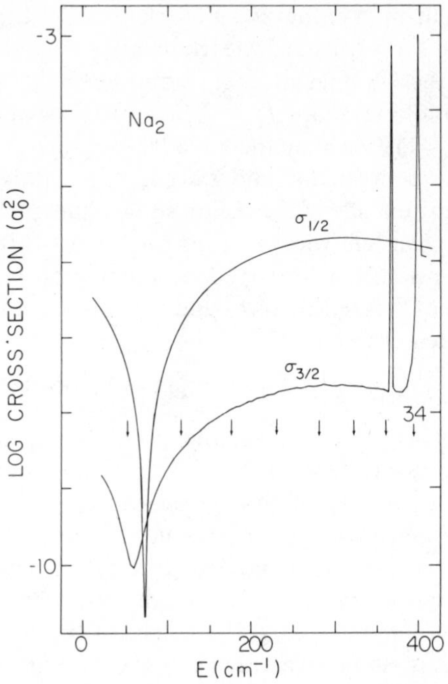

Figure 19. Photofragment cross section as a function of total energy from calculations of Struve et al. for one-photon dissociation of Na_2. The contribution to the cross section from a transition from $J_1 = 10$, $v_1 = 0$ of the $X\,^1\Sigma_g^+$ state to final angular momentum $J_2 = 9$ is shown in the figure. The maximum of the $B\,^1\Pi_u$ barrier is at 409 cm^{-1}. Single channel $B\,^1\Pi_u$ quasibound state energies are indicated by arrows.

photon absorption to final angular momentum $J_2 = 9$ is shown in Fig. 19. In real single-photon dissociation experiments, contributions to the observed cross section also arise from $J_2 = 10$ and 11. The cross section to both $j_{Na} = 1/2$ and 3/2 is seen to be vanishingly small until the excess dissociation energy is near the height of the $B\,^1\Pi_u$ barrier (407 cm^{-1} in the approximate potentials used in the calculation).

The positions of quasibound levels of the rotationless $B\,^1\Pi_u$ potential energy surface, shifted by the rotational energy for $J_2 = 9$, are indicated with arrows in Fig. 19. The two highest-lying quasibound levels are visible as resonance peaks in the calculated $j_{Na} = 3/2$ cross section. The fact that neither of these resonances are discernible in the calculated $j_{Na} = 1/2$ cross section indicates that nonadiabatic transitions between the $B\,^1\Pi_u$ state and electronic levels, that correlate with $j_{Na} = 1/2$, are negligible at these energies. Of course, these conclusions are dependent on the use of the $r \to \infty$ form of nonadiabatic coupling matrix elements at all r (Section III).

Quasibound levels at lower energies are not visible in our calculations as photodissociation resonances in either fine structure state cross section because the structureless contribution arising from $X\ ^1\Sigma_g^+ \rightarrow A\ ^1\Sigma_u^+$ oscillator strength, although very small, swamps the rapidly diminishing contribution from $X\ ^1\Sigma_g^+ \rightarrow B\ ^1\Pi_u$ flux as energy is decreased below $\sim 350\ \text{cm}^{-1}$. We speculate that the mechanism for appearance of $A\ ^1\Sigma_u^+$ flux in the $j_{Na} = 3/2$ cross section could be nonadiabatic transitions at the crossing of the $A\ ^1\Sigma_u^+$ and $b\ ^3\Pi_u$ surfaces, although it is doubtful whether this could be confirmed experimentally because the predicted cross sections are so small in this energy region.

Rotational effects predicted for NaH and CH^+ dissociation, such as strong Coriolis coupling and the occurrence of numerous resonance peaks arising from centrifugal barriers, are absent in Na_2. The reduced mass of Na_2 is an order of magnitude larger than in either of the hydrides, and therefore Coriolis coupling is diminished by this factor. The long-range RDI potentials combined with a weak centrifugal potential lead to very small centrifugal barriers with maxima at large R (e.g., $\sim 10^{-1}\ \text{cm}^{-1}$ centrifugal barrier maximum at $R = 4700$ bohr for $J_2 = 9$).

The calculations for Na_2 presented here only map out the general qualitative features of the photodissociation cross sections as a guide to future experiments, and this is done with the realization that fits to accurate experimental data will require future adjustments in the potentials, nonadiabatic couplings, and transition moment functions. The persistence of the RDI to very large interfragment separation complicates the calculation of transition amplitudes to magnetic sublevels m_{Na} of the sodium fine structure states. Although the different fine structure levels are not coupled by the RDI at large R, the same is not true of m_{Na} levels. This introduces the necessity to consider the dynamical effects of hyperfine coupling, the fluorescence from fragments while they are still interacting and the effects of retardation.

VII. DISCUSSION

The study of the dependence of fine structure level branching ratios, angular distributions, and polarization, produced from molecular photodissociation, is of interest as a probe of the interaction potentials between open shell atoms and of nonadiabatic collision dynamics. The treatment of diatomic molecule photodissociation reviewed in this article is a prelude to considering the fine structure dependence of photodissociation processes in polyatomic molecules. Rather little experimental information is currently available concerning molecular potentials generated from open shell atoms, especially the long-range attractive portions that are not readily accessed by spectroscopic methods. Molecular beam scattering techniques have provided important data when both atoms are closed shell and in some cases when one of the atoms is open shell; however, these experiments with two open shell atoms, perhaps with one in an electronically excited state, are not presently being performed. Photodissociation appears to be ideally suited to these purposes because it yields the fragment kinetic energy resolution provided by the narrow-band laser excitation. Furthermore, use of the two (or higher) photon excitation processes described here enables angular momentum selection which in semiclassical terms corresponds to impact parameter selection, a useful feature not generally available to crossed molecular beam methods.

Little theoretical effort has gone into the *ab initio* computation of the bond length dependence of spin-orbit and Coriolis matrix elements, especially those off-diagonal matrix elements which induce nonadiabatic transitions. Experimental data on nonadiabatic resonance positions and widths, whose general existence is predicted by us, could provide impetus for such calculations and for a strong interplay between theory and experiment. Absolute resonance positions are, of course, extremely sensitive to minute details of the potentials, but the widths of nonadiabatic resonances may, in part, reflect aspects of the off-diagonal couplings. Also, long range molecular interactions between open shell atoms have not been studied, and interesting questions arise concerning retardation corrections to resonant long-range dipole–dipole interactions. It would be interesting if low-energy photodissociation experiments could provide information concerning this problem.

Many polyatomic molecules dissociate to open shell fragments, so an analogous theory must apply to these systems. The basic features of the required theory follow directly. One example is the use of an "atomic" basis transformation diagonalizing the asymptotic Hamiltonian. The various high-energy recoil limits arise where dynamical calculations become unnecessary or greatly simplified to a single potential surface when oscillator strength is nonzero for only one of the possible intermediate molecular electronic states. Other details, introduced because of the greater complexity of the dynamics of the polyatomic half-collision, remain to be more fully explored. The experimental situation is, however, anticipated to be somewhat more complicated. Franck-Condon restrictions may preclude the investigation of the near-threshold region where nonadiabatic induced resonances are expected to reside. Even if this region is accessible, it may contain a much larger number of these resonances due to added quantum numbers such as those correlating with the product rotational levels. Hence, the individual nonadiabatic resonances may not easily be resolved.

Here we review the general theory of photodissociation in diatomic molecules where one or both of the fragments have nonvanishing electronic angular momentum quantum numbers and fine structure sublevels. Cases of one- and resonant two-photon dissociation are explicitly treated, and the many-resonant-photon situation follows analogously. The particular examples chosen do not involve curve crossings between states with different L-S atomic limits of one or both of the fragments, but the theory is easily extended to cover this important case by, at worst, including all molecular states dissociating to both participating sets of atomic limits.

The full quantum theory is developed because it is required for describing the low-energy resonances that occur by virtue of transitions due to nonadiabatic couplings that are present as the molecule breaks apart. Simple recoil appproximations are introduced in the high-energy limit, and other intermediate approximations would be useful to consider under certain circumstances. An important aspect of the theory is the derivation of general expressions for the differential cross sections for angularly resolved detection of photofragments and polarized fluorescence from excited photofragments produced by resonant photodissociation of a diatomic molecule. Further approximations are associated with simplifications in the calculation of the transition amplitudes appearing in these general quantum formulas or in the consideration of possible high angular momentum semiclassical limits. The general theory is also readily extended to the description of experiments using laser induced fluorescence to probe fragment polarization.

The general quantum theory is formulated in a density matrix formalism because of a number of advantages it affords. The lengthy expression for the differential cross section is decomposed into several smaller components, each with a clear physical significance. The formalism permits many extensions of the theory to be readily made merely by altering only one of these components, thus obviating a lengthy full derivation for each "new" case.

In this article, we have explored the utility of the density matrix approach in formulating approximation schemes which greatly simplify calculation of photodissociation cross sections. Often no single approximation describes the fragments at each step of the dissociation. Rather, approximations must be modified to reflect intramolecular coupling and dynamics at each stage of dissociation. In our formalism, the differential cross section is calculated by applying a series of tensor operators to the density matrix which describe the initial molecular ensemble. Each operator corresponds to a step in the photodissociation: resonant excitation of the molecule, promotion to electronic states that lead to fragmentation, and finally, collection and optical probe of photofragments. Distinct and appropriate physical approximations for each step of the photodissociation process are readily introduced using the density matrix formalism because each physical stage of the dissociation corresponds to a separate tensor operator in our expression for the differential cross section. This concept is fully developed in the context of photodissociation for the first time in this work. It should prove useful in describing dissociating molecule in which the coupling scheme changes as the fragments separate (e.g. from Hund's case (b) to (a) in light atom diatomics, or from case (a) to (c) for heavier diatoms), or when the dynamics are principally adiabatic in some regions along the internuclear coordinate and diabatic, or sudden, in other regions.

Calculations are provided for photodissociation in NaH, CH^+ and Na_2 using full close-coupled quantum calculations which are necessary in the low-energy region where resonances are present. Despite the fact that all three molecules dissociate to atoms in $2P + 2S$ states, their photodissociation dynamics are quite different. The NaH molecule has a repulsive $^1\Pi$ state which cannot provide resonances, so the resonance structure is rather simple and readily analyzed. Resonant long-range dipole–dipole interactions in Na_2 eliminate Coriolis induced shape resonances, but produce interesting ones due to tunneling through a rather large barrier. The presence of the attractive $^1\Pi$ state and perhaps the long-range ion–dipole interactions in CH^+ lead to the predicted occurrence of a large number of nonadiabatic induced resonances in qualitative accord with the currently available unanalyzed data. We hope that our calculations will be useful in aiding to unravel this interesting data despite the obvious limitations on the potentials and nonadiabatic matrix elements which have considerable effects on the computed resonances. The calculations are also useful to consider questions concerning the relative abundance of CH^+ and CH to C and C^+ in interstellar space.

The high-energy photodissociation dynamics is discussed in terms of the J-dependent and independent approximations where a full numerical evaluation of the photodissociation amplitudes can be avoided. The former approximation is appropriate when the relative fragment translational energy is much larger than the asymptotic splitting of atomic fine structure states. Curve crossing transitions and other short-distance couplings may be included at this stage if necessary.

The J-independent recoil approximation is valid under even more restrictive conditions, but it is used here to compute quantities like asymmetry parameters and branching ratios which are shown to often provide a signature of the participating intermediate molecular electronic states. Examples are provided for the simplest case where fragmentation proceeds along a single dissociative potential energy curve so no dynamical information is required.

Acknowledgments

This research is supported, in part, by NSF Grant CHE 83-17098 and a grant from the United States-Israel Binational Science Foundation. We are grateful to Carl Williams for assistance with calculations on CH^+, and the calculations for Na_2 have been performed in collaboration with Walter Struve.

References

1. Y. B. Band, K. F. Freed, and D. J. Kouri, *Chem. Phys. Lett.* **79**, 233 (1981).
2. Y. B. Band and K. F. Freed, *Chem. Phys. Lett.* **79**, 238 (1981).
3. W. M. Gelbart, *Annu. Rev. Phys. Chem.* **28**, 323 (1977).
4. K. F. Freed and Y. B. Band, in *Excited States*, Vol. 3, E. C. Lim, ed., Academic, New York, 1977.
5. H. Okabe, *Photochemistry of Small Molecules*, Wiley, New York, 1978.
6. S. R. Leone, *Adv. Chem. Phys.* **50**, 225 (1982).
7. M. Shapiro and R. Bersohn, *Annu. Rev. Phys. Chem.* **33**, 409 (1982).
8. S. J. Singer, K. F. Freed, and Y. B. Band, *J. Chem. Phys.* **79**, 6060 (1983).
9. S. J. Singer, Y. B. Band, and K. F. Freed, *Chem. Phys. Lett.* **91**, 12 (1982).
10. S. J. Singer, K. F. Freed, and Y. B. Band, *J. Chem. Phys.* **81**, 3091 (1984).
11. S. J. Singer, K. F. Freed, and Y. B. Band, *Chem. Phys. Lett.* **105**, 158 (1984).
12. U. Fano and J. M. Macek, *Rev. Mod. Phys.* **45**, 553 (1973).
13. J. P. Simons, in *Gas Kinetics and Energy Transfer 2*, senior reporters P. G. Ashmore and R. J. Donovan, The Chemical Society, London, 1977.
14. M. N. R. Ashford, M. T. Macpherson, and J. P. Simons, *Topics Curr. Chem.* **86**, 1 (1978).
15. R. C. Dunbar, in *Gas Phase Ion Chemistry*, M. T. Bowers, ed., Academic, New York, 1979, p. 181.
16. J. T. Moseley and J. Durup, *Annu. Rev. Phys. Chem.* **32**, 53 (1983).
17. J. T. Moseley in *Appl. Atomic Collision Phys. 5*, E. W. McDaniel, ed., Academic, New York, 1983, p. 269; *Adv. Chem. Phys.* (preprint).
18. T. E. Sharp, At. *Data* **2**, 179 (1971).
19. J. E. Mentall and P. M. Guyon, *J. Chem. Phys.* **67**, 3845 (1977).
20. R. S. Mulliken, *J. Am. Chem. Soc.* **86**, 1849 (1966).
21. G. Gabrielse and Y. B. Band, *Phys. Rev. Lett.* **39**, 697 (1977); Y. B. Band, *Phys. Rev.* **A19**, 1906 (1979).
22. P. Borrell, P. M. Guyon, and M. Glass-Maujean, *J. Chem. Phys.* **66**, 818 (1977).
23. L. C. Lee and D. L. Judge, *Phys. Rev.* **A14**, 1094 (1976).
24. E. J. Stone, G. M. Lawrence, and C. E. Fairchild, *J. Chem. Phys.* **65**, 5083 (1976).
25. R. J. Buenker and S. D. Peyerimhoff, *Chem. Phys. Lett.* **34**, 225 (1975); R. J. Buenker, S. D. Peyerimhoff, and M. Perić, *Chem. Phys. Lett.* **42**, 383 (1976); M. Yoshimine, K. Tanaka, H. Tatewaki, S. Okara, F. Sasaki, and K. Ohno, *J. Chem. Phys.* **64**, 2254 (1976); R. P. Saxon and B. Liu, *J. Chem. Phys.* **67**, 5432 (1978).
26. L. C. Lee, T. G. Slanger, G. Black, and R. L. Sharpless, *J. Chem. Phys.* **67**, 5602 (1977).
27. D. D. Konowalow, M. E. Rosenkrantz, and M. L. Olson, *J. Chem. Phys.* **72**, 2612 (1980).
28. E. W. Rothe, U. Krause, and R. Düren, *Chem. Phys. Lett.* **72**, 100 (1980); *J. Chem. Phys.* **72**, 5145 (1980).
29. E. K. Kraulings and M. L. Yanson, *Opt, Spectrosc.* **46**, 629 (1979).
30. D. L. Feldman and R. N. Zare, *Chem. Phys.* **15**, 415 (1976).
31. E. J. Bredford and F. Engelke, *Chem. Phys. Lett.* **75**, 132 (1980).
32. E. J. Bredford and F. Engelke in *Laser-Induced Processes in Molecules*, Vol. 6, K. L. Kompa and S. D. Smith, ed., Springer, Berlin, 1978; *J. Chem. Phys.* **71**, 1994 (1979).

33. J. McCormack and A. J. McCaffery, *Chem. Phys. Lett.* **64**, 98 (1979).
34. D. Eisel, D. Zevgolis, and W. Demtröder, *J. Chem. Phys.* **71**, 2005 (1979).
35. E. J. Sachs, J. Hinze, and N. H. Sabelli, *J. Chem. Phys.* **62**, 3367 (1975); *J. Chem. Phys.* **62**, 3377 (1975); *J. Chem. Phys.* **62**, 3384 (1975); *J. Chem. Phys.* **62**, 3389 (1975); *J. Chem. Phys.* **62**, 3393 (1975).
36. M. S. Child and R. B. Bernstein, *J. Chem. Phys.* **59**, 5916 (1973).
37. K. R. Wilson, in *Excited State Chemistry*, J. N. Pitts, Jr., ed., Gordon and Breach, New York, 1970.
38. R. K. Sander and K. R. Wilson, *J. Chem. Phys.* **63**, 4242 (1975).
39. D. M. Burde, R. A. McFarlane, and J. R. Wiesenfeld, *Phys. Rev.* **A10**, 1917 (1974).
40. R. S. Mulliken, *J. Chem. Phys.* **55**, 288 (1971).
41. T. G. Lindeman and J. R. Wiesenfeld, *Chem. Phys. Lett.* **50**, 364 (1977); T. G. Lindeman and J. R. Wiesenfeld, *J. Chem. Phys.* **70**, 2882 (1977).
42. A. B. Peterson and I. W. M. Smith, *Chem. Phys.* **30**, 407 (1978).
43. R. J. Oldman, R. K. Sander, and K. R. Wilson, *J. Chem. Phys.* **63**, 4252 (1975).
44. M. A. A. Clyne and M. C. Heaven, *J. Chem. Soc. Faraday Trans. II*, **74**, 1192 (1978).
45. M. A. A. Clyne and M. C. Heaven and E. Martinez, *J. Chem. Soc. Faraday Trans. II*, **76**, 177 (1980).
46. S. R. Leone and F. J. Wodarczyk, *J. Chem. Phys.* **60**, 314 (1974).
47. C. P. Hemenway, T. G. Lindemann, and J. R. Wiesenfeld, *J. Chem. Phys.* **70**, 3560 (1979).
48. S. L. Baughaum, H. Hofmann, S. R. Leone, and D. J. Nesbitt, *Faraday Discuss. Chem. Soc.* **67**, 306 (1979).
49. M. A. A. Clyne and M. C. Heaven, *J. Chem. Soc. Faraday Trans. II*, **76**, 49 (1980).
50. M. A. A. Clyne and I. S. McDermid, *J. Chem. Soc. Faraday Trans. II*, **74**, 1935 (1978); *J. Chem. Soc. Faraday Trans. II*, **76**, 1677 (1980).
51. L. N. Krasnoperov and V. N. Panfilov, *Kinet. Katal.* **20**, 540 (1979).
52. R. Bersohn, in *Alkalai Halide Vapors*, P. Davidovits and D. L. McFadden, Eds, Academic, New York, 1979.
53. R. D. Clear, S. J. Riley, and K. R. Wilson, *J. Chem. Phys.* **63**, 101 (1975).
54. T.-M. R. Su and S. J. Riley, *J. Chem. Phys.* **71**, 3194 (1979).
55. T.-M. R. Su and S. J. Riley, *J. Chem. Phys.* **72**, 1614 (1980).
56. P. P. Sorokin, N. S. Shiren, J. R. Lankard, E. C. Hammond and T. G. Kazyaka, *Appl. Phys. Lett.* **10**, 44 (1967).
57. J. C. White and D. Henderson, *Opt. Lett.* **7**, 204 (1982).
58. J. C. White and D. Henderson, *Opt. Lett.* **7**, 517 (1982).
59. W. Lüthy, P. Burkhard, T. E. Gerber, and H. P. Weber, *Opt. Comm.* **38**, 413 (1981).
60. G. Herzberg, *Molecular Spectra and Molecular Structure, Vol. 1, Spectra of Diatomic Molecules*, Van Nostrand, New York, 1950.
61. L. D. Landau and E. M. Lifshitz, *Quantum Mechanics—Non-Relativistic Theory*, Addison-Wesley, New York, 1965.
62. K. F. Freed, *J. Chem. Phys.* **45**, 1714 (1966).
63. K. F. Freed, *J. Chem. Phys.* **45**, 4214 (1966).
64. J. T. Hougen, *Calculation of Rotational Energy Levels and Rotational Line Intensities in Diatomic Molecules*, Natl. Bur. Stand. Monogr. 115 (1970).

65. A. Carrington, D. H. Levy, and T. A. Miller, *Adv. Chem. Phys.* **18**, 149 (1970).
66. D. H. Levy, *Adv. Mag. Res.* **6**, 1 (1973).
67. F. H. Mies, *Phys. Rev.* **A7**, 942 (1973).
68. J. C. Weisheit and N. F. Lane, *Phys. Rev.* **A4**, 171 (1971).
69. J. Launay and E. Roueff, *J. Phys.* **B10**, 879 (1977).
70. D. M. Brink and G. R. Satchler, *Angular Momentum*, 2nd ed., Clarendon, Oxford, 1968.
71. Throughout the chapter we adopt the conventions of Ref. 70 for Wigner rotation matrices, spherical harmonics, and all other angular momentum symbols.
72. Y.-N. Chiu, *J. Chem. Phys.* **42**, 2671 (1965).
73. Y.-N. Chiu, *J. Chem. Phys.* **45**, 2969 (1966).
74. J. K. Knipp, *Phys. Rev.* **53**, 734 (1938).
75. P. R. Fontana, *Phys. Rev.* **123**, 1871 (1961).
76. P. K. Fontana, *Phys. Rev.* **125**, 1597 (1962).
77. T. Y. Chang, *Rev. Mod. Phys.* **39**, 911 (1967).
78. R. S. Mulliken, *Rev. Mod. Phys.* **3**, 113 (1931).
79. R. S. Mulliken, *Rev. Mod. Phys.* **4**, 26 (1932).
80. R. S. Mulliken, *Phys. Rev.* **46**, 549 (1934).
81. R. S. Mulliken, *Phys. Rev.* **51**, 310 (1937).
82. R. S. Mulliken, *Phys. Rev.* **57**, 500 (1940).
83. R. S. Berry, in *Alkalai Halide Vapors*, P. Davidovits and D. L. McFadden, eds., Academic, New York, 1979.
84. W. Kolos, *Adv. Quantum Chem.* **5**, 99 (1970).
85. R. W. Numrich and D. G. Truhlar, *J.* Phys. *Chem.* **79**, 2745 (1975).
86. R. W. Numrich and D. G. Truhlar, *J. Phys. Chem.* **82**, 168 (1978).
87. J. H. Van Vleck, *Rev. Mod. Phys.* **23**, 213 (1951).
88. J. H. Van Vleck, *Phys. Rev.* **33**, 467 (1929).
89. W. G. Richards, H. P. Trivedi, and D. L. Cooper, *Spin-Orbit Coupling in Molecules*, Clarendon, Oxford, 1981.
90. P. S. Julienne and M. Krauss, *J. Mol. Spec.* **56**, 2701 (1975).
91. U. Fano and G. Racah, *Irreducible Tensorial Sets*, Academic, New York, 1959.
92. M. M. Lambropoulos and R. S. Berry, *Phys. Rev.* **A8**, 855 (1973).
93. J. C. Hansen, Ph.D. Dissertation, University of Chicago, 1979.
94. A. Omont, *Prog. Quantum Elec.* **5**, 69 (1977).
95. K. Chen and E. S. Yeung, *J. Chem. Phys.* **72**, 4723 (1980).
96. R.-L. Chien, *J. Chem. Phys.* **81**, 4023 (1984).
97. C. Cohen-Tanoudji and A. Kastler, *Prog. Optics* **5**, 3 (1966).
98. W. Happer, *Rev. Mod. Phys.* **44**, 169 (1972).
99. R. N. Zare, *Mol. Photochem.* **4**, 1 (1972).
100. R. J. van Brunt and R. N. Zare, *J. Chem. Phys.* **48**, 4304 (1968).
101. G. A. Chamberlain and J. P. Simons, *J. Chem. Soc. Faraday Trans. II* **71**, 2043 (1975).
102. M. J. MacPherson, J. P. Simons, and R. N. Zare, *Mol. Phys.* **38**, 2049 (1979).
103. G. W. Loge and R. N. Zare, *Mol. Phys.* **43**, 1419 (1981).

104. G. W. Loge and R. N. Zare, *Mol. Phys.* **47**, 225 (1982).
105. J. Vigué, P. Grangier, G. Roget, and A. Aspect, *J. Physique Letters* **42**, L-531 (1981).
106. J. Vigué, J. A. Beswick, and M. Broyer, *J. Physique* **44**, 1225 (1983).
107. R.-L. Chien, O. C. Mullins, and R. S. Berry, *Phys. Rev.* **A28**, 2078 (1983).
108. S. J. Singer, K. F. Freed, and Y. B. Band, *J. Chem. Phys.* **81**, 3064 (1984).
109. A. P. Yutsis, I. B. Levinson, and V. V. Vanagas, *Mathematical Apparatus of the Theory of Angular Momentum*, Israel Program for Scientific Translations, Jerusalem, 1962.
110. I. V. Hertel and W. Stoll, *Adv. Atom. Mol. Phys.* **13**, 113 (1978).
111. C. M. Greene and R. N. Zare, *Annu. Rev. Phys. Chem.* **33**, 89 (1982).
112. N. F. Mott and H. S. W. Massey, *The Theory of Atomic Collisions*, Clarendon, Oxford, 1965.
113. Y. B. Band and K. F. Freed, *Chem. Phys. Lett.* **28**, 328 (1974).
114. Y. B. Band and K. F. Freed, *J. Chem. Phys.* **63**, 3382 (1975).
115. M. D. Morse, K. F. Freed, and Y. B. Band, *J. Chem. Phys.* **70**, 3604 (1979).
116. M. D. Morse, K. F. Freed, and Y. B. Band, *J. Chem. Phys.* **70**, 3620 (1979).
117. M. D. Morse and K. F. Freed, *Chem. Phys. Lett.* **74**, 49 (1980).
118. M. D. Morse and K. F. Freed, *J. Chem. Phys.* **74**, 4395 (1981).
119. M. D. Morse and K. F. Freed, *J. Chem. Phys.* **78**, 6045 (1982).
120. M. D. Morse, Y. B. Band, and K. F. Freed, *J. Chem. Phys.* **78**, 6066 (1983).
121. K. C. Kulander and J. C. Light, *J. Chem. Phys.* **73**, 4337 (1980).
122. B. R. Johnson, *J. Chem. Phys.* **67**, 4086 (1977).
123. W. S. Struve, S. J. Singer, and K. F. Freed, *Chem. Phys. Lett.*, **110**, 588 (1984).
124. H. Sun, S. J. Singer, and K. F. Freed, (manuscript in preparation).
125. Y. B. Band, K. F. Freed, and D. J. Kouri, *J. Chem. Phys.* **74**, 4380 (1981).
126. W. N. Sams and D. J. Kouri, *J. Chem. Phys.* **51**, 4809 (1969).
127. W. N. Sams and D. J. Kouri, *J. Chem. Phys.* **51**, 4815 (1969).
128. Y. B. Band and B. M. Aizenbud, *Chem. Phys. Lett.* **77**, 49 (1981).
129. Y. B. Band and B. M. Aizenbud, *Chem. Phys. Lett.* **79**, 224 (1981).
130. S. J. Singer, K. F. Freed, and Y. B. Band, *J. Chem. Phys.* **77**, 1942 (1982).
131. M. Shapiro, *J. Chem. Phys.* **56**, 2582 (1972).
132. M. Shapiro, *Israel J. Chem.* **11**, 691 (1973).
133. M. Shapiro and G. G. Balint-Kurti, *J. Chem. Phys.* **71**, 1461 (1979).
134. W. P. Reinhardt, *Annu. Rev. Phys. Chem.* **33**, 223 (1982).
135. A. F. J. Siegert, *Phys. Rev.* **56**, 750 (1939).
136. O. Atabek and R. Lefebvre, *Chem. Phys.* **52**, 199 (1980).
137. O. Atabek and R. Lefebvre, *Phys. Rev.* **A22**, 1817 (1980).
138. O. Atabek and R. Lefebvre, *Chem. Phys.* **56**, 195 (1981).
139. O. Atabek and R. Lefebvre, *Chem. Phys.* **55**, 395 (1981).
140. U. Fano, *Rev. Mod. Phys.* **29**, 74 (1957).
141. S. Green, P. S. Bagus, B. Liu, A. D. McLean, and M. Yoshimine, *Phys. Rev.* **A5**, 1614 (1972).
142. M. Yoshimine, S. Green, and P. Thaddeus, *Astrophys. J.* **183**, 899 (1973).

143. The potential energy surfaces for CH^+ presented in Fig. 15 are constructed using singlet state curves inferred from experiment (kindly supplied to us prior to publication by H. Helm) and singlet-triplet state splittings from *ab initio* calculations.[141,142] The long-range portion of the potential energy curves is the ion-induced dipole potential as in Ref. 69.

144. P. C. Cosby, H. Helm, and J. T. Moseley, *Astrophys. J.* **25**, 52 (1980).

145. H. Helm, P. C. Cosby, M. M. Graff, and J. T. Moseley, *Phys. Rev.* **A25**, 304 (1982).

146. M. M. Graff, J. T. Moseley, J. Durup, and E. Roueff, *J. Chem. Phys.* **78**, 2355 (1983).

147. C. Williams, S. Singer, and K. F. Freed, (unpublished calculations).

148. A. Carrington, J. Buttenshaw, R. A. Kennedy, and T. P. Softley, *Mol. Phys.* **45**, 747 (1982).

149. A. Dalgarno, in *Atomic Processes and Applications*, P. A. Burke and B. L.Moisiwitch, eds., North-Holland, Amsterdam, 1976.

150. A. Giusti-Suzor, E. Roueff, and H. van Regenmorter, *J. Phys.* **B9**, 1021 (1976).

151. H. Abgrall, A. Giusti-Suzor, and E. Roueff, *Astrophys. J.* **207**, L69 (1976).

152. M. M. Graff, J. T. Moseley, and E. Roueff, *Astrophys. J.* **269**, 796 (1983).

153. E. I. Dashevskaya, A. I. Voronin, and E. E. Nikitin, *Can. J. Phys.* **47**, 1237 (1969).

154. M. Movre and G. Pichler, *J. Phys.* **B10**, 2631 (1977).

155. W. C. Stwalley, Y.-H. Uang, and G. Pichler, *Phys. Rev. Lett.* **41**, 1164 (1978).

156. P. Kusch and M. M. Hessel, *J. Chem. Phys.* **68**, 2591 (1978).

157. M. Lyyra and P. R. Bunker, *Chem. Phys. Lett.* **61**, 67 (1979).

158. N. Allard and J. Kielkopf, *Rev. Mod. Phys.* **54**, 1103 (1982).

159. K. Niemax and G. Pichler, *J. Phys.* **B8**, 179 (1975).

160. J. Huennekens and A. Gallagher, *Phys. Rev.* **A27**, 1851 (1983); *Phys. Rev.* **A28**, 238 (1983).

REDUCED DIMENSIONALITY THEORIES OF QUANTUM REACTIVE SCATTERING

JOEL M. BOWMAN*

The James Franck Institute
The University of Chicago, Chicago, Illinois

CONTENTS

I. INTRODUCTION

Theoretical approaches to chemical reactions have historically developed along two avenues. The older and more heavily used one was established

* Dr. Bowman's permanent address: Department of Chemistry, Illinois Institute of Technology, Chicago, Illinois 60616.

by transition state theory (TST).[1-3] In this theory the objective to directly relate properties of the potential energy surface to the thermal rate constant was realized by an astonishingly simple expression. This immensely popular theory is still predominantly used in its original form. A second, parallel track, begun relatively recently, emphasized the dynamical basis of chemical reactions. This work showed how the thermal rate constant was related to the collision cross section.[4,5] Soon after that relationship was known the dynamical basis of TST was investigated. Specifically, the cross section implied by TST was derived.[6-8]

At about the same time, quantum dynamical calculations were becoming feasible for model (i.e., collinearly constrained) reactions. These were very important calculations because they enabled unambiguous tests of TST to be made, albeit in a fictitious linear world. As a result of numerous comparisons between collinear exact quantum (CEQ) rate constants, reaction probabilities and wave functions with those given or implied by TST (in the collinear world) for the $H + H_2$ reaction in particular,[9] several shortcomings of TST became clear. First, tunneling is not described at all by TST, and simple one-dimensional corrections for tunneling were found to be inadequate. This meant that at room temperature and below, the TST rate constant was significantly (i.e., a factor of 10 or more) less than the accurate quantum rate constant. Second, as found numerically, the CEQ reaction probability could fall below (or not even reach) unity whereas the corresponding TST probability is a unit step function. Thus the TST rate constant exceeds the correct one at high temperatures. The exact classical reaction probability is qualitatively like the CEQ one.[10] That is, some trajectories that cross the transition state going in the direction of reactants to products are reflected by the repulsive part of the potential surface and re-cross the transition state in the direction back toward reactants. This re-crossing reduces the dynamical rate constant relative to the TST one, especially at higher temperatures where re-crossing becomes significant. These fundamental differences between collinear dynamical calculations and collinear TST have been confirmed in comparisons of recent three-dimensional accurate quantum and TST rate constants for $H + H_2$.[11] Thus tunneling and re-crossing effects have been shown to be significant differences between accurate dynamical and TST rate constants. Another obvious difference between dynamical theories and TST appears when considering rate constants and cross sections for quantum state-to-state processes which are beyond the scope of TST.

Given the insights gained from CEQ calculations of reaction probabilities and rate constants the challenge was to incorporate these somehow into three-dimensional TST. A number of approaches have been taken and these have been reviewed in several places.[12-16] An approach which we have taken and which is the main subject of this review was motivated by a study of CEQ and accurate three-dimensional rate constants for the $H + H_2$ reaction.[14] In this study the collinear and three-dimensional quantum transmission coefficients were compared for temperatures between 200 and 600 K. The transmission coefficient is just the ratio of the accurate quantum rate constant to the corresponding transition state theory rate constant. (The explicit mathematical expression will be given later.) The two did not agree very well; for example, the CEQ transmission coefficient was about two times smaller than the accurate three-dimensional one at 300 K. This was somewhat disappointing because the CEQ transmission coefficient contains effects of tunneling and re-crossing which for the collinearly dominated $H + H_2$ reaction should have been quite realistic.

In spite of this disappointing result we were stimulated to extend that study. Specifically, we obtained an expression for the reaction cross section implied by the rate constant given by the three-dimensional TST rate constant times the CEQ transmission coefficient.[17] Because that coefficient is defined for vibrational state-to-state processes the resulting cross section is also vibrationally state-to-state but averaged and summed over the rotational states of the reactant and product. In addition, we considered a simple, but very important, modification of the CEQ transmission coefficient.[17,18] This was motivated by the threshold behaviors of CEQ, coplanar and three-dimensional reaction probabilities for $H + H_2$; each higher dimensional threshold was shifted in energy by an amount approximately equal to the H_3 bending energy at the transition state from the threshold in the space of one less dimension.[11,19,20] Such a shift does occur in the above CEQ cross section, however; to account for it in a approximately dynamical way we treated the H_3 bending motion adiabatically. This results in a two-mathematical dimensional effective potential energy surface consisting of the collinear one plus the adiabatic bending eigenvalue which depends on the collinear configuration. The resulting collinear exact quantum reaction probabilities were labeled as CEQB to distinguish them from the CEQ ones.

The transmission coefficient obtained from the CEQB reaction probabilities was found to be in good agreement with the accurate three-dimensional one,[17] and thus by treating the bending motion adiabatically it appeared as though an accurate correction to three-dimensional TST was available through the CEQB transmission coefficient, at least for $H + H_2$. It should be noted that a similarly accurate transmission coefficient was obtained previous to our work by a simpler, one-dimensional reaction probability obtained for the effective potential described above.[21] Prior to that calculation, the CEQB transmission coefficient was calculated for $H + H_2$[22] (at least approximately) before three-dimensional accurate quantum results were available for comparison.

In this chapter, the CEQ and CEQB expressions for transmission coefficients, integral and differential cross sections will be derived from the Schrödinger equation. Previously, they were arrived at somewhat heuristically, starting with expressions for the transmission coefficients and three-dimensional rate constants and then deriving the cross sections implied by them. The results are all correct; nevertheless, it was felt that a strict derivation based on scattering theory should be given. The advantage of such a derivation is that it suggests new results and generalizations of the theory. Indeed a new, more rigorous model is obtained from this derivation. In the present context we shall refer to the CEQ, CEQB, and other related theories as reduced dimensionality theories of reactive scattering. These are derived in the next section where expressions for the rotationally averaged and summed but vibrational state-to-state integral cross section are derived. Following that, expressions for the corresponding transmission coefficient and rate constants are given and discussed. In Section IV the reduced dimensionality differential cross section is derived. Applications of the CEQ and CEQB theories to a number of reactions are reviewed in Section V, followed by a summary in Section VI. A postscript appears in Section VII.

This chapter is not intended to be a comprehensive review of reactive scattering. There are a number of excellent recent reviews[23-28] which together provide such a review.

II. REDUCED DIMENSIONALITY THEORIES

Several reduced dimensionality quantum approaches to three-dimensional $A + BC$ reactive scattering have been developed very recently. The spirit of these approaches is to treat a subset of all of the degrees of freedom dynamically and to use various decoupling approximations to treat the remaining degrees of freedom.[17,18,29-31] The model closest in spirit to ours is the earlier rotating linear model of Wyatt[32] and Connor and Child[33] and the extension by Walker and Hayes,[34] in which the ground state adiabatic bending energy is added to the potential. They term this the bending-corrected rotating linear model (BCRLM). Based on the derivation of our reduced dimensionality theories we obtain a new model which is close to the BCRLM and thus partially remove some of its ad hoc nature. Miller and Schwartz[35,36] and Skodje and Truhlar[37] have based reduced dimensionality theories on the reaction path Hamiltonian[38-40]. They have also suggested approximations to reintroduce coupling among all the degrees of freedom. Schwartz and Miller have obtained very good results for the zero partial wave of the $H + H_2$ reaction.[36] These methods have not been extended to the calculation of cross sections. Practically, only two degrees of freedom can be treated dynamically. However, from a decade of experience with strictly collinear, two-mathematical dimensional (2MD) calculations we can appreciate the substantial increase in richness that these calculations give relative to the one-mathematical dimensional treatments of reactive scattering that have existed for roughly 50 years. The reduction to a 2MD dynamical space is most straightforward for reactions with a collinear reaction path for the CEQB and BCRLM methods. Formally, there appears to be no simplification for collinear reaction paths in the more general reaction path approaches. These, however, pay the price of complexities and approximations which are not present in the CEQB and BCRLM methods.

The two degrees of freedom treated dynamically in the CEQB and BCRLM methods are those describing the vibrational motion of BC and the relative motion of A and BC with their relative orientations fixed. These degrees of freedom can be mapped exactly and most simply onto the corresponding ones for the products $AB + C$ or $AC + B$ for the collinear orientation. As a result we have restricted our attention thus far to reactions which prefer a collinear orientation (on energetic grounds). A consequence of this choice for the reduced dimensionality scattering space is that vibrational state-to-state transition amplitudes are calculated for reactive and nonreactive events, for example,

$$A + BC(v) \rightarrow AB(v') + C,\ AC(v') + B$$

An important issue we addressed (but which was not focused on in the original rotating linear model and hence in the BCRLM) is: What is the relationship between this reduced dimensionality amplitude and the full three-dimensional ones which of course contain the amplitudes for rotational transitions as well? The answer we gave, although somewhat heuristically derived, is correct as shown in the next section where a strictly dynamical approach is taken. This issue is also present in the reaction path approaches and still remains to be completely resolved.

In preparation for the next section and in anticipation of some of the results there the rotationally averaged and sum cross section will now be defined. Let the cross section for the reaction $A + BC(v, j, \Omega) \rightarrow AB(v', j', \Omega') + C$ be written at the total energy E as

$$Q_{vj\Omega \rightarrow v'j'\Omega'}(E) = \frac{\pi}{k_{vj}^2} \sum_{J=0} (2J + 1)|S^J_{vj\Omega \rightarrow v'j'\Omega'}(E)|^2 \tag{1}$$

where $S^J_{vj\Omega \rightarrow v'j'\Omega'}$ is the partial wave scattering matrix element, J is the total angular momentum quantum number, j, Ω (j', Ω') are the molecular rotational quantum number and its projection on some z-axis for the reactant (product)

$$k_{vj}^2 = 2\mu_{A,BC}(E - E_{vj})/\hbar^2 \tag{2}$$

and where $\mu_{A,BC}$ is the reduced mass of the $A + BC$ system and E_{vj} is the BC vibration-rotation energy eigenvalue.

We now wish to consider the following *cumulative* reaction probability

$$\bar{P}^J_{v \to v'}(E) = \sum_{j\Omega} \sum_{j'\Omega'} |S^J_{vj\Omega \to v'j'\Omega'}(E)|^2 f_j \tag{3}$$

in terms of which a rotationally averaged cross section $\bar{Q}_{v \to v'}(E)$ can be defined from Eqs. (1) and (3). It is

$$\bar{Q}_{v \to v'}(E) = \frac{\sum_{j\Omega} \sum_{j'\Omega'} f_j k_{vj}^2 Q_{vj\Omega \to v'j'\Omega'}(E)}{\sum_j f_j (2j+1) k_{vj}^2} = \frac{\pi}{\bar{k}_v^2} \sum_{J=0} (2J+1) \bar{P}^J_{v \to v'}(E) \tag{4}$$

where

$$\bar{k}_v^2 = \sum_j k_{vj}^2 (2j+1) f_j \tag{5}$$

and where the sum is over the open rotational states and f_j is the nuclear spin statistical factor. Equation (4) can be rewritten in terms of the degeneracy-averaged cross section

$$Q_{vj \to v'j'}(E) = \frac{1}{2j+1} \sum_{\Omega} \sum_{\Omega'} Q_{vj\Omega \to v'j'\Omega'}(E) \tag{6}$$

as follows

$$\bar{Q}_{v \to v'}(E) = \sum_j (2j+1) f_j k_{vj}^2 Q_{vj \to v'j'}(E)/\bar{k}_v^2 \tag{7}$$

It will be shown later that it is $\bar{P}^J_{v \to v'}$ which is obtained in the our reduced dimensionality theory. Some suggestions for ad hoc methods to obtain reduced dimensionality approximations to $S^J_{vj\Omega \to v'j'\Omega'}$ will also be made.

In the next section we derive a hierarchy of reduced-dimensionality approximations to $\bar{P}^J_{v \to v'}$ ending with the usual transition state theory expression for it. In our particular approaches we have been motivated to treat the dynamically excluded degrees of freedom within the spirit of transition state theory. How that is done in detail will be given in the next section. Such an approach can thus be dynamically justified and puts the excellent success of the theory on a firm theoretical foundation.

A. Hierarchy of Reduced Dimensionality Theories

We now derive a hierarchy of reduced dimensionality quantum theories of reactive scattering which are all based on an adiabatic treatment of the three-atom bending motion. Subsequent, additional approximations are then introduced from which the hierarchy results.

We start with the full three-dimensional quantum formulation for the $A + BC$ reaction in body-fixed coordinates[41,42] as given in detail by Schatz and Kuppermann.[43] The Schrödinger equation in terms of the mass-scaled Delves vectors $\mathbf{r}$ (the BC relative position vector) and $\mathbf{R}$ (the position vector of A to the center of mass of BC) is

$$\left[\frac{-\hbar^2}{2\mu}(\nabla_R^2 + \nabla_r^2) + V(\mathbf{R}, \mathbf{r}) - E\right]\Psi(\mathbf{R}, \mathbf{r}) = 0, \tag{8}$$

where

$$\mu = \left(\frac{m_A m_B m_C}{m_A + m_B + m_C}\right)^{1/2} \tag{9}$$

A standard partial wave decomposition of Ψ is done in terms of the good quantum numbers J and M, the total angular momentum and its projection on a space-fixed axis respectively, and then each component is expressed as

$$\Psi^{JM}(\mathbf{R}, \mathbf{r}) = \sum_{\Omega=-J}^{J} D^J_{M\Omega}(\phi, \theta, 0)\Psi^{J\Omega}(r, R, \gamma, \psi), \tag{10}$$

in the body-fixed coordinate system where the z-axis is along $\mathbf{R}$. The angles γ and ψ are the polar angles of $\mathbf{r}$ in the body-fixed system and Ω is the projection quantum number of $\mathbf{J}$ on the body-fixed z-axis. The resulting partial differential equations for the $\Psi^{J\Omega}$ have been given previously.[43] We shall consider the form of these in the so-called centrifugal sudden (j_z-conserving) approximation.[44,45] They are,

$$\left\{\frac{-\hbar^2}{2\mu}\left(\frac{1}{R}\frac{\partial^2}{\partial R^2}R + \frac{1}{r}\frac{\partial^2}{\partial r^2}r\right) + \frac{j_{op}^2}{2\mu r^2} + \frac{[J(J+1)\hbar^2 + j_{op}^2 - 2\Omega j_z \hbar]}{2\mu R^2} + V(r, R, \gamma) - E\right\}\Psi^{J\Omega}(r, R, \gamma, \psi) = 0 \tag{11}$$

where j_{op}^2 is the square of the angular momentum operator associated with **r**, expressed in terms of γ and ψ.

Instead of proceeding with the coupled channel expansion of $\psi^{J\Omega}$ in terms of eigenfunctions of j_{op}^2, we consider an adiabatic approximation to the γ-motion. Thus, we shall express $\Psi^{J\Omega}$ as

$$\Psi^{J\Omega}(r, R, \gamma, \psi) = \exp(i\Omega\psi)\xi_{n\Omega}(\gamma; r, R)U_{n\Omega}^J(r, R)/(rR\sqrt{2\pi}) \tag{12}$$

Inserting this into Eq. (11) we have

$$\left\{\frac{-\hbar^2}{2\mu}\left(\frac{1}{R}\frac{\partial^2}{\partial R^2}R + \frac{1}{r}\frac{\partial^2}{\partial r^2}r\right)\frac{-\hbar^2}{2I(r, R)}\left(\frac{\partial^2}{\partial\gamma^2} + \cot\gamma\frac{\partial}{\partial\gamma} - \frac{\Omega^2}{\sin^2\gamma}\right) + \frac{[J(J+1) - 2\Omega^2]\hbar^2}{2\mu R^2} + V(r, R, \gamma) - E\right\}\xi_{n\Omega}(\gamma; r, R)U_{n\Omega}^J/(rR) = 0 \tag{13}$$

where $I(r, R)$ equals $\mu^{-1}(r^{-2} + R^{-2})^{-1}$. As it stands, this equation is no less exact than Eq. (11). The adiabatic approximation is now made. Let $\xi_{n\Omega}(\gamma; r, R)$ be an eigenfunction of the r and R-fixed Schrödinger equation

$$\left[\frac{-\hbar^2}{2I(r, R)}\left(\frac{\partial^2}{\partial\gamma^2} + \cot\gamma\frac{\partial}{\partial\gamma} - \frac{\Omega^2}{\sin^2\gamma}\right) + V_b(\gamma, r, R) - \varepsilon_{n\Omega}(r, R)\right]\xi_{n\Omega}(\gamma; r, R) = 0 \tag{14}$$

where $V_b(\gamma, r, R)$ is

$$V_b(\gamma, r, R) = V(r, R, \gamma) - V(r, R, \gamma = 0) \tag{15}$$

where $V(r, R, \gamma = 0)$ is the full potential for the collinear reaction $A + BC$. [Note, $V(r, R, \gamma = \pi)$ would be the full potential for the collinear $A + CB$ reaction.] $\varepsilon_{n\Omega}(r, R)$ is the adiabatic bending eigenvalue. Making the adiabatic approximation now results in the following Schrödinger equation for the unknown function $U_{n\Omega}^J(r, R)$:

$$\left\{\frac{-\hbar^2}{2\mu}\left(\frac{\partial^2}{\partial R^2} + \frac{\partial^2}{\partial r^2}\right) + \frac{[J(J+1) - 2\Omega^2]\hbar^2}{2\mu R^2} + \varepsilon_{n\Omega}(r, R) + V(r, R, \gamma = 0) - E\right\}U_{n\Omega}^J(r, R) = 0 \tag{16}$$

Before proceeding, we note that for $J = 0$ ($\Omega = 0$), and ignoring the bending energy, the preceding equation looks like the standard Schrödinger equation for a two-mathematical dimensional collinear reaction.

As discussed in the previous section the quantity to be calculated in this reduced dimensionality theory is the partial wave cumulative reaction probability,

$$\bar{P}^J_{v\to v'} = \sum_{j\Omega}\sum_{j'\Omega'} |S^J_{vj\Omega\to v'j'\Omega'}|^2 \tag{17}$$

where $S^J_{vj\Omega\to v'j'\Omega'}$ is the full scattering matrix element in the helicity representation.[43] Recall also that the full scattering matrix encompasses both reactions $A + BC$ and $A + CB$, whereas the foregoing approximate approach treats these separately. This implies that the bending eigenfunctions are localized for either the ABC or ACB configurations. Referring now to Eq. (16), let us introduce the associated scattering matrix, $S^{Jn\Omega}_{v\to v'}$. In this adiabatic treatment n is a good quantum number in addition to J (which is rigorously a good quantum number) and Ω (which is a good quantum number within the j_z-conserving approximation). If it were our objective to obtain an approximation to the full scattering matrix $S^J_{vj\Omega\to v'j'\Omega'}$ it would be necessary to determine the correlation between the bending state $\xi_{n\Omega}(\gamma, r, R\to\infty)$ and the free-rotor states $|j\Omega\rangle$. Unfortunately, that is not a straightforward correlation.[46] However, because our objective is an approximation to the cumulative reaction probability $\bar{P}^J_{v\to v'}$, we can write

$$\bar{P}^J_{v\to v'} = \sum_{n\Omega} |S^{Jn\Omega}_{v\to v'}|^2 \tag{18}$$

That is, the summation over n and Ω spans the same space as the summation over Ω, Ω', j and j' in Eq. (17).

Equation (18) is the basis for a hierarchy of reduced dimensionality theories which we now discuss in detail.

The solutions to Eq. (16) with no further approximations are formally straightforward. For each n the two-mathematical dimensional equations are solved for each partial wave J. The Ω-quantum number (i.e., the bending angular momentum) is restricted to the range $-n$ to n in steps of 2.[46,47]

The equation for the ground bending state deserves special attention

$$\left[\frac{-\hbar^2}{2\mu}\left(\frac{\partial^2}{\partial R^2} + \frac{\partial^2}{\partial r^2}\right) + \frac{J(J+1)\hbar^2}{2\mu R^2} + \varepsilon_{00}(r, R) + V(r, R, \gamma = 0) - E\right] U^J_{00}(r, R) = 0 \tag{19}$$

This is because the bending eigenvalues $\varepsilon_{n\Omega}$ can be fairly large in the neighborhood of the saddle point of the potential $V(r, R, \gamma = 0)$ and for total energies E up to say several kcal/mole above the classical barrier height, excited bending states make a negligible contribution to $\bar{P}^J_{v\to v'}$. Thus, for this reason and for the sake of simplicity we shall base our subsequent discussion on Eq. (19) and make the generalization to include all bending states at the end of this section. (An analogous equation is given in terms of the product arrangement coordinates and the wavefunctions in each arrangement channel are matched smoothly along a matching line, usually $R = R'$.)

1. *Full Adiabatic Bend Theory*

Even for the ground bending state the solutions to Eq. (19) are fairly time-consuming to obtain because the 2MD equations must be re-solved for each partial wave J. The scattering boundary conditions associated with Eq. (19) are the usual ones and are expressed in terms of the asymptotic behavior of $U^J_{00}(r, R)$. For the reactants and products let $g^J_c(\tilde{R})$ and $g^J_p(\tilde{R}')$ be the radial wave functions where $\tilde{R}$ and $\tilde{R}'$ are the unscaled, physical, radial distances for the reactant and product arrangement channels. Thus, for each vibrational channel

$$g^J_c(\tilde{R}) \underset{\tilde{R}\to\infty}{\sim} V_v^{-1/2}\{\exp[-i(k_v\tilde{R} - J\pi/2)]\delta_{vv_c} - S^{J00}_{v\to v_c}(E)\exp[i(k_{v_c}\tilde{R} - J\pi/2)]\} \tag{20a}$$

$$g^J_p(\tilde{R}') \underset{\tilde{R}'\to\infty}{\sim} -V_v^{-1/2} S^{J00}_{v\to v'}(E)\exp[i(k_{v'}\tilde{R}' - J\pi/2)] \tag{20b}$$

where

$$V_v = \frac{\hbar k_v}{\mu_{A,BC}} \tag{21a}$$

$$k_{v_{(c)}} = [(2\mu_{A,BC}/\hbar^2)(E - E_{v_{(c)}})]^{1/2};\ k_{v'} = [(2\mu_{C,AB}/\hbar^2)(E - E_{v'})]^{1/2} \tag{21b}$$

and $S^J_{v\to v_c}$ is the nonreactive scattering matrix element. Note, we have used the fact that the ground state adiabatic bending energy vanishes asymptotically.

2. *CEQB Theory*

We have introduced by a different set of arguments an approximate solution to Eq. (19) which reduces the computational effort relative to the full adiabatic solution considerably.[17,18] We replace the centrifugal potential $J(J+1)\hbar^2/2\mu R^2$ by a constant $E^\ddagger_{J0}$ which is the value of the ABC rotational energy at the transition state configuration. The expression for $E^\ddagger_{J\Omega}$ in the general case is

$$E^{\ddagger}_{J\Omega} = \frac{[J(J+1) - 2\Omega^2]\hbar^2}{2\mu R^{\ddagger 2}} \tag{22}$$

where $R^{\ddagger}$ is the value of R at the ABC transition state. Clearly, it is at this point that the flavor of transition state theory is added to the dynamics. The rationalization for this replacement is that $E^{\ddagger}_{J\Omega}$ is assumed to be the maximum value of the centrifugal potential along the reaction path. The new 2MD equations to be solved for the ground bending state are

$$\left[\frac{-\hbar^2}{2\mu}\left(\frac{\partial^2}{\partial R^2} + \frac{\partial^2}{\partial r^2}\right) + \varepsilon_{00}(r, R) + V(r, R, \gamma = 0) - (E - E^{\ddagger}_{J0})U^{J}_{00}(r, R) = 0 \tag{23}$$

Because of its formal resemblance to the 2MD equation for the collinear reaction with the addition of the adiabatic bending energy we can relate the $U^{J}_{00}(r, R)$ wave functions to the 2MD ones as follows,

$$U^{J}_{00}(E) = U^{\mathrm{CEQB}}_{00}(E - E^{\ddagger}_{J0} | n = \Omega = 0) \tag{24}$$

where $U^{\mathrm{CEQB}}_{00}(E | n = \Omega = 0)$ is the solution to Eq. (23) for $J = 0$. The superscript CEQB has been used by us to signify that $U^{\mathrm{CEQB}}_{00}(E | n = \Omega = 0)$ is the wave function for the usual collinear (2MD) exact quantum system with an adiabatic treatment of the ABC bending motion. As a result of Eq. (24),

$$S^{J00}_{v \to v'}(E) = S^{\mathrm{CEQB}}_{v \to v'}(E - E^{\ddagger}_{J0} | n = \Omega = 0) \tag{25}$$

where $S^{\mathrm{CEQB}}_{v \to v'}(E | n = \Omega = 0)$ is the scattering matrix element associated with $U^{\mathrm{CEQB}}_{00}(E)$, the radial parts of which satisfy the following asymptotic boundary conditions:

$$g^{\mathrm{CEQB}}_{c}(\tilde{R}) \underset{\tilde{R} \to \infty}{\sim} V_v^{-1/2}\{\exp(-ik_v\tilde{R})\delta_{vv_c} - S^{\mathrm{CEQB}}_{v \to v}(E | n = \Omega = 0) \times \exp(ik_{v_c}\tilde{R})\} \tag{26a}$$

$$g^{\mathrm{CEQB}}_{p}(\tilde{R}') \underset{\tilde{R}' \to \infty}{\sim} -V_v^{-1/2} S^{\mathrm{CEQB}}_{v \to v'}(E | n = \Omega = 0) \exp(ik_{v'}\tilde{R}') \tag{26b}$$

Note that these equations differ from Eqs. (20a) and (20b) by the phase factors $\exp(iJ\pi/2)$ and $\exp(-iJ\pi/2)$ for the incoming and outgoing parts of the radial wave functions. The differences have important consequences for the CEQB differential cross section but are of no significance for the integral cross section which is independent of phase factors. Thus, we defer a discussion of the obvious modification of the CEQB boundary conditions to correct their phases until Section IV where differential cross sections are explicitly considered.

3. *CEQ Theory*

An additional simplification to Eq. (19) can be made, again within the spirit of transition state theory. It is to replace $\varepsilon_{00}(r, R)$ by $\varepsilon_{00}^{\ddagger}$, that is, the *ABC* bending energy at the transition state. Thus, Eq. (21) becomes

$$\left[\frac{-\hbar^2}{2\mu}\left(\frac{\partial^2}{\partial R^2} + \frac{\partial^2}{\partial r^2}\right) + V(r, R, \gamma = 0) - (E - E_{J0}^{\ddagger} - \varepsilon_{00}^{\ddagger})\right]U_{00}^{J}(r, R) = 0 \tag{27}$$

The scattering wavefunction U_{00}^{J} and its corresponding scattering matrix element can be directly related to those for the collinear exact quantum problem as follows,

$$U_{00}^{J}(E) = U^{\text{CEQ}}(E - E_{J0}^{\ddagger} - \varepsilon_{00}^{\ddagger}),\ S_{v \to v'}^{J00}(E) = S_{v \to v'}^{\text{CEQ}}(E - E_{J0}^{\ddagger} - \varepsilon_{00}^{\ddagger}) \tag{28}$$

The CEQ scattering boundary conditions are identical to Eq. (26a) and (26b) with $S_{v \to v'}^{\text{CEQB}}$ replaced by $S_{v \to v'}^{\text{CEQ}}$ and $S_{v \to v_c}^{\text{CEQB}}$ replaced by $S_{v \to v_c}^{\text{CEQ}}$.

4. *One-Dimensional Reaction Path Theory*

Equation (19) is a 2MD partial differential equation which requires a fairly extensive, though routine, computational effort. A great reduction in effort results if the variables r and R are replaced by reaction coordinates s and x,[38–40] the reaction path and the displacement orthogonal to it, respectively. If the vibrational x-motion is treated adiabatically the following equation is obtained from Eq. (19).

$$[T_s + V(s) + \varepsilon_v(s) - (E - E_{J0}^{\ddagger} - \varepsilon_{00}^{\ddagger})]U_{00v}^{J}(s) = 0 \tag{29}$$

where T_s is the kinetic energy operator for the s-motion (including curvature terms), $V(s)$ is the potential along s and $\varepsilon_v(s)$ is the adiabatic vibrational eigenvalue. For this equation the scattering matrix becomes

$$S_{v \to v'}^{J00}(E) = \delta_{v'v} S_v^{1D}(E - E_{J0}^{\ddagger} - \varepsilon_{00}^{\ddagger} - \varepsilon_v^{\ddagger}) \tag{30}$$

where $S_v^{1D}(E)$ is the scattering matrix corresponding to the one-dimensional

Schrödinger equation

$$[T_s + V(s) + \varepsilon_v(s) - E]U_v^{1D}(s) = 0 \tag{31}$$

Many important refinements to this one-dimensional reaction path approach have been made recently which take into account certain 2MD features of the potential energy surface.[48]

5. *Transition State Theory*

The full transition state solution to Eq. (19) follows by replacing the scattering matrix $S_v^{1D}(E)$ by a unit step function, $\theta(E - V_o - \varepsilon_v^\ddagger)$, where V_0 and $\varepsilon_v^\ddagger$ are the values of $V(s)$ and $\varepsilon_v(s)$ at the transition state. Thus,

$$S_{v \to v'}^{J00} = \delta_{v'v}\theta(E - V_0 - \varepsilon_v^\ddagger - E_{J0}^\ddagger - \varepsilon_{00}^\ddagger). \tag{32}$$

We have deferred up to this point giving an explicit definition of the transition state. Several definitions are of course possible. For example, the conventional transition state is the configuration corresponding to the saddle point of the potential surface $V(r, R, \gamma = 0)$. Another and better choice is to apply a classical variational criterion and determine the transition state by maximizing $V(s) + \varepsilon_v(s) + E_{J0}(s) + \varepsilon_{00}(s)$ with respect to s. This would be a bit cumbersome because of the need to re-optimize (in the general case) for each v, J, Ω, and n independently. A compromise between this full optimization and no optimization seems reasonable and indeed was suggested by Truhlar and co-workers. They chose to optimize the quantity $V(s) + \varepsilon_v(s)$.[15] A more general method to locate the optimum transition state has been given by Pollak and Pechukas.[49]

The complete expression for each approximation to $\bar{P}_{v \to v'}^J$ includes a summation over the bending states. As noted already we have adopted the standard small angle treatment of the bending motion[46,47] in which the quantum number Ω is restricted to the range $-n$ to n in steps of 2. This model treats each reaction path separately. Although it is not necessary to do so, we adopt this picture because it is the one usually used in transition state theory.

B. Correlation between Adiabatic Bending and Free-Rotor States

Thus far the only correlation that has been made between bending states $|n\Omega\rangle$ and free rotor states $|j\Omega\rangle$ is that Ω is a good quantum number, a strict consequence of the j_z-conserving assumption. This alone yields some qualitatively interesting results. It implies that in the energy range where the ground bending state $|n = \Omega = 0\rangle$ dominates the summation in Eq. (18) that

$$\bar{P}_{v \to v'}^J = \sum_j \sum_{j'} |S_{vj\Omega=0 \to v'j'\Omega'=0}^J|^2 \tag{33}$$

Thus transition probabilities involving nonzero Ω or Ω' should be very small relative to those for which both Ω and Ω' are zero. This result is in accord qualitatively with the coupled channel calculations of Schatz and Kuppermann for $H + H_2(v = 0)$ for the energies they considered, that is, $E \leqslant 0.7$ eV.[11] They took note of the propensity for $\Omega = 0$ to $\Omega' = 0$ transitions and offered an explanation based on the favorable, that is, collinear orientation that $|j\Omega = 0\rangle$-states have relative to $|j\Omega \neq 0\rangle$-states. This argument presupposes that the asymptotic free-rotor orientations are, roughly speaking, maintained throughout the collision. This is the perspective of the rotational sudden approximation[50,51] and although this perspective is quite different from the present adiabatic one it is interesting that both lead to the same qualitative conclusions.

As noted, a completely adiabatic correlation between $|n\Omega\rangle$ and $|J\Omega\rangle$ states is not straightforward and would not be very useful. The reason it would not be very useful is because there is strong nonadiabatic coupling due to the drastic changes in the nature of the free-rotor and hindered-rotor wave functions as R is varied from say $R^{\ddagger}$ to ∞. In fact the change is probably better described by a sudden rather than an adiabatic treatment. It is possible to incorporate certain aspects of that theory into the present one. We explore some of these possibilities now.

We may consider the $S^{Jn\Omega}_{v \to v'}$ to be related to the $S^{J}_{vj\Omega \to v'j'\Omega}$ by an orthogonal transformation, that is, in matrix notation,

$$\mathsf{S}^{J\Omega} = \mathsf{C}^t \mathscr{S}^{J\Omega} \mathsf{C} \tag{34}$$

where $(\mathsf{S}^{J\Omega})_{nm} = S^{Jn\Omega}_{v \to v'}\ \delta_{nm}$ and $(\mathscr{S}^{J\Omega})_{jj'} = S^{J}_{vj\Omega \to v'j'\Omega}$. It is straightforward to show that such a transformation implies Eqs. (17) and (18), by noting that

$$\sum_n |S^{Jn\Omega}_{v \to v'}|^2 = \mathrm{Tr}[\mathsf{S}^{J\Omega}(\mathsf{S}^{J\Omega})^t] \tag{35}$$

which from Eq. (34) can be rewritten as

$$\mathrm{Tr}[\mathsf{C}^t \mathscr{S}^{J\Omega} \mathsf{C}(\mathsf{C}^t \mathscr{S}^{J\Omega} \mathsf{C})^t] = \mathrm{Tr}[\mathsf{C}^t \mathscr{S}^{J\Omega}(\mathscr{S}^{J\Omega})^t \mathsf{C}] \tag{36}$$

which equals $\mathrm{Tr}[\mathscr{S}^{J\Omega}(\mathscr{S}^{J\Omega})^t]$ from the cyclic property of the trace operation. Thus $\mathrm{Tr}[S^{J\Omega}(S^{J\Omega})^t]$ equals $\mathrm{Tr}[\mathscr{S}^{J\Omega}(\mathscr{S}^{J\Omega})^t]$, that is,

$$\sum_n |S^{Jn\Omega}_{v \to v'}|^2 = \sum_j \sum_{j'} |S^{J}_{vj\Omega \to v'j'\Omega}|^2 \tag{37}$$

Clearly, both sides of Eq. (36) can then be summed over Ω to obtain Eq. (18).

The adiabatic correlation can be made by assuming that $(C)_{nj} = \delta_{nj}$, in which case

$$S^{J}_{vj\Omega \to v'j'\Omega} = \delta_{j'j}\delta_{jn}S^{Jn\Omega}_{v \to v'} \tag{38}$$

As implied previously, this correlation does not seem reasonable to us. First, accurate quantum calculations for the $H + H_2$ reaction definitely do not support it.[11,19,20] Second, on more general grounds the transition in going from a free to a hindered rotor is too drastic for an adiabatic picture to hold. It is more likely that the transition occurs rather abruptly at some value of R, R^*, and then for smaller or larger values of R^* the rotor wave function changes adiabatically. (The value of R^* would be greater than $R^{\ddagger}$.) This sudden picture implies that $C_{jn} = \langle j\Omega | n\Omega \rangle$ and so

$$S^{J}_{vj\Omega \to v'j'\Omega}(E) = \langle j'\Omega | n\Omega \rangle S^{Jn\Omega}_{v \to v'}(E) \langle n\Omega | j\Omega \rangle \tag{39}$$

This expression is similar in spirit to the Franck-Condon theory of reactions,[52-54] especially the one given by Schatz and Ross for three-dimensional reactions.[54] An important difference is that Eq. (39) is meant to be a *quantitative* approximation for the scattering matrix. Previous Franck-Condon theories contained an unknown electronic coupling interaction and so were capable of giving only relative final state distributions. The present expression, though intended to be quantitative, does share a problem with the Franck-Condon theories. Namely, it violates conservation of energy. In the multidimensional Franck-Condon theory this is mitigated by the decrease in the overlap of other components of the wave function as j' increases for j fixed. This can, in spirit, also be done in the present case by replacing Eq. (38) by

$$S^{J}_{vj\Omega \to v'j'\Omega}(E) = \langle j'\Omega | n\Omega \rangle S^{Jn\Omega}_{v \to v'}(E - \max(E_j, E_{j'})) \langle n\Omega | j\Omega \rangle \tag{40}$$

Thus, if either j or j' is energetically closed for the v to v' transition, $S^{J}_{vj\Omega \to v'j'\Omega}$ would be identically zero because $S^{Jn\Omega}_{v \to v'}$ vanishes if E is less than $\max(E_v, E_{v'})$. This ad hoc procedure has the disadvantage that Eq. (37) no longer exactly holds. A different ad hoc procedure to eliminate off-shell transitions is to restrict the matrix equation [Eq. (31)] to only energetically open values of j and j' and reorthogonalize the resulting nonorthogonal C-matrix. This would preserve Eq. (37).

Finally, and of importance to any ad hoc procedure to obtain $S^{J}_{vj\Omega \to v'j'\Omega}$ by a sudden method is the question of where to take the projections $\langle j'\Omega | n\Omega \rangle$ and $\langle n | j\Omega \rangle$. Schatz and Ross[54] found that projecting the free-rotor states onto the ground hindered-rotor state at the transition state gave too much rotational excitation. Their analysis of this suggests a value of R^* less than $R^{\ddagger}$

in agreement with our earlier analysis. By analyzing the dependence of $\langle n\Omega | j\Omega \rangle$ on R (and the analogous dependence for the product channel) it might be possible to determine where the switch from free to hindered-rotor motion occurs. At this point, these suggestions are to be regarded as speculative although perhaps worthy of investigation.

To summarize this section: A hierarchy of approximations to the partial wave cumulative reaction probability has been given. The expressions for this probability including the summation over bending states are

$$\bar{P}^J_{v \to v'}(E) = \begin{cases} \sum_{n\Omega} |S^{Jn\Omega}_{v \to v'}|^2 & (41) \\ \sum_{n\Omega} |S^{\text{CEQB}}_{v \to v'}(E - E^{\ddagger}_{J\Omega}|n\Omega)|^2 & (42) \\ \sum_{n\Omega} |S^{\text{CEQ}}_{v \to v'}(E - E^{\ddagger}_{J\Omega} - \varepsilon^{\ddagger}_{n\Omega})|^2 & (43) \end{cases}$$

There remains a question about how to explicitly treat the excited bending states asymptotically. (As already noted previously, the adiabatic eigenvalue for the ground bending state is assumed to vanish asymptotically.) There is no really satisfactory answer to this question without making some correlation with the correct asymptotic free-rotor states. Operationally we have simply done the following. We related the higher bending state probabilities to the ground bending state probability by an energy shift given by the difference between the $n\Omega$th bending state energy and the ground bending state energy. This essentially assumes that the adiabatic energy vanishes asymptotically for all the bending states. By making that assumption it is also possible to test the accuracy of the energy shift approximation. That was done and the results will be discussed in Section V.A.

Some new expressions for rotationally state-to-state scattering matrices and reaction probabilities were also given by considering the projection of the adiabatic bending wavefunction onto the free-rotor basis, in the spirit of previous qualitative Franck-Condon theories of reaction.

In the next section we present the rate constants implied by the expressions for the integral cross sections derived in this section followed by a section on differential cross sections.

III. THERMAL AND VIBRATIONAL STATE-TO-STATE RATE CONSTANTS

The rate constant implied by the rotationally averaged and summed integral cross section $\bar{Q}_{v \to v'}(E)$ has been given previously in terms of both the CEQB and CEQ approximations to $\bar{P}^J_{v \to v'}(E)$.[17] This rate constant is thermally averaged over initial rotational states and summed over final ones but vibrationally state-to-state. It is conveniently expressed in terms of transmission coefficients $\Gamma^{\text{CEQB}}_{v \to v'}$ and $\Gamma^{\text{CEQ}}_{v \to v'}$ and the conventional three-dimensional transition state theory rate constant $k^{\ddagger}$,

$$k^{\text{CEQ(B)}}_{v \to v'}(T) = \Gamma^{\text{CEQ(B)}}_{v \to v'} k^{\ddagger} \tag{44}$$

where in terms of the CEQ and CEQB reaction probabilities $P^{\text{CEQ}}_{v \to v'}(E_t)$ and $P^{\text{CEQB}}_{v \to v'}(E_t)$,

$$\Gamma^{\text{CEQ}}_{v \to v'} = \frac{\langle P^{\text{CEQ}}_{v \to v'}(E_t) \rangle}{\langle \theta(E_t + E_v - V_0 - \varepsilon^{\ddagger}_v) \rangle} \tag{45}$$

and

$$\Gamma^{\text{CEQB}}_{v \to v'} = \frac{\sum_n (n+1) \exp(-\varepsilon^{\ddagger}_{n\Omega}/kT) \langle P^{\text{CEQB}}_{v \to v'}(E_t | n\Omega) \rangle / \langle \theta(E_t + E_v - V_0 - \varepsilon^{\ddagger}_v - \varepsilon^{\ddagger}_{n\Omega}) \rangle}{\sum_n (n+1) \exp(-\varepsilon^{\ddagger}_{n\Omega}/kT)} \tag{46}$$

In these expressions the brackets $\langle\, \rangle$ represent the thermal integral over the initial translational energy E_t which equals $E - E_v$. For example,

$$\langle P^{\text{CEQ}}_{v \to v'} \rangle = \int_0^{\infty} \mathrm{d}E_t \exp(-E_t/kT) P^{\text{CEQ}}_{v \to v'}(E_t) \tag{47}$$

Recall that V_0, $\varepsilon^{\ddagger}_v$, and $\varepsilon^{\ddagger}_{n\Omega}$ are the potential, the symmetric stretch, and bending energies at the conventional transition state. Equation (46) is an explicit thermal average of the transmission coefficients for each bending state (of degeneracy $n + 1$) of the transition state. It is easy to show that if $P^{\text{CEQB}}_{v \to v'}(E_t | n\Omega)$ equals $P^{\text{CEQ}}_{v \to v'}(E_t - \varepsilon^{\ddagger}_{n\Omega})$ Eq. (46) reduces to (45). The proof is as follows. By assumption

$$\langle P^{\text{CEQB}}_{v \to v'}(E_t | n\Omega) \rangle = \int_0^{\infty} \mathrm{d}E_t \exp(-E_t/kT) P^{\text{CEQ}}_{v \to v'}(E_t - \varepsilon^{\ddagger}_{n\Omega}) \tag{47a}$$

$$= \exp(-\varepsilon^{\ddagger}_{n\Omega}/kT) \int_0^{\infty} \mathrm{d}E_t \exp(-E_t/kT) P^{\text{CEQ}}_{v \to v'}(E_t) \tag{47b}$$

where the fact that $P^{\text{CEQ}}_{v \to v'}(E_t)$ is zero for negative $-E_t$ was used to obtain Eq. (47b). Also,

$$\langle \theta(E_t + E_v - V_0 - \varepsilon_v^\ddagger - \varepsilon_{n\Omega}^\ddagger) \rangle = \exp(-\varepsilon_{n\Omega}^\ddagger/kT)\langle \theta(E_t + E_v - V_0 - \varepsilon_v^\ddagger) \rangle \tag{48}$$

So, by assumption and Eqs. (47b) and (48)

$$\langle P^{\text{CEQB}}_{v \to v'}(E_t|n\Omega) \rangle / \langle \theta(E_t + E_v - V_0 - \varepsilon_v^\ddagger - \varepsilon_{n\Omega}^\ddagger) \rangle$$
$$= \langle P^{\text{CEQ}}_{v \to v'}(E_t) \rangle / \langle \theta(E_t + E_v - V_0 - \varepsilon_v^\ddagger) \rangle \tag{49}$$

Inserting Eq. (49) into (46) and cancelling the bending partition functions gives the result that $\Gamma^{\text{CEQB}}_{v \to v'}$ equals $\Gamma^{\text{CEQ}}_{v \to v'}$. In practice the assumption that $P^{\text{CEQB}}_{v \to v'}(E_t|n\Omega)$ equals $P^{\text{CEQ}}_{v \to v'}(E_t - \varepsilon_{n\Omega}^\ddagger)$ is a fairly good one provided E_t is not in the deep tunneling region of $P^{\text{CEQB}}_{v \to v'}(E_t|n\Omega)$. A more accurate assumption is that

$$P^{\text{CEQB}}_{v \to v'}(E_t|n\Omega) = P^{\text{CEQB}}_{v \to v'}(E_t - \Delta\varepsilon_{n\Omega}^\ddagger | n = \Omega = 0) \tag{50}$$

where

$$\Delta\varepsilon_{n\Omega}^\ddagger = \varepsilon_{n\Omega}^\ddagger - \varepsilon_{00}^\ddagger \tag{51}$$

This leads to the very useful result [based on arguments similar to those used to obtain Eq. (49)]

$$\Gamma^{\text{CEQB}}_{v \to v'} = \langle P^{\text{CEQB}}_{v \to v'}(E_t | n = \Omega = 0) \rangle / \langle \theta(E_t + E_v - V_0 - \varepsilon_v^\ddagger - \varepsilon_{00}^\ddagger) \rangle \tag{52}$$

This expression, which requires only the CEQB reaction probability for the ground state bending motion gives quite accurate results for $\Gamma^{\text{CEQB}}_{v \to v'}$ and all of the CEQB rate constants given later were obtained using it.

The usual thermal rate constant $k(T)$ is obtained from $k^{\text{CEQB}}_{v \to v'}(T)$ by the standard expression

$$k(T) = \sum_v \exp(-E_v/kT) k^{\text{CEQB}}_{v \to v'}(T) / \sum_v \exp(-E_v/kT) \tag{53}$$

Thus, Eqs. (44)–(46), (52), and (53) are the transmission coefficients and rate constants implied by the reduced dimensionality rotational averaged and summed integral cross sections. In our original paper,[17] these expressions for the transmission coefficients and rate constants were promulgated and the expressions for the rotationally averaged and summed integral cross sections were derived from them.

In the next section we present expressions for the differential cross sections and comment on their relationship to the expression for the differential cross section given in the bending corrected rotating linear model.

IV. DIFFERENTIAL CROSS SECTIONS

Starting with Eq. (4) for the rotationally averaged and summed integral cross section, $\bar{Q}_{v \to v'}(E)$, we can ask what differential cross section when integrated over all scattering angles would yield Eq. (4) using Eq. (18) and any of the approximations to $S^{Jn\Omega}_{v \to v'}$. The straightforward result is

$$\bar{\sigma}_{v \to v'}(\theta) = \frac{1}{4\bar{k}_v^2} \sum_{n\Omega} \left| \sum_{J=0} (2J+1) S^{Jn\Omega}_{v \to v'} d^J_{\Omega\Omega}(\theta) \right|^2 \tag{54}$$

where $d^J_{\Omega\Omega}(\theta)$ is $D^J_{\Omega\Omega}(0, \theta, 0)$, the rotation matrix. For the particular case of the ground bending state,

$$\bar{\sigma}_{v \to v'}(\theta | n = \Omega = 0) = \frac{1}{4\bar{k}_v^2} \left| \sum_{J=0} (2J+1) S^{J00}_{v \to v'} P_J(\cos\theta) \right|^2 \tag{55}$$

where $\mathrm{d}^J_{00}(\theta)$ has been rewritten as $P_J(\cos\theta)$, the usual Legendre polynomial of order J. Consider how Eq. (55) could be obtained from the rigorous expression for the rotationally averaged and summed differential cross section. To do that, recall that the detailed reactive state-to-state differential cross section in the helicity representation is given by[43]

$$\sigma_{vj\Omega \to v'j'\Omega'}(\theta) = \frac{1}{4k_{vj}^2} \left| \sum_{J=0} (2J+1) S^J_{vj\Omega \to v'j'\Omega'} \mathrm{d}^J_{\Omega\Omega'}(\theta) \right|^2 \tag{56}$$

The usual degeneracy-averaged and summed cross section is

$$\begin{aligned} \sigma_{vj \to v'j'}(\theta) &= \frac{1}{2j+1} \sum_{\Omega\Omega'} \sigma_{vj\Omega \to v'j'\Omega'}(\theta) \\ &= \frac{1}{(2j+1)4k_{vj}^2} \sum_{\Omega\Omega'} \left| \sum_{J=0} (2J+1) S^J_{vj\Omega \to v'j'\Omega'} d^J_{\Omega\Omega'}(\theta) \right|^2 \end{aligned} \tag{57}$$

For those reactive systems and energies where the dominant Ω-transitions are $\Omega = \Omega' = 0$ then

$$\sigma_{vj \to v'j'}(\theta) = \frac{1}{(2j+1)4k_{vj}^2} \left| \sum_{J=0} (2J+1) S^J_{vj\Omega=0 \to v'j'\Omega'=0} P_J(\cos\theta) \right|^2 \tag{58}$$

Thus, the rotationally averaged and summed differential cross section [corresponding to Eq. (4) for the integral cross section] is by definition

$$\bar{\sigma}_{v \to v'}(\theta) = \frac{1}{k_v^2} \sum_{jj'} k_{vj}^2 f_j (2j+1) \sigma_{vj \to v'j'}(\theta) \tag{59}$$

$$= \frac{1}{4k_v^2} \sum_{jj'} f_j \left| \sum_{J=0} (2J+1) S^J_{vj\Omega=0 \to v'j'\Omega'=0} P_J(\cos\theta) \right|^2 \tag{60}$$

This expression is approximated by Eq. (54) for the case where $\Omega = 0$. Further, if the ground state bend dominates the summation in Eq. (54) then we have Eq. (55) as the approximation to Eq. (60).

Equation (55) can be used directly if $S^{J00}_{v \to v'}$ obtained from the full adiabatic scattering solutions to Eq. (19), is used. However, if the CEQB or CEQ approximations to $S^{J00}_{v \to v'}$ are used, a small modification must be done first. This is because of missing J-dependent phase factors in the CEQB and CEQ radial wavefunctions, as already noted in Section II; compare Eqs. (26a) and (26b) and (20a) and (20b). Fortunately, the problem is easily remedied. These sets of equations can be brought into the same form by multiplying Eq. (20a) and (20b) by $\exp(-iJ\pi/2)$ with the result that for the products

$$g_p^J(\tilde{R}') \underset{\tilde{R}' \to \infty}{\sim} -V_v^{-1/2} S^{J00}_{v \to v'} \exp(ik_{v'}\tilde{R}') \exp(-iJ\pi) \tag{61}$$

Comparing this equation with Eq. (26b) and using Eq. (25) we have that the CEQB approximation to $S^{J00}_{v \to v'}(E)$ is $S^{\text{CEQB}}_{v \to v'}(E - E^{\ddagger}_{J0} | n = \Omega = 0)(-1)^J$; that is, the missing phase factor is $(-1)^J$. Thus, the CEQB expression for $\bar{\sigma}_{v \to v'}(\theta | n = \Omega = 0)$ is

$$\bar{\sigma}^{\text{CEQB}}_{v \to v'}(\theta | n = \Omega = 0) = \frac{1}{4k_v^2} \left| \sum_{J=0} (2J+1) S^{\text{CEQB}}_{v \to v'}(E - E^{\ddagger}_{J0} | n = \Omega = 0) \right.$$
$$\left. \times (-1)^J P_J(\cos\theta) \right|^2 \tag{62}$$

[The CEQ approximation consists in replacing $S^{J00}_{v \to v'}(E)$ by $S^{\text{CEQ}}_{v \to v'}(E - E^{\ddagger}_{J0} - \varepsilon^{\ddagger}_{00})(-1)^J$.] Note that because $(-1)^J P_J(\cos\theta) = P_J[\cos(\pi - \theta]$, Eq. (62) can be rewritten as

$$\bar{\sigma}^{\text{CEQB}}_{v \to v'}(\theta | n = \Omega = 0) = \frac{1}{4k_v^2} \left| \sum_{J=0} (2J+1) S^{\text{CEQB}}_{v \to v'}(E - E^{\ddagger}_{J0} | n = \Omega = 0) \right.$$
$$\left. \times P_J[\cos(\pi - \theta)] \right|^2 \tag{63}$$

The expression for the differential cross section using the full adiabatic solution to Eq. (19) (which we have not yet implemented) is well defined physically and probably more realistic for the differential cross section than the CEQB and CEQ expressions because the full adiabatic $S^{J00}_{v \to v'}(E)$ contains more realistic J-dependent phase information than the CEQB and CEQ expressions. We have used the CEQB and CEQ expressions for the ground bending state differential cross section in a number of applications.[29,30,55] These will be reviewed in the next section.

Before concluding this section we briefly consider the relationship between the reduced dimensionality theory described here and the rotating linear model[32,33] and the bending corrected rotating linear model.[34] In these models a partial wave Hamiltonian is given by

$$H^l = \frac{-\hbar^2}{2\mu}\left[\frac{1}{\eta Q^2}\frac{\partial}{\partial v}\eta Q^2\frac{\partial}{\partial v} + \frac{1}{\eta Q^2}\frac{\partial}{\partial u}\frac{Q^2}{\eta}\frac{\partial}{\partial u}\right] + \frac{\hbar^2[l(l+1)+1]}{2\mu Q^2} + V(u, v) + E_0^b(u) \tag{64}$$

where l is the orbital angular momentum, u is a reference reaction path coordinate, v is the coordinate transverse to u, $E_0^b(u)$ is the ground state adiabatic bending energy evaluated on u, that is, for $v = 0$, η equals $1 + \kappa(u)v$, where $1/\kappa$ is the local radius of curvature of the reaction path, and

$$Q^2 = R^2 + r^2 \tag{64a}$$

where R and r were defined just before Eq. (8). In the rotating linear model E_0^b is not included in H^l.

Scattering solutions to $(H^l - E)\psi^l(u, v)$ are obtained and the differential cross section is given by

$$\sigma^{\text{BCRLM}}_{v \to v'}(\theta) = \frac{1}{4k^2_{vj=0}}\left|\sum_{l=0} (2l+1) S^l_{v \to v'} P_l(\cos\theta)\right|^2 \tag{65}$$

where $k^2_{vj=0}$ is given by Eq. (2).

There are a number of differences between Eqs. (64) and (65) and the analogous equations, Eq. (16) and Eq. (63), respectively. First, the centrifugal potentials are different; the one given in BCRLM does not vanish for zero angular momentum whereas the one in Eq. (16) does. Another difference is the use of $k^2_{vj=0}$ in Eq. (65) instead of the average squared wavevector of Eq. (5). This last difference arises because the differential cross section we have given is explicitly a rotationally summed and averaged quantity whereas the exact identity of the BCRLM differential cross section is still uncertain.[56] The recent applications of the BCRLM by Hayes and Walker[56] and Walker et al.[57] have been to qualitative studies of differential cross sections and their relationship to features of potential energy surfaces. Although qualitative, these are valid and very valuable applications of the model.

In the next section a number of applications of the CEQB and CEQ theories are presented. We consider the entire range of applications, from differential cross sections to thermal rate constants. The systems reviewed are: $H + H_2(v = 0, 1)$, $D + H_2(v = 0, 1)$, $F + H_2(v = 0)$, $F + HD(v = 0)$, $O(^3P) + H_2(v = 0, 1)$, and $O(^3P) + D_2(v = 0, 1)$.

V. APPLICATIONS OF THE CEQ AND CEQB THEORIES

In this section we review applications of the reduced dimensionality theories of differential and integral cross sections to a number of atom plus diatom systems. The organization is based on a given reaction.

A. $H + H_2(v = 0)$

The $H + H_2(v = 0)$ reaction was studied for its intrinsic historic interest but, primarily, because of the availability of accurate quantum calculations.[11,19,20] Rotationally averaged cross sections $\bar{Q}_{0\to 0}$ were calculated using both the CEQB and CEQ approximations. The ground and first excited bending states were used in the CEQB calculations. The potential energy surface used was the semiempirical one of Porter and Karplus[58] which has a classical barrier height, V_0, of 9.13 kcal/mole. This is probably too low by approximately 0.5–0.7 kcal/mole.[59,60] The choice of this surface was made because it was used in the accurate quantum calculations of Schatz and Kuppermann,[11] the results of which were used in comparisons with those of the CEQ and CEQB theories. This comparison is shown in Figs. 1 and 2. In Fig. 1 the accurate, CEQ, and CEQB $\bar{Q}_{0\to 0}(E)$ are plotted versus the total energy in the tunneling regime. (For the total energies considered in that plot, $H_2(v' = 0)$ is the only open vibrational state.) First, note the very good agreement between the accurate and CEQB cross sections even for very small values of the cross section. The CEQ cross section greatly underestimates the accurate result at low energies but becomes much more accurate for E greater than 14.0 kcal/mole. To understand the significant differences between the CEQB and CEQ results in the deep tunneling region, that is, for E much less than the CEQB adiabatic barrier, $V_0 + \varepsilon_v^\ddagger + \varepsilon_{00}^\ddagger$ of 15.0 kcal/mole, recall that the cumulative reaction probability is given by

$$\bar{P}^J_{v\to v'}(E) = \sum_{n\Omega} P^{\text{CEQB}}_{v\to v'}(E - E^\ddagger_{J\Omega}|n\Omega) \tag{66}$$

and

$$\bar{P}^J_{v \to v'}(E) = \sum_{n\Omega} P^{\text{CEQ}}_{v \to v'}(E - E^{\ddagger}_{J\Omega} - \varepsilon^{\ddagger}_{n\Omega}) \tag{67}$$

in the CEQB and CEQ approximations. For E less than 14.0 kcal/mole, $P^{\text{CEQB}}_{0\to 0}(E)$ is significantly larger than $P^{\text{CEQ}}_{0\to 0}(E - \varepsilon^{\ddagger}_{00})$ (although both are quite small). Clearly then the replacement of the bending ground state adiabatic eigenvalue by a simple energy shift $\varepsilon^{\ddagger}_{00}$ breaks down as an accurate approximation for total energies much less than the CEQB adiabatic barrier. This is reasonable because while the simple energy shift may be accurate for total energies just below the CEQB barrier it will break down for energies much below this barrier because the details of the barrier shape become more important. Evidently the CEQB barrier is thinner than the CEQ one. This is expected because the adiabatic bending energy is of course not a constant over the 2MD potential surface; it decays fairly rapidly to zero as the system moves into the reactant or product channels. Indeed the CEQ and CEQB approximations to $\bar{P}^J_{v\to v'}$ would be equal only if $\varepsilon_{00}(r, R)$ were a constant equal to $\varepsilon^{\ddagger}_{00}$.

A comparison of the CEQ and CEQB $\bar{Q}_{0\to 0}$ is given at higher energies in

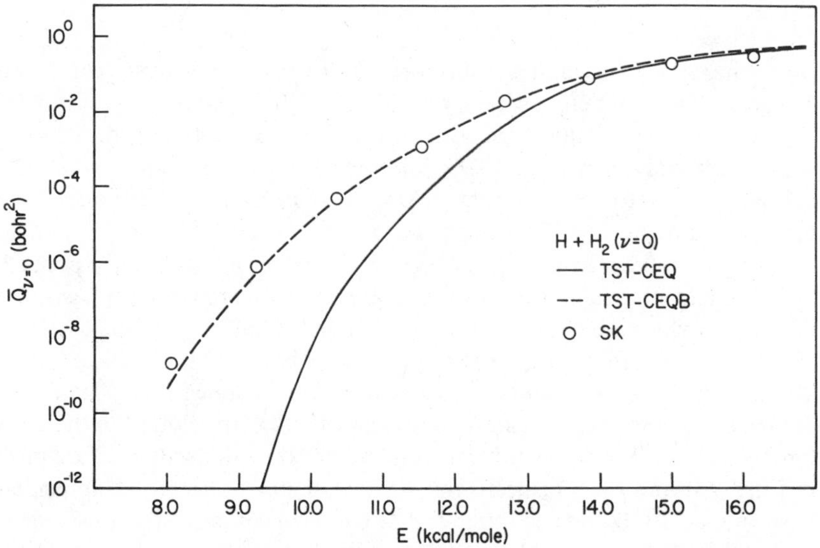

Figure 1. Comparison of reduced dimensionality CEQ and CEQB rotationally averaged cross sections with the accurate quantum ones of Schatz and Kuppermann (SK) (Ref. 11) emphasizing the tunneling region for $H + H_2(v = 0)$.

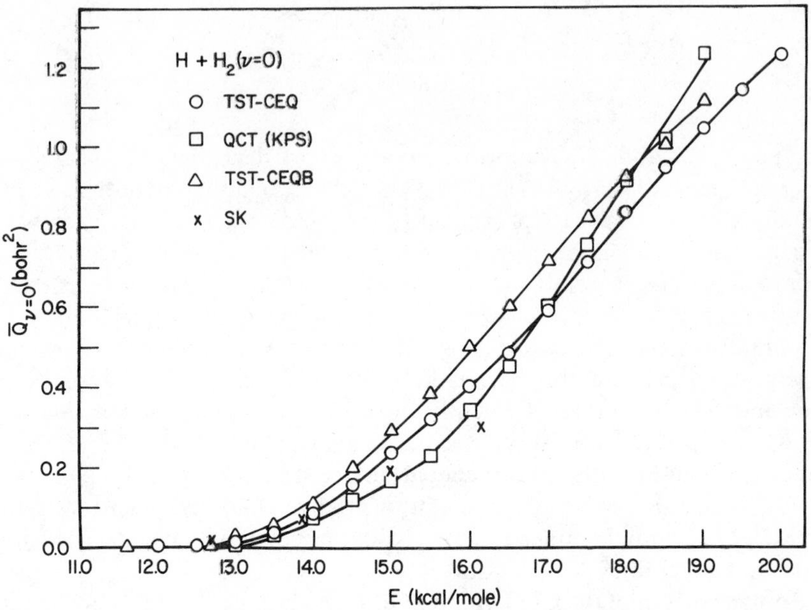

Figure 2. Comparison of CEQ and CEQB rotationally averaged cross sections with accurate quantum (SK) (Ref. 11) and quasiclassical trajectory KPS (Ref. 61) ones for $H + H_2(v = 0)$.

Fig. 2 where the quasiclassical trajectory (QCT) results of Karplus et al.[61] are also given. The CEQ cross sections are roughly 20% smaller than the CEQB ones which are generally larger than are QCT cross sections which have a sharp threshold energy of 13.0 kcal/mole. The accurate quantum cross section appears to be dropping below the CEQ and CEQB ones as E increases from roughly 15.0 kcal/mole.

It might appear from Fig. 2 that the CEQ cross section is perfectly adequate and that the results in Fig. 1 are shown merely to emphasize the differences (which are quite small on an absolute scale) between the CEQ and CEQB and accurate results. However, these differences are significant when considering the rate constant for the reaction. This is done in Fig. 3 where the CEQB, CEQ, and accurate thermal rate constants[11] are plotted in the usual semilog fashion. The relevant CEQB and CEQ expressions for $k(T)$ are given by Eqs. (46) and (45), respectively. As seen there is very good agreement between the CEQB and accurate results. This is of course just a consequence of the agreement between the CEQB and accurate $\bar{Q}_{0\to0}$ seen in Fig. 1. The important point is that there is significant disagreement between the CEQ and accurate results and on this basis Kuppermann, who first compared the

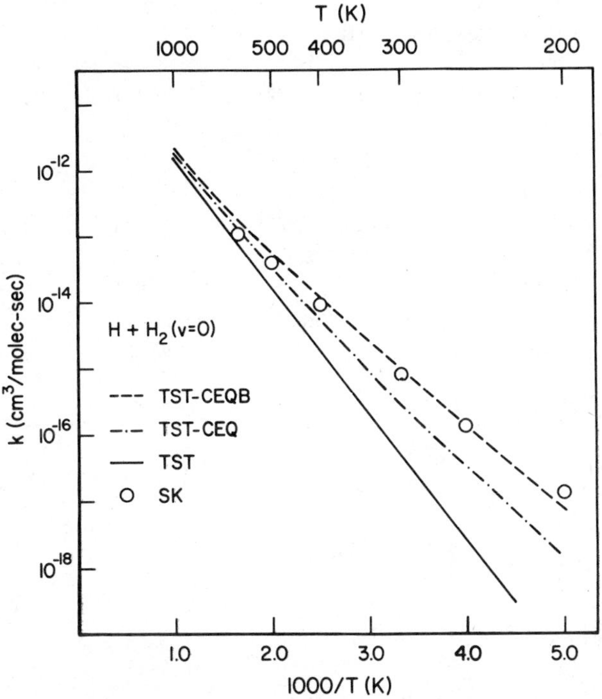

Figure 3. Thermal para to ortho rate constants for the $H + H_2(v = 0)$ reaction: conventional transition state theory (TST), reduced dimensionality CEQ and CEQB, and accurate quantum results of Schatz and Kuppermann (SK) (Ref. 11).

CEQ transmission coefficient correction to transition state theory to the accurate one,[14] concluded that the CEQ transmission coefficient was not accurate. However, the CEQB transmission coefficient is quite accurate. Garrett and Truhlar,[21] using the Marcus-Coltrin[62] one-dimensional tunneling path for a potential containing the ground state adiabatic bending eigenvalue obtained a rate constant of accuracy comparable to the CEQB result shown here. Clearly, then, the inclusion of the ground state bending energy in either one-dimensional or two-mathematical dimensional dynamics is essential to obtain an accurate thermal rate constant for $H + H_2(v = 0)$.

The accuracy of Eq. (50) was tested for this reaction by comparing the $\bar{Q}_{0\to 0}$ calculated using that equation and by explicitly calculating $P^{\text{CEQB}}_{v\to v'}(E - E^{\ddagger}_{J\Omega}|n\Omega)$ for the first two bending states and was found to be accurate to within 10% or less for the energies considered. In all the calculations reported next Eq. (50) was used when it was necessary to consider excited bending states.

We have also applied the CEQB expression for the rotationally averaged and summed differential cross section $\bar{\sigma}_{0\to 0}(\theta|n = \Omega = 0)$, given by Eq. (63), to this reaction for total energies between 0.6 and 0.95 eV.[55] There is currently considerable interest in reactive scattering resonances and their influence on the differential cross section (particularly for the $F + H_2$ reaction and the isotopic analogs of H_2 to be discussed later). The issue can also be addressed for $H + H_2(v = 0) \to H_2(v' = 0) + H$ on the PK2 surface where a weak resonance occurs at a total energy of 0.975 eV.[63] In Figs. 4 and 5 the CEQB partial wave cumulative reaction probabilities $\bar{P}^J_{0\to 0}$ (for the ground state bend state) and the corresponding differential cross section $\bar{\sigma}_{0\to 0}(\theta)$ are given for the total energies indicated. For E equal to 0.6 and

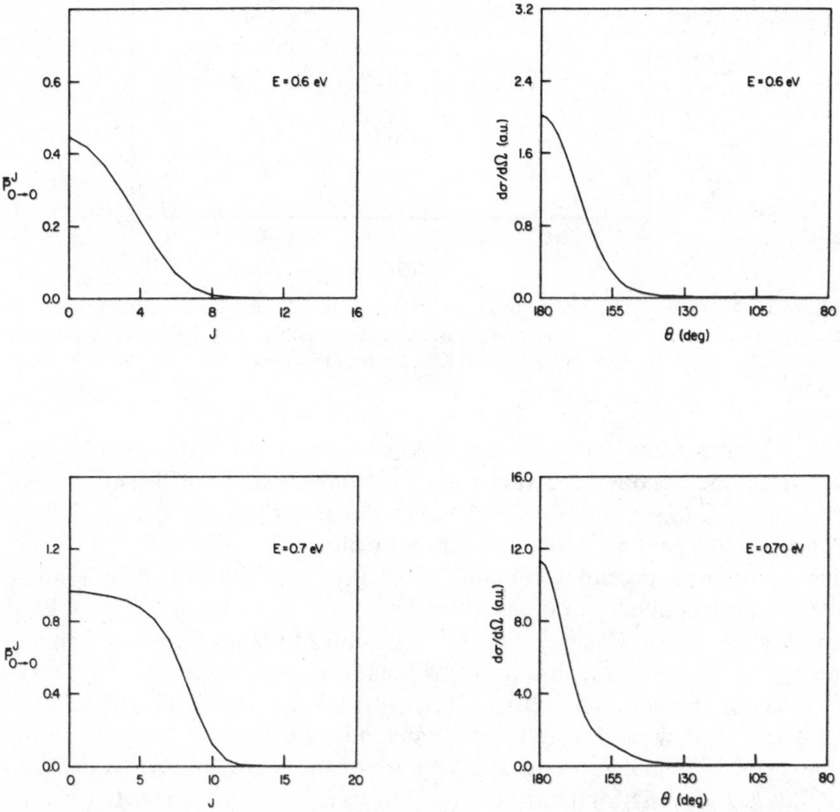

Figure 4. CEQB cumulative partial wave reaction probabilities versus the total angular momentum for $H + H_2(v = 0) \to H_2(v' = 0) + H$ and corresponding differential cross sections for total energies of 0.6 and 0.7 eV.

0.7 eV $\bar{P}^J_{0\to 0}$ is monotonically decreasing with J and the corresponding $\bar{\sigma}_{0\to 0}$'s are backward-peaked and quite smooth. The behavior of the differential cross sections are in *qualitative* agreement with the results of Schatz and Kupermann.[11] In contrast the $\bar{\sigma}_{0\to 0}(\theta)$ show interesting structure for the higher energies shown in Fig. 5. For E equal to 0.95 eV the $\bar{P}^J_{0\to 0}$ shows a broad maximum at nonzero J. This is a direct consequence of the weak resonance in $P^{\text{CEQB}}_{0\to 0}(E)$. The resonance energy for $P^{\text{CEQB}}_{0\to 0}(E)$ is 0.965 eV, in good agreement with the accurate quantum result.[63] As a result, $P^{\text{CEQB}}_{0\to 0}(E = 0.95\text{ eV})$ has already dropped from its maximum value at somewhat lower energies. Thus, from the CEQB expression for $\bar{P}^J_{0\to 0}$ for the ground state bend, Eq. (66), and for H_3 $E^\ddagger_J = 1.28 \times 10^{-3}\, J(J+1)$ eV we can easily predict that $\bar{P}^J_{0\to 0}$ will show a broad maximum at a nonzero value of J. Clearly, then, in this case the structure in $\bar{P}^J_{0\to 0}$ at $E = 0.95$ eV is due to the resonance. It would seem that the structure in differential cross section at this energy correlates with the resonance. However, $\bar{P}^J_{0\to 0}$ at 0.85 and 0.8 eV shows no structure, yet the corresponding differential cross section do show distinct structure. The 0.8 eV results especially would be difficult to ascribe to the 0.965 eV resonance which has a half-width of approximately 0.03 eV. The shift of the differential cross section to sidewise-peaking at 0.8 eV could be due in large part to the broad range of J-values where $\bar{P}^J_{0\to 0}$ is a maximum. These large J-values correspond physically to nonzero impact parameter collisions which tend to give rise to sidewise scattering. Thus, based on these results, there is some ambiguity in associating structure in the differential cross section with resonances. That is, resonances may be a sufficient but not necessary condition for structure in the differential cross section.

B. $H + H_2(v = 1)$

The CEQB and CEQ expressions for $\bar{Q}_{v\to v'}$ and $k(T)$ were applied to the $H + H_2(v = 1)$ reaction on the PK2 surface. This is an interesting reaction in several respects. First, it was found for the strictly collinear reaction that two vibrationally adiabatic barriers exist,[64–66] symmetrically placed about the saddle point toward the entrance and exit channels. And, significantly, the height of these two barriers exceeds the vibrationally adiabatic energy at the saddle point. Thus, for this reaction, the transition state is not at the saddle point, nor is it unique. There are two, mirror-image transition states located in the entrance and exit channels. This fact has direct consequences on the CEQ and CEQB theories. Both of these use data about the transition state to approximate $\bar{P}^J_{v\to v'}$. The CEQ theory uses the transition state bending and rotational energies, $\varepsilon^\ddagger_{n\Omega}$ and $E^\ddagger_{J\Omega}$, and the CEQB theory uses $E^\ddagger_{J\Omega}$. As mentioned in Section II, in three-dimensional variational transition state theory optimization of the transition state for each bending and rotational state of the transition state should be done. This is obviously a bit cumbersome and

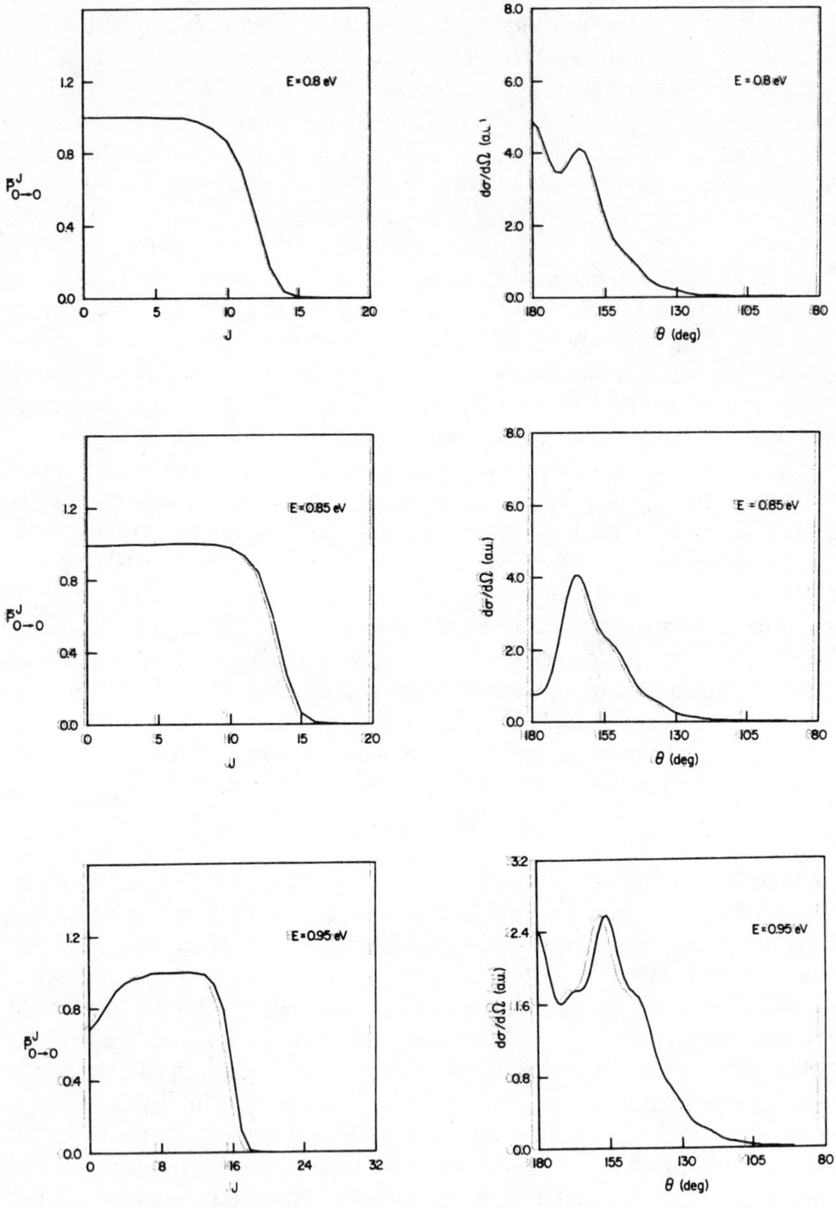

Figure 5. Same as Fig. 4 but for E equal to 0.8, 0.85, and 0.95 eV.

probably not necessary. Instead, it seems reasonable to optimize with respect to the transition state modes of highest frequency first and then to continue to the lower frequency modes in stages of increasing effort. This is because the high frequency modes are more important in relocating the transition state then the low-frequency ones. In the present case the rotational modes of the transition state are the lowest frequency ones and so we did not optimize with respect to them in the CEQB theory. In fact, we evaluated $E^{\ddagger}_{J0}$ at the saddle point rather than at the transition state optimized with respect to the symmetric stretch and bending modes. A simple estimate of the effect of this nonoptimization with respect to the rotational motion indicates that it is a small (i.e., 10% or less) effect over the temperature range considered. A large effect of nonoptimization does occur in the CEQ theory which did *not* use the $\varepsilon^{\ddagger}_{00}$ at the variational transition state but used the $\varepsilon^{\ddagger}_{00}$ at the saddle point. The magnitude of this can best be appreciated by examining the rate constant for the reaction summed over final vibrational states. This is shown in Fig. 6 where previous quantum sudden[67] and quasiclassical trajectory[68] results are also shown. The CEQ rate constant is substantially less than the CEQB one,

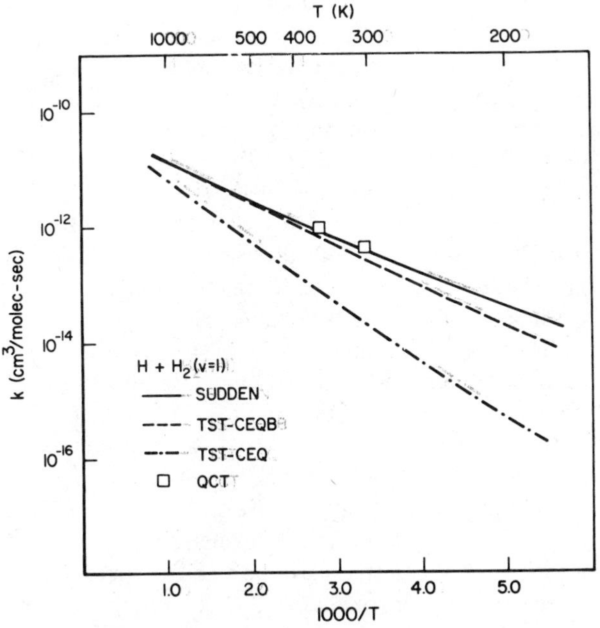

Figure 6. TST, reduced dimensionality CEQ and CEQB, rotational sudden approximation (SUDDEN) (Refs. 67 and 17), quasiclassical trajectory (QCT) (Ref. 68) thermal rate constants for the $H + H_2(v = 1)$ reaction using the PK2 potential.

due primarily to the use of the nonoptimized $\varepsilon_{00}^{\ddagger}$ in the CEQ theory. From the expression

$$\bar{P}_{0\to 0}^{J}(E) = P_{0\to 0}^{\text{CEQ}}(E - E_{J0}^{\ddagger} - \varepsilon_{00}^{\ddagger}) \tag{68}$$

we see that the CEQ approximation to $\bar{P}_{0\to 0}^{J}(E)$ will be an underestimate relative to the CEQB approximation [Eq. (66)] because the shift in E due to $\varepsilon_{00}^{\ddagger}$ is larger than it would be if the smaller variational $\varepsilon_{00}^{\ddagger}$ were used. This underestimation leads of course to a smaller cross section and rate constant. Note that by treating the bending motion adiabatically the CEQB theory automatically incorporates any variational effects that the bending motion may introduce. It also contains additional dynamical effects of the adiabatic bending motion, for example, tunneling modification as discussed for the $H + H_2(v = 0)$ reaction. The lesson here is that if the CEQ theory is to be used, variational effects should be taken into account especially in choosing $\varepsilon_{n\Omega}^{\ddagger}$. Although less important the same remark applies to both the CEQ and CEQB theories with respect to the rotational energies $E_{J0}^{\ddagger}$.

Returning to Fig. 6, we note that there is fairly good agreement between the CEQB and previous rotational sudden and quasiclassical trajectory rate constants, with, however, both the sudden and trajectory results greater than the CEQB ones. At 300 K the differences are about 80% of the CEQB results. Recent j_z-conserving, coupled-states calculations of this reaction by Schatz[69] allow a test of the accuracy of the CEQB calculations. This was done[31] and the results, given in a semilog plot in Fig. 7, show good agreement between the CEQB and CS results. (The CEQB calculations were done with the ground state bend only.) The crossing of the $1 \to 0$ and $1 \to 1$ curves at a total energy of 21.7 kcal/mole is quite interesting. We offer the following explanation for it. As already noted, there are two vibrationally adiabatic barriers for $H + H_2(v = 1)$. These are, however, strictly present only for the vibrationally adiabatic channel $H + H_2(v = 1) \to H_2(v' = 1) + H$. For the vibrationally nonadiabatic channel $H + H_2(v = 1) \to H_2(v' = 0) + H$ it can be argued that only the entrance channel barrier is relevant. That is, there is no exit channel barrier to form the $H_2(v' = 0)$ product. Thus, for total energies less than the $H + H_2(v = 1)$ adiabatic barrier, including the ground state bending energy, tunneling is primarily responsible for reaction. As a result, for these energies the probability to form $H_2(v' = 1)$ should be smaller than to form $H_2(v' = 0)$ because in the former case two barriers must be penetrated whereas in the latter case only one barrier is present. This adiabatic barrier height is approximately 23.0 kcal/mole, in good agreement with the energy where the $\bar{Q}_{1\to 0}$ and $\bar{Q}_{1\to 1}$ cross, 21.7 kcal/mol.

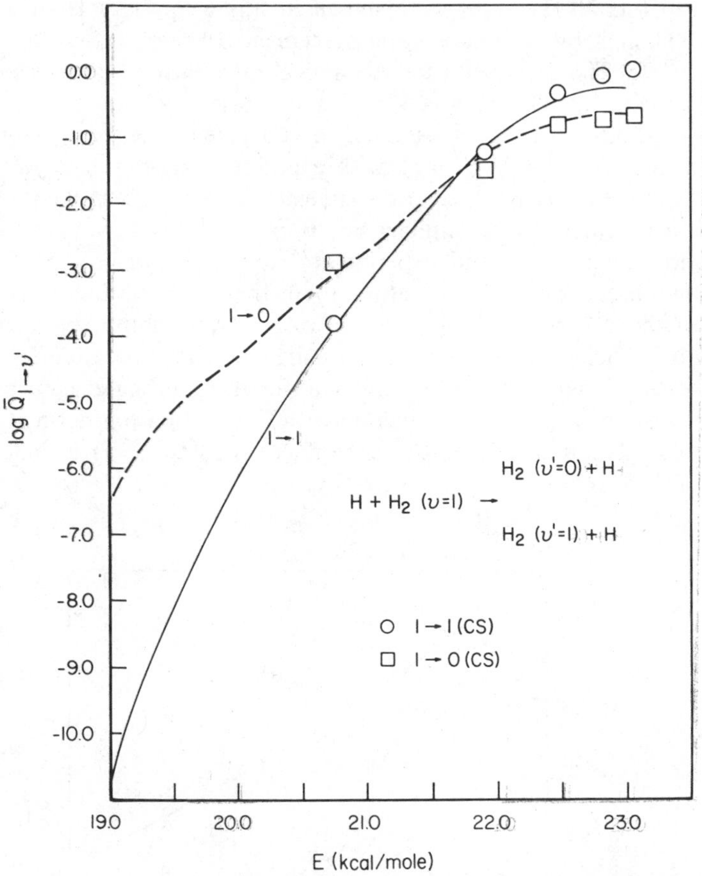

Figure 7. Semilog plot of $\bar{Q}_{1 \to v'}$ versus total energy for $H + H_2(v = 1) \to H_2(v') + H$ using the PK2 potential. The solid and dashed curves are the reduced dimensionality CEQB results and the circles and squares are the centrifugal sudden (CS) results of Schatz (Ref. 69).

C. $D + H_2(v = 0, 1)$

The CEQB expression for the vibrational state-to-state rate constant was used to obtain the temperature dependence of the rate constants for $D + H_2(v = 0)$ and $D + H_2(v = 1)$.[70] The objective in this study was to use the most accurate *ab initio* surface available, to calculate the rate constants using the CEQB theory and to compare with experiment. Based on the previous comparisons between the CEQB and accurate quantum calculations for $H + H_2(v = 0, 1)$ it was felt that the CEQB theory would yield rate constants which are accurate to 20% or so for a given potential surface. The potential surface used was the *ab initio* one of Liu and Siegbahn[59] as fit by Truhlar and Horowitz.[60] The results are shown in Fig. 8 for both $H_2(v = 0)$ and $H_2(v = 1)$. Generally, there is good agreement between theory and experiment[71,72] for $H_2(v = 0)$ with the theoretical rate constant consistently 15% below experiment. This may be significant as discussed later.

By contrast, there is substantial disagreement between theory and experiment[73] for $D + H_2(v = 1)$, with experiment approximately 10 times larger than theory. (Even greater disagreement exists between theory and an earlier experiment.[74]) A similar situation also exists between various theoretical calculations and experiment[75] for $H + H_2(v = 1)$.[76] This has generated discussion in the literature about the source of the disagreement between theory and experiment for $H_2(v = 1)$ with both experiment and theory being held responsible for the disagreement.[76,77] A resolution of this disagreement would also have to account for the good agreement between theory and experiment for the reaction with $H_2(v = 0)$. Considering this, the simplest explanation is that the experiment for $H_2(v = 1)$ is inaccurate.

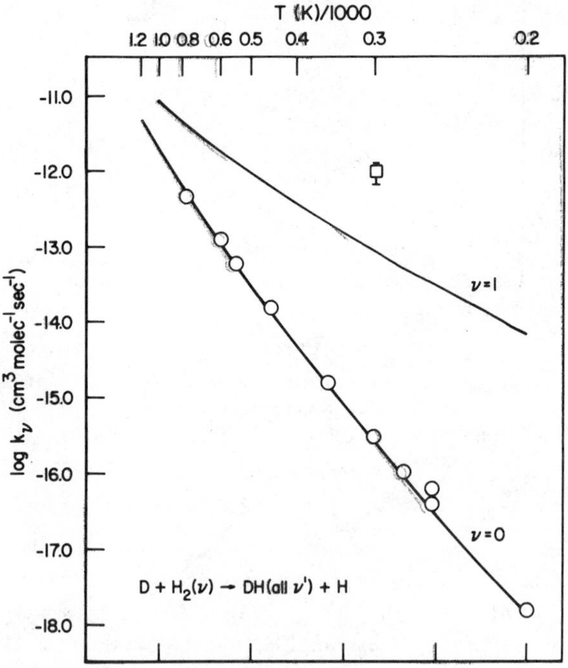

Figure 8. Comparison of reduced dimensionality CEQB rate constants using the LSTH potential (solid curves) with experiment for $D + H_2(v = 0, 1)$. Circles represent the $v = 0$ reaction (Ref. 71, 72) and the square represents the $v = 1$ reaction (Ref. 73).

Another way to rationalize the difference between $H_2(v = 0)$ and $H_2(v = 1)$ comparisons is to suppose that the fitted potential surface is more accurate at the saddle point, which is the transition state for the $H_2(v = 0)$ reaction, than at the vibrationally adiabatic barrier (the variational transition state) for the $H_2(v = 1)$ reaction which is located off the saddle point and towards the entrance channel. An error of roughly plus 1 kcal/mole in the $v = 1$ adiabatic barrier could result in a substantial change in the $D + H_2(v = 1)$ rate constant and possibly bring theory and experiment into good agreement.

Very recently, Liu has reported new *ab initio* CI results for the saddle point and asymptotic energies for $H + H_2$[78] and concludes that the barrier height at the saddle point is between 9.53 and 9.65 kcal/mole, lower by 0.15–0.27 kcal/mole than the LSTH barrier height. This small difference could lead to a slightly higher CEQB rate constant for the $D + H_2(v = 0)$ reaction, that is, an increase of roughly 20% at 300 K. Thus, the slightly lower CEQB rate constants for $D + H_2(v = 0)$ relative to experiment may be due to the slightly higher barrier in the LSTH surface.

D. $F + H_2(v = 0)$

The $F + H_2(v = 0)$ reaction has been studied intensively over the past 10 years, first because of its key role in the HF laser[79] and more recently because of molecular beam experiments[80] which have resolved interesting structure in the angular distribution of the product $HF(v')$. Numerous collinear exact quantum calculations have been reported for this reaction.[81,82] Most of these have used the semiempirical LEPS surface developed by Muckerman[83] (the so-called M5 surface). These calculations found that for total energies E less than 8.0 kcal/mole the major product is $HF(v' = 2)$. Further, the reaction probability $P_{0\to2}^{CEQ}$ shows a resonance at E equal to 6.54 kcal/mole. This resonance has been the subject of much analysis recently.[84–86] For the present discussion we merely wish to take note of it. For energies greater than 7.3 kcal/mole $P_{0\to3}^{CEQ}$ becomes substantial and increases with E while $P_{0\to2}^{CEQ}$ decreases with E. At $E \approx 8.26$ kcal/mole $P_{0\to3}^{CEQ}$ equals $P_{0\to2}^{CEQ}$.

The first coupled channel three-dimensional quantum study of this reaction was done by Redmon and Wyatt[87,88] who reported limited j_z-conserving calculations. They found for $F + H_2(v = 0,\ j = 0)$ that as E increased from 0.35 to 0.50 eV the partial wave reaction probabilities summed over final rotational states, that is, $P^J_{v=0,j=0\to v'=2,\mathrm{all}\,j'}$ and $P^J_{v=0,j=0\to v'=3,\mathrm{all}\,j'}$, displayed interesting behavior with J. The former probability peaked at increasing values of J as E increased whereas the latter one had its maximum for $J = 0$, in the above energy range. This structure was subsequently seen in calculations using the CEQ approximation to the cumulative reaction probabilities $\bar{P}^J_{0\to2}$ and $\bar{P}^J_{0\to3}$ (with the ground state bend).[29] The corresponding differential cross sections were also calculated.[29]

These are shown in Figs. 9 and 10 for E equal to 8.09 and 9.09 kcal/mole. At the lower energy both differential cross sections are smooth and backward-peaked. However, at the higher energy $\bar{\sigma}_{0\to 2}(\theta)$ shows structure and is sidwise-peaked, whereas $\bar{\sigma}_{0\to 3}(\theta)$ is still smooth and backward-peaked. The corresponding CEQ $\bar{P}^J_{0\to 2}$ and $\bar{P}^J_{0\to 3}$ also display quite different behavior with J for E equal to 9.09 kcal/mole, as seen in Fig. 11. $\bar{P}^J_{0\to 2}$ peaks at $J = 13$ whereas $\bar{P}^J_{0\to 3}$ peaks at $J = 0$. The analysis that was used to understand the $\bar{P}^J_{0\to 0}$ in the $H + H_2(v = 0)$ reaction in Subsection A can be used again to

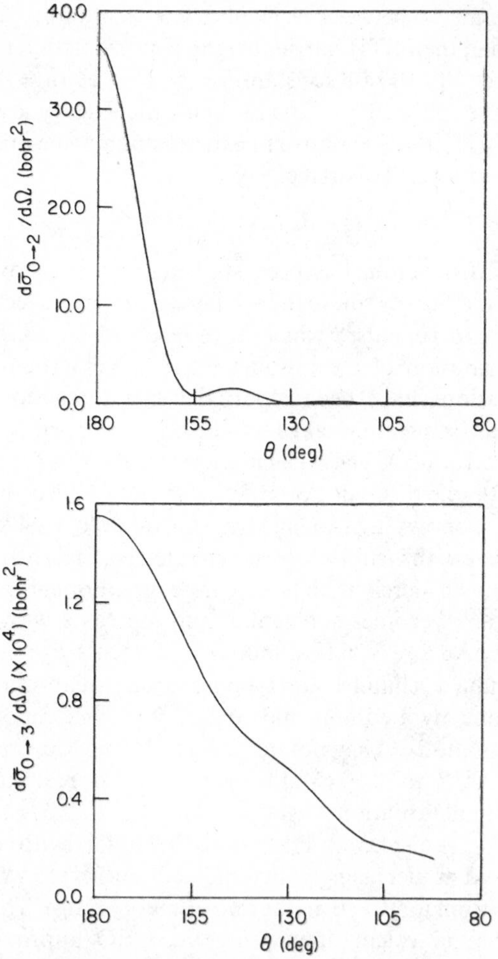

Figure 9. Reduced dimensionality CEQ differential cross sections for $F + H_2(v = 0) \to HF(v' = 2)$ and $HF(v' = 3) + H$ at a total energy E of 8.09 kcal/mole.

understand the peak in $\bar{P}^J_{0\to 2}$. At this total energy $\bar{P}^J_{0\to 2}$ peaks at $J = 13$ because $P^{\rm CEQ}_{0\to 2}(E - E^{\ddagger}_{J0} - \varepsilon^{\ddagger}_{00})$ is a maximum for $J = 13$. That maximum corresponds to the resonance maximum in $P^{\rm CEQ}_{0\to 2}(E)$, which occurs at $E = 6.54$ kcal/mole. A simple analytical expression for the J-values where $\bar{P}^J_{0\to 2}$ is a maximum based on the CEQ approximation to $\bar{P}_{0\to 2}$ can be given. It is based on Eq. (68). Let E_m be the resonance energy for $P^{\rm CEQ}_{0\to 2}$, then the value of J where $\bar{P}^J_{0\to 2}$ is a maximum at a given E is obtained from

$$B^{\ddagger}J(J+1) = E - E_m - \varepsilon^{\ddagger}_{00} \tag{69}$$

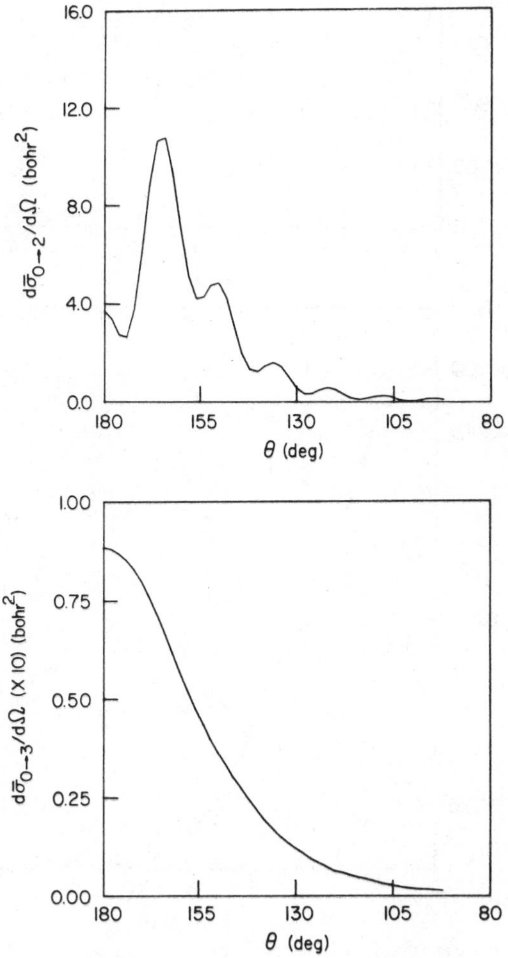

Figure 10. Same as Fig. 9 but for E of 9.09 kcal/mole.

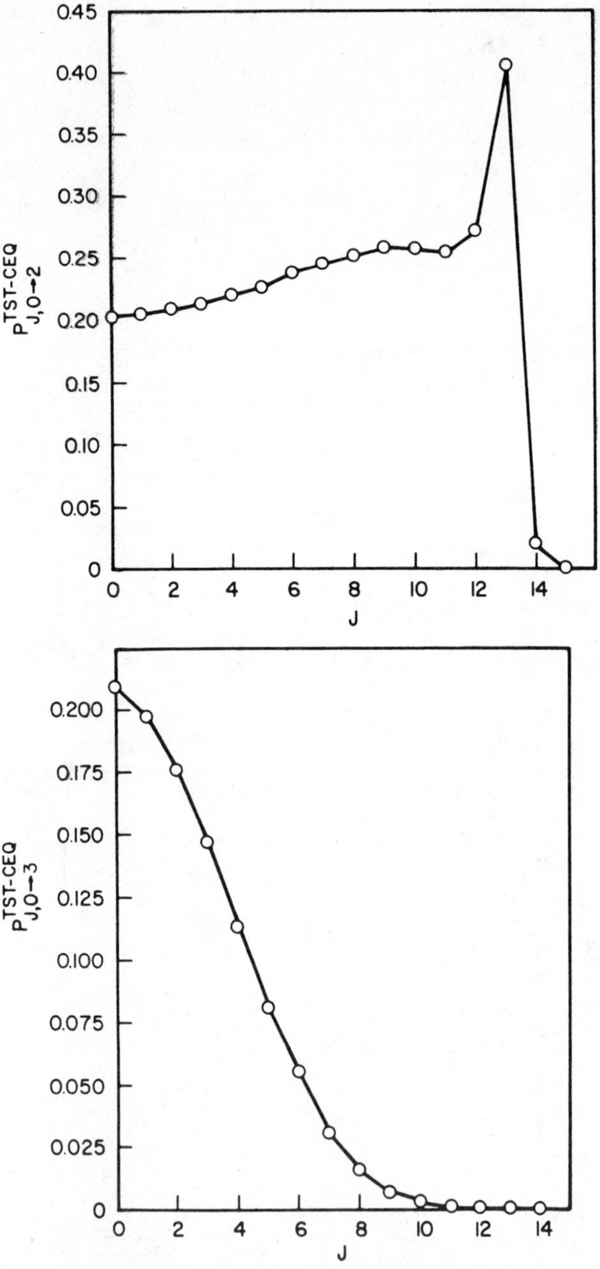

Figure 11. Reduced dimensionality CEQ partial wave reaction probabilities for $F + H_2(v = 0) \rightarrow HF(v' = 2)$ and $HF(v' = 3) + H$ at a total energy E of 9.09 kcal/mole.

Using the values appropriate for FH_2 at the saddle point of the M5 surface, $\varepsilon_{00}^{\ddagger} = 1.25$ kcal/mole, and $B^{\ddagger} = 6.67 \times 10^{-3}$ kcal/mole we find that for $E = 9.09$ kcal/mole, $J = 13$. (An identical expression with very similar constants has been given by Pollak and Wyatt,[89] based on an adiabatic model which is similar in spirit to the CEQ approximation to $\bar{P}^J_{v \to v'}$.) At $E = 8.09$ kcal/mole $\bar{P}^J_{0 \to 2}$ peaks near $J = 0$ and $\bar{P}^J_{0 \to 3}$ is a maximum at $J = 0$ and both corresponding differential cross sections are backward-peaked as already seen in Fig. 9. Thus, there is a predicted strong sensitivity in the behavior of the differential cross section with E for HF($v = 2$).

The CEQ approximation to $\bar{P}^J_{0 \to v'}$ and the differential cross sections should be regarded as semiquantitative because both our experience with CEQ/CEQB calculations on $H + H_2$ and comparison with the j_z-conserving calculations indicate that CEQ resonance features in reaction probabilities can be considerably broadened due to the adiabatic bending motion of the transition state. This has also been found in the recent BCRLM calculations of Hayes and Walker on the $F + H_2$, D_2, and HD systems.[56]

Differential cross sections for $F + H_2$ have also been calculated using the rotational infinite order sudden[90] and distorted wave Born approximations.[91] These also show sensitivity of the differential cross sections with E.

Finally, consider the rotationally averaged and summed integral cross sections using the CEQ approximation to $\bar{P}^J_{0 \to v'}$ for products HF($v' = 2$) and HF($v' = 3$). These are plotted in Fig. 12 versus the total energy E. As seen the HF($v' = 2$) product dominates the HF($v' = 3$) one for E less than 13 kcal/mole. However, the two cross sections are equal at 13.5 kcal/mole and for E above that value $\bar{Q}_{0 \to 3}$ rapidly exceeds $\bar{Q}_{0 \to 2}$. These results are in qualitative accord with rotational IOS calculations of Jellinek et al.[92] for the energies they considered (E less than 11.6 kcal/mole) and at lower energies with the j_z-conserving results of Redmon and Wyatt.[87]

In the next section we consider the F + HD reaction where again the CEQ approximation is used to obtain $\bar{P}^J_{0 \to v'}$ and differential cross sections.

E. F + HD($v = 0$)

A detailed analysis of the CEQ reaction probabilities, scattering matrices, time delays and eigenphase-shifts in the F + HD($v = 0$) → HF(v') + D, DF(v') + H reactions on the M5 surface was done recently by Kuppermann and Kaye.[85] They found and analyzed a narrow resonance in F + HD($v = 0$) → HF($v' = 2$) + D at a total energy of 0.245 eV with a width of 0.52×10^{-3} eV. By contrast the F + DH($v = 0$) → DF($v' = 3$) + H reaction has no resonance in the energy range examined up to 0.29 eV. (It does have a gentle maximum at $E = 0.28$ eV which was not attributed to a resonance.) The analysis just carried out for the $F + H_2$ reaction can be repeated for these reactions. The expectation is that $\bar{P}_{0 \to 2}$ for the reaction to form HF($v' = 2$)

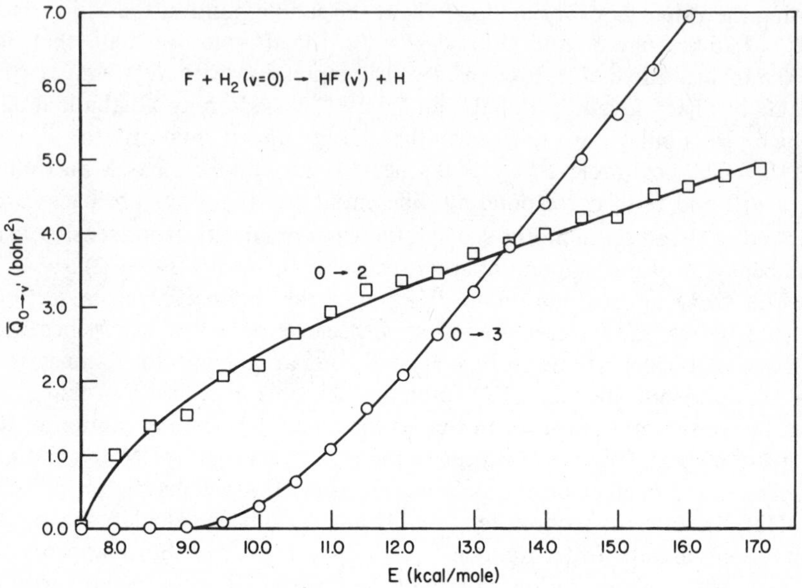

Figure 12. Reduced dimensionality CEQ cross sections for $F + H_2(v = 0) \rightarrow HF(v') + H$ versus the total energy E.

could exhibit a sharp maximum with J whereas $\bar{P}_{0\rightarrow 3}$ should be a smooth monotonically decreasing function of J. This is indeed borne out, at for example a total energy of 0.33 eV, as seen in Fig. 13. There the $\bar{P}^J_{0\rightarrow v'}$ and the corresponding differential cross sections are given. Note that the value of J where $\bar{P}^J_{0\rightarrow 2}$ is a maximum for F + HD is 14, as determined from Eq. (69). The corresponding differential cross section shows highly oscillatory structure and is both forward- and backward-peaked. Not surprisingly for this reaction, it has much of the character of $|P_{J=14}(\cos\theta)|^2$. This is because the $J = 14$ partial wave is the dominant one in the CEQ expression for the differential cross section. This is of course just a reflection of the narrowness of the CEQ resonance. An additional consequence of this sharp maximum in $\bar{P}^J_{0\rightarrow 2}$ is that, as E varies, the value of the dominant partial wave varies and the differential cross section would simply reflect the J-value where $\bar{P}^J_{0\rightarrow 2}$ is a maximum and take on the character of $|P_J(\cos\theta)|^2$. Again we caution that because these results, especially for resonant behavior, are based on the CEQ expressions for $\bar{P}^J_{v\rightarrow v'}$ and $\bar{\sigma}_{v\rightarrow v'}(\theta)$, they should be regarded as model results. Indeed we would anticipate a broadening of this resonance in the CEQB approximation.

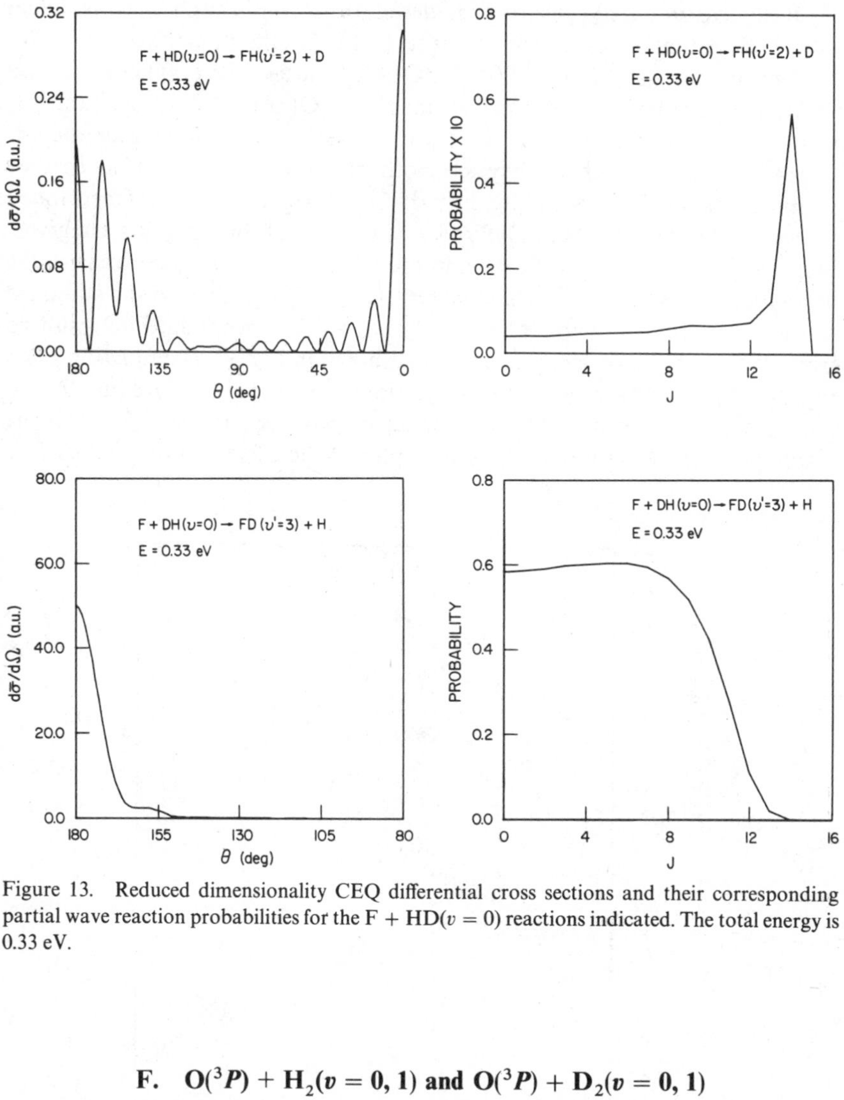

Figure 13. Reduced dimensionality CEQ differential cross sections and their corresponding partial wave reaction probabilities for the F + HD($v = 0$) reactions indicated. The total energy is 0.33 eV.

F. $O(^3P) + H_2(v = 0, 1)$ and $O(^3P) + D_2(v = 0, 1)$

CEQB and CEQ rate constants and reaction probabilities have been presented for the $O(^3P) + H_2(v = 0, 1)$ and $O(^3P) + D_2(v = 0, 1)$ reactions in a series of papers considering five potential energy surfaces.[93,94] Comparisons of these rate constants and reaction probabilities with analogous quasiclassical trajectory, transition state theory calculations, and experiment for each surface was done. Some of the highlights of this work are reviewed here.

Of the five potential surfaces considered, the *ab initio* surfaces of Walch et al.[95] and Schinke and Lester[96] appear to be the most accurate. We concentrate on the former, "MODPOLCI", surface here. There are two electronically adiabatic reactive surfaces for $O(^3P) + H_2$, of A' and A'' symmetry for noncollinear geometries. These surfaces are degenerate for collinear geometries where they are both of $^3\Pi$ symmetry and hence they exhibit a Renner-Teller coupling.[97] In the latest paper in the series mentioned and in a separate study the details of the inclusion of this coupling are given. Essentially, the major modification in the CEQ and CEQB theories is to the adiabatic bending energies which were calculated according to the standard perturbation theory treatment of the Renner-Teller coupling. The resulting ground state adiabatic bending energy is shown in Fig. 14 as a function of an angle ϕ which can be regarded as a reaction path coordinate; ϕ equal 0° and 90° correspond to reactants and products, respectively, and $\phi = 37.33°$ is the saddle point. An equipotential contour plot of the effective potential surface

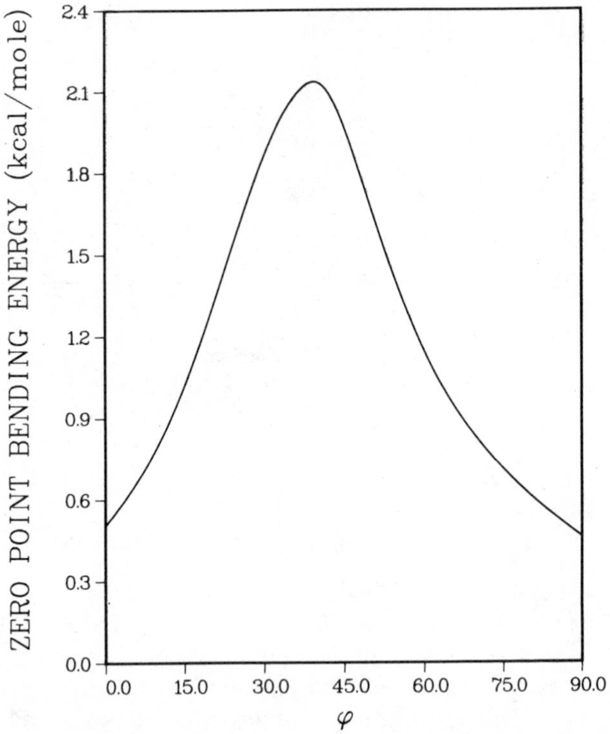

Figure 14. Adiabatic ground-state bending energy for $O(^3P) + H_2$ as a function of a reaction progress angle ϕ. ϕ equal to 0° and 90° corresponds to reactants and products, respectively.

consisting of the collinear MODPOLCI surface plus the ground state adiabatic bending energy is shown in Fig. 15 as a function of coordinates, Q_1 and Q_2, which are mass-scaled modifications of R and r, respectively. For $O(^3P) + H_2(v = 0, 1)$ and $O(^3P) + D_2(v = 0, 1)$ the skew angles between Q_1 and Q_2 are 46.7° and 48.2°, respectively. The barrier height of the collinear MODPOLCI surface is 12.58 kcal/mole.

CEQB reaction probabilities were calculated for total energies up to 30 kcal/mole for the H_2 reaction and 28 kcal/mole for the D_2 reaction. These were used to obtain transmission coefficients, Γ_0^{CEQB} and Γ_1^{CEQB}, for H_2 and D_2. The transmission coefficients are shown in Fig. 16. Before discussing that figure in detail, note that if transition state theory (at the saddle point) were exact, Γ_0^{CEQB} and Γ_1^{CEQB} would be unity at all temperatures. Note that variational versions of transition state theory locate the transition state at the

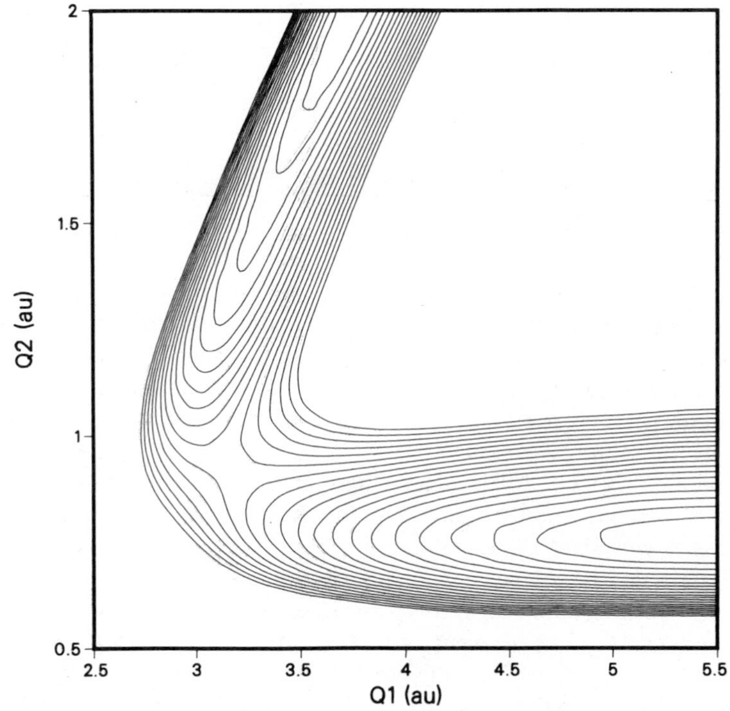

Figure 15. Equipotential contours of the effective potential surface for $O(^3P) + H_2$ consisting of the MODPOLCI *ab initio* surface plus the ground-state bending energy. Q_1 and Q_2 are mass-scaled coordinates proportional to r, the H_2 internuclear distance, and R, the distance of O to the center of mass of H_2, respectively. The energy contours are separated by 1.0 kcal/mole up to 20.0 kcal/mole.

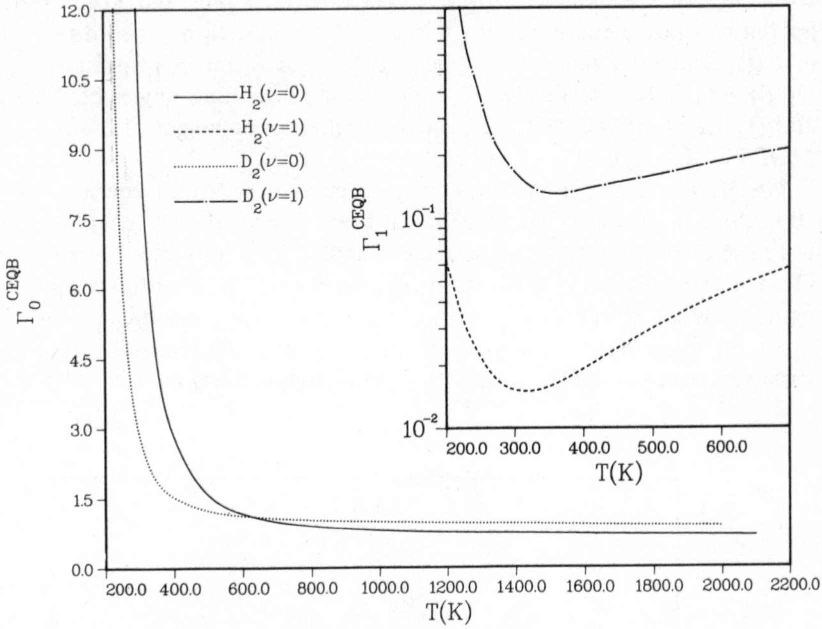

Figure 16. CEQB transmission coefficients Γ_0^{CEQB} and Γ_1^{CEQB} for $O(^3P) + H_2$ and $O(^3P) + D_2$ versus T.

maximum adiabatic barrier. For these reactions this barrier is, with the exception of the $O + H_2(v = 1)$ reaction, at the saddle point. When the CEQB transmission coefficient is used to obtain 3D rate constants, the choice of transition state is irrelevant as long as it is the same for the 3D TST rate constant it modifies.

As seen in Fig. 16 Γ_0^{CEQB} is greater than unity for T less than 600 K for the $H_2(v = 0)$ reaction and for T less than 900 K for the $D_2(v = 0)$ reaction. At 300 K it is considerably larger for H_2 than for D_2. This behavior is the expected result for isotopically different reactions. There is more tunneling for the small mass system, that is, H_2, and less re-crossing for the larger skew angle system, D_2 in the present case. The extent of recrossing becomes an important factor in the transmission coefficient at temperatures where a significant fraction of the reaction goes by the over-the-barrier route. Because of its slightly smaller skew angle the transmission coefficient for the $H_2(v = 0)$ reaction dips below the one for $D_2(v = 0)$ at 600 K and at 1600 K it is 22% smaller than the $D_2(v = 0)$ transmission coefficient.

The transmission coefficients for the vibrationally excited state reactions of H_2 and D_2 are also shown in Fig. 16. The transmission coefficients are qualitatively similar to each other and both are qualitatively different from the corresponding ones for the ground state reaction. First, consider the transmission coefficient for the $H_2(v = 1)$ reaction. It is substantially less than unity over the temperature range shown. This is not just a re-crossing effect. It is a consequence of the fact that the maximum adiabatic barrier, which controls the threshold behavior of P_1^{CEQB} is considerably greater than the adiabatic energy at the saddle point. There is of course some tunneling in this reaction and that is what is responsible for the upturn in the transmission coefficient at approximately 300 K.

Consider next the transmission coefficient for the $O + D_2(v = 1)$ reaction. It is considerably larger than the one for $O + H_2$ though still considerably less than unity. This result would be unexpected if the maximum adiabatic barrier were at the saddle point. However, this result is consistent with the maximum adiabatic barrier having a location and height somewhere between the values at the saddle point and the values for the maximum adiabatic barrier of $O + H_2$. Re-crossing, which also reduces the transmission coefficient, is significant because P_1^{CEQB} is always less than unity. At lower temperatures, tunneling causes the transmission coefficient to increase.

Re-crossing is also important for $O + H_2(v = 1)$. In fact, there is more re-crossing for this reaction than for its D_2 counterpart (again because of the slightly smaller skew angle for the H_2 reaction). However, while re-crossing is the dominant effect in D_2, for H_2 the shift in the adiabatic barrier off the saddle point has a comparable or even greater effect in reducing the transmission coefficient.

These transmission coefficients multiply the conventional 3D TST rate constant to obtain a final calculated rate constant. This CEQB rate constant can be directly compared to experiment and to the analogous CEQ rate constant. The resulting thermal and vibrationally excited rate constants are displayed as a function of inverse temperature in Fig. 17. The insert is a blowup of the high temperature region. The experimental results for these rate constants are also displayed. Most of the measurements come from a review by Westburg and Cohen,[98] but more recent work is also indicated, including the measurement of the vibrationally excited rate constant by Light.[99] For the thermal rate constant, the CEQB rate constant is somewhat higher than the CEQ one at all temperatures. The maximum difference occurs at the lowest temperatures where CEQB rate constants are about 60% higher than those for CEQ. The overall agreement between the CEQB and the experimental rate constants is very good. At temperatures of 2000 K or more, the calculated results pass through the cluster of experimental values, in contrast to the CEQ ones which appear to be too low. At the lower temperatures the calculated CEQB values may be slightly high.

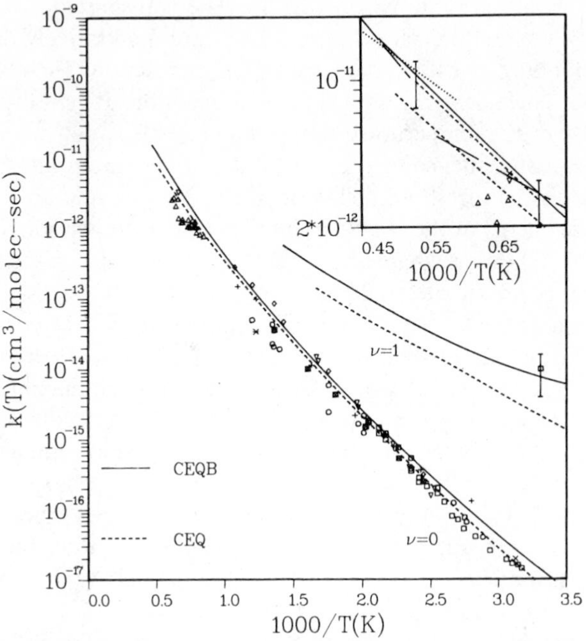

Figure 17. CEQB and CEQ rate constants for $O(^3P) + H_2(v = 0, 1)$. The insert is an expansion at high temperatures. Experimental results for $H_2(v = 0)$ are taken from Ref. 98 and for $H_2(v = 1)$ from Ref. 99.

For the vibrationally excited rate constant, the CEQB result is in better agreement than the CEQ one with the experimental rate constant. The CEQB transmission coefficients are larger than those for CEQ for two reasons: a decrease in bending energy because of a shift of the vibrationally adiabatic barrier from the saddle point toward the entrance channel, and an increase in tunneling at lower temperatures. As a result the CEQB rate constant decreases less rapidly with decreasing temperature and passes through the error bars of the one available experiment while the CEQ rate constant falls below the error bars.

The isotope effect in this reaction was also studied. Four measurements of the isotope effect on deuterium substitution are available[100–103] and the measured ratios of the H_2 to D_2 rate constant are shown in Fig. 18. All four measurements are for thermal rate constants. The error bars for the lower temperature measurements come from the published error bars on the individual rate constants and the assumption that the relative error in the ratio is the sum of the relative error in the numerator and the denominator.

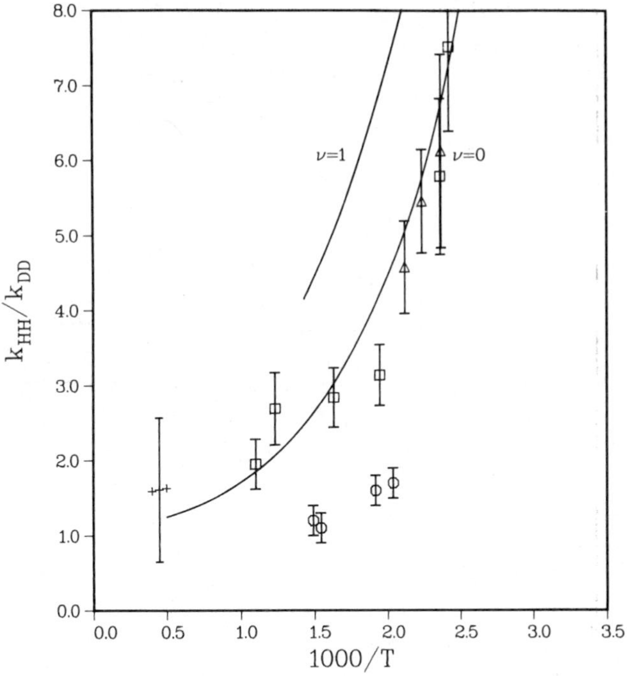

Figure 18. Ratio of CEQB and experimental rate constants k_{HH}/k_{DD}, plus signs from Refs. 100 and 104, circles from Ref. 101, squares from Ref. 102, and triangles from Ref. 103.

The highest temperature measurements were not published with error bars. However, in a subsequent report, the authors conservatively ascribe 30% errors to each rate constant in the ratio.[104] This makes the error bars on the ratio on the order of 60% by the same assumption used for the other measurements.

The calculated CEQB results for both the thermal and the vibrationally excited isotope effects are also displayed in Fig. 18. The agreement between theory and most of the experiments for the thermal result is quite good. The calculated results are 20% lower than the measured value at high temperatures but well within the rather generous error bars mentioned. At the lower temperatures the calculations are in very good agreement with the measurements of Westenberg and de Haas[102] and Presser and Gordon.[103] In particular, for temperatures below 500 K, both theory and experiment agree on a very steep temperature dependence. At intermediate temperatures (500–1000 K), the calculations are similar to the Westenberg and de Haas values with the measured values scattered about the theoretical curve. The experimental results of Clyne and Thrush[101] are qualitatively different at all temperatures from those of the calculations or other experiments.

The conventional 3D TST isotope effects are quite different from the CEQB ones. Although not shown in Fig. 18, the TST curve is nearly a straight line that passes through the high temperature values of Ref. 100 and through the value for the third highest temperature of Ref. 102. At low temperatures the TST isotope effects are much lower than the CEQB ones but at the highest temperatures shown they are greater than the CEQB ones. At low temperatures, the differences between the TST and CEQB isotope effects is due to the increase in the H_2 rate constant over the D_2 rate constant due to tunneling, which is only included in the CEQB calculations. Inclusion of the Wigner tunneling correction in the TST calculation will improve the low-temperature agreement, but the temperature dependence will still be too shallow. At the lowest temperature for which there is a measurement, the Wigner-corrected TST will still be too low and outside the experimental error bars. At high temperatures, the difference between CEQB and TST isotope effects is due to the differences in re-crossing in the H_2 and D_2 systems which is seen quantum mechanically and is consistent with skew angle arguments. Such re-crossing effects are not present in the TST calculations, with or without a Wigner correction.

The isotope effect for the vibrationally excited rate constant is calculated to be even larger than that for the thermal rate constant. At room temperature where the only vibrationally excited rate constant measurement has been made, the D_2 rate is predicted to be about 30 times slower. It is important to note that the conventional 3D TST isotope effect would have the D_2 rate 300 times slower, or a factor of 10 slower than the CEQB rate. The reason for this is that the vibrationally adiabatic barrier is considered in conventional TST to remain at the saddle point. With this restriction the vibrationally adiabatic barrier to reaction decreases much more rapidly for H_2 than for D_2 upon initial vibrational excitation. However, the vibrationally adiabatic barrier actually moves out into the reactant channel with increasing vibrational energy. As a consequence the vibrationally adiabatic energy at the barrier is more similar to the initial vibrational energy of the reactants and so the adiabatic barrier to reaction decreases in a more similar fashion for both H_2 and D_2. This effect is automatically included in the transmission coefficients to produce the less extreme isotope effects seen in Fig. 18. Variational TST would also describe this effect, qualitatively.

VI. SUMMARY AND OUTLOOK

We have reviewed a hierarchy of reduced dimensionality quantum theories of reactive scattering for atom–diatom systems. They are all based on treating the three-atom bending motion adiabatically, while treating the remaining two internal degrees of freedom in a fully coupled fashion. The highest level of approximation makes no further approximations and results in a set of two-mathematical dimensional partial differential equations for each partial wave J, the solutions of which yield scattering matrices and corresponding partial wave vibrational state-to-state, rotationally cumulative reaction probabilities, $\bar{P}^J_{v \to v'}(E)$. A substantial simplification of this fully adiabatic approximation results by replacing the J-dependent centrifugal potential by a constant value given by its value at the transition state, $E^{\ddagger}_{J\Omega}$. In the resulting theory, termed the CEQB theory, the vibrational state-to-state, rotationally cumulative partial wave reaction probability at the total energy E is given by the CEQB reaction probability at the energy $E - E^{\ddagger}_{J\Omega}$. In the next level of approximation the adiabatic bending energy $\varepsilon_{n\Omega}(r, R)$ is replaced by its value at the transition state, $\varepsilon^{\ddagger}_{n\Omega}$. In this level of approximation, termed the CEQ theory, the vibrational state-to-state, rotationally cumulative partial wave reaction probability at the total energy E is given by the CEQ reaction probability at the energy $E - E^{\ddagger}_{J\Omega} - \varepsilon^{\ddagger}_{n\Omega}$. It was then shown how further approximations lead to the familiar one-dimensional reaction path theory and finally to transition state theory itself. Thus, by starting from a rigorous formulation of reactive scattering, it was shown how a hierarchy of approximations could be developed leading finally to transition state theory.

The CEQB approximation was tested against accurate quantum rotationally averaged and summed integral cross sections and rate constants for the $H + H_2(v = 0)$ reaction and found to be quite accurate. Somewhat less accuracy was found for the CEQ approximation.

Applications of the CEQB and CEQ theories to $H + H_2(v = 0, 1)$, $D + H_2(v = 0, 1)$, $F + H_2(v = 0)$, $F + HD(v = 0)$, $O(^3P) + H_2(v = 0, 1)$, and $O(^3P) + D_2(v = 0, 1)$ were reviewed and the results, integral and differential cross sections and rate constants, were compared with other calculations and experiments.

This question of correlating the adiabatic bending states to the asymptotic free-rotor states was examined and some ad hoc suggestions were made to project the former onto the latter. The suggestion was based on the physically motivated idea that the change in character of the free-rotor wave functions to localized hindered-rotor wave function is abrupt in the atom–diatom separation distance. A good diagnostic criterion to characterize this change is clearly needed and awaits future work.

In fact, the correlation question is part of the issue surrounding the validity of the bending adiabatic approximation itself. Although there is ample numerical evidence that it is valid, at least in the vicinity of the reaction threshold energy, simple time-scale arguments would seem to invalidate the approximation. Those arguments compare the collision time with the frequency of the adiabatic bending motion and conclude that there is not enough time for even one period of bending motion. Taking note of this, Schatz, however, has argued that the uncertainty principle is a constraint on the dynamics that induces zero-point energy, time-scales notwithstanding.[105] This seems to suggest that only the ground bending state exists and that the existence of excited bending states must fall back (and fall down) on time-scale arguments. However, that might not necessarily follow either. Indeed there is substantial evidence from studies of coupled vibrational motion in bound states that the adiabatic approximation works surprisingly well for the ground and excited states even for cases where the frequencies are equal.[106–110] Clearly, in these cases time-scale arguments seem to be playing a smaller role than initially anticipated. Perhaps an explanation for this surprising success of the adiabatic approximation lies in an examination of the approximation within a time-independent framework. In that framework the validity of the adiabatic approximation depends on the variation of the adiabatic wave function with respect to the variables held fixed. For the adiabatic approximation to be valid that variation should be small relative to the variation of that part of the total wave function which depends explicitly on those variables. Obviously, the adiabatic approximation is exact for separable systems independent of the time scales. Similarly, even if the time scales appear to be favorable the approximation may break down if there is strong coupling between the slow and fast modes. So it would seem that both time scales and coupling strength need to be examined to assess the validity of the adiabatic approximation. And perhaps the former is a more important criterion than the latter.

A concrete and familiar example which illustrates the importance of coupling on adiabaticity is the effect of reaction-path curvature on vibrational nonadiabaticity. It is well known that curvature of the reaction path is largely responsible for vibrational nonadiabaticity.[111] Reaction models with no curvature show substantial vibrational adiabaticity in spite of time-scale arguments. Returning now to the question of adiabaticity in the bending motion we can ask about the magnitude of the coupling of that motion to the other degrees of freedom in the problem. For reactions with a collinear reaction path, such as we have considered, it can be argued that the bending motion is only weakly coupled to the two other degrees of freedom. That is simply due to the fact that along the reaction coordinate, γ remains constant, either 0 or π. Thus, on the basis of the weak coupling between γ and the reaction path we could argue for the adiabatic theory. On the other hand as mentioned already there is a substantial change in the character of the γ-wave function as it changes from a free-rotor function to a localized, oscillator-type wave function. If this transition is abrupt then it could be imagined that except where the transition occurs the adiabatic picture could be valid away from this localized region. This issue was already raised in connection with the correlation of bending eigenfunctions with asymptotic free-rotor states. Clearly, they are related and will receive more attention in the future.

Another question awaiting future work is the extension of the reduced dimensionality theories to more complex reactions. The idea that only a small number of degrees of freedom need to be explicitly coupled together certainly seems general. The details of how to implement that generally are just currently being worked on.[112] In certain cases it may be obvious which degrees of freedom to couple. As in the simpler atom–diatom case the question of the correlation of the hindered bending eigenfunctions to the asymptotic or near-asymptotic rotor states will arise. Thus, any progress made in the atom–diatom case will be of value in the general case.

Finally, their are many possible extensions of these ideas to any activated process and we hope to make contributions to some of these in the future.

VII. POSTSCRIPT

Very recently high-resolution crossed molecular beam experiments on the $F + H_2$, $F + D_2$, and $F + HD$ have been completed.[113,114] These indicate several important new features of these reactions which will require a revised potential energy surface and new dynamics calculations. Perhaps the most dramatic new result is that for $F + H_2$ the $HF(v' = 3)$ channel appears at the same energy as the $HF(v' = 2)$ one. This is in contradiction with the delayed onset of the former channel which was predicted using the semiempirical Muckerman 5 potential energy surface. In addition the $v' = 3$ channel shows substantial forward scattering, in qualitative disagreement with the results shown in Figs. 9 and 10 and seen in Refs. 56, 90, and 91. Some preliminary results on new potential surfaces have already begun to appear[57,114,115] and we can look forward to further theoretical work in the near future.

Acknowledgments

It is a pleasure to acknowledge the co-authors of the papers reviewed here. They are, Ki Tung Lee, Guan-Zhi Ju, Albert F. Wagner, and Robert B. Walker. I also thank John C. Light, William H. Miller, George C. Schatz, Donald G. Truhlar, and Robert E. Wyatt for stimulating discussions and correspondence.

The financial support of the National Science Foundation in the very early stages of this work is gratefully acknowledged. Most of the applications reviewed were supported in part by the Department of Energy, Office of Basic Energy Sciences, under contract DOEDE-AC02-81ER10900.

References

1. H. Pelzer and E. Wigner, *Z. Physik. Chem.* **B15**, 445 (1932).
2. H. Eyring, *J. Chem. Phys.* **3**, 107 (1935).
3. H. Johnston, *Gas Phase Reaction Rate Theory*, Ronald Press, New York, 1966.
4. M. A. Ellison and J. O. Hirschfelder, *J. Chem. Phys.* **30**, 1426 (1959).
5. J. Ross and P. Mazur, *J. Chem. Phys.* **35**, 19 (1961).
6. R. A. Marcus, *J. Chem. Phys.* **45**, 2138, 2630 (1966).
7. K. Morokuma, B. C. Eu, and M. Karplus, *J. Chem. Phys.* **51**, 5193 (1970).
8. S. H. Lin, K. H. Lau, and H. Eyring, *J. Chem. Phys.* **55**, 5657 (1971).
9. D. G. Truhlar and A. Kuppermann, *J. Am. Chem. Soc.* **73**, 1840 (1971).
10. See, for example, J. M. Bowman and A. Kuppermann, *Chem. Phys. Lett.* **12**, 1 (1971).
11. G. C. Schatz and A. Kuppermann, *J. Chem. Phys.* **65**, 4668 (1976).
12. P. Pechukas, in *Dynamics of Molecular Collisions*, Part B, W. H. Miller, ed., Plenum Press, New York, 1976, Chapter 6.
13. W. H. Miller, *Acc. Chem. Res.* **9**, 306 (1976).
14. A. Kuppermann, *J. Phys. Chem.* **83**, 171 (1979).
15. D. G. Truhlar and B. C. Garrett, *Acc. Chem. Res.* **13**, 440 (1980).
16. D. G. Truhlar, W. L. Hase, and J. T. Hynes, *J. Phys. Chem.* **87**, 2664, 5223 (1983).
17. J. M. Bowman, G.-Z. Ju, and K.-T. Lee, *J. Phys. Chem.* **86**, 2238 (1982).
18. J. M. Bowman, G.-Z. Ju, and K.-T. Lee, *J. Chem. Phys.* **75**, 5199 (1981).
19. A. B. Elkowitz and R. E. Wyatt, *J. Chem. Phys.* **62**, 2504 (1975).
20. R. B. Walker, E. B. Stechel, and J. C. Light, *J. Chem. Phys.* **69**, 2922 (1978).
21. B. C. Garrett and D. G. Truhlar, *Proc. Natl. Acad. Sci. U.S.A.* **76**, 4755 (1979).
22. E. M. Mortensen, *J. Chem. Phys.* **48**, 4029 (1968).
23. J. N. L. Connor, *Comp. Phys. Commun.* **17**, 117 (1979).
24. R. E. Wyatt, in *Atom-Molecule Collision Theory: A Guide for the Experimentalist*, R. B. Bernstein, ed., Plenum, New York, 1979.
25. R. B. Walker and J. C. Light, *Annu. Rev. Phys. Chem.*, **31**, 401 (1980).
26. G. C. Schatz, in *Potential Energy Surfaces and Dynamics Calculations*, D. G. Truhlar, ed., Plenum, New York, 1981.
27. M. Baer, *Adv. Chem. Phys.* **49**, 191 (1982).
28. M. Baer, in *Molecular Collision Dynamics*, J. M. Bowman, ed., Springer-Verlag, Berlin, Heidelberg, 1983.
29. J. M. Bowman, K.-T. Lee, and G.-Z. Ju, *Chem. Phys. Lett.* **86**, 5199 (1981).
30. K.-T. Lee and J. M. Bowman, *J. Phys. Chem.* **86**, 2289 (1982).
31. J. M. Bowman and K.-T. Lee, *Chem. Phys. Lett.* **74**, 363 (1983).
32. R. E. Wyatt, *J. Chem. Phys.* **51**, 3489 (1969).
33. J. N. L. Connor and M. S. Child, *Mol. Phys.* **18**, 653 (1970).
34. R. B. Walker and E. F. Hayes, *J. Phys. Chem.* **87**, 1255 (1983).
35. W. H. Miller and S. Schwartz, *J. Chem. Phys.* **77**, 2378 (1982).
36. S. D. Schwartz and W. H. Miller, *J. Chem. Phys.* **79**, 3759 (1983).
37. R. T. Skodje and D. G. Truhlar, *J. Chem. Phys.* **79**, 4882 (1983).
38. R. A. Marcus, *J. Chem. Phys.* **45**, 4493, 4500 (1966).

39. G. L. Hofacker, *Z. Naturforsch.* **A18**, 607 (1963).
40. W. H. Miller, N. C. Handy, and J. E. Adams, *J. Chem. Phys.* **72**, 99 (1980).
41. C. F. Curtiss, J. O. Hirschfelder, and F. T. Adler, *J. Chem. Phys.* **18**, 1638 (1950).
42. R. T Pack, *J. Chem. Phys.* **60**, 633 (1974).
43. G. C. Schatz and A. Kuppermann, *J. Chem. Phys.* **65**, 4642 (1976).
44. A. B. Elkowitz and R. E. Wyatt, *Mol. Phys.* **31**, 189 (1976).
45. A. Kuppermann, G. C. Schatz, and J. P. Dwyer, *Chem. Phys. Lett.* **45**, 71 (1977).
46. S. H. Harms and R. E. Wyatt, *J. Chem. Phys.* **62**, 3173 (1975).
47. See, for example, C. H. Townes and A. L. Schalow, *Microwave Spectroscopy*, McGraw-Hill, New York, 1955, Chapter 3.
48. B. C. Garrett and D. G. Truhlar, *J. Chem. Phys.* **79**, 4931 (1983).
49. E. Pollak and P. Pechukas, *J. Chem. Phys.* **70**, 325 (1979).
50. J. M. Bowman and K.-T. Lee, *J. Chem. Phys.* **72**, 5071 (1980).
51. V. Khare, D. J. Kouri, and M. Baer, *J. Chem. Phys.* **71**, 1188 (1979).
52. M. Berry, *Chem. Phys. Lett.* **27**, 73 (1974).
53. U. Halavee and M. Shapiro, *J. Chem. Phys.* **64**, 2826 (1976).
54. G. C. Schatz and J. Ross, *J. Chem. Phys.* **66**, 1037 (1977).
55. J. M. Bowman, K.-T. Lee, H. Romanowski, and L. B. Harding, in Proceedings of Symposium on *Electron-Molecule Scattering, van der Waal's Complexes, and Reactive Chemical Dynamics*, D. G. Truhlar, ed., ACS Symposium Series, 1984, Chap. 4.
56. E. F. Hayes and R. B. Walker, Reactive differential cross sections in the rotating linear model: Reactions of fluorine atoms with hydrogen molecules and their isotopic variants, *J. Phys. Chem*, submitted.
57. R. B. Walker, N. C. Blais, and D. C. Truhlar, Dependence of Reaction Attributes, Including Differential Cross Sections and Resonance Feature, on Changes in the Potential Energy Surface for the F + D_2 Reaction, *J. Chem. Phys.*, to be published.
58. R. N. Porter and M. Karplus, *J. Chem. Phys.* **40**, 1105 (1964).
59. B. Liu and P. Siegbahn, *J. Chem. Phys.* **68**, 2457 (1978).
60. D. G. Truhlar and C. J. Horowitz, *J. Chem. Phys.* **68**, 2466 (1978); *J. Chem. Phys.* **71**, 1514E (1979).
61. M. Karplus, R. N. Porter, and R. D. Sharma, *J. Chem. Phys.* **43**, 3259 (1965).
62. R. A. Marcus and M. E. Coltrin, *J. Chem. Phys.* **67**, 2609 (1977).
63. G. C. Schatz and A. Kuppermann, *Phys. Rev. Lett.* **35**, 1266 (1975).
64. B. C. Garrett and D. G. Truhlar, *J. Phys. Chem.* **83**, 1079 (1979).
65. E. Pollak, M. S. Child, and P. Pechukas, *J. Chem. Phys.* **72**, 1669 (1980).
66. E. Pollak, *J. Chem. Phys.* **74**, 5586 (1981).
67. J. M. Bowman and K.-T. Lee, *Chem. Phys. Lett.* **64**, 291 (1979).
68. H. R. Mayne, *Chem. Phys. Lett.* **66**, 487 (1979).
69. G. C. Schatz, *Chem. Phys. Lett.* **94**, 183 (1983).
70. J. M. Bowman, K.-T. Lee, and R. B. Walker, *J. Chem. Phys.* **79**, 3742 (1983).
71. O. N. Mitchell and D. J. LeRoy, *J. Chem. Phys.* **58**, 3449 (1973).
72. A. A. Westenberg and N. de Haas, *J. Chem. Phys.* **58**, 1393 (1967).
73. G. P. Glass and B. K. Chaturvedi, *J. Chem. Phys.* **78**, 3478 (1982).
74. M. Kneba, U. Wellhausen, and J. Wolfrum, *Ber. Bunsengs. Phys. Chem.* **83**, 940 (1979).

75. E. B. Gordon, B. I. Ivanov, A. P. Perminov, V. E. Balalaev, A. N. Ponomarev, and V. V. Filatov, *Chem. Phys. Lett.* **58**, 425 (1978).
76. See Ref. 26 for a review of previous calculations.
77. R. N. Zare and R. B. Bernstein, *Phys. Today* **33**, 43 (1980).
78. B. Liu, *J. Chem. Phys.* **80**, 581 (1984).
79. M. J. Berry, *J. Chem. Phys.* **59**, 6229 (1973).
80. R. K. Sparks, C. C. Hayden, K. Shobatake, D. M. Neumark, and Y. T. Lee, in *Horizons in Quantum Chemistry*, K. Fukui and B. Pullman, ed., Reidel, Dordrecht, 1980.
81. G. C. Schatz, J. M. Bowman, and A. Kuppermann, *J. Chem. Phys.* **63**, 674 (1975).
82. For a review of previous collinear quantum calculations see Ref. 23.
83. J. T. Muckerman, in *Theoretical Chemistry: Advances and Perspectives, 6A*, D. Henderson and H. Eyring, eds., Academic, New York, 1981.
84. A. Kuppermann, in *Potential Energy Surfaces and Dynamics Calculations*, D. G. Truhlar, ed., Plenum, New York, 1981.
85. A. Kuppermann and J. A. Kaye, *J. Phys. Chem.* **85**, 1969 (1981).
86. E. F. Hayes and R. B. Walker, *J. Phys. Chem.* **86**, 85 (1981).
87. M. J. Redmon and R. E. Wyatt, *Chem. Phys. Lett.* **63**, 209 (1979).
88. R. E. Wyatt, in *Horizons of Quantum Chemistry*, K. Fukui and B. Pullman, eds., Reidel, Dordrecht, 1980.
89. E. Pollak and R. E. Wyatt, *J. Chem. Phys.* **77**, 2689 (1982).
90. M. Baer, J. Jellinek, and D. J. Kouri, *J. Chem. Phys.* **78**, 2962 (1983).
91. R. W. Emmons and S. H. Suck, *Phys. Rev.* **A25**, 178 (1982).
92. J. Jellinek, M. Baer, V. Khare, and D. J. Kouri, *Chem. Phys. Lett.* **25**, 460 (1980).
93. K.-T. Lee, J. M. Bowman, A. F. Wagner, and G. C. Schatz, *J. Chem. Phys.* **76**, 3563, 3583 (1982).
94. J. M. Bowman, A. F. Wagner, S. P. Walch, and T. H. Dunning, Jr., *J. Chem. Phys.*, **81**, 1739 (1984).
95. S. P. Walch, T. H. Dunning, Jr., F. W. Bobrowitz, and R. Raffenetti, *J. Chem. Phys.* **72**, 406 (1980).
96. R. Schinke and W. A. Lester, Jr., *J. Chem. Phys.* **70**, 4893 (1979).
97. For a recent review, see J. M. Brown and F. Jorgensen, *Adv. Chem. Phys.* **52**, 117 (1983).
98. K. Westburg and N. Cohen, *J. Phys. Chem. Ref. Data* **12**, 531 (1983).
99. G. C. Light, *J. Chem. Phys.* **68**, 2831 (1978).
100. K. M. Pamidimukkala and G. B. Skinner, *J. Chem. Phys.* **76**, 311 (1982).
101. M. A. A. Clyne and B. A. Thrush, *Proc. Roy. Soc. (London)* **A5**, 544 (1963).
102. A. A. Westenberg and N. de Haas, *J. Chem. Phys.* **47**, 4241 (1967).
103. N. Presser and R. J. Gordon, private communication to be submitted to *J. Chem. Phys.*
104. K. M. Pamidimukkala and G. B. Skinner, Resonance Absorption Measurements of Atom Concentrations in Reacting Gas Mixtures. 8. Rate Constants for $O + H_2 \rightarrow OH + H$ and $O + D_2 \rightarrow OD + D$ from Measurements of O atoms in Oxidation of H_2 and D_2 by N_2O. 13th Shock Tube Symposium, p. 585, 1981.
105. G. C. Schatz, *J. Chem. Phys.* **79**, 5386 (1983).
106. K. M. Christoffel and J. M. Bowman, *J. Chem. Phys.* **74**, 5057 (1981).
107. M. Shapiro and M. S. Child, *J. Chem. Phys.* **76**, 6176 (1982).

108. Q. Zhi-Ding, Z. Xing-Guo, L. Xing-Wen, H. Kono, and S. H. Lin, *Mol. Phys.* **47**, 713 (1982).
109. G. S. Ezra, *Chem. Phys. Lett.* **101**, 259 (1983).
110. H. Romanowski and J. M. Bowman, *Chem. Phys. Lett.* **110**, 235 (1984).
111. See, for example, G. L. Hofacker and R. D. Levine, *Chem. Phys. Lett.* **9**, 617 (1971).
112. T. Carrington, Jr. and W. H. Miller, *J. Chem. Phys.* **81**, 3942 (1984).
113. D. M. Neumark, Ph.D. Dissertation, University of California, Berkeley, California, 1984.
114. D. M. Neumark, A. M. Wodtke, G. N. Robinson, and Y. T. Lee, in Proceedings of Symposium on *Resonances in Electron-Molecule Scattering, van der Waal's Complexes, and Reactive Chemical Dynamics*, D. G. Truhlar, Editor, ACS Symposium Series, 1984, Chap. 25.
115. E. F. Hayes and R. B. Walker, in Proceedings of Symposium on *Resonances in Electron-Molecule Scattering, van der Waal's Complexes, and Reactive Chemical Dynamics*, D. G. Truhlar, Editor, ACS Symposium Series, 1984, Chap. 26.

THE THEORETICAL INVESTIGATION OF THE ELECTRON AFFINITY OF CHEMICAL COMPOUNDS

G. L. GUTSEV

Institute of Chemical Physics (Branch), USSR Academy of Sciences, Chernogolovka, Moscow, USSR

and

A. I. BOLDYREV

Institute of Chemical Physics, USSR Academy of Sciences, Moscow, USSR

CONTENTS

I. INTRODUCTION

The electron affinity (*EA*) is an important molecular characteristic from both theoretical and practical points of view since one of the fundamental notions in chemistry—electronegativity—is defined by means of *EA*. Inductive effects in chemical bonding and electron donation properties are tightly related with *EA*. Energetical changes in redox reactions, ion–molecular reactions in upper layers of the atmosphere, and so on, depend on *EA* values. The concept of *EA* finds wide application in solid-state physics because energies of crystalline lattices may be calculated using the Born-Haber cycle if *EA*s of the constituents are known. The *EA* plays an important role in radiation damage and light-detection technologies, as well.

To determine *EA* values a great number of experimental methods are currently in use. There are the photodetachment method,[1,2] surface ionization technique,[3,4] electron impact fragmentation,[2] molecular phosodissociation,[2] equilibrium sublimation,[5] radiative electron attachment,[5] lattice energy measurements,[5] charge-transfer reactions,[6,7] photosensitivities ionization,[2] and other experimental methods.[8,13] The largest experimental success has been connected with development of the laser photodetachment technique, though its application is limited by the compounds that may be acquired in a sufficient concentration in gas phase. But this wide array of independent techniques has often given inconsistent and confusing results for molecular *EA*s. For example, the experimental *EA* for NO_2 is scattered between 1.6 and 4.6 eV, with one measurement claiming $EA(NO_2) > 3.613$ and other asserting $EA(NO_2) < 3.063$ eV.[6,7] Therefore, the quantum chemical calculations may serve as an alternative way to estimate *EA*s. This is sponsored by the considerable progress achieved recently in energy correlation calculations which essentially allows improvement of the accuracy of calculated *EA* values.

Undoubtedly, one of the important directions in both experimental and theoretical investigations is development of simple models which would allow prediction of whether *EA*s are large or small without any calculations and measurements at all. Of special interest are those systems which possess anomalously high *EA*s.

The *EA* of a molecule or radical is defined by the energy gain occurring when an electron is attached to a system. There may be defined two *EA*s: an adiabatic and a vertical one. The adiabatic *EA* is equal to the difference between total energies of a neutral system and the corresponding anion in their equilibriums:

$$EA_{\mathrm{ad}} = E(R_e) - E^{-}(R_e^{-}) \tag{1}$$

The vertical *EA* is equal to the difference between total energies of a neutral system and the anion in equilibrium geometry of the neutral system

$$EA_{\mathrm{vert}} = E(R_e) - E^{-}(R_e) \tag{2}$$

Moreover, for negatively charged systems, one may determine the first ionization potential (IP) which may also be vertical or adiabatic. The adiabatic IP is equal to the difference between total energies of a neutral system and the anion in their equilibriums:

$$IP_{\text{ad}} = -[E^{-}(R_e^{-}) - E(R_e)] \tag{3}$$

By definition, the adiabatic IP of an anion coincides with the adiabatic EA of the neutral system. The vertical IP is equal to the difference between energies of charged and neutral systems in the ion equilibrium geometry:

$$IP_{\text{vert}} = -[E^{-}(R_e^{-}) - E(R_e^{-})] \tag{4}$$

The vertical IP of an anion differs in the common case from the vertical EA of the neutral system. The difference between values of the vertical and adiabatic EAs and IPs is an adiabatic correction:

$$EA_{\text{ad}} = EA_{\text{vert}} + \Delta\, EA_{\text{ad}} \tag{5}$$

$$IP_{\text{ad}} = IP_{\text{vert}} + \Delta\, IP_{\text{ad}} \tag{6}$$

As it is seen from Eqs. (1)–(4), the EA is closely connected with the first IP; and the adiabatic IP_{ad} of an anion coincides with the adiabatic EA of the neutral system. Therefore, in the rest of this chapter we use results and conclusions obtained for IPs for which are collected much more experimental data and the theory is better developed. It should be underlined that all three experimental values EA_{ad}, EA_{vert}, and IP_{vert} are known practically for neither system. Nonetheless, they are equal to each other up to adiabatic corrections, and it should be pertinent to pay some attention to estimating values of these corrections.

From data obtained from the experimental photoelectron spectra (PES) of SF_6, C_6H_6, MoF_6, WF_6, and UF_6[9] it follows that the difference between IP_{ad} and IP_{vert} is small, as a rule 0.1–0.2 eV, and does not exceed 0.5 eV. In the investigation of PES for the SbF_5Cl, BrF_5, and IF_5 molecules[10] this difference has been also evaluated in 0.4–0.6 eV. The similar results are obtained for other molecules as well.[11]

The examples given above show the adiabatic corrections to be small (and it should be reasonable to suggest such values for EAs as well) and they do not exceed 0.5 eV as a rule. Therefore, in the cases where EAs are large and the accuracy in 0.5 eV is acceptable one needs to know only one of three values: EA_{ad}, EA_{vert}, or IP_{vert}. In such a case the EA calculation is somewhat simplified because geometries of radicals are frequently unknown, whereas for negative ions with closed shells one may use structure data for corresponding salts. Two levels in *EA* investigation. At the first level, one may obtain the crude *EA* estimations neglecting the differences between EA_{ad}, EA_{vert}, and IP_{vert} if the accuracy in 0.5–1.0 eV is acceptable to establish qualitative trends. At the second level, it should be necessary to take into account the adiabatic corrections to *EA*s and *IP*s, and conditions in which the experimental *EA* was obtained should be thoroughly examined when comparing the experimental and theoretical *EA* values. The second level is required to give the accuracy within 0.1–0.2 eV. At present this level appears to be achieved for small molecules and radicals only, whereas for large systems the modern theoretical methods require one to be restricted by the first level.

The literature on the subject includes several monographs[12,13] and reviews[1,2,5,8,14,15] dedicated to the *EA* investigation. But these works contain theoretical results for the simplest systems only, whereas comparatively large systems have been treated recently, and the accuracy of *EA* calculations for small compounds has grown sharply. Moreover, X_α methods, as well as *ab initio* ones, are now being used in *EA* calculations.

Therefore, it is of great interest to consider the state of the art in the modern theoretical *EA* calculations by both X_α and Hartree-Fock-Roothaan methods as well as by methods that include implicitly the correlation energy. On the other hand, it is equally of interest to present the new model notions based on the orbital approach which allow the qualitative *EA* estimations to be done. We pay particular attention to systems with large *EA* values because they meet a growing interest.

II. THE THEORETICAL METHODS OF *EA* CALCULATIONS

In this section we briefly discuss the theoretical methods of *EA* calculations with the primary aim of understanding the reliability of different approaches in current use. In this review we are concerned with *ab initio* and X_α methods only. *Ab initio* methods are the most reliable at present because an additional inclusion of correlation energy in framework of these methods allows one to calculate *EA*s with "experimental" accuracy. X_α methods are considered here because numerous calculations on the electronic structure of a great number of compounds have shown that these methods are highly effective for large systems and lead to the correct prediction of the *IP*s order almost in all cases studied up to now. Moreover, the calculated *IP* values did not differ from experimental ones by more than 1 eV, as a rule. Such an error may be acceptable in *EA* calculations for systems with large *EA* values. The semiempirical methods like EH, CNDO, and MNDO are intentionally excluded from consideration because they are not able to compete in accuracy and reliability with the *ab initio* methods for small molecules and radicals and with X_α methods for large systems.

A. The Hartree-Fock-Roothaan Method

We shall not consider in detail the theoretical foundations of the Hartree-Fock-Roothaan method, which is quite completely explained in a number of monographs[16-18] and reviews.[19-22] Let us briefly recall the main equations on which is based the *ab initio* calculational methods for understanding which sort of assumptions were made in one or another approach and which factors influence the accuracy of the results to be obtained.

The rigorous solution of the stationary Schrödinger equation is possible only for the hydrogen atom, the molecular hydrogen ion, and some other simplest systems. For complex molecular systems is widely used an approach connected with the solution of the Hartree-Fock-Roothaan (HFR) equations:

$$\sum_{\nu=1}^{N} c_{i\nu}(F_{\mu\nu} - \varepsilon_i S_{\mu\nu}) = 0, \qquad \mu = 1, 2, \ldots, N \tag{7}$$

$$F_{\mu\nu} = \int \chi_\mu(1)\left[-\tfrac{1}{2}\nabla_1^2 - \sum_A \frac{Z_A}{r_{1A}}\right]\chi_\nu(1)\, dV_1 + \sum_j \sum_\lambda \sum_\sigma c_{j\lambda} c_{j\sigma}[2(\mu\nu|\lambda\sigma) - (\mu\lambda|\nu\sigma)] \tag{8}$$

$$S_{\mu\nu} = \int \chi_\mu(1)\chi_\nu(1)\, dV_1 \tag{9}$$

$$(\mu\nu|\lambda\sigma) = \iint \chi_\mu^*(1)\chi_\nu^*(1)\frac{1}{r_{12}}\chi_\lambda(2)\chi_\sigma(2)\, dV_1\, dV_2 \tag{10}$$

These equations are obtained from the nonrelativistic Schrödinger equation by means of substitution of the Coulomb interelectronic interaction potential $\sum_{i,j} 1/r_{ij}$ by the approximate HFR one $\sum_i \sum_j \sum_\nu \sum_\mu \sum_\lambda \sum_\sigma c_{j\lambda} c_{j\sigma} c_{i\mu} c_{i\nu}[2(\mu\nu|\lambda\sigma) - (\mu\lambda|\nu\sigma)]$. Such a substitution allows one to reduce the multicenter problem of the electrons movement to an one-center problem of the movement of an electron in the average field of other electrons. The difference between rigorous nonrelativistic and HF energies is, by definition, the correlation energy. Solving HFR equations, the decomposition of numerical HF functions in analytic Slater (STO) or Gauss (GTO) basis functions is used (MO LCAO approximation). Therefore, quite accurate HF energies may be obtained if to use the large number of STOs or GTOs. The HFR equations are amenable for programming and there have been

TABLE I
Values of ε_{HOMO} Calculated Using the Ion (IP^K_{vert}) and Neutral Molecule (EA^K_{ad}) Equilibrium Geometries

Ion	Basis (STO)	IP^K_{vert} (in eV)	EA^K_{ad} (in eV)	References
OH^-	$(5s4p2d1f)_O + (3s1p)_H$	2.91	2.91	28
SH^-	$(6s5p1d)_S + (3s1p)_H$	2.41	2.42	28
CN^-	$(4s2p)_{C,N}$	5.21	5.14	29
BO^-	$(4s2p)_{B,O}$	3.04	3.14	29

compiled a great number of good quality programs[23-26] to solve the Eq. (7)–(10). Recently many methods to take into account the correlation energy have been elaborated, and in calculations with these elaborated techniques one approaches a rigorous solution of Schrödinger equation. These methods will be considered later, but now let us turn our attention to the HFR equations.

According to Koopmans theorem the one electron orbital energy ε_i in Eq. (1) equals the IP from the i-th MO if other MOs do not change, that is, do not relax, after detachment of an electron from the ith MO. This is a "frozen MO" approximation. Naturally, the simplest approach to the EA estimation in framework of the HFR method may be based on the orbital approach. For a negative ion with the closed shell structure it is reasonable to assume that

$$EA \approx \text{the first } IP = -\varepsilon_{HOMO} \tag{11}$$

where ε_{HOMO} is the energy of the highest occupied MO (HOMO). If ε_{HOMO} is calculated using the neutral molecule geometry then the first IP is the adiabatic Koopmans $IP(IP^K_{ad})$ which is equal to the EA^K_{ad} of the neutral system, and if the anion geometry is used then the first IP has sense of the vertical Koopmans $IP(IP^K_{vert})$. In many instances ε_{HOMO} changes only slightly when going from the neutral molecule equilibrium geometry to the anion one* (Table I). Therefore, to estimate the EA in the Koopmans approach one may use ε_{HOMO} of the anion at its equilibrium geometry. Before to discuss the quality of this approximation it is pertinent to consider dependence of ε_{HOMO} on the completeness and flexibility of the bases because a wide array of bases is suggested in the literature.

In Table II are given ε_{HOMO} for several negative ions calculated within

* This assertion is, of course, true only if the neutral molecule and its anion are stable to the monomolecular decay.

TABLE II

The Dependence of $-\varepsilon_{HOMO}$ (in eV) on the Basis Used[a]

Ion	STO-3G[b]	DZRS[b]	4-31G[b]	DZHD[b]	6-31G[b]	DZHD+P	DZHD+D[b]	DZHD+P+D	HF limit	References
H^-	−8.4	−0.6	−1.5	−0.6	−1.5	−0.6[b]	0.13	0.13[b]	0.13[b]	
F^-	−10.2	0.1	1.8		2.1		4.7	4.8[b]	4.92	27
OH^-	−6.3	−0.05	0.9	1.0				2.9[b]	2.91	27
SH^-	−5.2		1.1						2.42	27
CN^-	−1.2	3.79	3.9	4.54	4.1	4.6	4.8	5.0		31
BeH_3^-	1.7	4.0	3.8							32
BH_4^-	2.3	5.0	4.7	5.0		4.8		5.1[b]		32
AlH_4^-	1.5	5.4		5.4						32
NO_3^-	−0.3		6.0	6.5		5.7	6.7			30
BeF_3^-	2.2	8.1	8.7	9.6			9.7			33, 34
BF_4^-	2.7	9.2		10.9						33
MgF_3^-	0.1			9.2			9.6			33, 34

[a] STO-3G is the minimal basis in which are used three primitive GTO, contracted to one function, to represent each atomic shell orbital[35]; DZRS is the GTO basis of Roos and Siegbahn[36] and is of a double zeta quality (i.e., when each atomic shell orbital is represented by two optimized STOs[35]; 4-31G is the basis in which for each core state is used one contracted function consisting of four primitive GTOs and each valence AO is represented by two contracted functions, the first of which is composed of three primitive GTO and the second is of one GTO[37,38]; DZHD is the basis of Huzinaga and Dunning of a double zeta quality[39,40]; DZHD + P is the DZHD basis plus polarization functions (3*d* for Li through F and 2*p* for H atoms); HZHD + D is the DZHD basis extended by diffuse functions; DZHD + P + D is the DZHD + P basis extended by diffuse functions; HF limit is the basis within which nearly all the HF energy is to be obtained.

[b] Our calculations.

different bases beginning with the simplest STO-3G and up to bases of the near HF limit. As it is seen from Table II the minimal STO-3G basis leads to very crude values of ε_{HOMO}. The mean dimension bases (DZRS and 4-31 G) give better results, though the inaccuracy remains high. The 6-31G, DZHD, and DZHD + P bases which in the case of neutral molecules reproduce the majority of molecular properties on the HF limit level give quite good ε_{HOMO} estimations except for atomic ions. An extension of the DZHD basis by diffuse functions gives nearly HF limit values. The essential influence of diffuse functions in the case of the DZHD and DZHD + P bases is connected with the fact, that in calculations for ions are used bases which have been optimized for neutral atoms. To better describe the electronic structure of an ion it is necessary either to reoptimize the upper valence basis functions or to add the diffuse ones. Therefore, the results of calculations on ε_{HOMO} within the DZHD + D basis may be recommended as good estimations of *EA*s in the Koopmans approximation.

Table III contains the calculated EA^K and experimental values of *EA*s for some small radicals. It is worthwhile to notice first of all that calculated *EA*s may be both greater and smaller than experimental ones and are neither the upper nor lower bound for the latters. The inaccuracies in theoretical *EA*s may be quite large (up to 3 eV) and Koopmans approximation should be recommended for very crude *EA* estimations only.

When solving the HFR equations one obtains, apart from one-electron energies of occupied MO, the energies of vacant MOs as well. In principle the

TABLE III
The Calculated *EA*s Taken as the HOMO Energies of Negative Closed Shell Ions EA^K and Experimental *EA* Values

Radical	Theory	References	Experiment	References
F	4.92	27	3.448 ± 0.005	41
Cl	4.08	27	3.614 ± 0.001	41
OH	2.91	27	1.83 ± 0.04	41
SH	2.42	27	2.32 ± 0.01	41
NH_2	1.32	28	0.744 ± 0.022	13
PH_2	1.06	28	1.25 ± 0.03	13
BO	3.14	29	3.1 ± 0.1	42
CN	5.00	31	3.82 ± 0.02	41
BO_2	5.30	30	3.6 ± 0.1	42
NO_2	4.05	43a	3.10 ± 0.05	41
NO_3	6.72	43b	3.77 ± 0.25	8
PO_3	7.36	44	3.5	8
C_2H	3.33	45	3.73	41

TABLE IV
*EA*s Calculated as the Energies of Vacant MOs of the Neutral Closed Shell Molecules with Experimental Values

Molecule	Theory	References	Experiment	References
F_2[a]	−1.82		3.08 ± 0.1	41
Cl_2[a]	−0.74		2.38 ± 0.1	41
C_2	3.30	22	3.54	22
H_2O[b]	−1.01		1.78 − 1.82	41
SO_2	0.51	22	1.097	22
BF_3[a]	−3.6		2.65	41
O_3	1.44	22	1.9 − 2.2	22
P_2	−0.38	22	0.23 ± 0.23	22
SF_6[a]	2.21		0.3 − 0.75	119

[a] Our calculations, using the DZHD basis.
[b] The estimation within the DZHD + P + D basis.

one-electron energy of the lowest unoccupied MO (LUMO) may be considered as an *EA* estimation. However, this approximation is even more crude than the Koopmans approximation. As seen from the data of Table IV, in this case the approximation does not reach agreement with experimental values even in qualitative way. Moreover, the theoretical *EA*s have often the negative values.

The reasons of the orbital approach failure in the *EA* reproduction are the same as for *IP*s and were discussed early.[16,18] They are related with three assumptions on conservation of the relativistic energy (1), the MO composition (2), and the correlation energy (3) when going from an anion to the neutral molecule.

The first assumption should be fulfilled comparatively well for small systems because of smallness of valence MO energies in contrast to the core state energies. It may be demonstrated by the atomic *EA* example for which have been carried out both nonrelativistic and relativistic calculations (see Table V). As it is seen, the relativistic contributions to *EA*s are small and do not exceed ~0.05 eV. Hence, the first assumption is fulfilled quite well, but it is not so for the second one.

When detachment of an electron from a negative ion occurs, the MOs undergo severe alterations (see Table VI) and the corresponding changing in the total energy is named as the relaxation energy. To take into account this energy it is necessary two separate HFR calculations on the anion and the neutral to be done. We designate by "HFR" the *IP*s and *EA*s calculated accounting for the relaxation energy. Then, as in the case of the orbital

TABLE V

Relativistic Energies of Neutral Atoms and Negative Ions

Atom	State	E_{rel} (a.u.)	Ion	State	E_{rel} (a.u.)	References
Li	^2S	−0.00055	Li^-	^1S	−0.00055	46
B	^2P	−0.00610	B^-	^3P	−0.00605	46
C	^3P	−0.01379	C^-	^4S	−0.01369	46
N	^4S	−0.02740	N^-	^3P	−0.02720	46
O	^3P	−0.04943	O^-	^2P	−0.04917	46
F	^2P	−0.08288	F^-	^1S	−0.08242	46
Na	^2S	−0.2007	Na^-	^1S	−0.2009	47
Al	^2P	−0.4219	Al^-	^3P	−0.4214	47
Si	^3P	−0.5856	Si^-	^4S	−0.5846	47
P	^4S	−0.7927	P^-	^3P	−0.7917	47
S	^3P	−1.053	S^-	^2P	−1.052	47
Cl	^2P	−1.375	Cl^-	^1S	−1.373	47

TABLE VI

Percentage Decomposition of Valence MOs of ClF, ClF^+, ClF_2^-, and ClF_2 with Different Orbital Populations[a]

			ClF^+					ClF_2			
MO	AO[c]	ClF	$3\pi^b$	$7\sigma^b$	MO	AO	ClF_2^-	$6\sigma_g^b$	$3\pi_u^b$	$1\pi_g^b$	$4\sigma_u^b$
5σ	s	6	7	5	$4\sigma_g$	s	9	4	13	6	8
	p_σ	3	3	2		s'	90	95	86	92	91
	s'	89	87	91		p'_σ	1	1	1	2	1
	p'_σ	2	3	2	$3\sigma_u$	p_σ	5	5	5	4	4
6σ	s	84	85	93		s'	95	94	94	96	96
	s'	8	10	6	$5\sigma_g$	s	85	88	82	85	88
	p_σ	7	5	1		s'	10	8	14	8	9
7σ	s	7	5	2		p'_σ	5	4	4	8	3
	p_σ	31	39	11	$4\sigma_u$	p_σ	27	33	35	20	23
	s'	2	3	2		s'	3	3	4	3	3
	p'_σ	60	53	84		p'_σ	70	64	61	77	74
2π	p_π	13	12	6	$6\sigma_g$	s	6	4	5	9	4
	p'_π	87	88	94		p'_σ	94	96	95	91	96
3π	p_π	87	88	94	$2\pi_u$	p_π	39	14	40	35	26
	p'_π	13	12	6		p'_π	61	86	60	65	74
					$3\pi_u$	p_π	61	86	60	65	74
						p'_π	39	14	40	35	26

[a] From Ref. 48.

[b] Levels from which an electron is moved.

[c] Primed AOs are related to fluorine atoms, unprimed ones to the Cl cation.

approach, one may define vertical and adiabatic EAs and IPs:

$$EA_{\text{vert}}^{\text{HFR}} = E^{\text{HFR}}(R_e) - E^{-\text{HFR}}(R_e) \tag{12}$$

$$EA_{\text{ad}}^{\text{HFR}} = E^{\text{HFR}}(R_e) - E^{-\text{HFR}}(R_e^-) \tag{13}$$

$$IP_{\text{vert}}^{\text{HFR}} = -[E^{-\text{HFR}}(R_e^-) - E^{\text{HFR}}(R_e^-)] \tag{14}$$

$$IP_{\text{ad}}^{\text{HFR}} = -[E^{-\text{HFR}}(R_e^-) - E^{\text{HFR}}(R_e)] \tag{15}$$

Here, $EA_{\text{ad}}^{\text{HFR}} = IP_{\text{ad}}^{\text{HFR}-}$ as well.

Let us consider briefly the dependence of EA^{HFR} on a basis used in calculations. Table VII contains several examples of light molecules for which EA^{HFR} has been calculated within different bases. As seen from the table, the dependence of EA^{HFR} on the basis is not essential beginning with the DZHD basis, and corresponding EA values differ from the HF limit values within the DZHD basis by ~0.6 eV, within the DZHD+D basis by ~0.2 eV, within the DZHD+P basis by ~0.7 eV, and within the DZHD+P+D basis by ~0.2 eV. The inclusion in the bases of diffuse functions appears to be more important than that of polarization functions.

Using EA^{HFR} and EA^K it is possible to define the value of the relaxation energy as

$$\Delta R = EA^K - EA^{\text{HFR}} \tag{16}$$

The relaxation energy is negative and results, as a rule, in lowering EAs (Table VIII). Nonetheless, the more exact (from the theoretical point of view) EA^{HFR} values are in even worse agreement with the experimental values than in the case of EA^K.

The reason why EA^K values are in better agreement with experimental

TABLE VII
Dependence of EA^{HFR} on the Basis

Molecule	State	DZHD	DZHD+D	DZHD+P	DZHD+P+D	HF limit	References
C_2	$^2\Pi_u$	4.38	4.48	4.40	4.51	4.27	40
	$^2\Sigma_g^+$	4.10	4.23	4.08	4.21	4.10	40
	$^2\Sigma_u^+$	0.83	0.97	0.70	0.86	0.76	40
NH	$^2\Pi$	2.16	1.54	2.20	1.57	1.55	40
NH_2	$^2B_1$	1.68	0.95	1.73	1.03		49

TABLE VIII
Comparison of the Calculated EA_{vert}^{K} and E_{vert}^{HFR} Values with Experimental Values

Radical	EA_{vert}^{K} (eV)	EA_{vert}^{HFR} (eV)	References	EA_{vert}^{exp} (eV)	References
OH	2.91	−0.10	27	1.83	41
SH	2.42	1.21	27	2.3	41
F	4.92	1.36	27	3.445	41
Cl	4.08	2.57	27	3.613	41
CN	5.00	3.29	31	3.82	41

values is connected with neglecting, in this case, both the electronic relaxation and correlation energy changes, which possess different signs and may compensate each other, what is the case, for example, for OH, SH, and so on (see Table VIII). From data of Table VIII it is seen that adding an electron to a neutral system increases the electronic correlation energy and consequently the calculated $EA^{HFR+COR}$ value. The contributions to *EA*s from correlation are varied depending on three factors: (1) the increase of the correlation energy due to the extracorrelation of an additional electron with electrons of the neutral system; (2) the possible lowering of the correlation energy due to a decrease of the correlation energy between other electrons because in the ion the added electron occupies the space that was available for correlation in the neutral, and (3) the possible small increase in correlation due to the orbital deformation. Calculations have shown that contributions 1, 2, and 3 to the *EA* of OH are equal to +2.0, −0.6, and 0.2 eV, respectively[50] (in comparison with the experimental *EA* value 1.83 eV).

Therefore, for the reliable theoretical prediction of *EA*s by means of the quantum chemistry methods it is necessary to take into account both the relaxation and correlation energies.

B. The *EA* Calculations with Correlation

The *EA* calculations which are to take into account the relaxation and correlation energies are very cumbersome due to necessity of using large basis sets (contrary to the IP calculations, which require much smaller bases). It is connected with a diffuse character of the electronic cloud in anions and smallness of the *EA* values themselves. If, from the calculational point of view, accounting for the relaxation energy is a relatively simple problem, that is, it is sufficient to calculate the electronic structures of an anion and the molecule, then accounting for electronic correlation changings is much more complex one. The difficulties arising in such calculations were demonstrated in the *EA* calculations for atoms[51] in which have been used very large basis sets and lengthy expansions within the configuration interactions formalism. These computations have accounted for 95% of the correlation energy, however, for *EA*s it gives 83% of the correlation contributions only and the remaining part of unaccounted for correlation results in errors of 0.1–0.3 eV.

Up to now many approaches allowing inclusion of the electronic relaxation (ΔR) and correlation (ΔE_{corr}) energies have been developed to calculate *IP*s, *EA*s and other electronic characteristics. Among them the most known are techniques based on Green's functions (GF),[22,52] many-body Rayleigh-Schrödinger perturbation theory (PT),[18-20] the equation of motion method (EOM),[53,54] the usual second-order term in expansion of the self-energy part,[55] the ordinary Rayleigh-Schrödinger perturbation theory (PT),[56] the superoperator technique,[57] the electron propogator technique (EP),[58] the pair natural orbital configuration interaction method (PNO-CI), and the independent electron pair and coupled electron pair approximations (IEPA and CEPA).[59,60]

The more frequently used method to account for the correlation energy in the *ab initio* calculations is the method of configuration interactions (CI) in which the wave function of a system is represented by a linear combination of Slater's determinants describing various electronic configurations:

$$\Psi_l = \sum_{k=1}^{M} A_{kl}\phi_k \tag{17}$$

where M is the number of configurations (for more details see Refs. 61–63). Thus, the computational problem is to solve the system of inhomogeneous equations:

$$\sum_k A_{kl}(H_{kl} - E_l\,\delta_{kl}) = 0 \tag{18}$$

In the Hartree-Fock calculations for the ground state configuration ϕ_0, one obtains energies of both occupied and vacant MOs. The total number of the vacant MOs is defined by the basis size used. The CI wave function is constructed as a linear combination of determinants corresponding to removing electrons from occupied MOs and placing them at MOs vacant in the ground state. Configurations arising when one electron is excited are called singly excited configurations, when two electrons—doubly excited, and so on. In such an approach the correlation energy is accounted for by means of the virtual excitations from one, two, and so on, MOs. The A_{kl} coefficients are determined by the variational method whose application to the total energy functional leads to Eq. (18). In the CI approach the results obtained depend on both the number of basis AOs and the number of determinants included in the expansion Eq. (17). In principle, this method would give an exact solution, provided the complete basis were used. However, in real calculations, one is usually restricted to a relatively small number of MOs. Moreover, even using medium-size bases it is difficult to construct the full CI function, that is, one in which expansion Eq. (17) included all the possible excitations.

In the literature are described several limited CI approaches, and almost none of the calculations accounts for the configurations that correspond to core electron excitations. Their omission is based on an assumption that the correlation energy of core states is not changed when valence ionization or attachment of an electron occurs. These approaches succeeded in selecting the most important configurations and in accounting for the correlation energy quite completely using relatively short CI expansions.

Wide application has found the other group of CI approaches in which the full range of particular configurations [i.e., all the doubly excited (CI-D) or all the single *and* doubly excited (CI-SD) configurations, etc. are included in Eq. (17)]. The success of these approaches is due to the fact that the electron correlation has mainly the pair character. Indeed, CI-D and CI-SD calculations accounted for 85–95% of the valence electron correlation energy.[18-20] Sometimes, to reduce a number of determinants in CI-D calculations, the HF orbitals are transformed to the pair natural orbital representation (PNO). This approach is called PNO CI (details in Refs. 59 and 60).

The other variational method incorporating correlation is the self-consistent multiconfigurational method (MC-SCF),[64,65] but it is not widespread due to enormous calculational efforts required. This method differs from the CI method by variations in Eq. (18) of the MO LCAO coefficients in each particular configuration in Eq. (17) apart from the A_{kl} coefficients. This procedure leads to equations

$$\sum_i c_{i\mu}[F_{\mu v}(c_{i\mu}, A_{kl}) - \varepsilon S_{\mu v}] = 0 \tag{19}$$

$$\sum_k A_{kl}[H_{kl}(c_{i\mu}) - E_l\, \delta_{kl}] = 0 \tag{20}$$

The orbital approach is not used in the MC-SCF method, unlike the CI method, where it is tacitly assumed that MOs do not change after an excitation of an electron. Due to enormous computations required in the MC-SCF calculations, a number of determinants in Eq. (17) is to be small. However, owing to optimization of MOs in each configuration it accounts for much correlation using a small number of determinant functions. The MC-SCF-CI approach,[66,67] which begins with a solution of the MC-SCF equations using several most important determinants and then constructing for them a CI matrix, is more flexible but is too cumbersome.

It should be mentioned that in applications of the CI, MC-SCF, and MC-SCF-CI methods it is necessary to use large basis sets which are to be not worse than DZHD + P + D because to describe the excited MOs it is important to use the polarization and diffuse functions. All these methods allow calculation of EA_{vert}, EA_{ad}, and IP_{vert} with accuracy in several tenths of an eV, if a quite large basis and lengthy CI expansion are used.

The CI, MC-SCF, and MC-SCF-CI methods are variational and therefore present the lower boundaries for energy. However, when dealing with EAs, one calculates the neutral molecule and its anion. The bases published in the literature are suited to describe neutral systems and this circumstance results in better accounting for the correlation energy for the former. Thus one would calculate a lower boundary for EA as well.

Apart from these approaches in EA calculations, nonvariational methods are used, to which may be attributed various variants of the perturbation theory, the equation of motion, Green's functions, and some others. These techniques may be better understood from the point of view of the PT approach and we shall briefly discuss its main peculiarities.

The PT technique is based on the fact that knowing an exact solution of the Schrödinger problem for some model Hamiltonian $H^{(0)}$:

$$H^{(0)}\phi^{(0)} = E^{(0)}\phi^{(0)} \tag{21}$$

which differs from the exact Hamiltonian by a small perturbation λW, one may solve the Schrödinger equation with the full Hamiltonian:

$$H\psi = E\psi \tag{22}$$

Both the wave function ψ and total energy E are functions of the λ parameter and may be expanded in series by λ:

$$\psi = \phi^{(0)} + \lambda\phi^{(1)} + \lambda^2\phi^{(2)} + \cdots \tag{23}$$

$$E = E^{(0)} + \lambda E^{(1)} + \lambda^2 E^{(2)} + \cdots \tag{24}$$

Substitute Eqs. (23) and (24) into Eq. (22). To satisfy Eq. (22), coefficients of terms containing λ in equal powers on both sides of the equation have to be equal. It leads to the infinite system of equations:

$$H^{(0)}\phi^{(0)} = E^{(0)}\phi^{(0)} \tag{25}$$

$$(H^{(0)} - E^{(0)})\phi^{(1)} = E^{(1)}\phi^{(0)} - W^{(1)}\phi^{(0)} \tag{26}$$

$$(H^{(0)} - E^{(0)})\phi^{(2)} = E^{(2)}\phi^{(0)} + E^{(1)}\phi^{(1)} - W^{(1)}\phi^{(1)} \tag{27}$$

Expanding $\phi^{(1)}$ in the series with respect to a complete set of $\phi_i^{(0)}$:

$$\phi^{(1)} = \sum_{i=1}^{\infty} c_i \phi_i^{(0)} \tag{28}$$

and substituting this expansion into Eq. (26) following by multiplication of both sides by $\phi_i^{(0)*}$ and integrating, one obtains

$$E^{(1)} = \int \phi^{(0)} W^{(1)} \phi^{(0)} \, \mathrm{d}V \tag{29}$$

In the same manner may be obtained the corrections $E^{(2)}$, $E^{(3)}$, and so on. In quantum chemical calculations as an initial $H^{(0)}$ is usually chosen the HFR Hamiltonian H^{HFR} for which the zero and first corrections $E^{(0)}$ and $E^{(1)}$ do vanish. The correlation is usually chosen as a perturbation. Then the higher-order corrections are

$$E^{(2)} = \langle \phi^{(0)} | W Q_0 W | \phi^{(0)} \rangle \tag{30}$$

$$E^{(3)} = \langle \phi^{(0)} | W Q_0 W Q_0 W | \phi^{(0)} \rangle \tag{31}$$

$$E^{(4)} = \langle \phi^{(0)} | W Q_0 W Q_0 W Q_0 W | \phi^{(0)} \rangle - \langle \phi^{(0)} | W Q_0 W | \phi^{(0)} \rangle \langle \phi^{(0)} | W Q_0 Q_0 W | \phi^{(0)} \rangle \tag{32}$$

where Q_0 is a projector, that is, an idempotent operator. Recently programs have been developed incorporating the correlation energy up to the fourth order.[23,68] Numerous calculations have shown that in the second order PT accounts for more than 95% of the valence correlation energy for the molecular ground state configurations except for the quasidegenerate cases. The example of such an exception is the Be atom whose 2*s* and 2*p* shell energies are very close. It should be noted that PT is the nonvariational method and may overestimate E_{corr} though on average not more than 5% in the second order. The third-order PT corrections improve E_{corr} in the right direction, decreasing an inaccuracy up to 2% except for quasidegenerate cases, where the inaccuracy remains up to 5%. The inclusion of the fourth-order PT terms allows accounting for 99% of the valence correlation energy.

To calculate *EA*s using PT one need not do separate calculations on a neutral system and its anion. Instead, in PT there is possible to extract the varying part of the correlation energy. Therefore it is sufficient to do one run only.[18,20] In the case of *IP*s this approach gave quite good results,[69] though it has not been applied yet to *EA* calculations. Undoubtedly, PT is one of the simplest ways to include correlation and should find a wide application in *EA* calculations.

Among other theoretical methods of EA calculations with inclusion of correlation and relaxation effects, the equation-of-motion method[53,54] is widely used. In this method an operator Q_λ^+ is introduced that connects the wave function of an excited state $|\lambda\rangle$ with that of the ground state $|g\rangle$:

$$|\lambda\rangle = Q_\lambda^+ |g\rangle \tag{33}$$

Using the fact that both $|\lambda\rangle$ and $|g\rangle$ are eigensolutions to the same Hamiltonian H:

$$H|g\rangle = E_g|g\rangle \tag{34}$$

$$H|\lambda\rangle = E_\lambda|\lambda\rangle \tag{35}$$

one may obtain an expression for the equation of motion:

$$[H, Q_\lambda^+]|g\rangle = (E_\lambda - E_g)Q_\lambda^+ |g\rangle \tag{36}$$

We shall not derive the equation for EA calculations, because it has been in detail described in Ref. 54. The final result is:

$$EA_{\text{vert}} = -\varepsilon_{N+1} - \sum_{\alpha<\beta,m} \frac{|\langle N+1m|\alpha\beta\rangle|^2}{E_{\alpha\beta}^{m} + \varepsilon_{N+1}} + \sum_{N+1<n,\alpha} \frac{|\langle N+1\alpha|N+1n\rangle|^2}{E_{\alpha}^{N+1,n} - \varepsilon_{N+1}} + \sum_{N+1<m<n,\alpha} \frac{|\langle N+1\alpha|mn\rangle|^2}{E_{\alpha}^{mn} - \varepsilon_{N+1}} \tag{37}$$

where ε_{N+1} is the orbital energy of the anion HOMO, expressions in brackets are matrix elements, energies E_α^{mn}, $E_{\alpha\beta}^m$, and $E_\alpha^{N+1,n}$ are expressed in terms of matrix elements and orbital energies of the HFR equations. The first sum gives the approximate changes in E_{corr} of parent's N electrons coursed by the extra electron. The second sum represents an approximation to the HF ion-neutral energy difference and accounts for the charge redistribution on calculated EAs while the third sum approximates E_{corr} of an electron occupying ϕ_{N+1}. To derive Eq. (37), a number of approximations[54] have been used which are equivalent in terms of CI to an inclusion of all double excitations or accounting for the PT second- and third-order corrections. Therefore, the EOM should give ~95% of the valence correlation.

The other approach applied successfully in the EA and IP calculations is the Green's function method. The one-particle Green's function $G(\omega)$ possess poles at ω values that are equal to the EA or IP values (up to signs). To find the poles of $G(\omega)$ it is necessary to solve the well-known Dyson equation

$$G = G^{(0)} + G^{(0)}MG \tag{38}$$

which connects the Green's function with the self-energy part $M(\omega)$. Here $G^{(0)}$ is a free particle Green's function with

$$G_{ij}^{(0)} = \delta_{ij}(\omega - \varepsilon_i) \tag{39}$$

Considering the diagonal part of the self-energy only as it has usually been done in numerical calculations, one may obtain an expression to determine EAs:

$$EA = -\varepsilon_k - M_{kk}(\varepsilon_k) \tag{40}$$

In derivation of this expression an anion A^- *is* assumed to be the reference system. Then ε_k is the orbital energy of the anion HOMO and $M_{kk}(\varepsilon)$ is expressed in terms of matrix elements of the HF Hamiltonian and its orbital energies. In the literature is used the expansion of $M(\omega)$ up to the third order. The most intensive applications this approach finds in FRG[22,52] where a number of computer programs has been developed for its implementation.

According to the Brillouin theorem contributions from one-electron single excitations to the total energy are small, and the main contribution gives double excitations. The contributions from excitations of higher orders are also small. The importance of accounting for double excitations to include correlation effects are used in IEPA and CEPA methods.[59,60]

In terms of CI, the IEPA wave function is

$$\psi_{AB} = \Phi_0 + \sum_{R<S} C_{AB}^{RS}\Phi_{AB}^{RS} \tag{41}$$

where Φ_0 and Φ_{AB}^{RS} are Slater's determinants describing the ground and excited configurations. In IEPA, one CI run is carried out separately for each pair of the occupied spin orbitals A and B. The procedure calculates the ε_{AB} energy increments that are to be used in estimations of the electronic correlation between A and B MOs. Suggesting each pair to be independent from other pairs, one obtains the correlation energy as a sum of the pair contributions:

$$E_{\text{corr}} = \sum_{A,B} \varepsilon_{AB} \tag{42}$$

This approach may considerably overestimate the correlation energy (up to 30%). Contrary to IEPA, CEPA represents one of the most successfully applied methods to calculate E_{corr}. The CEPA equations are

$$\langle \Phi_{AB}|H|\Phi_0\rangle + \sum_{C<D} \sum_{T<U} \langle \Phi_{AB}^{RS}|H|\Phi_{CD}^{TU}\rangle\, d_{CD}^{TU} = d_{AB}^{RS}\langle \Phi_0|H|\Phi_0\rangle + d_{AB}^{RS}\varepsilon_{AB} \tag{43}$$

$$\varepsilon_{CD} = \sum_{T<U} \langle \Phi_0|H|\Phi_{CD}^{TU}\rangle d_{CD}^{TU} \tag{44}$$

There equations are to be solved by iterations starting, for example, from the IEPA coefficients C_{AB}^{RS}. In terms of PT CEPA may be classified as a method which accounts for double excitations fully in the second, third and partially fourth order of PT. Therefore, its accuracy should be not worse than 5%. Both methods have been used in EA calculations. Table IX contains the results of EA_{vert}, EA_{ad} and IP_{vert} estimations for a number of molecules obtained in calculations with inclusion of the correlation energy. As it is seen EAs calculated by various methods are nearly the same and stable with respect to the choice of a method, except for PNO-CI. These estimations agree with the experimental EAs within 0.2–0.3 eV.

As a summary it may be said that the 95% limit of the valence correlation energy accounted for may be obtained in CEPA, MC and CI if one takes into account all the singly and doubly excited configurations, in PT, GF, and EOM methods when all terms up to the third order are considered provided the quite large and flexible basis sets are used.

For readers interested in more details in the comparison of various methods with correlation which are used in the modern EA calculations it may be recommended the recent review of Freed et al.[14]

C. X_α Methods

In the recent time wide application of the electronic structure calculations has used the so-called X_α methods.[85,86] The most widespread of them are the scattered wave (SW-X_α)[87] and discrete variational (DV-X_α)[88] X_α methods. These methods have been devised to solve the Hartree-Fock-Slater (HFS)[88] or Dirac-Slater equations[89,90] (nonrelativistic and relativistic variants, respectively). Because the EA calculations to which we pay attention in this work have been carried out using nonrelativistic programs only, we shall consider the one-electron HFS equations only, which may be written down as

$$\hat{h}\psi_i(1) = \left(-\tfrac{1}{2}\nabla^2 - \sum_A \frac{Z_A}{\mathbf{r}_{1A}} + V_c + V_{XC} \right) \psi_i(1) = \varepsilon_i \psi_i(1) \tag{45}$$

TABLE IX
EA Values Calculated Taking into Account Correlation Energy

Molecule	Method of calcula-tions	EA_{ad}	IP_{vert}	EA_{vert}	EA_{ad}^{exp}	IP_{vert}^{exp}	EA_{vert}^{exp}	References
LiH	CEPA	0.26						50
	PNO-CI	0.26						50
	CI	0.32	0.34	0.31				70
	MC-SCF	0.28						71
	EOM	0.30						72
BeH	CEPA	0.48						50
	PNO-CI	0.27						50
	EOM	0.77	0.79	0.76	0.74			73
BH	CEPA	0.03						50
	PNO-CI	−0.11						50
	EOM	0						74
CH	CEPA	1.04						50
	PNO-CI	0.95						50
OH	CEPA	1.51						50
	PNO-CI	1.27						50
	PT		2.07				1.825	75
	EOM		1.76					76
F	PT	3.68						75
	CI	3.12					3.448	51
Li_2	EOM	0.90	1.06	0.80				77
C_2	GF	3.60		3.59			3.54	22
	CI	3.30						81
BO	EOM		2.88	2.81				29
CN	EOM		3.71	3.69				29
	CI		3.85					78
	SCEP		3.92					78
HC_2	CI	2.14	2.09	2.17		3.75 ± 0.5	2.21 ± 0.4	79
O_3	GF	2.17		1.67			1.9 − 2.2	22
LiF	EOM	0.46						72
BeO	EOM	1.77						72
NaH	CEPA	0.31						50
	PNO-CI	0.29						50
	MC-SCF	0.28						71
	CI	0.37						80
	EOM	0.36						72
MgH	CEPA	0.83						50
	PNO-CI	0.70						50
AlH	CEPA	0.03						50
	PNO-CI	−0.06						50
SiH	CEPA	1.13						50
	PNO-CI	1.08						50

TABLE IX
(Continued)

Molecule	Method of calcula-tion	EA_{ad}	IP_{vert}	EA_{vert}	EA_{ad}^{exp}	IP_{vert}^{exp}	EA_{vert}^{exp}	References
PH	CEPA	0.76						50
	PNO-CI	0.58						50
SH	CEPA	2.12						50
	PNO-CI	1.99						50
Cl	PT	3.66			3.613			76
P_2	GF	0.30		0.18	0.3 ± 0.5			22
SO_2	GF	0.93		0.69	1.097			22
CsH	MC-SCF	0.36						71
CH_3	CI	−0.4		−0.4				82
Be_2	PT	0.3						83
NH_2	EP	0.61	0.61			0.779 ± 0.037		28
PH_2	EP	0.85	0.85			1.271 ± 0.010		28
Na_2	CI	0.47	0.54	0.42				84
K_2	CI	0.51	0.56	0.47				84
Rb_2	CI	0.51	0.56	0.49				84

where the first term in $\hat{h}$ is the kinetic energy operator, the second and third are Coulomb potentials of the nuclear attractions and interelectronic interactions, respectively, and the fourth is the exchange-correlation potential. The most frequently used for V_{XC} form reads

$$V_{X_\alpha}(1) = -3\alpha \left[\frac{3}{8\pi} \rho(1) \right]^{1/3} \tag{46}$$

where the exchange parameter α is chosen within an interval (2/3, 1). The electronic density is defined by

$$\rho(\mathbf{r}) = \sum_i n_i |\psi_i(\mathbf{r})|^2 \tag{47}$$

where n_i are the occupation numbers of the ith MOs. In statistical methods and in X_α, in particular, they may be fractional.

The way of solving Eq. (45) defines the representation of the one-electron MOs. In the SW-X_α methods, molecular space is divided into three types:

spheres centered at all the atoms of a system considered, the region outside the Watson sphere which comprises the cluster of atomic spheres, and the interstitial region. Inside each atomic sphere the electronic density (and hence the potential) is spherically averaged and the solution reduces to a one-dimensional integration of the Schrödinger equation of the atomic type. In the interstitial region the electronic density is volume averaged and the solutions to Eq. (45) are plane waves. Outside the Watson sphere the charge density is spherically averaged and the solution reduces again to a one-dimensional radial Schrödinger equation. Boundary conditions at the sphere surfaces lead to the secular equations which depend on the energy E to be determined. The potential in SW-X_α, therefore has a muffin-tin shape that may lead to a bad description of some molecular characteristics. Moreover, the one-electron MOs possess various representations in all three types of regions. This makes the well-developed MO LCAO technique difficult to use, in particular, within the framework of a Mulliken population analysis.

On contrary, DVM-X_α is free of restrictions on the shape of the charge density and potential, and ψ_is are presented by linear combinations of symmetry-adapted and orthogonalized basis functions ϕ_j (the usual MO LCAO representation):

$$\chi_i^\kappa = \sum_j C_{ji}^\kappa \phi_j \tag{48}$$

where $\kappa = (k, \lambda)$ designates the λth row of the kth irreducible representation of the correspondent molecular point group. The total wave function is a Slater determinant constructed of MOs ψ_i

$$\psi_i = \sum_k C_{ki} \chi_i \tag{49}$$

Because the most known today theoretical EA estimations for large systems are obtained using DVM, let us consider this method in more details.

In the DV approach the usual variational Raylay-Ritz procedure based on a requirement to the total energy functional to be stationary under arbitrary variations of the total wave function is replaced by the requirement to the error functionals to be minimal at a discrete set of sample points

$$\min_{\mathbf{r}_k} \langle \psi_i(\mathbf{r}) | \hat{h} - \varepsilon | \psi_j(\mathbf{r}) \rangle \tag{50}$$

The fulfilment of this requirement leads to the usual secular equations:

$$HC = ESC \tag{51}$$

The matrix elements are integral sums

$$H_{ij} = \sum_k w_k \chi_i(\mathbf{r}_k) \hat{h} \chi_j(\mathbf{r}_k) \tag{52}$$

$$S_{ij} = \sum_k w_k \chi_i(\mathbf{r}_k) \chi_j(\mathbf{r}_k) \tag{53}$$

where w_k are positive weights and r_k are sample points.[92,93] The choice of the integration schemes has been discussed in Ref. 94 and the question about what number of sample points should be preferred in the Diophantine "closed"[92] and "open"[93] integration schemes has been investigated in Ref. 95.

Whereas evaluating the matrix element for $-(1/2)\nabla^2$, $\Sigma(Z_A/r_{1A})$, and V_{X_α} represents no problem, the evaluation of these of V_C requires estimation of a great number of three-center integrals of the nuclear attraction type

$$V_C(\mathbf{r}_k) = \int \frac{\rho(\mathbf{r})}{|\mathbf{r} - \mathbf{r}_k|} \mathrm{d}\mathbf{r} = \sum_{\mu,\nu} P_{\mu\nu} \int \frac{\chi_\mu^*(\mathbf{r})\chi_\nu(\mathbf{r})}{|\mathbf{r} - \mathbf{r}_k|} \mathrm{d}\mathbf{r} \tag{54}$$

To calculate them quickly the electronic density is projected either onto sets of analytical fit functions[94,95] or onto sets of one-center charge distributions centered at nuclear sites.[96]

$$\rho(\mathbf{r}) = \sum_i \sum_A a_i^A f_i^A(\mathbf{r}) \tag{55}$$

The a_i^A coefficients are defined by means of the least square fitting.[94,95] The choice of a procedure to fit the density defines the possible bases to be used in calculations and the structure of the computer program on the whole. In Ref. 94 we have developed a scheme to find optimal fit function sets using the gradient method of the nonlinear functional optimization. This approach allows the use as basis functions both analytical and numerical HF or HFS functions, together with their arbitrary combinations. The accuracy of calculations (in the framework of the X_α approach itself) depends on the quality of the approximation Eq. (55) and the choice of the basis sets only and is guided.

When approximating the charge density it is sufficient usually to include in the fit function sets functions of the spherical symmetry only. The extension of such sets by the functions with an angular dependence results in changes of *IP*s not exceeding 1 eV.[91,97,98] Therefore, in most cases the inclusion in the approximation Eq. (55) of the spherical functions only is sufficient.

Up to now, by DVM a great number of calculations was carried out beginning with small diatomic molecules[95,99] and up to $Re_3(CO)_{12}$[96] and large clusters simulating solids.[100] Practically in all the cases studied the discrepancy between experimental and calculated valence *IP*s did not exceed 1 eV and on average is equal to 0.5 eV or even less. Moreover, to reach such an accuracy it is sufficient to use the near minimal bases of the numerical HF or HFS atomic functions.[99,101,102]

Let us consider the question of the inclusion of correlation and relaxation within X_α methods which was discussed previously for *ab initio* methods. In the X_α approach, Koopman's theorem is no longer valid because $\varepsilon_i = \partial E_{X_\alpha}/\partial n_i$ where E_{X_α} is the total statistical X_α energy. This relationship allows calculations of the excitation energies, *EA*s, *IP*s, and Auger transition energies[103] using the Slater transition state concept.[86] Here *EA*s are calculated as the first *IP*s of anions, that is, when $n_{\text{HOMO}} = 1/2$ in the transition state. Evaluated in such a way *IP*s are equal to differences of total energies of the initial and final states up to third derivatives of the energy with respect to occupation numbers. These derivatives are usually small for systems near their equilibrium.[86] Therefore, the relaxation effects are automatically included in the transition state calculations.

It is known that the X_α methods predict the correct level orderings even in those cases where the *ab initio* ΔSCF one (in which relaxation is taken into account) fails. The question of accounting for correlation in the X_α methods has been discussed in the literature[104] but up to now is open. The X_α exchange itself is the first term of the expansion of the exchange-correlation potential in the Kohn–Sham–Hohenberg density functional theory[105,106] under some assumptions on its shape. In recent years much work has been done in search of further terms in the expansion of V_{XC} and to estimate nonlocal correlation effects.[107,108] A good review of directions in this field was presented, for example, in Ref. 109.

D. Comparison of Methods for *EA* Calculation

The discussion just presented on nonempirical methods to calculate *EA*s of molecules and radicals allows some conclusions on the possibilities of the approaches considered. The maximal accuracy in *EA* calculations is attained in the *ab initio* methods with correlation. It should be noted that to achieve an agreement between calculated and experimental *EA*s within 0.1–0.3 eV it is necessary to use the bases not worse than double zete with inclusion of diffuse and polarization functions as well as to account for all the terms of the perturbation theory up to the fourth order or, equivalently, all the singly and doubly excited configurations in the CI expansions. However, due to enormous calculational efforts required, such calculations are possible at present for systems containing 30–50 electrons only.

To calculate *EA*s of large systems, the less accurate HFR methods in the Koopman's approximation and X_α methods are in current use. From these

TABLE X
EA (in eV) Calculated by HFR in Koopman's Approximation and by DV-X_α Method

Radical	Calculated		Experimental	References
	HFR	DVM		
BO_2	5.3[a]	3.6	3.6±0.13	101, 8
AlO_2	4.9[a]	3.4	4.11	101, 8
NO_3	5.7[a]	3.6	3.77±0.25 3.68±0.2	101, 80, 12, 13
PO_3	7.4[a]	4.4	3.5	101, 8
BeF_3	9.7	4.5		34, 101
BF_4	10.9	6.2		34
MgF_3	9.6	3.8		34, 101
ClO_4	8.6[a]	6.3	5.82, 6.2	101, 41, 110
AlF_4	11.1	6.1		34

[a] This work, the DZHD + *d* functions of central atom basis.

the preference should be given to the latter because their inaccuracy in *EA* estimations is 0.5–1.0 eV as a rule, compared to 3–4 eV of the former (see Table X).

For systems possessing large *EA*s the inaccuracy in 0.5–1.0 eV may be considered as acceptable. From the other hand, the differences between experimental *EA* estimations obtained by various methods are of 1–2 eV.[1-13] Therefore, at present the X_α methods are the most reliable in estimations of *EA*s of large systems, and we shall mainly use the results of X_α calculations. Obtained by these methods *EA* estimations of the MX_k systems together with experimental values are shown in Table XI.

III. THE RELATIONSHIP BETWEEN *EA* AND ELECTRONIC STRUCTURES OF MOLECULES AND RADICALS

Apart from obtaining quantitative estimations, the main interest in the theoretical and experimental *EA* investigations is the construction of simple models that allow identification of molecules or radicals possessing high or low *EA* values without calculations or measurements at all. Of special interest are systems with anomalous values of *EA*s and *IP*s for which were formulated such simple representations.[101,131,132] Let us consider the arguments they are based on in more detail.

To develop the model representations it is natural to use an orbital approach which is widely applied for the interpretation of electronic, photo,

TABLE XI
EA Estimations and Experimental Values

Compound	Experimental	Ref.	SW-X_α	Ref.	DVM HF(DZ)	Ref.
BO_2	3.6 ± 0.2	8			4.6(3.6)	101
AlO_2	4.1 ± 0.2	8			3.7(3.4)	101
NO_3	3.68 ± 0.2	12				
	3.77 ± 0.25	13			4.0(3.6)	101
PO_3	3.5	8			5.3(4.4)	101
ClO_4	5.8	41	6.9	112	6.9(6.3)	101
	6.2	110				
ClO_3	2.83 ± 0.02	41				
	3.96					
BrO_3	3.22 ± 0.02	41				
IO_3	5.5 ± 0.7	41				
BeF_3					4.1(4.5)	101
MgF_3					3.2(3.8)	101
BF_4					6.2(6.2)[a]	
AlF_4					6.1(6.2)[a]	
SiF_5					6.4(6.3)	101
PF_6					6.8(6.6)	101
SF_6	0.3–0.75	12, 13	0.5, 0.7	113, 114	0.3(0.6)	116
SeF_6	3.0 ± 0.2	111	2.9	113	2.1(3.0)	116
TeF_6	3.34 ± 0.2	111	2.6	113	1.2(2.2)	116
TiF_6					7.5	117
VF_6					6.7	117
CrF_6					5.0	117
MnF_6					5.9	117
FeF_6					7.0	117
CoF_6					6.8	117
NiF_6					6.9	117
CuF_6					6.1	117
ZnF_6					5.8	117
ZrF_6					7.1	117
MoF_6	$\geqslant 5.14$, 3.6 ± 0.2, $\geqslant 5.36$, $\geqslant 4.5$, 5.77	119, 118, 120, 124, 125	4.5–4.8	114	3.2	117
TcF_6					4.3	117
RuF_6					5.2	117
RhF_6					6.1	117
PdF_6					6.1	117
AgF_6					6.4	117
CdF_6					4.8	117
AsF_6					7.6[a]	
SbF_6	6.0	8			7.3[a]	
HfF_6					8.8	121
TaF_6					8.4	121

TABLE XI
(Continued)

Compound	Experimental	Ref.	SW-X_α	Ref.	DVM HF(DZ)	Ref.
WF_6	3.51 ± 0.1, $\geqslant 4.9$, 3.7, 5.1 ± 0.5	8, 124, 123, 119	4.3–5.1	114	3.5	121
ReF_6			5.3–6.3	114	4.8	121
OsF_6			6.2–7.3	114	6.0	121
IrF_6	>4.34, $\geqslant 5.1 \pm 0.5$	41, 119	7.0–8.3		7.2	121
PtF_6	5.1–7.9	8	7.6–9.1	114	7.4	121
AuF_6			8.4–10.1	114	8.1	121
HgF_6					5.8	121
UF_6	4.9 ± 0.5, >5.0, $\geqslant 3.6$, $\geqslant 4.3$, 5.1, 4.85 ± 0.25	126–130	5.0	114		
$BeCl_3$					4.5(4.2)	122
BCl_4					4.9(4.5)	122
$MgCl_3$					4.5(4.5)	122
$AlCl_4$					5.8(5.0)	122
$SiCl_2$					4.9(3.7)	122
PCl_6					4.3(3.7)	122
$MgCl_3$					4.5(4.5)	122
$AlCl_4$					5.8(5.0)	122
$SiCl_5$					4.9(4.2)	122
PCl_6					4.3(3.7)	122
SCl_6					3.4	116
$SeCl_6$					4.3	116
$TeCl_6$					3.5	116
$SbCl_6$					5.6[a]	
$AsCl_6$					5.4[a]	
TcF_8					5.8	137
MnF_8					6.7	137
Al_2F_7					7.5	137
P_2F_{11}					8.6	137
As_2F_{11}					8.6	137
V_2F_{11}					8.0	137
ScF_4					4.8	137
MoO_4					6.2	138
TcO_4					4.9	138
RuO_4					3.1	138
RhO_4					3.8	138
PdO_4					4.7	138
AgO_4					4.4	138
CdO_4					3.4	138
CrO_4					6.1	138
MnO_4					5.0	138
FeO_4					3.4	138

TABLE XI
(Continued)

Compound	Experimental	Ref.	SW-X_α	Ref.	DVM HF(DZ)	Ref.
CoO_4					4.2–4.5	138
NiO_4					4.7	138
CuO_4					4.4	138
ZnO_4					3.9	138
MnF_3	4.36	8				
FeF_3	4.25	8				
FeF_4	5.40	115				
MnF_4	5.25	115				
MnF_2	4.36 ± 0.15	115				
PtF_4	5.22 ± 0.2	115				
CrF_4	3.6 ± 0.3	8				
$TiCl_4$	2.88 ± 0.15	8				
WF_5	$\geqslant 1.8 \pm 0.3$	8				
MoF_5	$\geqslant 3.5$, 3.63 ± 0.2	8				
UF_5	2.4, 3.30 ± 0.16	8				
	4.0 ± 0.4					
UF_7	$\leqslant 5.5$	8				
FeO_2	2.85	8				
CrO_3	2.38 ± 0.52	8				
MoO_3	2.58 ± 0.41	8				
WO_3	3.64 ± 0.41	8				
ReO_3	2.85 ± 0.55	8				
ReO_4	4.45	8				

[a] Values from this work.

and X-ray spectra and which follows Woodward-Hoffman rules. According to this approach the orbital energies should have the following order: ε(bonding) > ε(nonbonding) > ε(antibonding) for MOs composed of AOs of the same type. Therefore, it is natural to relate the *EA* value with the structure of the lowest unoccupied MO(LUMO) of a neutral system, which accepts an extra electron, as a rule, when the anion forms, or to relate the first *IP* value of an anion with the structure of its highest occupied MO(HOMO). Respectively, *IP*s should decrease in the series *IP*(bonding) > *IP*(nonbonding) > *IP*(antibonding), that is, the *EA* value depends on the type of anion HOMO.

The second important factor influencing the molecular *EA* is the *EA* of atoms that constitute the molecule. The larger the atomic *EA*s, the larger an *EA* may possess the compound. Therefore, the cases are possible when the

HOMO of an anion possesses an antibonding character but, due to the large *EA*s of ligand atoms, the molecule will possess the large *EA*, which may be larger *EA*s of molecules with nonbonding or even bonding HOMOs but whose atomic *EA*s are small.

The third important factor influencing the *EA* of a molecule is the extent of the delocalization of an extra electron. The relationship between delocalization and the *EA* value may be elucidated using simple arguments of Lowe.[133] According to them, the smaller the part of an electron $\delta\bar{e}$ is added to an atom, the easier is the task for the atom to retain it. In the limiting case when $\delta\bar{e} \rightarrow 0$ the first *IP* of an anion would approach the first *IP*s of neutrals, which are quite large. Thus, the *EA* should increase if an extra electron were delocalized over a large number of atoms constituting a molecule.

Apart from these factors influencing the relationship between the *EA* and peculiarities of the electronic structure, the *EA* value depends on some other specific factors which will be considered later. However, given the three main factors just discussed, it is possible to formulate conditions that should satisfy the systems with anomalous *EA* or *IP* values. Next we shall consider in detail these extreme cases and possible consequences of the anomalous *EA* and *IP* values for chemistry.

A. Systems with High *EA* Values

Among atoms of the periodic table the maximal *EA*s are associated with halogens (3.0–3.6 eV). Due to collective effects, molecules, and polyatomic radicals may possess *EA*s exceeding that of halogens; such systems have been called superhalogens.

Two problems arise in the investigations of systems with anomalous *EA* values. The first one is the necessity to identify those molecules that might be considered as superhalogens, and the second one is to determine the maximal possible *EA* values in molecular systems and in what specific cases the maximal *EA* might be realized.

Let us first apply the ideas given in Section III to systems of the MX_k type. For the *EA* of these systems to be quite large, the ligands *X* must be electronegative atoms with large *EA*s, for example, the fluorine or oxygen atoms. Further, keeping in mind the order ε(bond.) > ε(nonbond.) > ε(antibond.), it would be desirable that the LUMO be a bonding MO. However, if the ligands are halogens or oxygens (in general, if ligand atoms possess many electrons), the LUMO is always either nonbonding or antibonding. If the LUMO is nonbonding, the *EA* of MX_k should be greater than the *EA* of the ligand atoms *X* due to high atomic *EA*s and the delocalization effects.

It should be noted that even in those cases where the HOMO is antibonding by symmetry but the contribution of the central atom AOs is small to the LUMO (i.e., it is practically nonbonding), the *EA* of MX_k may be large. Apart from the three main factors mentioned previously, which may lead to high *EA*s, in MX_k the central atom will possess a positive charge. This charge would provide an additional stabilization of an extra electron which is delocalized over ligands, in particular, if the ligands are halogens or oxygens. The calculations[101] on the systems consisting of the F^- ion and a positive point charge displaced by 3.0 a.u. from F^- have shown that the increase of the point charge in +0.1 e leads to the *EA* increase of the F atom in 1.0 eV. Therefore, the larger the charge of *M*, the larger value may approach the *EA*.

The foregoing notions may be reformulated in terms of localized MOs. If in the MX_k^- ion with electronegative ligands the central atom *M* possesses no lone pair (LP) or the MX_k^- radical possesses no unpaired electrons localized at *M* (i.e., all the valence electronic density is distributed over localized MOs of bonds and LPs of ligands), then an extra electron will be delocalized over electronegative ligands, and the first IP of the anion MX_k^- should be large. On the contrary, if the central atom possesses LPs or unpaired electrons and the ionization involves these electrons, then the first IP of MX_k^- should be relatively small due to small *EA*s of *M*.

It should be noted that the electronic structure of MX_k where *M* is a transition metal differs from that of MX_k where *M* is a nontransition element. In MX_k, where *k* is the maximal formal valency of *M*, the extra electron is accepted by an antibonding MO, as a rule. If *M* is a nontransition atom, then the geometry of MX_k^- undergoes distortions so as to an extra electron was formed a LP of *M* (or, in radicals is an unpaired electron of the central atom). For example, if an electron is added to the linear BeF_2 molecule, then it occupies an antibonding MO and BeF_2^- undergoes an angular distortion to decrease the antibonding influence of this MO. In this case an extra electron becomes an unpaired electron of Be.

The opposite situation is observed in the MX_k molecules and radicals of transition metals whose geometry does not change, as a rule, after the occupation of the antibonding HOMOs. For example, in the series of hexafluoride anions from WF_6^- to AuF_6^-, the consecutive filling up the antibonding $4t_{2g}$ HOMO occurs; however, their geometry remain undistorted.

Therefore, it is expedient to consider the cases separately when *M* is a nontransition element and when *M* is a transition metal.

B. Superhalogens among the *sp*-Element Compounds

According to ideas presented, in the MX_k systems where *k* is less than or equal to the maximal formal valency of the central atom *M*, an extra electron will be accepted by either an antibonding or nonbonding MO of the central atom (LP of M). In both cases EA should be relatively small due to the small EA of M, and such systems will possess low EAs. Indeed, both the results of *ab initio* calculations with correlation (Table X) and the experimental data confirm this conclusion. In the MX_{k+1} radicals (k is the maximal formal valency of M) an extra electron occupies a nonbonding MO of ligands in which the central atom AOs do not enter by symmetry and its delocalization over ligands possessing high EAs results in a high EA of MX_{k+1}.

According to results of calculations on the electronic structure of BeF_3, BF_4, MgF_3, AlF_4, SiF_5, PF_6,[101] $BeCl_3$, BCl_4, $MgCl_3$, $AlCl_4$, $SiCl_5$, and PCl_6[122] by the DV-X_α method, the EAs of these radicals exceed 3.5 eV, and all of them may be considered as superhalogens. The analogous assignment is valid for systems $MX_{(k+1)/2}$ where X is a divalent atom and k is the maximal formal valency of M, because the calculated EAs of NO_3, PO_3, and ClO_4[101] exceed 3.5 eV as well.

To underline the importance of high-ligand EAs for high EAs of MX_{k+1} let us consider radicals whose HOMOs are bonding. It seemed to be in this case that IPs from HOMOs of MH^-_{k+1} should be sufficiently large. However, the estimations of the first IPs in the Koopman's approximation,, which are probably overestimated by 1.0–1.5 eV, for BeH_3^-, BH_4^-, MgH_3^-, and AlH_4^- (Table II) are lesser IPs from HOMOs of the corresponding MF^-_{k+1} anions[32, 34] whose HOMOs are nonbonding. The reason is apparently connected with the low EA of hydrogen atoms.

Thus it is possible to identify a number of systems of *sp*-elements that should be attributed to superhalogens. These are radicals MX_{k+1} where X is a halogen and $MX_{(k+1)/2}$ where X is oxygen. Of course, some other systems may possess sufficiently high EAs. The examples of such molecules represent the SeF_6 and TeF_6 molecules[116] possessing the MX_k type. An extra electron in MX_k^- (k is the maximal formal valency of M) occupies the antibonding MO, and the EA of such systems should be low. This is the case for SF_6 whose EA is equal to 0.3–0.75 eV.[12, 13] However, the EAs of SeF_6 and TeF_6 are quite large and equal 3.0 ± 0.2 and 3.34 ± 0.2 eV, respectively.[111] The quite large EA of these molecules is related to the high polarity of the M–F bonds. Consequently, though the LUMOs are antibonding with respect to the central atom–ligand interaction the real contribution of the central atom AOs to the LUMO is small. The LUMOs of SeF_6 and TeF_6 are practically nonbonding ligand MOs, which results in the high molecular EA. It is pertinent to note that the SeF_6^- and TeF_6^- ions as well as "metallic" octahedra MoF_6^- and WF_6^-, whose extra electron occupies the antibonding HOMOs, retain the undistorted octahedral configurations, too.[134, 135] The difference in the EA value in the series of isoelectronic molecules SF_6–SeF_6–TeF_6 is closely related to chemical properties of the chalcogen hexafluorides.

Apart from SeF_6 and TeF_6, some other systems of the MX_k type, where M

is an atom of the fourth, fifth and farthest periods, may possess quite large *EA*s. However, without calculations or experimental measurements it is not possible to know beforehand what the *EA* of these quite large systems would have to be for these systems to be considered as superhalogens.

The *EA* of MX_{k+1} should increase with increasing k due to delocalization of an extra electron over a larger number of ligands. Indeed, in series MgF_3 ($EA \sim 3.8$ eV)–AlF_4 ($EA \sim 6.1$ eV)–SiF_5 ($EA \sim 6.3$ eV)–PF_6 ($EA \sim 6.6$ eV) and AlO_2 ($EA \sim 3.4$ eV)–PO_3 ($EA \sim 4.4$ eV)–ClO_4 ($EA \sim 6.3$ eV) the *EA*s increase from 3.8 to 6.6 eV in the first case and from 3.4 to 6.3 eV in the second. Therefore, by analogy, the extreme *EA* values should be expected from MX_{k+1} with $k > 5$, however it should be mentioned that the increase in k may lead to a decrease in *EA*. In particular, in the series $MgCl_3$ ($EA \sim 4.5$ eV)–$AlCl_4$ ($EA \sim 5.2$ eV)–$SiCl_5$ ($EA \sim 4.4$ eV) – PCl_6 ($EA \sim 4.0$ eV) the *EA* decreases due to the strong repulsion of ligands.[122] The latter is connected with the fact that the HOMO of MX_{k+1} is nonbonding with respect to the central atom–ligand interaction but the HOMO possesses an antibonding character at the same time with respect to ligand–ligand interactions. When k is sufficiently large the mutual repulsion of ligands will prevail over delocalization effects and it will lead to low *EA* values. At optimal values of k the ligand–ligand antibonding character is small and does not prevail over delocalization of an extra electron that leads to high *EA*s. For the time being, such an optimal alignment of antibonding factors and stabilization is known for AsF_6 and SbF_6 which possess the highest *EA* values among superhalogens of the *sp*-elements.

C. Superhalogens among the Transition Metal Systems

In transition metals compounds of the MX_k type filling up the antibonding HOMO does not lead to geometry distortions of the high symmetry configuration of these systems, as a rule, that may be related with peculiarities in the *EA* behaviour of these systems. Let us consider the dependence of the *EA* value on the electronic structure in the transition metal hexafluorides. The consideration is begun with the 5*d* metal hexafluorides because the peculiarities of their electronic structure have the most clear form. In Table XII are given the valence *IP*s, in Table XIII are one-electron ground-state energies and compositions of the upper MOs of the e_g and t_{2g} symmetries, and in Table XIV are Mulliken population analysis data for the 5*d* metal hexafluorides.

It is seen that in the Hf–Hg series the transition from filling up the nonbonding purely ligand t_{1g} HOMO(Hf, Ta) to filling up the antibonding HOMO with respect to the central atom–ligand interaction of the t_{2g} symmetry (W–Au) and then to filling up the next antibonding $6e_g$ MO(Hg) is realized. The data of Table XII shows the following peculiarities in the

TABLE XII
Valence *IP*s (in eV) of the 5*d* Metal Hexafluoride Anions[a]

		HfF_6^-	TaF_6^-	WF_6^-	ReF_6^-	OsF_6^-	IrF_6^-	PtF_6^-	AuF_6^-	HgF_6^-
MO	Spin[b]	$(t_{1g}\alpha)^2$	$(t_{1g})^6$	$(t_{2g}\alpha)^1$	$(t_{2g}\alpha)^2$	$(t_{2g}\alpha)^3$	$(t_{2g}\beta)^1$	$(t_{2g}\beta)^2$	$(t_{2g})^6$	$(e_g\alpha)^1$
$6e_g$	H									4.5
$4t_{2g}$	L			2.2	3.5	4.7	5.9	6.1	6.8	7.6
	H						4.5	5.6		7.2
$1t_{1g}$	L	7.5	7.1	7.3	7.1	7.6	7.4	7.2	6.9	7.3
	H	6.8		6.9	6.7	6.6	6.6	6.7		7.0
$9t_{1u}$	L	7.3	7.3	7.7	7.6	7.7	7.9	7.5	7.3	8.0
	H	6.8		7.3	7.2	7.1	7.1	7.1		7.3
$2t_{2u}$	L	8.0	7.5	7.5	7.8	8.0	7.9	7.7	7.4	7.8
	H	7.2		7.3	7.3	7.2	7.1	7.2		7.5
$8t_{1u}$	L	8.3	8.0	8.1	8.4	8.7	8.6	8.5	8.2	9.0
	H	7.5		7.9	7.9	8.0	8.0	8.1		8.5
$8a_{1g}$	L	8.6	8.7	8.8	9.0	9.5	9.3	9.2	9.1	10.5
	H	8.2		8.7	8.7	8.8	8.9	9.0		9.5
$3t_{2g}$	L	9.9	10.1	10.5	11.3	12.1	12.2	12.4	13.4	16.3
	H	9.3		10.0	10.1	10.2	10.9	12.0		15.7
$5e_g$	L	10.3	10.6	11.0	11.5	12.5	12.2	12.5	13.6	17.2
	H	9.9		10.6	10.7	10.9	11.4	12.1		17.0

[a] From Ref. 121.
[b] H and L designate transitions to the high and low spin multiplicity final states of the MF_6 neutral systems.

valence *IP* behavior when going along the series. First, the first *IP* changes nonmonotonously; its value drops sharply when going from TaF_6^- to WF_6^- then grows steadily to AuF_6^- and drops again in HgF_6^-. Second, *IP*s from the "ligand" band $1t_{1g}$–$8a_{1g}$ do not change practically in the series. Third, *IP*s from the bonding $3t_{2g}$ and $5e_g$ MOs increase almost monotonously, although the change in their energies when going along the series is less that of 5*d* AOs (see Table XV).

Such behavior of *IP*s may be understood if the composition and symmetry of corresponding *MO*s is considered. The HOMOs of the Hf and Ta hexafluoride anions are nonbonding ligand *MO*s and the *EAs* of MF_6 are large due to delocalization of an extra electron over six strongly electronegative fluorines in accordance with notions about *sp*-superhalogens. It is pertinent to note that the *EA*s of these hexafluorides are close to the *EA*s of the most electronegative superhalogens among *sp*-superhalogens, namely, AsF_6 and SbF_6. When going to WF_6 the *EA* drops; that is connected with beginning the filling up of the $4t_{2g}$ MO which is antibonding and the gap between energies of $4t_{2g}$ and the next $1t_{1g}$ MO is ~4 eV. Because $4t_{2g}$ is antibonding it

TABLE XIII

One-electron Ground State Energies (in eV) and Composition of the Upper e_g and t_{2g} MOs for the 5d Metal Hexafluoride Anions[a]

MO	HfF_6^-					TaF_6^-					WF_6^-				
	$\varepsilon_i(\alpha)$	$\varepsilon_i(\beta)$	%M	$5d_M$	$2p_F$	$\varepsilon_i(\alpha)$	$\varepsilon_i(\beta)$	%M	$5d_M$	$2p_F$	$\varepsilon_i(\alpha)$	$\varepsilon_i(\beta)$	%M	$5d_M$	$2p_F$
$6e_g$	8.1	8.1	4	0.40	0.05	7.6		21	0.72	0.04	6.8	7.1	45	−0.96	0.12
$4t_{2g}$	6.0	5.9	78	−1.16	0.33	3.3		74	−1.05	0.33	1.8	2.4	68	−1.00	0.33
$3t_{2g}$	−6.0	−6.3	22	0.35	0.40	−6.6		29	0.40	0.38	−6.9	−6.6	35	0.46	0.36
$5e_g$	−6.4	−6.6	13	0.46	0.25	−7.0		20	0.48	0.23	−7.2	−7.1	24	0.50	0.22
	ReF_6^-					OsF_6^-					IrF_6^-				
$6e_g$	5.5	6.1	57	−1.08	0.17	4.1	5.0	55	−1.03	0.20	3.2	3.7	52	−0.98	0.21
$4t_{2g}$	0.3	1.5	61	−0.99	0.36	−1.0	0.5	52	−0.90	0.38	−1.6	−0.8	47	−0.90	0.39
$3t_{2g}$	−7.3	−6.7	43	0.57	0.33	−7.8	−6.9	51	0.66	0.30	−8.0	−7.4	57	0.70	0.28
$5e_g$	−7.6	−7.3	29	0.58	0.22	−8.1	−7.5	35	0.62	0.20	−8.3	−7.9	39	0.65	0.19
	PtF_6^-					AuF_6^-					HgF_6^-				
$6e_g$	2.4	2.2	48	−0.91	0.22	0.4		36	−0.78	0.25	−0.9	−0.4	25	0.66	0.26
$4t_{2g}$	−1.9	−2.3	40	−0.77	0.40	−3.2		24	−0.59	0.44	−4.0	−3.8	16	0.46	0.45
$3t_{2g}$	−8.2	−8.5	64	0.76	0.25	−9.4		80	0.88	0.18	−12.1	−12.0	90	0.93	0.12
$5e_g$	−8.5	−8.7	44	0.69	0.18	−9.5		58	0.79	0.15	−11.7	−11.4	64	0.83	0.12

[a] From Ref. 121.

TABLE XIV
Mulliken Populations of AOs, Charges on Atoms (Q_A) and Overlap Populations ($Q_{M\text{-}F}$) of the 5*d* Metal Hexafluoride Anions and *EA*s of the 5*d* Metal Hexafluorides[a]

	HfF_6^- t_{1g}^5	TaF_6^- t_{1g}^6	WF_6^- t_{2g}^1	ReF_6^- t_{2g}^2	OsF_6^- t_{2g}^3	IrF_6^- t_{2g}^4	PtF_6^- t_{2g}^5	AuF_6^- t_{2g}^6	HgF_6^- e_g^1
6*s*	0.29	0.37	0.40	0.43	0.48	0.50	0.55	0.64	0.58
$5de_g$	0.83	1.07	1.29	1.47	1.67	1.88	2.15	2.68	3.37
M $5dt_{2g}$	1.29	1.84	2.71	3.58	4.40	5.21	5.98	6.29	6.34
6*p*	0.28	0.40	0.42	0.46	0.49	0.51	0.52	0.57	0.63
2*s*	1.93	1.93	1.93	1.93	1.93	1.93	1.93	1.94	1.93
$2p_{x,y}$	1.79	1.83	1.83	1.84	1.86	1.87	1.90	1.95	1.94
$2p_z$	1.84	1.78	1.75	1.71	1.67	1.63	1.58	1.48	1.52
Q_F	−0.39	−0.39	−0.36	−0.34	−0.33	−0.315	−0.32	−0.30	−0.34
Q_M	1.33	1.36	1.19	1.06	0.96	0.89	0.91	0.83	1.06
$Q_{M\text{-}F}$	0.093	0.136	0.127	0.110	0.092	0.071	0.049	0.017	0.078
EA (in eV)	8.8	8.4	3.5	4.8	6.0	7.2	7.4	8.1	5.8

[a] From Ref. 121.

might be expected that its further occupation along the series will lead to decreasing the *EA*. However, as it is seen from Table XIV there is observed a monotonous increase in the *EA* value when going along the series which approaches ~7 eV in AuF_6' and the gap between $4t_{2g}$ and $1t_{1g}$ disappears.

Let us consider the composition of the $4t_{2g}$ MO in more detail. According to data of Table XIII its character changes from essentially antibonding (WF_6^-) to practically nonbonding (AuF_6^-). With the decrease in the antibonding character the energy of this MO decreases and approaches energies of the nonbonding ligand MOs ($1t_{1g}$, $9t_{1u}$, $2t_{2u}$). The *EA* of AuF_6 is close to that of TaF_6 in which an extra electron fills up the purely ligand by symmetry $1t_{1g}$ MO. When going from AuF_6^- to HgF_6^- an extra electron fills up the next antibonding $6e_g$ MO that is accompanied by lowering the *EA*. The latter is related mainly with the large $6e_g$—$4t_{2g}$ splitting by the ligand field because the $6e_g$ MO is essentially a ligand one. Thus, according to Table XIII the contribution of 5*d* AOs to $6e_g$ is ~25% and coincides practically with that of 5*d* AOs to $4t_{2g}$ of AuF_6^-. In HgF_6^- (in comparison with AuF_6^-) there is observed an increase of the *M*-F overlap population.

It is of interest to relate the *EA* behavior in the series of the 5*d* metal hexafluorides with the behavior of energies of 5*d* atomic levels. When going along the series from Hf to Hg the energy of the 5*d* AO increases (Table XV); that, in turn, leads to lowering energies of the $3t_{2g}$ and $5e_g$ MOs, with its transformation mostly to the nonbonding atomic 5*d* levels. In other terms,

TABLE XV

One-Electron Energies of the *nd* and $(n + 1)s$ HF Atomic Orbitals of the *nd* Metals for Their High Orbital Momentum Configurations[a]

Configuration	$nd^3(n+1)s^1$	$nd^4(n+1)s^1$	$nd^5(n+1)s^1$	$nd^6(n+1)s^1$	$nd^7(n+1)s^1$	$nd^8(n+1)s^1$	$nd^9(n+1)s^1$	$nd^{10}(n+1)s^1$	$nd^{10}(n+1)s^2$
atom	Ti	V	Cr	Mn	Fe	Co	Ni	Cu	Zn
3*d*	7.44	8.73	10.16	10.42	11.07	11.76	12.47	13.36	21.29
4*s*	5.58	5.83	6.04	6.18	6.27	6.35	6.43	6.50	7.96
atom	Zr	Nb	Mo	Tc	Ru	Rh	Pd	Ag	Cd
4*d*	6.82	8.18	9.74	10.30	11.23	12.25	13.30	14.62	20.78
5*s*	5.60	5.87	6.06	6.08	6.05	6.03	6.01	5.98	7.21
atom	Hf	Ta	W	Re	Os	Ir	Pt	Au	Hg
5*d*	6.70	8.01	9.50	10.02	10.90	11.88	12.92	14.18	19.43
6*s*	5.72	6.00	6.21	6.21	6.16	6.10	6.06	6.00	7.10

[a] From Refs. 117 and 121.

the 5*d* electrons tend to localize near the central atom that leads to decreasing the 5*d* AO population in the upper $4t_{2g}$ and $6e_g$ MOs and to approaching $4t_{2g}$ the ligand valance band. This conclusion is confirmed by the decrease in overlap populations in the series WF_6^-–AuF_6^- and by an increase of $Q_{M\text{-}F}$ when going to HgF_6^- since in the latter the 5*d* electrons are enforced to fill up partially the antibonding $6e_g$ MO.

The main peculiarities of the electronic structure of the 4*d* metal hexafluorides are the same as for the 5*d* metal hexafluorides (compare Tables XII and XVI), nonetheless there are some differences in the electronic structure of the isoelectronic species.

First, the valence ligand band of the 4*d* metal hexafluorides is somewhat displaced toward the highest energies (0.3–0.5 eV) in comparison with the 5*d* case. This result is consistent with the known experimental data on *IP* values for MoF_6 and WF_6,[9] according to which $\Delta IP(WF_6\text{–}MoF_6) \approx 0.3$ eV. Second, the HOMO of the t_{2g} type in MoF_6^- is placed 1.3 eV closer to the valence band in comparison with that of WF_6^-. This leads to the larger *EA* value in MoF_6 (by ~ 1.0 eV). The calculated $\Delta IP(4t_{2g}\text{–}1t_{1g})$ of WF_6^- and $\Delta IP(3t_{2g}\text{–}1t_{1g})$ of MoF_6^- differences (1.3 eV) are consistent with the difference between

TABLE XVI
One-Electron Energies of the 4*d* Metal Hexafluoride Anions Calculated in the Appropriate Transition States[a]

MO	Spin	ZrF_6^-	NbF_6^-	MoF_6^-	TcF_6^-	RuF_6^-	RhF_6^-	PdF_6^-	AgF_6^-	CdF_6^-
$5e_g$	α									4.8
$3t_{2g}$	α			3.2	4.3	5.2	5.1	5.9	6.7	6.9
	β						5.6	6.0		7.0
$1t_{1g}$	α	6.5	6.8	7.0	6.3	6.1	6.1	6.1	6.4	6.1
	β	6.8		7.0	6.6	6.7	6.5	6.5		6.3
$7t_{1u}$	α	6.7	7.1	7.2	6.9	6.6	6.6	6.5	6.8	6.3
	β	6.8		7.2	7.1	7.1	7.0	6.9		6.7
$1t_{2u}$	α	7.0	7.1	7.3	6.9	6.6	6.5	6.4	6.5	6.5
	β	7.3		7.4	7.2	6.9	6.8	6.6		6.8
$6t_{1u}$	α	7.2	7.6	8.0	7.7	7.5	7.5	7.5	7.7	7.6
	β	7.6		8.0	7.9	7.9	7.8	7.7		7.9
$7a_{1g}$	α	7.9	8.3	8.5	8.5	8.3	8.2	8.2	8.5	8.4
	β	8.1		8.6	8.6	8.5	8.4	8.3		8.9
$2t_{2g}$	α	9.1	9.4	10.0	9.7	9.6	10.2	11.2	13.7	18.0
	β	9.4		10.3	10.4	10.7	10.8	11.5		18.4
$4e_g$	α	9.6	10.1	10.4	10.4	10.4	10.7	11.5	14.3	16.8
	β	9.8		10.6	10.8	11.0	11.1	11.6		16.9

[a] From Ref. 117.

energies of the first optical transition in MoF_6 and WF_6 $[\Delta\nu_1(WF_6)-\Delta\nu_1(MoF_6)] = 1.33$ eV.[136] The lowering of t_{2g} in MoF_6^- in comparison with WF_6^- is connected with a less antibonding character of this MO that, in turn, is caused by the larger energy of the 4*d* AO of Mo (Table XV).

As well as in the case of the 5*d* metal hexafluorides the antibonding character of the t_{2g} HOMO decreases when this HOMO is filling up in the series of the 4*d* metal hexafluorides. However, in the latter case the decrease is more rapid and this results in a more rapid approach of the HOMO to the ligand valence band. Moreover, in AgF_6^- (whose $3t_{2g}$ HOMO is fully occupied) the energy of this MO falls below energies of the pure ligand $1t_{1g}$ and $1t_{2u}$ MOs. It should be noted that for AuF_6^- [121] such an inversion of levels was not observed. The MO order in AgF_6^- contradicts the traditional notions about the MO order in dependence on their character: ε(bond.) > ε(nonbond.) > ε(antibond.). However, the reasons are clear after a close analysis of the nodal structure of the upper valence MOs presented schematically in Fig. 1. As can be seen from Fig. 1, the $3t_{2g}$ MO is an antibonding MO of the π type with respect to the central atom–ligand interaction and a bonding MO of the σ type with respect to the ligand–ligand interactions; the t_{1g} MO is σ-antibonding and purely ligand by symmetry; the $7t_{1u}$ and $1t_{2u}$ MOs are of the π type, and if $7t_{1u}$ contains some contribution of the central atom, then $1t_{2u}$ is an antibonding and purely ligand MO. When going from MoF_6^- to AgF_6^- the contribution of the central atom to the $3t_{2g}$ decreases sharply and in AgF_6^- all the $3t_{2g}$–$1t_{2u}$ band may be considered as purely ligand and all the central atom–ligand interactions in this band may be neglected. In such a case, according to traditional notions, one may expect the following sequence of MOs: $3t_{2g}$ (σ-bond.) > $7t_{1u}$(π-bond.) > $1t_{2u}$(π-antibond.) > $1t_{2g}$ (σ-antibond.). Exactly this MO sequence occurs for AgF_6^- and CdF_6^-; that is, the electronic structure of the end of the 4*d* metal hexafluorides is determined by purely ligand interactions.

The main peculiarities in the *EA* behavior and one-electron energies in the 3*d* metal hexafluorides (Table XI) are the same as in the case of the 4*d* and 5*d* metal ones (Tables XII, XVI, and XVII). In particular, the antibonding $2t_{2g}$ HOMO of CrF_6^- is placed only 1.5 eV above the ligand MOs, whereas the corresponding $3t_{2g}$ and $4t_{2g}$ MOs in MoF_6^- and WF_6^- are 3.8 and 5.1 eV above their ligand MOs, respectively. Under further filling up of the $2t_{2g}$ MO in the 3*d* anions, its energy approaches rapidly the energies of ligand MOs and the inversion of the $2t_{2g}$ and $1t_{1g}$ MOs is already observed in FeF_6^-, whereas the analogous inversion of the $3t_{2g}$ and $1t_{1g}$ MOs was observed in the end of the 4*d* metal hexafluorides only. Such behavior of the $2t_{2g}$ MO is connected with its lesser antibonding character due to the increase in 3*d* AO energies of the first transition row atoms. The decrease in the antibonding character of the HOMO with an increase of the *nd* energies is also observed

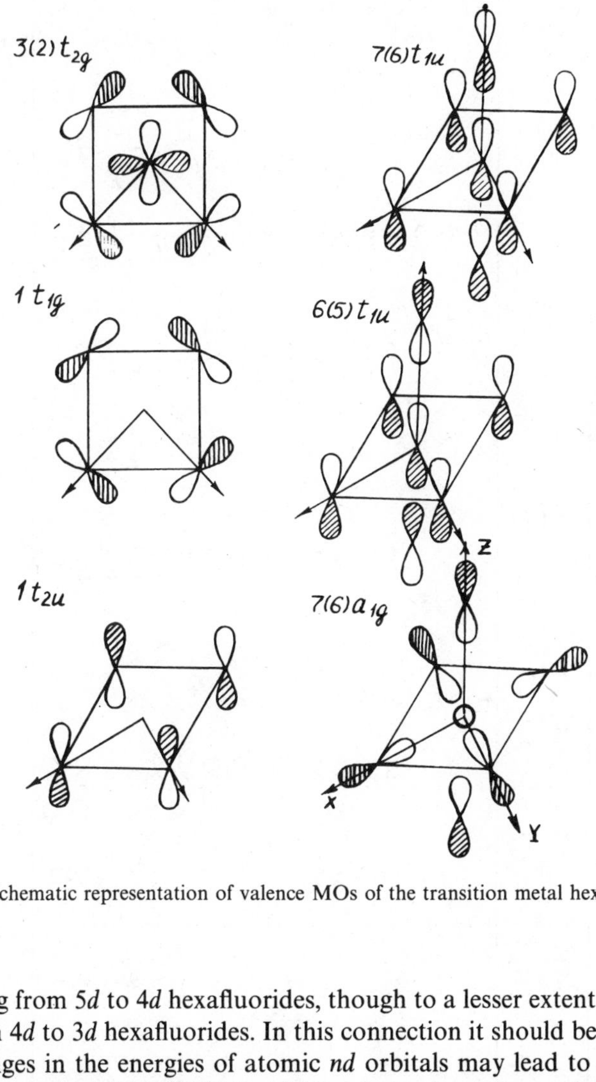

Fig. 1. Schematic representation of valence MOs of the transition metal hexafluorides.

when going from 5*d* to 4*d* hexafluorides, though to a lesser extent than when going from 4*d* to 3*d* hexafluorides. In this connection it should be noted that small changes in the energies of atomic *nd* orbitals may lead to quite large changes of energies of the HOMOs of the t_{2g} type. A small contribution of the central atom AOs to the HOMO in the 3*d* metal hexafluorides leads to high *EA* values (6–7 eV) for all the series except for, probably, CrF_6, whose *EA* is also quite large (5.0–5.5 eV).

The analysis of the relationship of the *EA* values to the electronic structure just presented reveals the essential difference between *nd* metal hexafluorides and *sp*-atoms fluorides, and we may consider all the *nd* metal hexafluorides as superhalogens. The dependence of *EA*s on the electronic structure is quite

TABLE XVII

One-Electron Energies of the $3d$ Metal Hexafluoride Anions Calculated in the Appropriate Slater Transition States[a]

MO		TiF_6^-	VF_6^-	CrF_6^-	MnF_6^-	FeF_6^-		CoF_6^-		NiF_6^-	CuF_6^-	ZnF_6^-
		$t_{1g}{}^5$	$t_{1g}{}^6$	$t_{2g}{}^1$	$t_{2g}{}^2$	α^3	$\alpha^3\beta^1$	$\alpha^2\beta^2$	$\alpha^3\beta^1$	$t_{2g}{}^5$	$t_{2g}{}^6$	$e_g{}^1$
$4e_g$												5.8
$2t_{2g}$	α			5.0	5.9	7.0	5.7	6.8	6.4	6.7	7.0	7.6
	β								7.5	6.9		7.7
$1t_{1g}$	α	6.7	6.7	6.5	6.6	6.5	6.6	7.0	6.9	6.7	6.1	6.2
	β	7.1		6.5	6.8	6.8	6.7		7.3	6.9		6.4
$6t_{1u}$	α	7.1	7.0	6.8	7.0	7.0	7.0	7.5	7.4	6.9	6.3	6.5
	β	7.4		6.9	7.0	7.1	7.1		7.6	7.0		6.7
$1t_{2u}$	α	7.3	7.3	7.1	7.2	7.1	7.2	7.6	7.5	7.2	6.6	6.8
	β	7.7		7.1	7.3	7.4	7.3		7.8	7.5		7.0
$5t_{1u}$	α	8.1	8.2	8.2	8.4	8.4	8.5	9.0	9.0	8.4	8.0	8.2
	β	8.5		8.2	8.5	8.6	8.6		9.3	8.6		8.6
$6a_{1g}$	α	8.6	8.7	8.6	9.2	9.1	9.1	9.8	9.8	9.1	8.7	9.2
	β	8.8		8.7	9.3	9.2	9.2		9.9	9.2		9.7
$1t_{2g}$	α	9.5	9.9	9.6	10.1	9.6	10.0	10.9	10.6	10.7	12.6	17.8
	β	9.8		10.3	10.4	11.1	10.4		11.4	11.2		17.9
$3e_g$	α	9.8	10.0	9.8	10.3	10.2	10.1	11.0	10.9	10.4	12.3	16.9
	β	9.9		10.3	10.7	11.4	10.6		11.4	11.0		16.9

[a] From Ref. 117.

complicated. Nonetheless, it may be stated that among 4*d* and 5*d* metal hexafluorides the largest *EA*s will belong to those whose HOMOs are occupied fully or nearly fully (e.g. NbF_6, TaF_6, AgF_6 and AuF_6), and all the 3*d* metal hexafluorides possess high *EA*s.

The other group of anions that are widespread in crystalline salts are the MX_4^- tetraanions, where X = F, O, and Cl. Therefore it is of interest to investigate their *EA* behavior and to study the relationship of the *EA* values to peculiarities in the electronic structure of the tetraanions. Let us analyze this relationship, for example, in tetraoxyanions. In Table XVIII are presented the orbital energies of the 4*d* metal tetraoxyanions calculated in the appropriate Slater transition states. The entries of this table indicate the following peculiarities in the valence *IP* behavior when going along the series. First, the first *IP* changes nonmonotonously; its value drops when going from TcO_4^- to RuO_4^-, then increases in the RuO_4^-–AgO_4^- series, and again drops when going to CdO_4^-. Second, *IP*s from the $1t_1$, $8t_2$, and $7a_1$ MOs of the ligand character do not practically change along the series as well as from the $6t_2$ and $6a_1$ MOs composed mainly of O 2*s* states. Third, *IP*s from the bonding $7t_2$ and 2*e* MOs grow monotonously when going along the series. On the whole, the peculiarities in the electronic structure of the tetraoxyanions are the same as for the hexafluoride anions. The main difference is the order of the filling up of MOs along the series. It begins with the HOMO of the *e* type and the filling up the antibonding t_2 HOMO begins with CdO_4^-. Additionally, the *EA* of tetraoxides and the energy jump when filling up the antibonding HOMOs are essentially lower than in the case of the *nd* metal hexafluorides. The changes in the upper valence *IP*s along the series of the 3*d* metal tetraoxianions are analogous to those of the 4*d* metal ones. However, as in the case of the 3*d* and 4*d* metal hexafluorides the *EA* values themselves are larger and the energy jumps are smaller when filling up the next antibonding MO (see Table XI). On the whole, the *EA*s of tetraoxides are large (3.5–5.0 eV) and all of them, (except for possibly RuO_4), may be considered as superhalogens.

The same analysis of the *EA* dependence on the electronic structure might be done for the MF_5 pentafluorides, MF_4 tetrafluorides, MX_3 oxides and fluorides, difluorides, dioxides, and so on. In this review we shall not consider all these cases, and point out only that the high *EA*s belong to not only the MX_{k+1} superhalogens, where k is the maximal formal valency of M and X is a monovalent ligand, but also the MX_n systems, whose antibonding HOMO in the corresponding anion is fully occupied.

D. A Search of Systems with Maximal Possible *EA*s

The theoretical investigations on the *EA* of superhalogens just presented and some experimental estimations show that chemical compounds may

TABLE XVIII

One-Electron Energies (in eV) of the 4d Metal Tetraoxianions and Charges on Atoms Obtained Using the Bond Length $R(M - O) = 1.79$ Å[a]

MO	MoO_4^-		TcO_4^-	RuO_4^-		RhO_4^-		PdO_4^-		AgO_4^-	CdO_4^-	
Configuration:	$t_1{}^5$		$t_1{}^6$	$3e^1$		$3e^2$		$3e^3$		$3e^4$	$9t_2{}^1$	
Spin:	α	β	$\alpha = \beta$	α	β	α	β	α	β	$\alpha = \beta$	α	β
$9t_2$											3.4	
$3e$				3.1		3.8		4.1	3.5	4.4	4.4	4.3
$1t_1$	5.6	4.8	4.9	4.8	4.5	5.1	4.2	4.8	4.2	4.7	4.5	4.3
$8t_2$	5.8	5.4	5.5	5.5	5.2	5.7	5.0	5.5	5.1	5.4	5.5	5.0
$7a_1$	6.2	5.8	6.2	6.0	5.8	6.0	5.6	5.8	5.5	5.8	6.3	5.5
$7t_2$	8.3	8.0	8.7	9.0	8.8	9.3	9.2	10.2	10.1	12.4	15.6	15.3
$2e$	8.3	7.9	8.7	9.0	8.7	10.0	9.3	10.7	10.5	13.6	17.8	17.8
$6t_2$	20.9	20.3	20.5	20.4	20.1	20.7	19.9	20.4	20.1	20.4	20.6	20.1
$6a_1$	21.1	20.5	20.6	20.6	20.2	20.7	20.0	20.5	20.1	20.5	21.0	20.6
Charges on Atoms												
Q_M	0.80		0.95	0.85		0.72		0.60		0.73	0.91	
Q_O	−0.45		−0.49	−0.46		−0.43		−0.40		−0.43	−0.48	

[a] From Ref. 138.

possess very high *EA*s which are considerably greater than *EA*s of the halogen atoms. Naturally, two questions arise: what maximal possible *EA*s may possess superhalogens and among what systems they should be searched for. In accordance with presented ideas such systems should contain as ligands a maximal number of fluorine atoms, because in anions of such systems the energy of the HOMO would tend to that of neutral fluorides due to delocalization of an extra electron over a large number of ligands. Thus, the first *IP* value of neutral fluorides may serve as an upper bound for maximal possible *EA* values. According to experimental data for the series of fluorides such as SF_6, WF_6, SiF_4, CF_4,[9] and many others, the upper *IP*s of fluorides are within 15–17 eV; therefore the value of 17 eV may be considered as an upper bound for *EA*s of chemical compounds.

Let us outline some ways to search systems for maximal *EA*s. One direction is the increase of a number of ligands in the MF_{k+1} mononuclear systems. However, with the increase in k the destabilization effects in the HOMO due to a strong interligand repulsion may prevail over stabilization effects of delocalization and may not lead to the further increase in *EA* as it takes place in the case of the *sp*-chlorides (see Section III.B). The second direction to increase the number of ligands is to go to polynuclear superhalogens such as Al_2F_7 and Sb_2F_{11}. In these systems the interligand repulsion does not increase with increasing in k, and due to the maximal delocalization of an extra electron their *EA*s may approach the maximal value. Of course, the ideas presented are only qualitative and point out the trends in the *EA* changes only. To obtain quantitative estimations it is necessary to calculate the *EA* in every case.

To establish trends in the *EA* changes when the number of ligands increases let us use results of the DVM calculations on the electronic structure of the MnF_8^- and TcF_8^- anions, that is, mononuclear MF_{k+1}^- anions, and $Al_2F_7^-$, $P_2F_{11}^-$, $As_2F_{11}^-$, and $V_2F_{11}^-$ anions (binuclear $M_2F_{2k+1}^-$ anions) of Ref. 137.

In the series of the mononuclear fluorides ScF_4–VF_6–MnF_8 the *EA* grows sharply when going from ScF_4 (4.8 eV) to VF_6 (6.7 eV) and does not practically change when going to MnF_8 (6.7 eV), and decreases even when going from NbF_6 (6.8 eV) to TcF_8 (5.8 eV). This behavior indicates that an optimal combination of stabilization and destabilization factors for attaining the largest *EA*s occurs in the hexacoordinated fluorides. The further increase of the number of ligands does not lead to increasing the *EA* because of delocalization due to the increase in the interligand repulsion. The maximal possible *EA* seems to be 7–8 eV in mononuclear compounds. That value was obtained in the DVM calculations for TaF_6 and AuF_6.[121]

When going from mono- to binuclear complexes, a systematical increase in the *EA* by 1–2 eV is observed. Thus, for example, in series AlF_4–Al_2F_7,

PF_6–P_2F_{11}, AsF_6–As_2F_{11}, and VF_6–V_2F_{11} the *EA* increases from 6.1 to 7.5 eV, from 6.7 to 8.6 eV, and from 6.7 to 8.0 eV, respectively. Keeping in mind these results one may conclude that the maximal *EA* of binuclear systems is 9–10 eV. In particular, such *EA*s may be associated with Ta_2F_{11} and Au_2F_{11}. Thus, it may be concluded that the transition from mono- to polynuclear complexes is a possible way to search systems with maximal possible *EA*s. Such polynuclear superhalogens with *EA*s exceeding by more than two times the *EA*s of the halogen atoms should be called hyperhalogens. They are of great importance in obtaining new classes of salts with nontraditional cations. The use of complex polynuclear hyperhalogens seems to be a way to syntheses of stable salts with the Ar^+ cations.

E. Peculiarities in the Chemical Properties of Superhalogens

The presence of the large *EA* in a molecule leads to many distinctive features of such molecules. It was pointed out previously for SCl_6, $SeCl_6$, and $TeCl_6$, as an example, that high *EA*s of the two latter molecules result in the existence of the $SeCl_6^{2-}$ and $TeCl_6^{2-}$ anions, which are not ordinary from the traditional point of view.

The high *EA* allows one to obtain in principle new salts with nontraditional cations such as $O_2^+[PtF_6]^-$, $XeF_5^+[PtF_6]^-$, and $Xe_2F_{11}^+[AuF_6]^-$. The O_2, XeF_5, and XeF_{11} molecules possess high *IP*s; therefore to obtain salts with their cations it is necessary to use superhalogens with anomalously high *EA*s such as PtF_6 and AuF_6. Thus, superhalogens open wide possibilities for synthesis of new nontraditional salts with both organic and inorganic cations.

The other unusual property of systems containing superhalogen groups is based on the high values of the first *IP*s of superhalogen anions as well. It is well known that chemical bonds of compounds in gas phase dissociate in a homolitic way with formation of a pair of neutral residuals in most cases. However, it is not so for molecules containing superhalogens. Indeed, the first *IP*s of alkali atoms from Cs to Li are placed within 3.9–5.4 eV, whereas the *EA*s of superhalogens may exceed these values, and in such cases the heterolitic dissociation may be preferable. For example, the Cs–AlF_4 bond will dissociate in Cs^+ and AlF_4^- but not in Cs and AlF_4. Such a heterolitic dissociation permits the low-temperature molecular plasmas possessing high electrical conductivity, using ordinary heating. The published experimental data for gas phase mixtures of $NaAlCl_4$ and $KAlCl_4$ with $AlCl_5$–Al_2Cl_6[139] confirm a quite high (for gases) electrical conductivity from 10^{-11} to 10^{-6} Ohm^{-1} cm^{-1} at temperatures from 500 to 900 K. The partial pressure of ions in the gas has been estimated in 10^{-8} atm under partial pressure of $NaAlCl_4$ equaling 0.1 atmosphere.

F. Systems with Low *IP* Values

After experimental and theoretical detection of the large group of molecular systems possessing high *EA*s (superhalogens), the other question arises: what minimal values may attain *IP*s of neutral molecules and radicals, that is, *EA*s of the corresponding singly charged cations. Among atoms of the Periodic Table the alkali atoms possess the lowest first *IP*s; therefore, the systems whose *IP*s are lower those of alkali atoms should be named superalkalies.

From ideas presented in the beginning of this chapter it follows that for the first *IP* of a neutral system of the ML_{k+1} type to be small it is necessary that the HOMO to be antibonding with respect to central atom–ligand interaction and, too, ligand atoms should possess low values of the first *IP*s. In this case, under formation of the ML_{k+1} compound, the electrons of atoms will occupy an antibonding HOMO, and their energy of bonding in ML_{k+1} will be lower than that in the isolated *L* atom. Therefore, it is reasonable to search superalkalies among the ML_{k+1} systems where *L*s are atoms of alkali metals. To support this suggestion, calculations have been made on the electronic structure of radicals of such a type by DVM.[140] The calculated values of valence *IP*s and Mulliken charges on atoms for linear Li_2F, Li_2Cl, Na_2F, Na_2Cl, Cs_2F, and Cs_2Cl are given in Table XIX; for triangular Li_3O, Li_3S, Na_3O, and Na_3S, in Table XX; for tetrahedral Li_4N, Li_4P, Na_4N, and Na_4P, in Table XXI.

The calculations are carried out using the equilibrium geometry of cations; that is, the first *IP* values of radicals ML_{k+1} given in the tables are equal to the vertical *EA* values of the ML_{k+1}^{+} cations. Table XXII contains the LCAO

TABLE XIX
Valence *IP*s (in eV), Mulliken Charges on Atoms Calculated by DVM within the Double Zeta Basis of Clementi[140] for Linear Superalkalies

Compound	Li_2F	Li_2Cl	Na_2F	Na_2Cl	Cs_2F[a]	Cs_2Cl
Level						
$1\sigma_g$	3.6	3.6	3.7	3.7[b]	2.6	2.3
$1\pi_u$	12.6	10.1	10.5	9.0	7.4	7.2
$1\sigma_u$	13.3	11.6	10.5	9.9	5.7	6.5
Charges on Atoms						
Q_M	−0.53	−0.56	−0.6	−0.6	−0.15	−0.8
Q_L	+0.26	+0.28	+0.3	+0.3	+0.08	+0.4

[a] The electronic structure of Cs_2F and Cs_2Cl is calculated within the numerical HF basis extended by external STOs: $6p_{Cs}$ ($\xi = 2.0$) and $3d_L$ ($\xi = 1.0$).
[b] The experimental value is 4.14 ± 0.2 eV.[144]

TABLE XX
Electronic Characteristics of the Triangular Superalkalies Calculated by DVM within a Double Zeta Basis[a]

Compound		Li_3O	Li_3S	Na_3O	Na_3S
Valence	$1a'_1$	3.4	3.2	3.5	3.3
IP	$1a''_2$	6.4	6.2	4.7	5.9
(in eV)	$1e'$	8.0	7.7	6.3	6.4
Charges on Atoms					
Q_M		−0.88	−0.90	−0.98	−0.85
Q_L		+0.29	+0.30	+0.32	+0.28

[a] From Ref. 140.

TABLE XXI
Electronic Characteristics of the Tetrahedral Superalkalies Calculated by DVM within a Basis of the Double Zeta Functions[a]

Compound		Li_4N	Li_4P	Na_4N	Na_4P	NH_4[b]
Valence	$1a_1$	3.6	3.0	3.7	3.2	4.1
IP	$1t_2$	5.9	5.7	4.8	4.9	18.1
(in eV)	$2a_1$	15.8	12.7	14.6	11.8	29.1
Charges on Atoms						
Q_M		−1.07	−0.88	−0.95	−0.81	−0.89
Q_L		0.27	0.22	0.24	0.20	0.22

[a] From Ref. 140.
[b] The results of calculations within a double zeta basis extended by 3*s*, 3*p* and 3*d* STOs with exponents of 0.5. The use of the "pure" double zeta bases leads to low values of the first *IP*. In particular, within Clementi's basis has been obtained the value in 2.6 eV.[140]

coefficients of valence MOs obtained in a double zeta basis for Li_2F, Li_3O, and Li_4N. The valence MOs of other radicals of the corresponding type possess the analogous structure. As is seen, the double occupied MOs are composed of the central atom AOs mainly, whereas the upper half filled HOMOs are of ligand AOs. The HOMO of radicals are antibonding with respect to the central atom–ligand interaction, and in accordance with traditional notions its energy should be lower than the energies of both the central atom and the ligands. Indeed, it is seen from Tables XVIII–XX that *IP*s of radicals are lower than those of alkali metal atoms in all the cases considered. Therefore, these ML_{k+1}^+ cations should be attributed to superal-

TABLE XXII
LCAO Coefficients of the Valence MOs for Li_2F, Li_3O, and Li_4N.[140]

Compound	Level	Ligand AOs			Central metal AOs				
		$2s$	$2s'$	$2p$	$2s$	$2s'$	$2p$	$2p'$	$3d$
Li_2F	$4\sigma_g$	0.054	−0.643	0.167	0.048	0.173			−0.032
	$1\pi_u$			0.118			0.335	0.689	
	$2\sigma_u$		−0.140	−0.143			0.328	0.655	
Li_3O	$4a_1'$	0.055	−0.459	0.139	0.114	0.136			0.070
	$1a_2''$			0.156			0.265	0.662	
	$2e'$	0.009	−0.280	−0.174			0.274	0.612	0.033
Li_4N	$4a_1$	0.038	−0.381	0.086	0.092	0.309			
	$2t_2$	−0.007	0.234	−0.080			0.191	0.564	0.043
	$3a_1$	0.005	−0.008	−0.025	0.533	0.483			

kalies since in the corresponding radicals the extra electron is less bound in comparison with alkali metal atoms.

Radicals considered here are selected to trace the dependence of *IP*s on the substitution of *L* and *M* along the subgroup as well as on the increase of the number of ligands along the period. The substitution of Li by Na leads to small changes of *IP*s possessing no pronounced trends. However, the substitution of Na by Cs in the Na_2F–Cs_2F and Na_2Cl–Cs_2Cl series leads to essential decreasing *IP*s. The substitution of ligands by heavier ones along the subgroup appears to result in decreasing the *EA* of the corresponding anion, though relative changes are small in general. When the number of ligands in the series of the Li_2F^+–Li_3O^+–Li_4N^+ type is increased, the *EA* decreases on the whole, though the *EA* changes are small as before.

It should be noted that the fulfillment of the second condition for the *EA* to be small is unnecessary. If the HOMO is highly antibonding then the energy of bonding of the extra electron in ML_{k+1} may be small in spite of a high *EA* of the L^+ cation.

The other type of molecular radicals with low values of the first *IP*s may present the Rydberg radicals ML_{k+1} in which the extra electron (with respect to the corresponding cation) fills up a Rydberg AO of the central atom. Due to high energies of atomic Rydberg states, *IP*s from these AOs should be small in these cases. An example of such systems is the ammonium radicals NR_4 where R is an organic residual. For the simplest NH_4 radical there are reliable *ab initio* calculations with geometry variations separately for NH_4^+ and NH_4 and that account for correlation energy within the MC-SCF approach in extended by Rydberg AOs bases.[141] These calculations have shown that the HOMO of NH_4 is composed mainly of a Rydberg AO of N and the first IP_{ad} of NH_4 (i.e., EA_{ad} of NH_4^+) is equal to 4.3 eV. This value is in a good agreement with the known experimental estimations in 4.6 $\pm$ 0.2 eV[142] and 4.73 $\pm$ 0.06 eV.[143] The other NR_4 radicals in which R possesses electron-donor properties should have yet lower *IP*s and they appear to be attributed to alkalies or superalkalies.

The low *EA* of the superalkaline cations points out the possibility of syntheses of new salts in which anions may be molecular systems with low *EA*s. The first salts of this type (e.g., $Li_3O^+NO^-$ and $Li_3O^+NO_2^-$) are already synthesized and the structure analysis confirms the existence in them of Li_3O^+ + cations.[145,146] The superalkali cations are able to form salts in those cases when a formation of analogous salts with alkali metals may be unfavorable due to relatively high *IP* values or to steric obstacles. The superalkali cations together with superhalogen anions appear to give the lowest energy of dissociation into ions, which is of a great interest to physics of low-temperature molecular plasmas (see Section III.D).

Recently, using results of the *ab initio* calculations, a new class of so-called hyperlithium compounds such as OLi_4, OLi_5, CLi_5, CLi_6 has been detected,[147-150] in which the formal valency of the central atom exceeds the maximal classic valency. According to calculations the first *IP*s of these compounds are relatively small and they may be probably attributed to superalkalies as well. It should be noted, however, that the question of the existence and energetics of the hyperlithium compounds requires further theoretical and experimental investigations..

IV. CONCLUSIONS

Based on data presented in this work, one may conclude that great progress was achieved in the theoretical estimation of the *EA* of molecules and radicals lately. For systems with a small number of electrons the calculational methods incorporating the electronic correlation permit determination of the *EA* with accuracy frequently exceeding the experimental method. Additionally, these methods give opportunity to determine all three values of the *EA*, namely, EA_{ad}, EA_{vert}, and IP_{vert} of the corresponding anion and to study *EA*s of those molecules and radicals that are difficult to acquire in a sufficient concentration for experimental investigations. Therefore, for small systems, the theoretical calculations of the *EA*s are in essence the computer experiments for their determinations. The rapid growth of computational possibilities indicates that in the near future such methods will be applicable for sufficiently large systems. From the other hand, on the basis of calculations it is possible to relate the *EA* value to peculiarities of the electronic structure of compounds under consideration. In particular, such an analysis indicates two large classes of chemical compounds, those possessing anomalously high *EA*s (superhalogens) and those possessing anomalously low first *IP*s (superalkalies). Both classes are of great interest both for theoretical chemistry due to extensions of traditional notions about electropositivity and electronegativity of chemical groups, and for practical applications, such as for syntheses of new chemical compounds, explanations of the electrical conductivity in vapors of inorganic salts, and so on. It may be expected that new theoretical notions about the electronic structure of superhalogens and superalkalies will be developed and attract the wide attention of experimentalists.

References

1. L. M. Branscomb, *Atomic and Molecular Processes*. D. R. Bates ed., Academic Press, New York, 1962.
2. R. S. Barry, *Chem. Rev.* **69**, 533 (1969).
3. P. P. Sutton and J. E. Meyer, *J. Chem. Phys.* **3**, 20 (1935).
4. A. L. Garragher, F. M. Page, and R. C. Wheller, *Discuss. Faraday Soc.* **37**, 203 (1964).
5. B. Moiseiwetsch, *Adv. Mol. Phys.* **1**, 61 (1964).
6. W. A. Chupka, J. Berkowitz, and D. Gutman, *J. Chem. Phys.* **55**, 2724 (1971).
7. W. A. Chupka, J. Berkowitz, and D. Gutman, *J. Chem. Phys.* **55**, 2733 (1971).
8. L. N. Sidorov, *Usp. Khimii* **51**, 625 (1982).
9. L. Karlsson, L. Mattsson, R. Jadrny, T. Bergmark, and K. Siegbahn, *Phys. Scripta* **14**, 230 (1976).
10. R. L. Dekock, B. R. Higginson, and D. R. Lloyd, *Discuss Faraday Soc.* **54**, 84 (1972).
11. D. W. Turner, C. Barker, A. D. Barker, and C. R. Brundle, *Molecular Photoelectron Spectroscopy*, Wiley-Interscience, London, 1970.
12. B.. M. Smirnov, *Negative Ions*, Atomizdat, Moscow, 1978.
13. H. S. W. Massey, *Negative Ions*, 3rd ed., Cambridge University, London, 1976.
14. M. F. Herman, K. F. Freed, and D. L. Yeager, *Adv. Chem. Phys.* **48**, 1–69 (1981).
15. C. F. Bunge and A. V. Bunge, *Int. J. Quant. Chem.* **12S**, 345 (1978).
16. H. F. Schaefer, III, *The Electronic Structure of Atoms and Molecules. A Survey of Rigorous Quantum Mechanical Results*, Addison-Wesley, Massachusetts, 1972.
17. R. McWeeny and B. T. Sutcliffe, *Methods of Molecular Quantum Mechanics*, Academic Press, New York, 1969.
18. P. Carsky and M. Urban, *Lecture Notes in Chemistry. Ab initio Calculations. Methods and Applications in Chemistry*. Springer-Verlag, Berlin–New York–London, 1980.
19. R. J. Bartlett, *Annu. Rev. Phys. Chem.* **32**, 359 (1981).
20. I. Hubac and P. Carsky, *Topp. Carr. Chem.* **75**, 97 (1978).
21. C. W. McCurdy, T. N. Rescigno, D. L. Yager, and V. McKoy, in *Modern Theoretical Chemistry*, Vol. 3, H. F. Schaefer, III, ed., Plenum Press, New York, 1977, p. 339.
22. W. von Niessen, L. S. Cederbaum, and W. Domcke, in *Excited States in Quantum Chemistry*, C. A. Nicolaides, D. R. Beck, eds., Reidel, Dorchcht, Holland, 1979.

23. GAUSSIAN-80, J. S. Binkley, R. Whiteside, R. Krishnan, R. Seeger, H. B. Schlegel, D. J. De Frees, S. Topiol, L. R. Kahn, and J. A. Pople, *QCPE*, **13**, 406 (1981).
24. HONDO-5, M. Dupuis, J. Rys, and H. F. King, *Resour. Comput. Chem. Software Cat.*, Vol. 1, Progr. N QHO2, 1980.
25. MOLECULE, J. Almlof, USIP Report 74-29, University of Stockholm, 1974.
26. POLYATOM/2, J. W. Moscowitz and L. C. Snyder, in *Modern Theoretical Chemistry*, Vol. 3, H. F. Schaefer, III, ed., Plenum Press, New York, 1977, p. 339.
27. P. E. Cade, *J. Chem. Phys.*, **47**, 2390 (1967).
28. J. V. Ortiz and Y. Ohrn, *Chem. Phys.* **77**, 548 (1981).
29. K. M. Griffing and J. Simons, *J. Chem. Phys.* **64**, 3610 (1976).
30. S. P. Konovalov and V. G. Solomonik, *Zh. Fiz. Khimii* **57**, 636 (1983).
31. P. R. Taylor, G. B. Bacskay, N. S. Hush, and A. C. Harley, *J. Chem. Phys.* **70**, 4481 (1979).
32. A. I. Boldyrev, L. P. Sukhanov, and O. P. Charkin, *Koord. Khim.* **8**, 430 (1982).
33. A. I. Boldyrev, V. G. Zakzhevskii, and O. P. Charkin, *Koord. Khim.* **8**, 437 (1982).
34. A. I. Boldyrev and O. P. Charkin, *Koord. Khim.* **8**, 618 (1982).
35. E. Clementi, Tables of Atomic Wave Functions, a Supplement to *IBM J. Res. Develop.* **9**, 2 (1965).
36. B. Roos and P. Siegbahn, *Theor. Chim. Acta* **17**, 209 (1970).
37. R. Ditchfield, W. J. Hehre, and J. A. Pople, *J. Chem. Phys.* **54**, 724 (1971).
38. W. J. Hehre and J. A. Pople, *J. Chem. Phys.* **56**, 4233 (1972).
39. T. H. Dunning, *J. Chem. Phys.* **53**, 2823 (1970); *J. Chem. Phys.* **55**, 3958 (1971); *J. Chem. Phys.* **55**, 716 (1971).
40. T. H. Dunning and P. J. Hay, in *Methods of Electronic Theory*, H. F. Schaefer, III, ed., Plenum, New York, 1977, p. 1.
41. L. V. Gurvich, G. V. Karachevtsev, V. N. Kondratjev, Y. A. Lebedev, V. A. Medvedev, V. K. Potapov, and Y. S. Hodeev, *Destruction Energies of Chemical Bonds. Ionization Potentials and Electron Affinities*, Nauka, Moscow, 1974, p. 226.
42. R. D. Srivastava, O. M. Uy, and M. Farber, *Trans. Faraday Soc.* **67**, 2941 (1971).
43. [a]Our calculations within the $(9s5p1d/4s2p1d)_N + (9s5p/4s2p)_O$ basis; [b]Our calculations within the $(9s5p/4s2p)_{N,O} + DF_{N,O}$ basis.
44. Our calculations within the $(12s9p1d/6s4p1d)_P + (9s5p/4s2p)_O$ basis.
45. Our calculations within the $(9s5p1d/4s2p1d)_C + (4s1p/2s1p)_H$ basis + $DF_{C,H}$
46. E. Clementi and A. D. McLean, *Phys. Rev.* **A133**, 419 (1964).
47. E. Clementi, A. D. McLean, D. L. Raimondi, and M. Yoshimine, *Phys. Rev.* **A133**, 1274 (1964).
48. A. E. Smolyar, A. I. Boldyrev, O. P. Charkin, N. M. Klimenko, and V. I. Avdeev, *J. Struct. Chem.* (USA) **17**, 188 (1976).
49. J. Bell, *J. Chem. Phys.* **69**, 3879 (1979).
50. P. Rosmus and W. Meyer, *J. Chem. Phys.* **69**, 2745 (1978).
51. F. Sasaki and M. Yoshimine, *Phys. Rev.* **A9**, 26 (1974).
52. L. S. Cederbaum and W. Domcke, *Adv. Chem. Phys.* **36**, 205 (1977).
53. C. W. McCurdy, T. N. Rescingno, D. L. Yager, and V. McKoy, in *Modern Theoretical Chemistry*, Vol. 3, H. F. Schaefer, III, ed., Plenum Press, New York, 1977, p. 339.
54. J. Simons and W. D. Smith, *J. Chem. Phys.* **58**, 4899 (1973).

55. G. Hohlneicher, F. Ecker, and L. S. Cederbaum, in *Electron Spectroscopy*, D. H. Shirley, ed., North-Holland, Amsterdam, 1972.
56. D. R. Chong, F. G. Herring, and D. McWilliams, *J. Chem. Phys.* **61**, 78 (1974).
57. O. Goscinski and B. Lukman, *Chem. Phys. Lett.* **7**, 573 (1970).
58. H. Sun and K. F. Freed, *J. Chem. Phys.* **76**, 5051 (1982).
59. W. Kutzelnigg, in *Methods of Electronic Structure Theory*, H. F. Schaefer, III, ed., Plenum Press, New York, 1977.
60. W. Meyer, in *Methods of Electronic Structure Theory*, H. F. Schaefer, III, ed., Plenum Press, New York, 1977.
61. I. Shavitt, in *Modern Theoretical Chemistry*, Vol. 3, H. F. Schaefer, III, ed., Plenum Press, New York, 1977.
62. P. S. Bagus, B. Liu, A. D. McLean, and M. Yoshimine, in *Computational Methods for Large Molecules and Localized States in Solids*, F. Herman, A. D. McLean, and R. H., Nesbet, eds., Plenum Press, New York, 1973.
63. R. J. Buenker and S. D. Peyerihoff, *Theor. Chim. Acta* **35**, 33 (1974).
64. R. McWeeny and R. Steiner, *Adv. Quant. Chem.* **2**, 93 (1965).
65. A. C. Wahl and G. Das, *Adv. Quant. Chem.* **5**, 261 (1970).
66. F. P. Billingsley, *J. Chem. Phys.* **62**, 864 (1975); *J. Chem. Phys.* **63**, 2267 (1975).
67. M. Hackmeyer, *Int. J. Quant. Chem.* **8**, 783 (1974).
68. M. Urban, V. Kello, J. Noga, and I. Cernusak, POLYCOR (unpublished).
69. I. Hubac and M. Urban, *Theor. Chim. Acta* **45**, 185 (1977).
70. B. Liu, K. O-Ohata, and K. Kirby-Docken, *J. Chem. Phys.* **67**, 1985 (1977).
71. A. M. Karo, M. A. Gardner, and J. R. Hiskes, *J. Chem. Phys.* **68**, 1942 (1978).
72. K. D. Jordan, K. M. Griffing, J. Kenney, E. L. Andersen, and J. Simons, *J. Chem. Phys.* **64**, 4730 (1976).
73. J. Kenney and J. Simons, *J. Chem. Phys.* **63**, 592 (1975).
74. K. M. Griffing and J. Simons, *J. Chem. Phys.* **62**, 535 (1975).
75. L. S. Cederbaum, K. Schonhammer, and W. von Niessen, *Phys. Rev.* **A15**, 833 (1977).
76. D. Smith, T. Chen, and J. Simons, *Chem. Phys. Lett.* **27**, 499 (1974).
77. E. Anderson and J. Simons, *J. Chem. Phys.* **64**, 4549 (1976).
78. P. R. Taylor, G. B. Bacskay, N. S. Hush, and A. C. Hurley, *J. Chem. Phys.* **70**, 4481 (1979).
79. K. Vasudevan and F. Grein, *J. Chem. Phys.*, **68**, 1418 (1978).
80. R. E. Olson and B. Lieu, *J. Chem. Phys.* **73**, 2817 (1980).
81. M. Dupuis and B. Liu, *J. Chem. Phys.* **73**, 337 (1980).
82. D. S. Marynick and D. A. Dixon, *Proc. Natl. Acad. Sci. USA* **74**, 410 (1977).
83. K. D. Jordan and J. Simons, *J. Chem. Phys.* **67**, 4027 (1977).
84. H. Partridge, D. A. Dixon, S. P. Walch, C. W. Bauschlicher, and J. L. Gole, *J. Chem. Phys.* **79**, 1859 (1983).
85. J. C. Slater, *The Self-Consistent Field for Molecules and Solids*, Vol. 4, McGraw-Hill, New York, 1974.
86. J. C. Slater, *Adv. Quant. Chem.* **6**, 1 (1972).
87. K. H. Johnson, *Adv. Quant. Chem.* **7**, 143 (1973).
88. D. E. Ellis and G. S. Painter, *Phys. Rev.* **B2**, 2887 (1970).

89. T. Ziegler, J. G. Snijders, and E. J. Baerends, *J. Chem. Phys.* **74**, 1271 (1981).
90. D. D. Koelling, D. E. Ellis, and R. J. Bartlett, *J. Chem. Phys.* **65**, 3331 (1975).
91. E. J. Baerends and P. Ros, *Chem. Phys.* **8**, 412 (1975).
92. H. Conroy, *J. Chem. Phys.* **47**, 5307 (1967).
93. C. B. Haselgrove, *Mat. Comp.* **15**, 323 (1961).
94. G. L. Gutsev and A. A. Levin, *Chem. Phys.* **51**, 459 (1980).
95. E. L. Baerends, D. E. Ellis, and P. Ros, *Chem. Phys.* **2**, 41 (1973).
96. B. Delly and D. E. Ellis, *J. Chem. Phys.* **76**, 1949 (1982).
97. G. L. Gutsev and A. E. Smoljar, *Chem. Phys.* **56**, 189 (1981).
98. P. J. M. Geurts, J. W. Gosselink, A. Avoird van der, E. J. Baerends, and J. G. Snijders, *Chem. Phys.* **46**, 133 (1980).
99. G. L. Gutsev and A. A. Levin, *J. Struct. Chem.* (USA) **20**, 771 (1979).
100. V. A. Gubanov, E. Z. Kurmaev, and D. E. Ellis, *J. Phys.* *C*14, 5567 (1981).
101. G. L. Gutsev and A. I. Boldyrev, *Chem. Phys.* **56**, 277 (1981).
102. A. Rosen, D. E. Ellis, H. Adachi, and F. W. Averill, *J. Chem. Phys.* **65**, 3629 (1976).
103. G. M. Mikhailov, G. L. Gutsev, and Ju. G. Borod'ko, *Chem. Phys. Lett.* **96**, 70 (1983).
104. D. R. Salanub, S. H. Lamson, and R. P. Messmer, *Chem. Phys. Lett.* **85**, 430 (1982).
105. P. Hohenberg and W. Kohn, *Phys. Rev.* **136**, 864 (1964).
106. W. Kohn and L. J. Sham, *Phys. Rev.* **A140**, 1133 (1965).
107. L. Wilk and S. H. Vosko, *J. Phys.* **C15**, 2139 (1982).
108. J. P. Perdew and A. Zunger, *Phys. Rev.* **B23**, 5048 (1981).
109. D. D. Koelling, *Rep. Progr. Phys.* **44**, 141 (1981).
110. P. Prins, *J. Chem. Phys.* **61**, 2580 (1974).
111. R. N. Compton and C. D. Cooper, *J. Chem. Phys.* **59**, 4140 (1973).
112. J. G. Norman, Jr., *Mol. Phys.* **31**, 1191 (1976).
113. M. Boring, *Chem. Phys. Lett.* **46**, 242 (1977).
114. J. E. Bloor and R. E. Sherrod, *J. Am. Chem. Soc.* **102**, 4333 (1980).
115. L. N. Sidorov, *Int. J. Mass. Spectr. Ion Phys.* **39**, 311 (1981).
116. G. L. Gutsev and A. P. Klyagina, *Chem. Phys.* **75**, 243 (1983).
117. G. L. Gutsev and A. I. Boldyrev, *Mol. Phys.* **11**, 1 (1984).
118. L. N. Sidorov, A. V. Borshchevsky, E. B. Rudny, and V. D. Butsky, *Chem. Phys.* **71**, 145 (1982).
119. R. N. Compton, P. W. Reinhardt, and C. D. Cooper, *J. Chem. Phys.* **68**, 2023 (1978).
120. J. Burgess, I. H. Haigh, R. D. Peacock, and P. Taylor, *J. Chem. Soc. Dalton Trans.*, 1064 (1974).
121. G. L. Gutsev and A. I. Boldyrev, *Chem. Phys. Lett.* **101**, 441 (1983).
122. G. L. Gutsev and A. I. Boldyrev, *Chem. Phys. Lett.* **84**, 352 (1981).
123. H. Dispert and K. Lacmann, *Chem. Phys. Lett.* **45**, 311 (1977).
124. B. P. Martur, E. W. Rothe , and G. P. Reck, *J. Chem. Phys.* **67**, 377 (1977).
125. I. D. Webb and E. R. Bernstein, *J. Am. Chem. Soc.* **100**, 483 (1978).
126. N. Bartlett and M. K. Iha, *Chem. Commun.* **168** (1966).
127. L. N. Sidorov, E. V. Skokan, M. I. Nikitin, and I. D. Sorokin, *Int. J. Mass. Spectr. Ion Phys.* **35**, 215 (1980).

128. T. Mallouk, B. Desbat, and N. Bartlett, *J. Fluor. Chem.* **21**, 88 (1982).

129. J. L. Beauchamp, *J. Chem. Phys.* **64**, 929 (1976).

130. A. G. Pyatenko, A. V. Gusarov, and L. N. Gorokhov, *Zh. Fiz. Khimii.* **54**, 1906 (1982).

131. G. L. Gutsev and A. I. Boldyrev, *Russ. J. Inorg. Chem.* **26**, 2353 (1981).

132. G. L. Gutsev and A. I. Boldyrev, *Russ. J. Inorg. Chem.* **26**, 2557 (1981).

133. J. P. Lowe, *J. Am. Chem. Soc.* **99**, 5557 (1977).

134. G. A. Tsigdinos and F. W. Moore, *Annu. Rep. Inorg. Gen. Synth.*, 1973. Wiley, New York, 1974, p. 157.

135. J. Burgess and R. D. Peacock, *J. Fluor. Chem.* **10**, 479 (1977).

136. R. McDiarmid, *Chem. Phys. Lett.* **76**, 300 (1980).

137. G. L. Gutsev and A. I. Boldyrev, *Chem. Phys. Lett.* **108**, 250 (1984).

138. G. L. Gutsev and A. I. Boldyrev, *Chem. Phys. Lettt.* **108**, 255 (1984).

139. B. W. Dowing, *J. Phys. Chem.* **75**, 1260 (1971).

140. G. L. Gutsev and A. I. Boldyrev, *Chem. Phys. Lett.* **92**, 262 (1982).

141. H. Cardy, D. Liotard, A. Dargelos, and E. Poquet, *Chem. Phys.* **77**, 287 (1983).

142. B. W. Williams and R. F. Porter, *J. Chem. Phys.* **73**, 5598 (1980).

143. G. I. Gellene, D. A. Cleary, and R. F. Porter, *J. Chem. Phys.* **77**, 3471 (1982).

144. K. I. Peterson, P. D. Dao, and A. W. Castleman, Jr., *J. Chem. Phys.* **79**, 777 (1983).

145. J. Jensen, *Angew. Chem.* **B88**, 412 (1976).

146. J. Jensen, *Angew. Chem.* **B89**, 5567 (1977).

147. E. D. Jemmis, J. Chandrasekhar, E. U. Wurthwein, P. V. R. Schleyer, J. W. Chinn, F. J. Lando, R. J. Langow, B. Luke, and J. A. Pople, *J. Amer. Chem. Soc.* **104**, 4275 (1982).

148. P. V. R. Schleyer, E. U. Wurthwein, and J. A. Pople, *J. Am. Chem. Soc.* **104**, 5839 (1982).

149. P. V. R. Schleyer, E. U. Wurthwein, E. Kaufmann, T. Clark, and J. A. Pople, *J. Am. Chem. Soc.* **105**, 5930 (1983).

150. N. M. Klimenko, V. G. Zakzhevskii, and O. P. Charkin, *Koord. Khim.* **8**, 903 (1982).

SCATTERING THEORY IN SUPERSPACE

C. GEORGE, F. MAYNÉ and I. PRIGOGINE*

Service de Chimie-Physique II, Université libre de Bruxelles, Brussels, Belgium

CONTENTS

PRELIMINARY

There is today a growing interest in irreversible processes at the macroscopic level, and, with it a renewed interest in the microscopic formulation of irreversibility.

---

* Dr. Prigogine is also affiliated with the Center for Studies in Statistical Mechanics, The University of Texas at Austin, Texas.

Curiously, the second law appears to be in contradiction with all fundamental theories (be it classical or quantum) in which the entropy in terms of the distribution function in phase space (or the density operator) is conserved as a consequence of the unitary character of the dynamical evolution. Consequently irreversibility, as expressed through the second law, has generally been considered as coming from approximations or even as being subjective in character.

In contrast, in our approach, we seek to formulate the law of entropy increase as a fundamental dynamical principle and we study the modifications in the conceptual structure of classical and quantum mechanics implied by this point of view.

We have shown that irreversibility is the result of a symmetry breaking of the time-reversal invariance of the dynamical group that governs the evolution of complex systems. As a consequence of this symmetry breaking, the dynamical group is "realized" as a dissipative semigroup.

The physical origin of this symmetry breaking is a limitation on physically realizable states or observables. Not all initial conditions permitted by dynamics can be physically prepared or observed in nature, but *only a subset of them, which is asymmetrical with respect to the two directions of time.*

This asymmetry is implemented through a suitable (necessarily nonunitary) transformation Λ of states ρ

$$^{p}\rho = \Lambda^{-1}\rho$$

such that the evolution operator of the transformed states $^{p}\rho$ is a dissipative semigroup. The transformation Λ is a nonlocal, time symmetry-breaking transformation, leading from a unitary description to a contractive semigroup, which incorporates the second law of thermodynamics. Its explicit form depends on the dynamics itself. As the transformed density operator $^{p}\rho$ incorporates the second law of thermodynamics, we have for $^{p}\rho$ what we call a "physical representation." In this representation, matter as described by $^{p}\rho$, has an intrinsic arrow of time, which is propagated by the dissipative semigroup.

Previous publications have dealt with both classical and quantum cases, as in these two situations irreversibility originates in slightly different contexts. As discussed elsewhere, irreversibility can only appear when the Liouvillian operator has a continuous spectrum. For quantum systems, one must therefore consider the limit to large systems. A compact presentation of the quantum mechanical situation has been given elsewhere. In Reference 1, to which we shall frequently refer as I, some of these delicate questions related to the large volume limit were mentioned but not discussed in detail. For this reason, we find it useful to give here a rather complete presentation of a relatively simple case, that of potential scattering. Already here we encounter the basic questions of the microscopic theory of irreversible processes.

The first is the question of the subset of states which is compatible with the second law. In our formulation, this question is settled by studying the domain of the transformation operator Λ (or its inverse Λ^{-1}).

As is well known, quantum scattering theory leads to two integral equations, whose solutions $|k^{\pm}\rangle$ contain, in addition to plane waves, outgoing or incoming spherical waves. It is generally stated that, based on physical intuition, only one stationary solution, say $|k^{+}\rangle$, corresponds to actual experiments, in which an incoming plane wave is scattered to produce an outgoing spherical wave.

Our theory permits indeed to exclude $|k^{-}\rangle$ as the result of the formulation of the second law acting as a selection principle. This result is in agreement with previous results.[1,2] In short, for unstable systems we have to exclude states corresponding to a "controlled" future. In our terminology,[1] such states require precollisional correlations of arbitrary range, while the second law as a selection principle states that only transient precollisional correlations can be prepared or observed in nature. This gives to the second law a highly intuitive meaning. However the interest of our approach is far from being exhausted by the study of the second law as a selection principle.

In scattering theory, a fundamental role is played by concepts such as scattering cross sections. While the calculation of cross sections in the simple cases considered here presents no special problem, their incorporation into a dynamical theory as generators of motion can only be performed in terms of a nonunitary transformation theory such as mediated by our transformation operator Λ. This paper deals exclusively with the scattering problem as met in two body scattering or in dilute gases ("Boltzmann" problem). More generally, our method permits the extraction of information about "irreversible behavior" starting with the traditional quantum mechanical Schrödinger equation or field equations.

Though we shall also frequently refer to,[1] this paper is intended to present a *self-contained* introduction to our approach. This accounts for its length. Readers familiar with the traditional scattering theory will find most of the calculations presented here quite elementary.

I. INTRODUCTION

As is well known, quantum theory can be formulated both in Hilbert space using wave functions or in what we call "superspace" using density operators. In the first case, the time evolution is expressed through the Schrödinger equation, in terms of the Hamiltonian operator H, in the second through the

Liouville-von Neumann equation, involving the "superoperator" $L = H \times 1 - 1 \times H$, which operates on "superspace." With L a difference of factorized superoperators,* the evolution superoperator e^{-iLt} takes also the factorized form, $e^{-iHt} \times e^{iHt}$. Therefore it might seem that no new information can be obtained by going from the description in Hilbert space to the description in superspace. However, this is no more so when irreversible processes are included in the description. We have then to introduce *nonfactorizable* superoperators.[3]

Indeed as was shown more recently,[4,5,6] if there exists at all a possibility of enlarging the algebra of observables in order to include entropy, the corresponding observable can only be defined as a nonfactorizable superoperator.

Superspace plays therefore a central role in all formulations of the second law of thermodynamics as applicable to quantum systems, because irreversibility thus leads to observables which cannot be included in the Hilbert space formulation. Inversely, irreversibility implies that not all operators in Hilbert space can be considered as observables (in addition to the usual restrictions, regarding, e.g., hermiticity, superselection rules, etc.). However, the introduction of an observable corresponding to entropy is severely restricted. There is no such observable if H has a purely discrete spectrum, or a continuous but bounded spectrum.

The introduction of nonfactorizable superoperators leads to drastic changes in the basic formulation of quantum theory. Pure and mixed states have then to be treated on an equal footing, because the application of a nonfactorizable superoperator to a pure state transforms it into a mixture.

Both in classical and quantum theory irreversibility can appear only when the description in terms of trajectories or of wave functions ceases to be physically meaningful as the consequence of some inherent instability of motion.

The introduction of nonfactorizable superoperators means that the simple correspondance between descriptions in Hilbert space and in superspace is broken.

The reason for this breakdown is that in both classical and quantum mechanics, irreversibility leads to "non locality". In classical mechanics, this non locality appears in phase space.[7]

As compared to classical theory, quantum theory is obviously already a "nonlocal" theory, but an extended definition of locality can be restored through the use of the Hilbert space. The additional nonlocality in quantum

* When confusion is possible, operators O on Hilbert space may be denoted by $\hat{O}$, and superoperators M on superspace by M. Numbers and matrices may stay unmarked. A *factorized* superoperator F can be written $\mathsf{F} = \hat{G} \times \hat{D}$, this notation means that F acting on an operator O gives $\mathsf{F}\hat{O} = \hat{G}\hat{O}\hat{D}$.

theory forces us to consider states in the product superspace which do not reduce to "points" (the pure states) in Hilbert space.[7]

This situation will be studied here on the simple example of potential scattering. The starting point is the Schrödinger equation which corresponds to a unitary evolution. The problem however is to extract from this equation the physical information included in quantities such as cross sections and lifetimes which are finally to be incorporated in the entropy production.

The technique that enables us to do so is the theory of starunitary transformations which is an appropriate extension of the standard transformation theory of quantum mechanics. There are of course conditions for this noncanonical transformation theory to be applicable in quantum theory: it always requires some asymptotic elements, such as large volumes, large number of degrees of freedom, large times, and so on.

As has already been noticed, we indeed need to take the limit of a continuous spectrum. However, this requires great care because by taking from the start an infinite volume would suppress the effects of scattering which we precisely want to retain. We therefore have to develop specific asymptotic methods valid for a large but not strictly infinite volume.

We want to illustrate our general scheme considering first the problem of two-body scattering and then the Boltzmann problem for a dilute gas. Potential scattering is associated with the Hamiltonian, in standard second quantization notations (the relation with two-body scattering is recalled in Appendix A):

$$H = \sum \omega_k b_k^+ b_k + \sum V_{kk'} b_k^+ b_{k'} \tag{1}$$

If L^3 is the volume of the quantization box, $V_{kk'}$ is assumed to be of order L^{-3}. At some later stage, the limit $L^3 \to \infty$ will be considered.

The simplicity of this model originates from the fact that the number of excitations is conserved in time. The vacuum state $|0\rangle$, the one-particle states $\{|k\rangle\}$, the two-particle states $\{|kk'\rangle\}, \ldots$ evolve separately. Similarly, we may independently study the contributions to the density operator arising from $|0\rangle\langle 0|$, $|0\rangle\langle k|$, $|k\rangle\langle 0|$, $|k\rangle\langle k'|, \ldots$ (we note the corresponding contributions $\rho_{00}, \rho_{0k}, \rho_{k0}, \rho_{kk'}, \ldots$). As could be expected, the evolution of the set ρ_{0k} (or ρ_{k0}) is isomorphic to that of the wave function, but already the study of $\rho_{kk'}$ reveals new features due to the role of scattering which describes an irreversible process (see our previous work).[8,9,10]

The reason for the appearance of these new features, including the introduction of nonfactorizable superoperators, is very simple: as is well known, the discussion of scattering in terms of wave functions in the limit of large systems involves appropriate analytical continuation.[11]

This is also true for the study of the density operator. However, analytic continuation for the density operator, which is a quadratic functional of the wave function, has to be performed with special care. We cannot simply take the square of the wave function, because the analytic continuation of a product is generally different from the product of the analytic continuations of the factors.

This observation is also related to a remark which was, we believe, first made by Van Hove:[12,13] scattering has to be considered as the result of "constructive interference" of the two wave functions whose time evolutions therefore cannot be treated independently, because they give the time evolution of the density matrix. We may consider this question from a more fundamental point of view, involving the very meaning of time in quantum theory.[14] The time dependence of averages in the Schrödinger representation comes from the wave function. As expectation values of observables depend on wave functions in a quadratic way, we get a "double" time dependence in these averages. However this *double* time dependence has to be reordered into a *single* sequence when processes such as collisions, which are at the origin of irreversible processes, are to be introduced.

In other words, analytical continuation has to be performed directly on the level of the density matrix itself. The regularization procedure[15] for doing it is simple and will be recalled in Section III of this chapter.

Let us now briefly sketch the content of this chapter. In Section II, we indicate the difficulties that arise in the traditional scattering theory (some aspects of this theory are reviewed in Appendix B) when we go from the Möller operators in Hilbert space to superoperators. In Section III, we introduce briefly the basic concepts that appear in the study of the dynamics of superspace (resolvent, creation and destruction fragments, definition of subdynamics, transformation theory, etc.) as well as the rules for analytical continuation. In Sections IV to VI, we apply these rules to the superspace description of the sets ρ_{k0} (or ρ_{0k}) and ρ_{kk}, $\rho_{kk'}$, of the density matrix. We show that the evolution of ρ_{k0} (or ρ_{0k}) is indeed described in terms of a unitary transformation. Its normal form corresponds to a diagonal form of H. In contrast, the evolution of the set ρ_{kk} and $\rho_{kk'}$ leads to a nonunitary ("starunitary") evolution. Its normal form is now one in which its infinitesimal generator of motion is directly related to the cross sections. In short, concepts such as cross sections are obviously foreign to the description of the time evolution in terms of a unitary transformation. On the other hand, the importance of these nonunitary concepts (cross sections, lifetimes) is such that we have to seek to include them in a consistent description starting with the traditional formalism of quantum theory.

While we limit ourselves in this chapter to the simple Hamiltonian [Eq. (1)], our conclusions extend to more general situations in field theory. In fact, the transition to superspace seems to us the natural way for extracting information about irreversible processes associated with given fields (such as treated in quantum electrodynamics).

Moreover the microscopic theory of the second law of thermodynamics as a selection principle is beautifully illustrated in scattering theory as our formulation shows that scattering states involving incoming spherical waves transformed into "controlled" outgoing plane waves are excluded. The analogy with the result obtained earlier,[16] using "fibers" in classical theory of K-flows, is striking: contracting fibers are excluded as all paths corresponding to these fibers tend to a common future (a single point in phase space). Similarly here the scattering states that are excluded are such that they would tend in the distant future to a single value of momentum (as labeled by the unperturbed Hamiltonian).

The importance of excluding "nonphysical" solutions in scattering theory and more generally in all problems involving classical or quantum fields has been repeatedly emphasized in the literature. The restriction of the solutions to "retarded potentials," although of foremost importance, is generally only justified by technical considerations, such as the inability of the experimentalists to realize appropriate initial conditions. It is therefore important that we now give a rigorous formulation to such ad hoc procedures.

For instance, when discussing radiation damping, Landau and Lifchitz, in their well-known textbook on field theory, exclude "self-accelerating" solutions. They note that such solutions would lead to physically absurd results. In their opinion, these contradictions have to be traced back to the difficulty of defining consistently an electrodynamical mass. This however does not explain why the time inverse of a permitted solution would be prohibited.

Note that in earlier days of quantum mechanics the emphasis by Planck and Einstein was often on the thermodynamical aspects of fields. With the advent of quantum field theory the emphasis moved to dynamical aspects. However with the discovery of the instability of most elementary particles[17] and that of the residual black body radiation,[18] the situation has again changed and we have to look for the origin of irreversibility in the frame of field theory.

Independently of field theory, with the renewed interest in irreversible processes,[19] we need new procedures to perform the transition from small quantum systems, presenting quasi-periodic motion, to large systems that exhibit irreversible behavior.

Since the formulation of quantum field theory, attempts have been made repeatedly to introduce some nonlocality (e.g., in the form of discrete space–time lattices) to eliminate some of the difficulties of standard field theory. From this point of view it is interesting that irreversibility introduces indeed a form of nonlocality. Irreversibility on one side and space–time description on the other are closely related problems as this appears already so clearly in the macroscopic theory.

II. SCATTERING THEORY*

The traditional scattering theory is formulated in the Hilbert space of wave functions. The state vector $|\psi(t)\rangle$ of the system obeys the Schrödinger equation

$$i\frac{\partial}{\partial t}|\psi(t)\rangle = H|\psi(t)\rangle \tag{2}$$

For a finite range potential, the decomposition of the Hamiltonian H as a sum of a kinetic part H_0 and a potential part V

$$H = H_0 + V \tag{3}$$

is quite unambiguous. This corresponds to the decomposition in terms of projectors P and Q as used in *I*, Section II.

The general solution of Eq. (2) is easily expressed in the eigenbasis of the H operator. From this point of view, scattering theory reduces to the problem of finding the eigenstates of H, in the unperturbed basis $\{|k\rangle\}$ such that

$$H_0|k\rangle = \omega_k|k\rangle \tag{4}$$

Equivalently we may diagonalize H, that is, introduce operators U such that

$$U^{-1}HU = H_0 \tag{5}$$

The operators U are unitary in the absence of bound states. (For simplicity, we shall assume in the following that the potential does not produce any bound state. The extension to potentials with a finite number of bound states can easily be performed.[20])

With appropriate boundary conditions, Eq. (5) leads to the well-known Möller operators $U_\pm$. Their action on the vectors $|k\rangle$ of the unperturbed basis gives the in- or out-eigenstates of H:

$$|k^\pm\rangle = U_\pm|k\rangle \tag{6}$$

* See for the notations Appendix B which presents a survey of the formal scattering theory.

As known,

$$(U_{\pm})_{kk'} = \delta_{kk'} + \frac{t^{\pm}_{kk'}(\omega_{k'})}{\pm i\varepsilon + \omega_{k'} - \omega_k} \tag{7a}$$

$$(U^{-1}_{\pm})_{kk'} = \delta_{kk'} + \frac{t^{\mp}_{kk'}(\omega_k)}{\mp i\varepsilon + \omega_k - \omega_{k'}} \tag{7b}$$

where $t(z)$ is the transition operator or briefly the t-operator (for details, see Appendix C). Note that as the result of the limiting procedure to large volumes, $|k^{\pm}\rangle$ or $U_{\pm}$ contain distributions.

The physical meaning of the scattering states is discussed in every textbook on scattering theory (see especially Ref. 11). The in-states $|k^{+}\rangle$ describe states with an incoming particle in a plane wave state and an outgoing particle in a (spherical, e.g.) scattered wave state. The out-states $|k^{-}\rangle$ contain an outgoing plane wave and an incoming spherical wave.

From the knowledge of the stationary states [Eq. (6)], the time evolution of the state $\psi(t)$ can be obtained.

Note however that, to define cross sections, one has still to proceed to some ad hoc handling of divergent contributions which appear when $\psi(t)$ is squared.[11]

An alternative formulation of scattering, as already mentioned, is directly in terms of density matrices.

The evolution of the system is then described by the Liouville-von Neumann equation

$$i\frac{\partial}{\partial t}\rho = L\rho \tag{8}$$

where L is the commutator with the Hamiltonian

$$L = [H, \ \ ]_- \tag{9}$$

We shall first show that, if one limits oneself to factorizable unitary superoperators, one encounters divergencies, which have the same origin as those met when one derives cross sections by the usual squaring procedure.

Consider the superoperator $\mathscr{U}^{-1}$ diagonalizing H

$$H_0 = U_{+}^{-1} H U_{+} = \mathscr{U}_{+}^{-1} H \tag{10}$$

Its matrix elements are

$$\mathscr{U}^{-1}_{+k_1k_2k_3k_4} = (U_{+}^{-1})_{k_1k_3}(U_{+})_{k_4k_2} \tag{11}$$

and one can immediately check that, for instance, the elements $\mathscr{U}^{-1}_{+kkk'k'}$ are ill-defined. Indeed, using Eq. (7), one finds

$$\mathscr{U}^{-1}_{+k'k'kk} = (U^{-1}_{+})_{k'k}(U_{+})_{kk'} = \frac{t^{-}_{k'k}(\omega_{k'})}{-i\varepsilon + \omega_{k'} - \omega_{k}} \cdot \frac{t^{+}_{kk'}(\omega_{k'})}{i\varepsilon + \omega_{k'} - \omega_{k}}$$

$$= \frac{1}{\varepsilon} \cdot \pi\delta(\omega_{k} - \omega_{k'})|t^{+}_{kk'}(\omega_{k})|^{2} \tag{12}$$

The origin of the divergence is quite clear: in the time behavior of $|\psi(t)|^2$, there are interferences between the evolutions of the wave-function $\psi(t)$ and its complex conjugate $\psi(t)^{cc}$. These cannot be properly taken into account when limiting procedures, as involved in the definition of the Möller operators $U_{\pm}$, are performed on $\psi(t)$ and $\psi(t)^{cc}$ separately.

This difficulty also shows up, for instance, in expressions such as

$$\varphi^{\pm} = \sum_{k} g(k)|k^{\pm}\rangle\langle k^{\pm}| \tag{13}$$

Because the $|k^{\pm}\rangle$ are eigenstates of the Hamiltonian H, one would expect this quantity to be invariant, whatever the function $g(k)$. Following Eq. (8), an invariant φ should satisfy

$$L\varphi = 0 \tag{14}$$

However, the foregoing φ's [Eq. (13)] fail to satisfy this relation. Indeed it can be shown (see Section VIII) that

$$(L\varphi^{\pm})_{kk} = g(k)[t^{\pm}_{kk}(\omega_{k}) - t^{\mp}_{kk}(\omega_{k})] \pm 2\pi i \sum_{k_1} \delta(\omega_{k} - \omega_{k_1})g(k_1)|t^{\pm}_{kk_1}(\omega_{k})|^{2} \tag{15}$$

which vanishes only in the special case of a function $g(k)$ depending only on the energy $g(k) = g(\omega_k)$ as the result of the optical theorem Eq. (C.13).

For sure, the r.h.s. of Eq. (15) vanishes if we take into account the volume dependence and let the volume go to infinity. However, this merely shows that $\varphi^{\pm}$ is an invariant of motion if the effect of scattering is neglected. This of course we cannot accept because we are precisely interested in the effect of scattering on the dynamical evolution.

Such difficulties show that the transition from the Hilbert space description [Eq. (2)] to the superspace description [Eq. (8)] is indeed more complex than could be expected at first.* It is this transition we want to discuss now in more detail.

* The mathematical difficulties come from the presence of distributions in Eq. (7). However, in addition they express a basic physical fact, the emergence of a new type of description (a nonunitary one) on the level of superspace.

III. DYNAMICS OF CORRELATIONS, SUBDYNAMICS AND NONUNITARY TRANSFORMATION THEORY

We present here, for the convenience of the reader, a compact formulation of the general theory[21] as far as it is needed. However no proofs are given because all results used later are derived in the subsequent sections.

Our starting point in the description of the evolution in superspace is the Liouville-von Neumann equation [Eq. (8)]. Its formal solution can be expressed in terms of the following Laplace transform:

$$\rho(t) = \frac{1}{2\pi i} \int_{\tilde{\gamma}} \mathrm{d}z\, \mathrm{e}^{-izt} \frac{1}{z - L} \rho(0) \tag{16}$$

where $\tilde{\gamma}$ is a contour, from right to left, above the singularities of the resolvent $(z - L)^{-1}$.

In the case of a finite system, the unit superoperator

$$1 = \frac{1}{2\pi i} \int_{\tilde{\gamma}} \mathrm{d}z \frac{1}{z - L} \tag{17}$$

can be easily decomposed into a complete set of (super)projectors, each arising from a (isolated) pole of the resolvent. In going to the limit of a continuous spectrum, an asymptotic procedure has to be introduced in order to take into account the fact that some of the poles in Eq. (16) or (17) may coalesce into a cut.

The solution of the Liouville-von Neumann equation for a given element of the density matrix in the free basis is, from Eq. (16) (for the notations, see Appendix A):

$$\rho_{\alpha\beta}(t) = \frac{1}{2\pi i} \int_{\tilde{\gamma}} \mathrm{d}z\, \mathrm{e}^{-izt} \sum_{\gamma\delta} \langle\alpha\beta| \frac{1}{z - L} |\gamma\delta\rangle \rho_{\gamma\delta}(0) \tag{18}$$

where the matrix elements of the Liouvillian L are given by

$$L_{\alpha\beta\alpha\gamma} = H_{\alpha\gamma}\delta_{\delta\beta} - H_{\delta\beta}\delta_{\alpha\gamma} \tag{19}$$

The states $|\alpha\beta\rangle$ are eigenstates of the unperturbed Liouville superoperator L_0 (associated to the unperturbed Hamiltonian H_0)

$$L_0 = [H_0, \quad]_- \tag{20a}$$

with eigenvalues $(\omega_\alpha - \omega_\beta)$. Similarly, one defines an interaction Liouvillian

$$\delta L = [V, \ \ ]_- \tag{20b}$$

Among these states $(\alpha\beta)$ which can be connected through δL, one distinguishes a vacuum state and correlation states. Correlation states are classified according to the minimum number of steps [Eq. (20b)] necessary to reach them from the vacuum.

Usually, the vacuum state consists of all the diagonal elements $\rho_{\alpha\alpha}$. The "monocorrelations" are the off-diagonal elements $\rho_{\alpha\beta}$ which are obtainable in one step from the vacuum, the "bicorrelations" are obtained in two steps, and so on. We obtain in this way correlations of higher and higher degree.

Using these definitions, it may be shown that, in the limit of a continuous spectrum, in contributions coming from a given state, the vacuum $\rho_{\alpha\alpha}$ for instance, connected to itself, only (irreducible) sequences of correlations of higher degrees have to be considered. We call this important result the "theorem in dynamics of correlations."[22]

Let us now turn to the regularization procedure already mentioned which is needed to give a meaning to Eq. (18) in the limit of a continuous spectrum. This procedure is often called the $i\varepsilon$-rule.[15]

Before we state this rule, let us first write Eq. (18) in its perturbative form:*

$$\rho_{\alpha\beta}(t) = \frac{1}{2\pi i}\int dz\, e^{-izt} \sum_{\gamma\delta}\sum_{n:0}^{\infty} \langle\alpha\beta| \frac{1}{z - L_0}\left(\delta L \frac{1}{z - L_0}\right)^n |\gamma\delta\rangle \rho_{\gamma\delta}(0) \tag{21}$$

Each term in this expansion is characterized by a sequence of correlated states, with possible repetitions. To obtain all contributions to $\rho_{\alpha\beta}(t)$, we have to deal with all singularities due to the vacuum and correlation states. Consider one of these states, say v. If the distinguished state v appears only once, the singularity is a simple pole.

When taking the residue at this pole, the $i\varepsilon$-rule states that the other propagators, corresponding to less-correlated states, are to be treated $-i\varepsilon$ (that is to z in the propagator has to be added a small negative imaginary part) and $+i\varepsilon$ when they are more correlated. States of the same degree of correlation have to be treated $-i\varepsilon$ if they appear at the right of the distinguished propagator, or $+i\varepsilon$ if they appear at the left. The $i\varepsilon$-rule is thus a regularization procedure corresponding to a "chronological" ordering. The equivalence with classical procedures in simple systems will be studied in Section IV (see also Appendix D).

In the case of a multiple pole, the same rule applies. Note that there is no ambiguity as a result of the theorem in dynamics of correlations, which as

* For a nonperturbative extension of the $i\varepsilon$-rule in the Friedrichs model, see Ref. 23 and Appendix F on potential scattering.

already mentioned says that between the same correlation appear only correlations of higher degree (which are thus treated $+i\varepsilon$).

It is very important to notice that our regularization procedure differs radically from usual analytical continuation which in the context of superspace has been discussed elsewhere.[24,25] Here we first write the matrix elements of the resolvent for a large (but finite) system, introduce the $i\varepsilon$-rule, and then go to the limit of an infinite system; in the usual techniques, which till now have been applied only to much simpler situations (such as discrete poles), one first goes to the limit of infinite systems and then proceeds to a suitable contour deformation.

The expression in the formal solution Eq. (21) which appears at the right of the distinguished propagator is called the destruction fragment $\mathscr{D}$ (in the case of a multiple pole, at the right of the first appearance of the propagator). Similarly, the creation fragment $\mathscr{C}$ is the expression at the left (of the last appearance) of the distinguished propagator. This terminology originates from the decomposition of the resolvent of L in the form

$$\frac{1}{z-L} = \mathscr{C}(z)\frac{1}{z - PLP - \psi(z)}\mathscr{D}(z) + \mathscr{P}(z) \tag{22}$$

where P symbolizes in general the projection on a given class of correlation states, here the ν-state. For instance, P^0 acting on ρ would select the vacuum component $\rho_0 \equiv \{\rho_{\alpha\alpha}\}$. The complementary projection on all other correlation states is denoted by Q. In contrast to Eq. (2.4) in Ref. I, we have included the P-parts into the definition of $\mathscr{C}(z)$ and $\mathscr{D}(z)$.

With the use of the $i\varepsilon$-rule, a complete set of superprojectors Π can be constructed. Their matrix elements are

$$\overset{\nu}{\Pi}_{\alpha\beta\gamma\delta} = \frac{1}{2\pi i}\int_{\gamma_\nu}' dz \left(\frac{1}{z-L}\right)_{\alpha\beta\gamma\delta} \tag{23}$$

γ_ν specifying the contour around the singularity corresponding to the selected ν-state. The prime on the integration sign recalls that the $i\varepsilon$-rule has been used in the transition to the continuous spectrum.

The basic properties of these projectors Π^ν are:[15]

$$\overset{\nu}{\Pi}{}^2 = \overset{\nu}{\Pi}; \qquad \overset{\nu}{\Pi}\,\overset{\nu'}{\Pi} = 0 \qquad \nu' \neq \nu \tag{24}$$

$$\sum_\nu \overset{\nu}{\Pi} = 1 \tag{25}$$

$$\left[\overset{\nu}{\Pi}, L\right]_- = 0 \tag{26}$$

These properties will be discussed and derived in the case studied here in Sections IV and V. As the result of the property of Eq. (26), the time evolution is split into a complete set of independent subdynamics.

Let us now summarize briefly our starunitary (nonunitary) transformation theory, based on the complete set of projectors Π^{ν}'s.[15] Its basic aim is the construction of a superoperator Λ which leads from ρ, satisfying the Liouville-von Neumann equation, to ${}^{p}\rho$, which corresponds to a semigroup description, including the second law:

$$ {}^{p}\rho = \Lambda^{-1}\rho \tag{27} $$

It is this transformation that breaks the time symmetry and introduces the nonlocality already mentioned in the Introduction. Whenever such a Λ exists, it is a nonfactorizable superoperator. As in the new representation, ${}^{p}\rho$ incorporates the second law of thermodynamics, we call it a physical representation.

The general method of construction of Λ starting from the Π's has been discussed elsewhere.[15] Let us summarize the essential steps. The starting point is the set of relations

$$ \Pi = \Lambda P \Lambda^{-1} \tag{28} $$

(when possible we drop from now on the upper index on the projectors).

To use this relation, it is convenient to express Π in the form:[21]

$$ \Pi = CAD \tag{29} $$

In Eq. (29), A stands for the $P - P$ component of Π:

$$ A = P\Pi P \tag{30} $$

and C and D are respectively defined by

$$ CA = \Pi P \tag{31} $$

$$ AD = P\Pi \tag{32} $$

The C and D are closely related to the creation and destruction superoperators in Eq. (22) (see also Section IV).

The components of the Λ and Λ^{-1} superoperators are then written as

$$ \Lambda P = C\chi \tag{33} $$

$$ P\Lambda^{-1} = \chi^{*}D \tag{34} $$

where χ is the $P - P$ component of Λ:

$$\chi = P\Lambda P \tag{35}$$

and χ^* that of Λ^{-1}:

$$\chi^* = P\Lambda^{-1}P \tag{36}$$

As the result of Eqs. (28) and (29), χ and χ^* have to satisfy the relation

$$\chi\chi^* = A \tag{37}$$

The projectors Π are obtained using our regularization procedure. The determination of χ may require supplementary conditions that will be formulated later in this section.

However, in the problems discussed in this chapter, χ, and therefore these supplementary conditions, plays no role (for reasons of order of magnitude in the volume).

That Λ^{-1} is the inverse of Λ:

$$\Lambda^{-1}\Lambda = 1 \tag{38}$$

insures the projection properties [Eq. (24)] of the Π's:

$$\overset{\nu}{\Pi}\overset{\nu'}{\Pi} = \Lambda\overset{\nu}{P}\Lambda^{-1}\Lambda\overset{\nu'}{P}\Lambda^{-1} = \Lambda\overset{\nu}{P}\overset{\nu'}{P}\Lambda^{-1} = \overset{\nu}{\Pi}\,\delta_{\nu\nu'} \tag{39}$$

Similarly, the right inverse property

$$\Lambda\Lambda^{-1} = 1 \tag{40}$$

leads to the completeness relation [Eq. (25)] of the Π's:

$$\sum_{\nu}\overset{\nu}{\Pi} = \sum_{\nu}\Lambda\overset{\nu}{P}\Lambda^{-1} = \Lambda\Lambda^{-1} = 1 \tag{41}$$

as the set of orthogonal P-projections can be made complete.

In the new representation defined by Eq. (27), the time evolution is described by

$$i\,\partial_t{}^P\rho = \phi^P\rho \tag{42}$$

with the new generator of motion ϕ:

$$\phi = \Lambda^{-1} L \Lambda \tag{43}$$

It will be shown later that the $\phi_{\alpha\alpha\beta\beta}$ elements are precisely the cross sections. For this reason in the representation Eq. (27), Eq. (42) incorporates irreversible processes in contrast with Eq. (8).

From Eqs. (26), (28), and (43), one derives the important property of ϕ:

$$[\phi, P]_- = 0 \tag{44}$$

This so-called block-diagonality property of ϕ implies that the various P-components of the physical density operator ${}^p\rho$ evolve independently:

$$i\,\partial_t\, P^p\rho = P\phi P^p\rho \tag{45}$$

So far, we have studied the evolution of the matrix elements of the density operator. This corresponds to the Schrödinger picture. We can equivalently study the evolution of the operators (Heisenberg picture).

The average value of an operator O has to have the same value in both representations

$$\begin{aligned}\langle O\rangle_t &= (O, \rho(t)) = (O, \mathrm{e}^{-iLt}\,\rho)\\ &= (\mathrm{e}^{iL^+t}\,O, \rho) = (O(t), \rho)\end{aligned} \tag{46}$$

The superoperator L being hermitian, the evolution of the operator $O(t)$ is indeed given by the Heisenberg equation for observables

$$i\frac{\mathrm{d}}{\mathrm{d}t}\,O(t) = -LO(t) \tag{47}$$

On the other hand, one can define the Heisenberg-Schrödinger conjugation (noted by a prime (′) and also called the L-inversion):

$$i\frac{\partial}{\partial t}\,\rho(t) = L\rho(t); \qquad i\frac{\mathrm{d}}{\mathrm{d}t}\,O(t) = L'O(t) \tag{48}$$

with obviously,

$$L' = -L \tag{49}$$

If one focuses on the evolution of observables instead of that of density operators, as previously, one has to consider $e^{-iL't}$ instead of e^{-iLt}. It would seem that one has only to consider the same expressions as above, except for the replacement of L by $L' = -L$. However, it is necessary to adapt our regularization procedure when applied to observables. It turns out that the only modification of the $i\varepsilon$-rule that has to be introduced concerns correlations with the same degree.[23] In this case, with respect to a specified propagator, propagators at its left have to be considered as less correlated (and treated $-i\varepsilon$) and at its right as more correlated (and treated $+i\varepsilon$). Using this rule, one obtains the Λ'-transformation, which applies to observables.

Then with ${}^{p}O$ defined by

$$ {}^{p}O = \Lambda'^{-1} O \tag{50} $$

one obtains

$$ \langle O \rangle = (O, \rho) = ({}^{p}O, {}^{p}\rho) \tag{51} $$

We shall apply the L-inversion in Sections IV and V.

The starunitarity property of Λ

$$ \Lambda^{*} = \Lambda'^{+} = \Lambda^{-1} \tag{52} $$

insures the validity of Eq. (51). Note that when $\Lambda' \neq \Lambda$ the transformation is indeed no longer unitary.

In Eq. (52) the adjoint superoperator Λ^{+} appears. Its matrix elements are defined through

$$ (\Lambda^{+})_{\alpha\beta\gamma\delta} = (\Lambda_{\gamma\delta\alpha\beta})^{cc} \tag{53} $$

Another property that the transformation has to satisfy is the adjoint-symmetry, which insures that hermiticity of states (and observables) is preserved by the transformation to the physical representation:

$$ \Lambda^{a} = \Lambda \tag{54} $$

with the definition

$$ (\Lambda^{a})_{\alpha\beta\gamma\delta} = (\Lambda_{\beta\alpha\delta\gamma})^{cc} \tag{55} $$

As we have seen, the Λ-transformation is not entirely determined by the rules of analytical continuation. Specially the $P-P$ components of Λ we have denoted by χ [see Eq. (35)] remain undetermined. A supplementary condition which generalizes the usual diagonalization can be introduced.[26] This condition involves the energy superoperator $\mathscr{H}$ defined as half the anticommutator with the Hamiltonian operator:

$$\mathscr{H} = \tfrac{1}{2}[H, \ ]_+ \tag{56}$$

which is shown to commute with all projectors Π defining the subdynamics:

$$[\mathscr{H}, \Pi]_- = 0 \tag{57}$$

The condition is that the Λ-superoperator diagonalizes $\mathscr{H}$ in the unperturbed basis. In other words, the Λ-transform of $\mathscr{H}$,

$$^{p}\mathscr{H} = \Lambda^{-1}\mathscr{H}\Lambda \tag{58}$$

satisfies the condition

$$^{p}\mathscr{H}_{\alpha\beta\gamma\delta} = {}^{p}\mathscr{H}_{\alpha\beta\alpha\beta}\delta_{\alpha\gamma}\delta_{\beta\delta} \tag{59}$$

In the particular case of the vacuum elements, one requires that

$$^{p}\mathscr{H}_{\alpha\alpha,\beta\beta} = \Omega_\alpha\delta_{\alpha\beta} \tag{60}$$

where the Ω_α are the physical energies.

The importance of this condition rests on the fact that, as L and $\mathscr{H}$ commute, one has also

$$[\phi, {}^{p}\mathscr{H}]_- = 0 \tag{61}$$

which, together with Eq. (44) where P is the projector on the vacuum, insures that the collision superoperator $P\phi P$ strictly conserves the physical energies, defined as the eigenvalues of $P{}^{p}\mathscr{H}P$.

We have summarized so far some of the general properties of the Λ-transformation. The aim of this chapter is to construct this transformation explicitly for the potential scattering problem (see also Refs. 8, 9) as well as for the Boltzmann problem corresponding to a dilute gas.

IV. UNITARY SECTORS OF THE POTENTIAL SCATTERING PROBLEM

We shall now apply the methods just described to the problem of potential scattering, for which the Hamiltonian has the form of Eq. (1) (in the second quantization formalism).

In general, a state of the system can be represented by a linear combination in the basis $|0\rangle$, $|k\rangle$, $|kk'\rangle, \ldots$ of the Fock space of 0, 1, 2,... particles generated by the free fields

$$|\psi\rangle = a_0|0\rangle + \sum_k c_k|k\rangle + \sum_{kk'} d_{kk'}|kk'\rangle + \ldots \tag{62}$$

In the present case, the interaction conserves the number of particles. This implies an uncoupling of the sectors corresponding to different numbers of particles.

The density matrix ρ will in general have elements not only between states of the same sector but also between states belonging to different sectors. We shall limit ourselves to the study of ρ_{00}, ρ_{10}, ρ_{01}, ρ_{11}, which are uncoupled due to the decomposition in sectors.

ρ_{00} is constant in time.

ρ_{10} evolves inside the (1, 0) sector in superspace. We shall show that, as expected, its evolution is isomorphic to that of the wave function, as

$$\rho_{k0} \sim \langle k|\psi\rangle\langle\psi|0\rangle = c_k a_0 \sim c_k \tag{63}$$

Similarly,

$$\rho_{0k} \sim c_k^* \tag{64}$$

is isomorphic to the complex conjugated wave function, and evolves in the (0, 1) sector.

The (1, 1) components ρ_{kk} and $\rho_{kk'}$ evolve independently from ρ_{0k} and ρ_{k0}. They are the components of the density operator for the one particle states.

First we deal with the sectors (1, 0) and (0, 1).

The evolution of the ρ_{k0} component is given by the Liouville-von Neumann equations:

$$i\frac{\partial}{\partial t}\rho_{k_10} = L_{k_10k_10}\rho_{k_10} + \sum_{k_2} L_{k_10k_20}\rho_{k_20} \tag{65}$$

Using the definition of the matrix elements of the Liouville operator [Eq. (19)], which gives in the present case

$$L_{k_10k_10} = \omega_{k_1} \tag{66a}$$

$$L_{k_10k_20} = V_{k_1k_2} \tag{66b}$$

One obtains from Eq. (65) the explicit form

$$i\frac{\partial}{\partial t}\rho_{k_1 0} = \omega_{k_1}\rho_{k_1 0} + \sum_{k_2} V_{k_1 k_2}\rho_{k_2 0} \tag{67}$$

This is indeed isomorphic to the time evolution of the wave function in the unperturbed basis

$$|\psi(t)\rangle = \sum_{k} c_k(t)|k\rangle \tag{68}$$

as

$$i\frac{\partial}{\partial t} c_{k_1}(t) = \omega_{k_1} c_{k_1}(t) + \sum_{k_2} V_{k_1 k_2} c_{k_2}(t) \tag{69}$$

In the limit of a continuous spectrum, the orders of magnitude in L^3 of the different contributions play an important role in retaining in a consistent way the dominant contributions. Obviously,

$$\omega_k \sim O(1); \quad V_{kk'} \sim O\left(\frac{1}{L^3}\right) \tag{70}$$

In addition, we assume that ρ_{k0} is of order $L^{-3/2}$ (this corresponds to the usual wave function normalization)

$$\rho_{k0} \sim O\left(\frac{1}{L^{3/2}}\right) \tag{71}$$

The formal solution of Eq. (65), is [see Eq. (18)]

$$\rho_{k_1 0}(t) = \frac{1}{2\pi i}\int_{\pm} \mathrm{d}z\, \mathrm{e}^{-izt} \sum_{k_2} \langle k_1 0| \frac{1}{z - L} |k_2 0\rangle \rho_{k_2 0} \tag{72}$$

To analyze this solution, we start with the perturbation expansion for the resolvent of L:

$$R_{k_1 0 k_2 0} = \langle k_1 0| \frac{1}{z - L} |k_2 0\rangle = \sum_{n:0}^{\infty} \langle k_1 0| \frac{1}{z - L_0}\left(\delta L \frac{1}{z - L_0}\right)^n |k_2 0\rangle \tag{73}$$

We have introduced in Section III a classification of correlations according to the number of steps necessary to reach them from the "vacuum". In the (1, 0) sector all components ρ_{k0} are obviously of the same type. They all belong to the vacuum state for this sector.

As

$$\langle k_1 0| \frac{1}{z - L_0} |k_2 0\rangle = \frac{1}{z - \omega_{k_1}} \delta_{k_1 k_2} \tag{74}$$

and

$$\langle k_1 0| \delta L |k_2 0\rangle = V_{k_1 k_2} \tag{75}$$

a typical term appearing in Eq. (73) takes the form

$$\sum_{k'_1 k'_2 \ldots} \frac{1}{z - \omega_{k_1}} V_{k_1 k'_1} \frac{1}{z - \omega_{k'_1}} V_{k'_1 k'_2} \frac{1}{z - \omega_{k'_2}} \cdots \frac{1}{z - \omega_{k'_3}} V_{k'_3 k'_4} \frac{1}{z - \omega_{k'_4}} V_{k'_4 k_2} \frac{1}{z - \omega_{k_2}}$$

Let us first note that situations in which the same vector k would occur more than once in this expression are ruled out by the theorem in dynamics of correlations we stated in Section III. Indeed such a term (consider for instance $k'_1 = k'_4$) would be negligible in the limit of a large volume ($L^3 \to \infty$), because one summation would then be lost.

Using Eqs. (74) and (75), one can rewrite Eq. (73) as:

$$R_{k_1 0 k_2 0}(z) = \sum_{k'} \mathscr{C}_{k_1 0 k' 0}(z) \frac{1}{z - \omega_{k'}} \mathscr{D}_{k' 0 k_2 0}(z) \tag{76}$$

where $\mathscr{C}_{k_1 0 k' 0}(z)$ is

$$\mathscr{C}_{k_1 0 k' 0}(z) = \delta_{k_1 k'} + \frac{1}{z - \omega_{k_1}} \left\{ V_{k_1 k'} + \sum_{k'_1} V_{k_1 k'_1} \frac{1}{z - \omega_{k'_1}} V_{k'_1 k'} + \cdots \right\} \tag{77a}$$

or using Eq. (C.7)

$$\mathscr{C}_{k_1 0 k' 0}(z) = \delta_{k_1 k'} + \frac{t_{k_1 k'}(z)}{z - \omega_{k_1}} \tag{77b}$$

Similarly $\mathscr{D}_{k'0k_20}(z)$ is given by

$$\mathscr{D}_{k'0k_20}(z) = \delta_{k'k_2} + \frac{t_{k'k_2}(z)}{z - \omega_{k_2}} \tag{78}$$

Due to the fact that the state $k'0$ can appear only once, the $\mathscr{C}_{k_1k'0}(z)$ and $\mathscr{D}_{k'0k_20}(z)$ are regular $z = \omega_{k'}$.

In the terminology introduced in Section III, $\mathscr{C}$ and $\mathscr{D}$ are called respectively a creation and a destruction fragment, with respect to the state $(k', 0)$.*

From Eq. (76), one sees that all the contributions to the temporal evolution of ρ_{k_10} [Eq. (72)] can be expressed as a sum of contributions arising from the pole $z = \omega_k$.

Let us now single out one of these contributions from the state $(k, 0)$. Equation (76) is then written as

$$R_{k_10k_20} = \mathscr{C}_{k_10k0}(z)\frac{1}{z - \omega_k}\mathscr{D}_{k0k_20}(z) + \mathscr{P}_{k_10k_20}(z) \tag{79}$$

where the irreducible part, $\mathscr{P}_{k_10k_20}(z)$, corresponds to all the contributions that do not go through the intermediate state $k0$ and thus has no singularity in $z = \omega_k$. (It is Eq. (76) but the sum $\Sigma_{k'}$ is replaced by $\Sigma_{k' \neq k}$).

The contribution to $\rho_{k_10}(t)$ [Eq. (72)] arising from the pole $z = \omega_k$ (a contribution we shall note $\rho^{(k0)}_{k_10}(t)$) is now easily evaluated using the $i\varepsilon$-rule of Section III. All the states $k_10, k'_10, \ldots$ appearing at the left of $k0$, that is, in the creation fragment $\mathscr{C}_{k_10k0}$, have to be considered as more correlated and thus treated $+i\varepsilon$ when the residue in $z = \omega_k$ is taken.

On the contrary, the states $k_20, \ldots$ at the right of $k0$, that is, in the destruction fragment $\mathscr{D}_{k0k_20}$, are to be considered as less correlated and treated $-i\varepsilon$.

This gives†

$$\overset{(k0)}{\rho_{k_10}}(t) = \left(\overset{(k0)}{\Sigma}(t)\rho(0)\right)_{k_10} = C_{k_10k0}\, e^{-i\omega_k t} \sum_{k_2} D_{k0k_20}\rho_{k_20}(0) \tag{80}$$

* When necessary, this may be emphasized using an upper index $(k'0)$. However in $\mathscr{C}^{(k'0)}$ and $\mathscr{D}^{(k'0)}$ there is no possible confusion because the lower right index in $\mathscr{C}$ and the lower left index in $\mathscr{D}$ represent this state.

† The notation $\Sigma^{(k0)}(t)$ ought not be confused with a summation sign on the vector k: Σ_k. The upper index $(k0)$ indicates which singularity has been retained.

where

$$C_{k_1 0 k 0} = \lim_{z : \omega_k + i\varepsilon} \mathscr{C}_{k_1 0 k 0}(z) = \delta_{k_1 k} + \frac{t^+_{k_1 k}(\omega_k)}{i\varepsilon + \omega_k - \omega_{k_1}} \tag{81a}$$

Note that this expression is nothing but the $k_1 k$ element of the Möller matrix U_+ [see Eq. (7) and, for more details, the Appendix B]. Thus

$$C_{k_1 0 k 0} = (U_+)_{k_1 k} \tag{81b}$$

Similarly, one has

$$D_{k 0 k_2 0} = \lim_{z : w_k - i\varepsilon} \mathscr{D}_{k 0 k_2 0}(z) = \delta_{k k_2} + \frac{t^-_{k k_2}(\omega_k)}{-i\varepsilon + \omega_k - \omega_{k_2}} \tag{82a}$$

or

$$D_{k 0 k_2 0} = (U_+^{-1})_{k k_2} \tag{82b}$$

The evolution of $\rho_{k_1 0}(t)$ is then given [see Eqs. (76) and (72)] by:

$$\rho_{k_1 0}(t) = \sum_k \overset{(k0)}{\rho}_{k_1 0}(t) = \sum_k \left(\overset{(k0)}{\Sigma}(t) \rho(0) \right)_{k_1 0} \tag{83}$$

We give in Appendix D the result that one obtains in the usual quantum mechanical treatment, starting with the resolvent corresponding to the Hamiltonian H. We show that, in this case, our regularization procedure coincides with the results reached traditionally by standard procedures.

We are now able to make contact with the general transformation theory as outlined in the preceeding section.

Indeed, the zero time limit of the $\Sigma^{(k0)}(t)$ operator defines the subdynamics projector $\Pi^{(k0)}$:

$$\overset{(k0)}{\Pi}_{k_1 0 k_2 0} = \lim_{t \to 0} \overset{(k0)}{\Sigma}_{k_1 0 k_2 0}(t) \tag{84}$$

with [see Eq. (80)]

$$\overset{(k0)}{\Pi}_{k_1 0 k_2 0} = C_{k_1 0 k 0} D_{k 0 k_2 0} \tag{85}$$

Let us now verify that the $\Pi^{(k0)}$'s indeed satisfy the general relations of Eqs. (24)–(26). This is done in Appendix E using the explicit forms of C and D [Eqs. (81) and (82)] as well as the properties of the t-operator [Eqs. (C.6)–(C.15)]. Let us summarize some basic results.

1. The completeness of the $\Pi^{(k0)}$'s [see Eq. (25)]:

$$\sum_k \overset{(k0)}{\Pi}_{k_1 0 k_2 0} = \delta_{k_1 k_2} \tag{86}$$

is a consequence of Eq. (C.14).

2. In a similar way, the orthogonality and idempotence

$$\overset{(k0)}{\Pi}\;\overset{(k'0)}{\Pi} = \overset{(k0)}{\Pi}\,\delta_{kk'} \tag{87}$$

immediately results from Eq. (C.11).

3. The commutation relation with the Liouville operator

$$\left[\overset{(k0)}{\Pi}, L\right]_- = 0 \tag{88}$$

is a consequence of the definition of the t-operator [Eq. (C.6)]. It is indeed shown in Appendix E that

$$\left(\overset{(k0)}{\Pi}\,L\right)_{k_1 0 k_2 0} = \left(L\,\overset{(k0)}{\Pi}\right)_{k_1 0 k_2 0} = C_{k_1 0 k 0}\,\omega_k D_{k0k_2 0} \tag{89}$$

The properties of Eqs. (86)–(88) are general properties which have their correspondents in other sectors. However, in this sector, we have an additional property which fails to be satisfied in the sector (1, 1), namely that $\Pi^{(k0)}$ is hermitian:

$$\overset{(k0)}{\Pi}{}^{+} = \overset{(k0)}{\Pi} \tag{90}$$

Indeed, using Eq. (C.8) and the definition Eq. (53) of the adjoint of a superoperator, one immediately obtains the relations

$$\begin{aligned}(C^+)_{k0k_2 0} = (C_{k_2 0 k 0})^{cc} &= \delta_{kk_2} + \frac{t^-_{kk_2}(\omega_k)}{-i\varepsilon + \omega_k - \omega_{k_2}} \\ &= D_{k0k_2 0}\end{aligned} \tag{91}$$

This then leads to Eq. (90).

The Λ superoperators are obtained from the Π's through the relation Eq. (28) or Eqs. (33) and (34).

Note that, in the present sector,

$$A_{k0k0} = \frac{1}{2\pi i}\int_{z:\omega_k} dz\langle k0|\frac{1}{z - L_0}|k0\rangle = 1 \tag{92}$$

hence $\chi_{k0k0} = 1$ and $\chi^*_{k0k0} = 1$. Therefore, in this case, we have simply

$$\Lambda_{k_1 0k0} = C_{k_1 0k0} \tag{93}$$

$$(\Lambda^{-1})_{k0k_2 0} = D_{k0k_2 0} \tag{94}$$

As shown in Section III, the invertibility property of the Λ in the sector considered so far:

$$\Lambda\Lambda^{-1} = \Lambda^{-1}\Lambda = 1 \tag{95}$$

is a consequence of the properties [Eqs. (86) and (87)] of the $\Pi^{(k0)}$'s.

Furthermore, the relation in Eq. (89) directly implies that

$$(\Lambda^{-1}L\Lambda)_{k_1 0k_2 0} = \omega_{k_1}\delta_{k_1 k_2} = (L_0)_{k_1 0k_2 0} \tag{96}$$

This last relation indicates that the Λ transformation in this sector does not lead to any dissipative behavior. This could also have been immediately inferred from Eq. (91), which shows that in this sector Λ is unitary

$$(\Lambda^{-1})_{k0k'0} = (\Lambda^{+})_{k0k'0} \tag{97}$$

However, as will be shown in Section V, the unitarity is lost in sector (1, 1).

The remaining question is how to insure the invariance [Eq. (51)]. Using the $i\varepsilon$-rule (see Section III), we obtain the L-inverted superoperators by changing L into $-L$ (i.e., ω_k into $-\omega_k$, $V_{kk'}$ into $-V_{kk'}$) and replacing everywhere $i\varepsilon$ by $-i\varepsilon$, because all states in the (1, 0) sector are of the same degree of correlation. Consider as an example [see Eq. (81)]

$$\begin{aligned} C'_{k_1 0k0} &= \delta_{k_1 k} + \frac{1}{-i\varepsilon - \omega_k + \omega_{k_1}} \\ &\quad \times \left(-V_{k_1 k} + \sum V_{k_1 k'_1}\frac{1}{-i\varepsilon - \omega_k + \omega_{k'_1}}V_{k'_1 k} + \cdots\right) \\ &= \delta_{k_1 k} + \frac{t^{+}_{k_1 k}(\omega_k)}{i\varepsilon + \omega_k - \omega_{k_1}} = C_{k_1 0k0} \end{aligned} \tag{98}$$

As the same holds for other elements in this sector, Λ is invariant under the L-inversion

$$(\Lambda')_{k_1 0 k 0} = \Lambda_{k_1 0 k 0} \tag{99}$$

Hence, together with Eq. (97), it proves that the Λ-transformation restricted to the sector (1, 0) is indeed starunitary [see Eq. (52)].

The subdynamics (0, k) in the sector (0, 1) is described in Appendix E in a completely similar way.

When the elements of Λ obtained are compared with Eqs. (93) and (94) one immediately sees that

$$\Lambda_{0k_1 0k} = (\Lambda_{k_1 0 k 0})^{cc} \tag{100}$$

This relation, using the definition of the associated superoperator [Eq. (55)], means that

$$\Lambda = \Lambda^a \tag{101}$$

Therefore hermiticity is preserved by the Λ-transformation.

We have treated the sectors (1, 0) and (0, 1) using the $i\varepsilon$-rule with the chronological ordering.

In these sectors, the difference between chronological and antichronological ordering is of no consequence. With the choice of the antichronological ordering, all expressions obtained so far, of course, change, but the modifications turn out to consist simply in the overall replacement of U_+ by U_- (see Appendix E).

However, this is no longer so in sector (1, 1), which includes scattering, because the two orderings lead to the two different semigroups, one oriented toward the future, the other toward the past. The chronological ordering is the one adapted to the description toward the future.[28]

In the unitary sectors, in contrast, there is no dissipativity, hence no criterion available for adopting one chronological ordering rather than the opposite one.

Let us now turn to the the sector (1, 1), in which the description of the scattering process takes place.

V. THE STARUNITARY SECTOR (1, 1)

In this sector, the density matrix has two kinds of elements: diagonal elements ρ_{kk} and off-diagonal elements $\rho_{kk'}$.

In accordance with our classification of correlations, one considers the diagonal states as the (degenerate) vacuum, and the off-diagonal elements as monocorrelations (which may be reached in a single step from the vacuum).

In the evolution equation [Eq. (21)] two types of propagators appear: the vacuum propagator

$$\langle kk| \frac{1}{z - L_0} |kk\rangle = \frac{1}{z} \tag{102}$$

and the monocorrelation propagators

$$\langle k_1 k_2| \frac{1}{z - L_0} |k_1 k_2\rangle = \frac{1}{z - \omega_{k_1} + \omega_{k_2}} \tag{103}$$

The $z = 0$ singularity will give rise to the Π^0-subdynamics and the $z = \omega_{k_1} - \omega_{k_2}$ poles to various correlation subdynamics $\Pi^{(k_1 k_2)}$. Formally,

$$\overset{0}{\Pi} = \frac{1}{2\pi i} \int_{\gamma_0}' dz \frac{1}{z - L} \tag{104}$$

$$\overset{(k_1 k_2)}{\Pi} = \frac{1}{2\pi i} \int_{\gamma_{12}}' dz \frac{1}{z - L} \tag{105}$$

γ_0 being a contour around $z = 0$, γ_{12} around $z = \omega_{k_1} - \omega_{k_2'}$ and the prime on the integral sign to remind that the residues have to be evaluated with the use of the $i\varepsilon$-rule.

For Π^0, this rule is very simple to apply. The pole at $z = 0$ corresponds to the vacuum state. All other states are more correlated, and so the residues are to be treated $+i\varepsilon$. A further simplifying remark is that

$$\overset{0}{\Pi}_{kkkk} = A_{kkkk} = \frac{1}{2\pi i} \int_{\gamma_0} dz \langle kk| \frac{1}{z - L_0} |kk\rangle = 1 \tag{106a}$$

and

$$\overset{0}{\Pi}_{kkk'k'} = A_{kkk'k'} \sim O\left(\frac{1}{L^6}\right) \tag{106b}$$

so that the dominant contribution of $\Pi^0_{k_1 k_2 kk}$ reduces to $C_{k_1 k_2 kk'}$ and that to

$\overset{0}{\Pi}_{kkk_1k_2}$ to the corresponding $D_{kkk_1k_2}$ [see Eqs. (30)–(32)]. For instance,

$$\begin{aligned}\overset{0}{\Pi}_{k_1kkk} = C_{k_1kkk} &= \frac{1}{2\pi i}\int_{\gamma_0} dz \sum_{n:0}^{\infty} \langle k_1k| \frac{1}{z-L_0}\left(\delta L \frac{1}{z-L_0}\right)^n |kk\rangle \\ &= \frac{1}{2\pi i}\int_{\gamma_0} dz \left\{\delta_{k_1k} + \frac{1}{z-\omega_{k_1}+\omega_k}\left(V_{k_1k} + \sum V_{k_1k_2}\right.\right. \\ &\quad \times \left.\left.\frac{1}{z-\omega_{k_2}+\omega_k} V_{k_2k+\cdots}\right)\right\}\frac{1}{z} = \frac{1}{2\pi i}\int_{\gamma_0} dz\, \mathscr{C}_{k_1kkk}(z)\frac{1}{z} \\ &= \delta_{k_1k} + \frac{1}{i\varepsilon-\omega_{k_1}+\omega_k}\left(V_{k_1k} + \sum V_{k_1k_2}\frac{1}{i\varepsilon-\omega_{k_2}+\omega_k} V_{k_2k+\cdots}\right)\end{aligned}$$

The summation is easily performed using Eq. (C.7) and gives

$$C_{k_1kkk} = \delta_{k_1k} + \frac{t^+_{k_1k}(\omega_k)}{i\varepsilon+\omega_k-\omega_{k_1}} \tag{107}$$

Similarly one obtains

$$C_{kk_2kk} = \delta_{kk_2} + \frac{t^-_{kk_2}(\omega_k)}{-i\varepsilon+\omega_k-\omega_{k_2}} \tag{108}$$

and $(k \neq k_1 \neq k_2 \neq k)$

$$C_{k_1k_2kk} = \frac{t^+_{k_1k}(\omega_k)}{i\varepsilon+\omega_k-\omega_{k_1}} \cdot \frac{t^-_{kk_2}(\omega_k)}{-i\varepsilon+\omega_k-\omega_{k_2}} \tag{109a}$$

The proof of Eq. (109a) is given in Appendix F.
The expressions Eq. (107)–(109) can be written in the compact form

$$C_{k_1k_2kk} = \left(\delta_{k_1k} + \frac{t^+_{k_1k}(\omega_k)}{i\varepsilon+\omega_k-\omega_{k_1}}\right)\left(\delta_{kk_2} + \frac{t^-_{kk_2}(\omega_k)}{-i\varepsilon+\omega_k-\omega_{k_2}}\right) \tag{109b}$$

Similarly, one obtains for the elements of the destruction fragment

$$D_{kkk_1k_2} = \left(\delta_{kk_1} + \frac{t^+_{kk_1}(\omega_k)}{i\varepsilon+\omega_k-\omega_{k_1}}\right)\left(\delta_{kk_2} + \frac{t^-_{k_2k}(\omega_k)}{-i\varepsilon+\omega_k-\omega_{k_2}}\right) \tag{110}$$

For the sake of completeness, let us also consider the diagonal-to-diagonal elements of the Π^0 operator, which are evaluated in Appendix F: their expression

$$\overset{0}{\Pi}_{kkk'k'} = A_{kkk'k'} = \frac{1}{2\pi i} \int_{\gamma_0} \mathrm{d}z \frac{1}{z^2} \sum_{n:1}^{\infty} \langle kk| \, \delta L \left(\frac{1}{z - L_0} \, \delta L \right)^n |k'k'\rangle \quad (111a)$$

involves a double pole in $z = 0$. It leads to

$$A_{kkk'k'} = \tfrac{1}{2}[|t^+_{kk'}(\omega_k)|^2 + |t^+_{kk'}(\omega_{k'})|^2] \cdot \left[\frac{1}{(i\varepsilon + \omega_k - \omega_{k'})^2} + \frac{1}{(-i\varepsilon + \omega_k - \omega_{k'})^2} \right] + i\pi \, \delta(\omega_k - \omega_{k'}) \left[\left(\frac{\partial}{\partial \omega_k} t^+_{kk'}(\omega_k) \right) \times t^-_{k'k}(\omega_k) - t^+_{kk'}(\omega_k) \left(\frac{\partial}{\partial \omega_k} t^-_{k'k}(\omega_k) \right) \right] \quad (111b)$$

The elements of the Π^0 projector are easily obtained as in all propagators corresponding to monocorrelations, we have to add $+i\varepsilon$.

In the case of the monocorrelation subdynamics $\Pi^{(kk')}$, the states are either the vacuum (which is treated $-i\varepsilon$), or monocorrelations. The chronological ordering determines then if these are to be treated $+i\varepsilon$ or $-i\varepsilon$. In the destruction fragment $\mathscr{D}_{kk'k_1k_2}$ all monocorrelations are at the right of the distinguished correlation (kk'). In accordance with the $i\varepsilon$-rule of Section III, they are treated as "less correlated," as is the vacuum. The quantity $-i\varepsilon$ has therefore to be added.

$$D_{kk'k_1k_2} = \frac{1}{2\pi i} \int_{z:\, -i\varepsilon + \omega k - \omega k'} \mathrm{d}z \frac{1}{z - \omega_k + \omega_{k'}} \times \langle kk'| \sum_{n:0}^{\infty} \delta L \left(\frac{1}{z - L_0} \, \delta L \right)^n |k_1k_2\rangle \frac{1}{z - \omega_{k_1} + \omega_{k_2}} \quad (112)$$

By recurrence (see Appendix F), one obtains $((k_1k_2) \neq (kk'))$

$$D_{kk'k_1k_2} = \left(\delta_{kk_1} + \frac{t^-_{kk_1}(\omega_k)}{-i\varepsilon + \omega_k - \omega_{k_1}} \right) \left(\delta_{k'k_2} + \frac{t^+_{k_2k'}(\omega_{k'})}{i\varepsilon + \omega_{k'} - \omega_{k_2}} \right) \quad (113)$$

This result is also valid in the case where $k_1 = k_2$, as the vacuum is also treated $-i\varepsilon$.

However, for the creation fragments, the vacuum and the monocorrelations are treated differently, and thus the cases $k_1 \neq k_2$ and $k_1 = k_2$ have to be considered separately.

As the result, one gets $(k_1 \neq k_2; (k_1 k_2) \neq (kk'))$

$$C_{k_1 k_2 kk'} = \frac{1}{2\pi i} \int_{z: i\varepsilon + \omega k - \omega k'} \mathrm{d}z \frac{1}{z - \omega_{k_1} + \omega_{k_2}} \langle k_1 k_2 | \sum_{n:0}^{\infty} \times \delta L \left(\frac{1}{z - L_0} \delta L \right)^n | kk' \rangle \frac{1}{z - \omega_k + \omega_{k'}}$$

$$= \left(\delta_{kk_1} + \frac{t^+_{k_1 k}(\omega_k)}{i\varepsilon + \omega_k - \omega_{k_1}} \right) \left(\delta_{k'k_2} + \frac{t^-_{k'k_2}(\omega_{k'})}{-i\varepsilon + \omega_{k'} - \omega_{k_2}} \right) \tag{114}$$

When $k_1 = k_2$, the vacuum propagator $1/z$ has to be treated $-i\varepsilon$, giving $(-i\varepsilon + \omega_k - \omega_{k'})^{-1}$, while all the other monocorrelations are treated $+i\varepsilon$, as for $C_{k_1 k_2 kk'}$.

This leads to

$$C_{kkkk'} = \frac{t^-_{k'k}(\omega_{k'})}{i\varepsilon + \omega_{k'} - \omega_k} \tag{115}$$

$$C_{k'k'kk'} = \frac{t^+_{k'k}(\omega_k)}{-i\varepsilon + \omega_k - \omega_{k'}} \tag{116}$$

$$C_{k_1 k_1 kk'} = \frac{1}{-i\varepsilon + \omega_k - \omega_{k'}} \left(\frac{1}{i\varepsilon + \omega_{k_1} - \omega_{k'}} + \frac{1}{i\varepsilon + \omega_k - \omega_{k_1}} \right) \times t^+_{k_1 k}(\omega_k) t^-_{k'k_1}(\omega_{k'}) \tag{117}$$

All these results are easily checked in the lowest orders in perturbation.

As we have seen in Section III, the transformation superoperator is obtained using Eqs. (33)–(36).

In the present case,

$$\Lambda_{kkkk} = (\Lambda^{-1})_{kkkk} = \chi_{kkkk} = 1 \tag{118}$$

Using Eqs. (33)–(36) and Eq. (118), we now present a list of equations giving the explicit form of the transformation superoperator. We include also expressions for Λ^{-1}. A proof that

$$\Lambda\Lambda^{-1} = \Lambda^{-1}\Lambda = 1 \tag{119}$$

is given later in this section.

$$\Lambda_{k_1k_2kk'} = \left(\delta_{k_1k} + \frac{t^+_{k_1k}(\omega_k)}{i\varepsilon + \omega_k - \omega_{k_1}}\right)\left(\delta_{k'k_2} + \frac{t^-_{k'k_2}(\omega_{k'})}{-i\varepsilon + \omega_{k'} - \omega_{k_2}}\right) \tag{120}$$

where $k_1 \neq k_2$; $k = k'$ or $k \neq k'$.

$$\Lambda_{kkkk'} = \frac{t^-_{k'k}(\omega_{k'})}{i\varepsilon + \omega_{k'} - \omega_k} \tag{121}$$

$$\Lambda_{k'k'kk'} = \frac{t^+_{k'k}(\omega_k)}{-i\varepsilon + \omega_k - \omega_{k'}} \tag{122}$$

$$\Lambda_{k_1k_1kk'} = \frac{-1}{-i\varepsilon + \omega_k - \omega_{k'}}\left(\frac{1}{i\varepsilon - \omega_{k'} + \omega_{k_1}} + \frac{1}{i\varepsilon + \omega_k - \omega_{k_1}}\right) \times t^+_{k_1k}(\omega_k)t^-_{k'k_1}(\omega_{k'}) \tag{123}$$

$$\Lambda^{-1}_{kkk_1k_2} = \left(\delta_{kk_1} + \frac{t^+_{kk_1}(\omega_k)}{i\varepsilon + \omega_k - \omega_{k_1}}\right)\left(\delta_{kk_2} + \frac{t^-_{k_2k}(\omega_k)}{-i\varepsilon + \omega_k - \omega_{k_2}}\right) \tag{124}$$

$$\Lambda^{-1}_{kk'k_1k_2} = \left(\delta_{kk_1} + \frac{t^-_{kk_1}(\omega_k)}{-i\varepsilon + \omega_k - \omega_{k_1}}\right)\left(\delta_{k'k_2} + \frac{t^+_{k_2k'}(\omega_{k'})}{i\varepsilon + \omega_{k'} - \omega_{k_2}}\right) \quad \text{for } (k' \neq k) \tag{125}$$

In the preceding list, only the diagonal-to-diagonal elements $\Lambda_{kkk'k'}$ and $\Lambda^{-1}_{kkk'k'}$ are omitted. Their determination requires supplementary conditions, such as Eq. (59), which we discuss in Appendix G.

The physical representation for the scattering problem is completely defined by the matrix elements of the Λ operator, as given by Eqs. (120)–(125).

Indeed, the order of magnitude of the elements of the density matrix corresponding to a wave packet are

$$\rho_{kk} \sim O\left(\frac{1}{L^3}\right) \quad \text{and} \quad \rho_{kk'} \sim O\left(\frac{1}{L^3}\right) \tag{126}$$

Therefore, from Eq. (27), one has

$$\begin{aligned} {}^p\rho_{kk} &= \rho_{kk} + \sum_{k_1k_2}{}' (\Lambda^{-1})_{kkk_1k_2}\rho_{k_1k_2} \\ &= \sum_{k_1k_2}\left(\delta_{kk_1} + \frac{t^+_{kk_1}(\omega_k)}{i\varepsilon + \omega_k - \omega_{k_1}}\right)\left(\delta_{kk_2} + \frac{t^-_{k_2k}(\omega_k)}{-i\varepsilon + \omega_k - \omega_{k_2}}\right)\rho_{k_1k_2} \end{aligned} \tag{127}$$

and

$$\begin{aligned} {}^{p}\rho_{kk'} &= \rho_{kk'} + \sum_{k_1k_2}{}' (\Lambda^{-1})_{kk'k_1k_2}\rho_{k_1k_2} \\ &= \sum_{k_1k_2}\left(\delta_{kk_1} + \frac{t^{-}_{kk_1}(\omega_k)}{-i\varepsilon + \omega_k - \omega_{k_1}}\right)\left(\delta_{k'k_2} + \frac{t^{+}_{k_2k'}(\omega_{k'})}{i\varepsilon + \omega_{k'} - \omega_{k_2}}\right)\rho_{k_1k_2} \end{aligned} \tag{128}$$

One sees that the $kk\,k'k'$ elements of the Λ^{-1}-transformation do not play any role here, because they give contributions of order L^{-6} [see Eq. (106b) and (126)] and may be neglected.

Let us recall that the $kk\,k'k'$ elements of the transformation, when expressed as products of Möller operators, are divergent (see Section II and Appendix E). It is therefore interesting to verify that, in contrast, using our regularization procedure, the $kk\,k'k'$ elements of Λ and Λ^{-1} become well defined. This is done in Appendix H.

The proof that Λ^{-1} is indeed the inverse of Λ proceeds in two steps. The proof of Eq. (38) follows from the explicit expressions of Eqs. (120)–(125) as well as the relations in Eqs. (C.5) and (C.11). Equation (40) also results from the relations in Eqs. (120)–(125) and (C.17). The proofs are sketched in Appendix G.

Let us emphasize that Λ is not an unitary transformation, because

$$\Lambda^{-1} \neq \Lambda^{+} \tag{129}$$

As an example, the element $\Lambda^{-1}_{kkk_1k}$ [Eq. (124)] obviously differs from $\Lambda^{+}_{kkk_1k}$ [see also Eq. (53)]:

$$\Lambda^{+}_{kk\,k_1k} = \Lambda^{cc}_{k_1kkk} = \delta_{k_1k} + \frac{t^{-}_{kk_1}(\omega_k)}{-i\varepsilon + \omega_k - \omega_{k_1}} \tag{130}$$

Let us notice that the nonunitary elements of the Λ transformation are those involving the vacuum state. This can be proved in a straightforward way by taking the adjoint of these elements, using Eq. (53) and showing that, for instance,

$$\Lambda^{+}_{kk\,k_1k_2} \neq \Lambda^{-1}_{kk\,k_1k_2} \tag{131}$$

Instead, the basic property of the Λ-transformation is now the starunitarity [Eq. (52)].

$$\Lambda^{-1} = \Lambda'^{+} \tag{132}$$

To calculate the L-inverse of an element of Λ, one has to change L into $-L$ and invert the ordering for correlations of the same degree from chronological to antichronological. Let us consider an example (see also Section IV).

To obtain $\Lambda'_{k'kkk}$ one has only to change in $\Lambda_{k'kkk}$, L_0 into $-L_0$ and δL into $-\delta L$ because all intermediate states are "more correlated" than the vacuum state [contrast this example with Eq. (98)]. This leads to

$$\Lambda'_{k_1kkk} = C'_{k_1kkk} = \delta_{k_1k} + \frac{t^{-}_{k_1k}(\omega_k)}{-i\varepsilon + \omega_k - \omega_{k_1}} \tag{133}$$

and from the definition of the adjoint, one obtains

$$(\Lambda'^{+})_{kkk_1k} = \Lambda'^{cc}_{k_1kkk} = \delta_{k_1k} + \frac{t^{+}_{kk_1}(\omega_k)}{i\varepsilon + \omega_k - \omega_{k_1}} \tag{134}$$

which is indeed $\Lambda^{-1}_{kkk_1k}$ [Eq. (124)].

The starunitarity [Eq. (52)] can be verified in this way for all matrix elements. We always have to take care of the inversion of the chronological ordering when states of equal degree of correlation are involved.

Note that the elements of the Λ-transformation which do not involve the vacuum state "look" unitary, as

$$\Lambda'_{k_1k_2kk'} = \Lambda_{k_1k_2kk'} \tag{135}$$

The adjoint-symmetry property of Λ [see Eq. (54)],

$$\Lambda^a = \Lambda \tag{136}$$

expressing that the hermitian character of the transformed operator ρ is preserved, can also be verified.

Note that the nonunitarity (starunitarity) of the Λ-transformation is not in contradiction with the unitarity of the S-matrix [Eq. (B.19)].

Indeed, the relations in Eqs. (C.11) and (C.15) which enable us to prove that Λ^{-1} is the inverse of Λ, precisely express the unitarity of the Møller wave operators [Eq. (B.12)].

One way of obtaining the collision operator ϕ, describing the evolution in the physical representation, is using the definition in Eq. (43), taking into account that the matrix elements of L are given by Eq. (19).

For instance, the $kkk'k'$ elements of ϕ are given by

$$\phi_{kkk'k'} = \sum \Lambda^{-1}_{kkk_1k_2} L_{k_1k_2k_3k_4} \Lambda_{k_3k_4k'k'} \tag{137}$$

and one obtains

$$\phi_{kkk'k'} = 2\pi i\ \delta(\omega_k - \omega_{k'})|t^+_{kk'}(\omega_k)|^2 \tag{138}$$

Similarly,

$$\phi_{kkkk} = 2\pi i\ \mathrm{Im}\ t^+_{kk}(\omega_k) \tag{139}$$

and

$$\phi_{kk'kk'} = (\omega_k - \omega_{k'}) \tag{140}$$

Equation (138) is precisely the cross section for the process $k \to k'$, and Eq. (139) the forward scattering cross section. The (nonnegative) cross sections appear thus quite directly here as the elements of the transform of the Liouville operator L through our nonunitary, nonfactorizable transformation Λ and as the generators of the evolution of the physical entities defined through the Λ-transformation.

As mentioned previously, most of the elements of the Λ-transformation look unitary [see Eq. (135)]. However, in the relation [Eq. (137)] between ϕ and L, it is precisely the specifically nonunitary elements of Λ, those involving the vacuum state, which play the major role to obtain cross sections as transforms of the Liouville operator.

VI. PHYSICAL REPRESENTATION

The physical representation for the potential scattering problem is completely defined by the matrix elements of the Λ-transformation that we have obtained in Sections IV and V. In order to investigate it in more detail, let us consider, for simplicity, the case of the wave packet, with the orders of magnitude of the elements of the density operator given by Eqs. (71) and (126).

We have shown that the Λ-transformation, when restricted to the sectors (1, 0) and (0, 1), is the usual unitary transformation diagonalizing H (see Appendix E).

The evolution in the unperturbed basis of the amplitude $c_k(t)$, given by Eq. (D.6), is expressed as a superposition of oscillations.

For the elements of the sector (1, 0), the physical representation is defined by

$$
\begin{aligned}
{}^{p}\rho_{k0} &= \sum \Lambda^{-1}_{k_0 k_{10}} \rho_{k_1 0} \\
&\propto c_k + \sum_{k_1} \frac{t^{-}_{kk_1}(\omega_k)}{-i\varepsilon + \omega_k - \omega_{k_1}} c_{k_1} = c_k^{(-)}
\end{aligned} \tag{141}
$$

[compare this expression with Eq. (B.38)].

Their evolution thus reduces to independent oscillations [see Eq. (B.34)]

$$
{}^{p}\rho_{k0}(t) \propto c_k^{(-)}(t) = \mathrm{e}^{-i\omega_k t} c_k^{(-)}(0) \tag{142}
$$

Similarly, in the (0, 1) sector, one obtains, in the physical representation, the complex conjugated expressions.

Let us notice that if we had chosen the antichronological ordering, the corresponding results would have been Eq. (D.7) instead of Eq. (D.6), and

$$
{}^{(p')}\rho_{k0} \propto c_k + \sum_{k_1} \frac{t^{+}_{kk_1}(\omega_k)}{i\varepsilon + \omega_k - \omega_{k_1}} c_{k_1} = c_k^{(+)} \tag{143}
$$

instead of Eq. (141), with

$$
c_k^{(+)}(t) = \mathrm{e}^{-i\omega_k t}\, c_k^{(+)}(0) \tag{144}
$$

[These expressions may again be compared with Eqs. (B.24) and (B.38)].

Let us now turn to the (1, 1) sector. For the case of the wave packet, the matrix elements of the density operator in the unperturbed basis are

$$
\rho_{kk} = c_k c_k^{*}; \qquad \rho_{kk'} = c_k c_{k'}^{*} \tag{145}
$$

Then, in the physical representation, defined by Eqs. (127) and (128), using the expressions in Eqs. (141) and (143), one can write

$$
{}^{p}\rho_{kk} = c_k^{(+)} c_k^{(+)cc} \tag{146a}
$$

$$
{}^{p}\rho_{kk'} = c_k^{(-)} c_{k'}^{(-)cc} \tag{146b}
$$

This already shows that the ${}^{p}\rho$ arising from a pure case is no longer a pure case, as a consequence of the nonfactorization of the Λ-transformation (see Appendix G).

In addition, the time-variation of ${}^{p}\rho_{kk}$ is given by Eq. (45)

$$i\,\partial_t\,{}^{p}\rho_{kk} = \sum_{k'} \phi_{kkk'k'}\,{}^{p}\rho_{k'k'} \tag{147}$$

where the infinitesimal generator of motion $\phi_{kkk'k'}$ given by Eq. (138), is clearly nonfactorizable.

The solution of Eq. (147) is

$${}^{p}\rho_{kk}(t) = {}^{p}\rho_{kk}(0) + t\sum_{k'} 2\pi\;\delta(\omega_k - \omega_{k'})|t^{+}_{kk'}(\omega_k)|^2\,{}^{p}\rho_{k'k'}(0) \tag{148}$$

hence the form Eq. (146a) is not maintained in the course of time.

The very notion of a pure case is thus lost when the time evolution is described in terms of irreversible processes such as collisions.

The irreversible character of the collision processes is also manifest through the study of appropriate Lyapunov functions.[27,28]

In the Π^0 subdynamics, one can introduce the Lyapunov functional

$$\overset{(0)}{\Omega} = \sum_{k} |{}^{p}\rho_{kk}|^2 \quad : 0(L^{-3}) \tag{149}$$

for which one can prove an H-theorem. Indeed, using Eqs. (147) (138), (139), and (C.13),

$$\begin{aligned} \partial_t\overset{(0)}{\Omega} &= 2\sum_{k} {}^{p}\rho_{kk}\,\partial_t\,{}^{p}\rho_{kk} \\ &= -2i\sum_{kk'} {}^{p}\rho_{kk}\phi_{kkk'k'}({}^{p}\rho_{k'k'} - {}^{p}\rho_{kk}) \end{aligned} \tag{150}$$

leads immediately to

$$\partial_t\overset{(0)}{\Omega} = -\tfrac{1}{2}\sum_{kk'} 2\pi\;\delta(\omega_k - \omega_{k'})[|t^{+}_{kk'}(\omega_k)|^2 + |t^{+}_{k'k}(\omega_k)|^2]({}^{p}\rho_{kk} - {}^{p}\rho_{k'k'})^2 \tag{151}$$

This is obviously nonpositive and therefore shows the monotonous decrease of $\Omega^{(0)}$ until ${}^{p}\rho_{k'k'} = {}^{p}\rho_{kk}$ for k and k' such that $\omega_{k'} = \omega_k$.

In the correlation subdynamics $\Pi^{(kk')}$, the evolution is nondissipative, because by inserting Eq. (140) into Eq. (45), one gets

$$i\,\partial_t\,{}^{p}\rho_{kk'} = (\omega_k - \omega_{k'})\,{}^{p}\rho_{kk'} \tag{152}$$

This is manifest at the level of the $\Omega^{(1)}$ function which can be associated with the monocorrelations

$$\overset{(1)}{\Omega} = \sum_{kk'} |{}^P\rho_{kk'}|^2 \quad : 0(1) \tag{153}$$

as, from Eq. (152),

$$\partial_t \overset{(1)}{\Omega} = 0 \tag{154}$$

The same holds for the subdynamics of the (1, 0) and (0, 1) sectors, because

$$\sum_k |{}^P\rho_{k0}|^2 \quad : 0(1) \tag{155}$$

also remains constant in time, as a consequence of Eq. (142).

The decrease of the Lyapunov function, $\partial_t\Omega$, through Eq. (151), is therefore entirely due to the collision process.

VII. THE BOLTZMANN EQUATION

The relation between the problem of potential scattering and that of a dilute gas (Boltzmann problem) is quite immediate. Indeed when one studies the approach to equilibrium for dilute gases, one starts by taking into account only two-body collisions. As is well known, the two-body scattering when treated in the center of mass reference frame is identical to potential scattering. The single variable k which has been introduced in this work plays the role of the momentum transfer in two-body collisions while the total momentum is being conserved (see Appendix A).

The single variable k is thus replaced by a couple of variables k_1 k_2 such that their sum is conserved and, for instance, Eq. (147) is replaced by

$$\begin{aligned} i\,\partial_t\, {}^P\rho_{k_1k_2k_1k_2} &= \phi_{k_1k_2k_1k_2k_1k_2k_1k_2}\, {}^P\rho_{k_1k_2k_1k_2} \\ &\quad + \sum \phi_{k_1k_2k_1k_2k_1'k_2'k_1'k_2'}\, {}^P\rho_{k_1'k_2'k_1'k_2'} \end{aligned} \tag{156}$$

with $k_1 + k_2 = k_1' + k_2'$ and [compare with Eq. (138)]

$$\phi_{k_1k_2k_1k_2k_1'k_2'k_1'k_2'} = 2\pi i\,\delta(\omega_{k_1} + \omega_{k_2} - \omega_{k_1'} - \omega_{k_2'})|t^+_{k_1k_2k_1'k_2'}(\omega_{k_1} + \omega_{k_2})|^2 \tag{157}$$

This expression has to be viewed as the evolution of two arbitrary particles within a N-body system. The one-body distribution function is then derived by the classic procedure through integration over the momentum of a "second particle". One obtains in this way the master equation

$$\partial_t {}^P\rho_{k_1k_1}(t) = \sum_2 \sum_{\substack{k_2 \\ k_1'k_2'}} 2\pi\,\delta(\omega_{k_1'} + \omega_{k_2'} - \omega_{k_1} - \omega_{k_2})|t^+_{k_1k_2k_1'k_2'}(\omega_{k_1} + \omega_{k_2})|^2$$
$$\times [{}^P\rho_{k_1'k_2'k_1'k_2'} - {}^P\rho_{k_1k_2k_1k_2}] \tag{158}$$

When in addition the usual factorization assumption

$$\rho_{k_1k_2k_1k_2} = \phi(k_1)\phi(k_2) \tag{159}$$

is introduced, one obtains the quantum analog of the Boltzmann equation:

$$\partial_t {}^P\phi(k_1) = N \sum_{\substack{k_2 \\ k_1'k_2'}} 2\pi\,\delta(\omega_{k_1'} + \omega_{k_2'} - \omega_{k_1} - \omega_{k_2})|t^+_{k_1k_2k_1'k_2'}(\omega_{k_1} + \omega_{k_2})|^2$$
$$\times [{}^P\phi(k_1'){}^P\phi(k_2') - {}^P\phi(k_1){}^P\phi(k_2)] \tag{160}$$

The factor N arises from the summation on the particle 2 in Eq. (158).

Now, in this case, the usual assumptions made on the order of magnitude in L^3 are

$$\rho_{k_1k_2k_1k_2} \sim 0(L^{-6}) \tag{161}$$

$$\rho_{k_1k_2k_1'k_2'} \sim 0(L^{-9}) \quad \text{if} \quad k_1 + k_2 = k_1' + k_2' \tag{162}$$

These orders of magnitude (see Ref. 13) insure the preservation of the extensive or intensive character of the macroscopic observables and are conserved by the evolution.

With these orders of magnitude, one has

$${}^P\rho_{k_1k_1} \sim 0(L^{-3}) \tag{163}$$

and the evolution $\partial_t {}^P\rho_{k_1k_1}$ is now of order $0(L^{-3}) \times c$, where c is the concentration, which is kept constant in the thermodynamical limit ($L^3 \to \infty, N \to \infty$) replacing in that problem the infinite volume limit that we have considered here.

The feature which emerges from this brief discussion of N-body systems is that the way we have treated the limit to infinite volume in potential scattering, keeping only dominant contributions in L^{-3}, leads, in the case of gases, to retain the first correction due to finite concentrations.

The main difference with the potential scattering or two-body scattering lies in the assumption in Eq. (162). Indeed starting with a wave packet in the two-body collisions $\rho_{k_1k_2k'_1k'_2}$ would have been also of order L^{-6}.

The dilute gas situation obviously does not correspond to a pure case. Indeed, as we have seen, as the result of collisions, the notion of wave function is lost. Moreover, if the initial correlations $\rho_{k_1k_2k'_1k'_2}$ were of order L^{-6}, these correlations would be damped through the collisions with other particles. New correlations would be formed through the creation fragment of order L^{-9}.

Once this situation is realized, $\Omega^{(0)}$ becomes dominant with respect to $\Omega^{(1)}$ and corresponds to Boltzmann's entropy. Let us again emphasize that Boltzmann's kinetic description is incompatible with a description in terms of pure states.

VIII. THE SECOND LAW AS A SELECTION PRINCIPLE

The microscopic formulation of the second law as a selection principle was already studied in Refs. 1 and 2. However, the present chapter illustrates the physical meaning of this selection. We mentioned already in the Introduction that scattering theory leads to two types of solutions $|k^+\rangle$ and $|k^-\rangle$ which contain asymptotically radially outgoing and incoming waves, respectively. For the description in Hilbert space, both solutions are necessary and play essentially symmetric roles. However, as noticed in every textbook on scattering theory, only $|k^+\rangle$ states are realized in actual experiments. We want now to investigate this basic difference between $|k^+\rangle$ and $|k^-\rangle$ from the thermodynamical point of view.

Let us start with the generalized master equation (GME) [Eq. (2.6) in Ref. 1], obtained from Eq. (16), using the decomposition Eq. (22) of the resolvent of L, taking for P the projector P^0 on the vacuum state $\rho_0 = \{\rho_{kk}\}$:

$$\partial_t \rho_0 = \int_0^t \mathrm{d}t' \psi(t - t')\rho_0(t') + P\mathscr{D}(t)Q\rho(0) \tag{164}$$

$\psi(t)$ and $\mathscr{D}(t)$ are, respectively, the inverse Laplace transforms of the collision and destruction fragments $\psi(z)$ and $\mathscr{D}(z)$ of Section III.

It is clear from Eq. (164) that the long-time behavior is closely related to the initial correlations. Only if the initial (precollisional) correlations disappear asymptotically for $t \to +\infty$, that is, when

$$\lim_{t \to +\infty} P\mathscr{D}(t)Q\rho(0) = 0 \tag{165a}$$

do we have to expect universality in the approach to equilibrium.

Note that the violation of Eq. (165a), or equivalency of

$$\lim_{t \to +\infty} \int dz\, e^{-izt} P\mathscr{D}(z)Q\rho(0) = 0 \tag{165b}$$

implies that the integrand is singular at the origin $z = 0$.

We have therefore to show that for an "out-state" involving $|k^-\rangle$ (containing asymptotically incoming spherical waves) the condition in Eq. (165) is indeed violated whereas it is not for an "in-state" involving $|k^+\rangle$. We have also to show that this distinction has a persistent character (see Ref. 1, Sec. 5).

This is done using the projectors defining the asymptotic subdynamics. Indeed, when ρ is such that the condition in Eq. (165) is satisfied, then $\Pi\rho$ converges and $\Lambda^{-1}\rho$ is well defined. Conversely, a divergent $\Pi\rho$ implies that ρ violates the condition of Eq. (165).

Let us consider the states

$$\varphi^{(\pm)} = \sum_{k_1} g(k_1)|k_1^\pm\rangle\langle k_1^\pm| \tag{166}$$

with $g(k_1) \sim 0(L^{-3})$ [Eq. (13)].

We first observe that their matrix elements in the unperturbed basis are

$$\varphi_{kk}^{(\pm)} = g(k) \tag{166a}$$

$$\varphi_{k_1k_2}^{(\pm)} = \frac{1}{\mp i\varepsilon + \omega_{k_1} - \omega_{k_2}} h \tag{166b}$$

where h is a regular function of ω_{k_1} and ω_{k_2}:

$$\begin{aligned} h = h(k_1, k_2) = g(k_1)t_{k_1k_2}^{\mp}(\omega_{k_1}) + g(k_2)t_{k_1k_2}^{\pm}(\omega_{k_2}) + \sum g(k)t_{k_1k}^{\pm}(\omega_k)t_{kk_2}^{\mp}(\omega_k) \\ \times \left(\frac{1}{\pm i\varepsilon + \omega_k - \omega_{k_1}} - \frac{1}{\mp i\varepsilon + \omega_k - \omega_{k_2}}\right) \end{aligned} \tag{167}$$

We now study the action of the projector Π^0 on such states: $\Pi^0\varphi^{(\pm)}$. Let us consider the diagonal elements

$$\left(\overset{0}{\Pi}\, \varphi^{(\pm)}\right)_{kk} = \sum_{k_1k_2} D_{kkk_1k_2}\varphi_{k_1k_2}^{(\pm)} \tag{168}$$

From Eq. (110), we see that for $k_1 \neq k_2$,

$$D_{kkk_1k_2} = \frac{1}{i\varepsilon - \omega_{k_1} + \omega_{k_2}} d \tag{169}$$

where d is a regular at $\omega_{k_1} = \omega_{k_2}$

$$d = d(k, k_1, k_2) = t^{+}_{kk_1}(\omega_k)\delta_{kk_2} - t^{-}_{k_2k}(\omega_k)\delta_{kk_1} + t^{+}_{kk_1}(\omega_k)t^{-}_{k_2k}(\omega_k) \times \left(\frac{1}{i\varepsilon + \omega_k - \omega_{k_1}} - \frac{1}{-i\varepsilon + \omega_k - \omega_{k_2}}\right) \tag{170}$$

With the Eqs. (169) and (166), one obtains for Eq. (168) a convergent result with $\varphi^{(+)}$ and a divergent one with $\varphi^{(-)}$. Indeed $(\Pi\varphi^{(+)})_{kk}$ contains $\Sigma_{k_1k_2}\ 1/(-i\varepsilon + \omega_{k_1} - \omega_{k_2})^{-2}hd$ whereas $(\Pi\varphi^{(-)})_{kk}$ contains $(1/i\varepsilon)\ \Sigma\ 2\pi i\ \delta(\omega_{k_1} - \omega_{k_2})hd$.

The divergence clearly results from the presence in $\varphi^{(-)}_{k_1k_2}$ [Eq. (166b)], of the distribution $(1/i\varepsilon + \omega_{k_1} - \omega_{k_2})$, whereas no trouble arises from the presence in $\varphi^{(+)}_{k_1k_2}$ of the distribution $(1/-i\varepsilon + \omega_{k_1} - \omega_{k_2})$ which differs from the first one by the change of sign in front of $i\varepsilon$.

Now, in the potential scattering problem, $D_{kkk_1k_2}$ [Eq. (110)] is simply the limit for $z \to +i0$ of $\mathscr{D}_{kkk_1k_2}(z)$. Thus the foregoing results also mean that

$$\lim_{z \to +i0} zP\mathscr{D}(z)Q\varphi^{(+)} = \lim_{\varepsilon \to 0} i\varepsilon PDQ\varphi^{(+)} = 0 \tag{171}$$

so that $\varphi^{(+)}$ satisfies the condition of Eq. (165), whereas

$$\lim_{z \to +i0} zP\mathscr{D}(z)Q\varphi^{(+)} = \lim_{\varepsilon \to 0} i\varepsilon PDQ\varphi^{(-)} = \text{finite} \tag{172}$$

and $\varphi^{(-)}$ violates the condition in Eq. (165).

In accordance with the formulation of the second law as a selection principle, the $\varphi^{(-)}$-states have to be excluded from the class of physically acceptable states.

Let us notice that the opposite would be true, had we chosen the other semigroup: the exclusion principle therefore fixes the choice of the semigroup.

In a completely similar way, one can show that, due to the presence of the $|k^-\rangle$ vector in the two following states

$$\varphi^{\mathrm{I}} = \sum g(k)|k^+\rangle\langle k^-| \tag{173a}$$

$$\varphi^{\mathrm{II}} = \sum g(k)|k^-\rangle\langle k^+| \tag{173b}$$

none is an acceptable state, because

$$\lim_{z \to +i0} = P\mathscr{D}(z)Q\varphi^{\text{I,II}} \neq 0 \tag{174}$$

The requirement of Eq. (165) that precollisional correlations are not persistent, or equivalently that $\Pi\rho$ and $\Lambda^{-1}\rho$ are well defined can thus be used as a criterion to distinguish between the in- and out-states and to exclude the latter as physically nonrealizable.

From the traditional scattering theory, it would seem that the states in Eqs. (166) and (173) are invariants of motion.

An invariant is any operator φ that commutes with the Hamiltonian operator, that is, such that

$$L\varphi = 0 \tag{175}$$

However, one has to be very precise about the meaning to give to this equality. Indeed, we have seen that, in order to take properly into account the scattering processes, we have to retain in Eq. (8) for the variation of diagonal elements of the density operator $\partial_t \rho_{kk'}$ terms which are of order L^{-3} with respect to ρ_{kk} itself. These terms in the strict limit of large volume would have given a vanishing contribution.

For the discussion of the invariants, their orders of magnitude in the volume will also play an important role. Localized wave packets obviously cannot be invariant. Only declocalized states, such that

$$\rho_{kk} \sim 0(L^{-3}); \qquad \rho_{kk'} \sim 0(L^{-6}) \tag{176}$$

are to be considered and they have to satisfy Eq. (175) up to order $0(L^{-6})$.

If we now calculate the diagonal elements of $L\varphi$:

$$(L\varphi)_{kk} = \sum_{k'} (V_{kk'}\varphi_{k'k} - V_{k'k}\varphi_{kk'}) \tag{177}$$

and insert the Eqs. (166a) and (166b), using Eq. (C.6), we obtain Eq. (15). Thus the states $\varphi^{(\pm)}$ [Eq. (166)] are in general not invariants of the motion when the collision processes are retained.

On the other hand, one can prove in a completely similar way that the operators φ^{I} and φ^{II} [Eq. (173)] are truly invariant. Because of Eq. (174), however, these invariants belong, following the classification we introduced previously (see Ref. 29) to the category of "singular" invariants.

The "regular" invariants of potential scattering take the form

$$\varphi_{kk} = f(\omega_k)$$

$$\varphi_{kk'} = \sum_{k_1} C_{kk'k_1k_1} f(\omega_{k_1})$$

with

$$f(\omega_k) \sim 0(L^{-3}) \tag{178}$$

and this corresponds to a homogeneous, isotropic state, the "equilibrium" distribution.

As repeatedly emphasized, our microscopic formulation of the second law states that not all initial conditions are acceptable. Only states going to equilibrium in the future are physically realizable or observable.

This is certainly not the case for singular invariants. On the one hand, their occurrence shows that there indeed exist in the conventional formulation states that have to be excluded, and on the other hand, their form shows [see Eq. (173)] that the selection principle can only be formulated in the superspace and not at the level of Hilbert space.

For all excluded states, the off-diagonal elements $\rho_{k_1k_2}$ contain a denominator such that Eq. (165) is violated [as an example, consider $\varphi^{(-)}$ in Eq. (166)]. They would evolve in time toward a singular invariant.

We intend to develop further the question of singular invariants in a separate publication.

IX. CONCLUDING REMARKS

The study of the potential scattering problem presented here shows that the traditional formulation of quantum mechanics cannot claim to be complete, because it is unable to distinguish between $|k^+\rangle$ states as realized in actual experiments, and $|k^-\rangle$ states which are neither realized in laboratory experiments nor observed in nature. It is certainly not satisfactory to blame this difference on our technical inabilities. If so, we could as well claim that gravitation is only a practical matter which will be solved once an antigravitational device will have been invented.

We therefore consider that quantum theory requires an extension to superspace to be able to formulate a selection principle that distinguishes permitted motions from excluded ones.

We have already mentioned in Ref. 1 some general consequences of the inclusion of irreversibility as a basic dynamical principle. We are now in position to make more precise some of the general conclusions of that paper.

Our general formalism rests on the decomposition of the Hamiltonian H using complementary projectors P and Q:

$$H = PH + QH \tag{179}$$

Here we have made the "natural" choice [Eq. (3)], corresponding to the separation of energy into a kinetic part and a potential part. The choice is unambiguous in this problem.

A more refined decomposition, involving a complete set of projectors P has proved to be necessary for the explicit construction of the various subdynamics (see Section III). However, the decomposition into P and Q (without further decomposition into subdynamics) is sufficient for the discussion of the asymptotic properties of the system.

The simplicity of the model considered enables us to go beyond a perturbative approach which is the starting point for our regularization procedure (see Section III). Resummations can be explicitly performed and connected with a convolutive approach (see also Ref. 23).

The basic object of our theory is the Λ-transformation, which as we have shown, is nonunitary and nonfactorizable. This Λ-transformation replaces the traditional diagonalization of L, starting from the diagonalization of H and using a factorizable unitary transformation. Indeed, as we have shown (see Section II), the diagonalization of L fails because it involves ill-defined products of distributions. In contradistinction, the form of Λ that we reach through our method is well defined and leads directly to cross sections as Λ-transforms of the Liouville superoperator [Eq. (137)].

It has also been emphasized that the nonunitarity and nonfactorizability of Λ, which are the properties that authorize the definition of an entropy observable, are also responsible for the loss of the notion of wave function. The Λ-transform of a pure state is no longer a pure state, and, moreover, the collision processes destroy any initially pure state. This appears as a severe limitation of the very concept of wave function and consequently of the superposition principle.

Once the physical L-inversion has been given a precise mathematical meaning, one is able to prove that the obtained Λ-transformation is starunitary [see Eq. (52)]. We then go from the unitary group of evolution to two different contractive semigroups, one defined using Λ, the other using its L-inverse, Λ'.

The selection principle, according to which "to be acceptable, states must evolve toward equilibrium in the future," then fixes the choice of the semigroup, hence of the Λ-transformation. It excludes, as physically unrealizable, all states that do not belong to the domain of Λ^{-1}.

In this way, only transient precollisional correlations are admitted. As it involves the nonfactorizable Λ-superoperator, this selection principle cannot be formulated in Hilbert space.

We have shown that the exclusion of persistent correlations amounts also to the exclusion of singular invariants.

In the scattering problem, in order to retain the collision processes, one is led to keep contributions of order $0(L^{-3})$, which, in the limit of an infinite volume, would be strictly vanishing. As discussed briefly in Section VII, this is validated by the fact that, in the case of the dilute gas (the Boltzmann's problem), these are precisely the terms that will give rise, in the thermodynamic limit, to contributions proportional to the concentration.

These systems are, however, still *too simple* to exhibit some important features present in the general formalism.

For instance, the physical state ${}^{P}\rho$ has an intrinsic broken time-symmetry.

However, this could only lead to physical predictions when we have to go to higher orders in higher approximation, such as in the three-body collision problem, to be dealt with in a forthcoming publication.

Also the nonfactorizability of Λ does not permit formulation of the renormalization problem in Hilbert space, but in superspace. Again the potential scattering problem is too simple to study this problem (see Ref. 26).

In conclusion, as stated in Ref. 1, "the microscopic content of the second law appears as expressing a limit to observations and manipulations, exactly as does the second law on the macroscopic level. Correlations cannot be controlled to an extent that they undo the effect of collisions. The probabilistic interpretation of entropy, which presupposes a direction of time, becomes only possible as a consequence of this negative statement".

These considerations bring us into the heart of the famous Ritz-Einstein controversy, about whether the second law implies the principle of "retarded waves" (Ritz's viewpoint) or follows from the use of probabilities (Einstein's viewpoint). We have shown that the second law both selects retarded solutions *and* leads as a result to a probabilistic description.

Acknowledgments

The authors thank M. de Haan, F. Henin, and B. Misra for useful discussions. This work was supported in part by the Instituts Internationaux de Physique et de Chimie, fondés par E. Solvay, and the Robert A. Welch Foundation, Houston, Texas.

APPENDIX A. POTENTIAL SCATTERING

One considers the scattering problem in a system with two particles labeled 1 and 2. The Liouville-von Neumann equation for the density operator $\rho(1, 2)$ of the system is

$$i\,\partial_t\,\rho(1, 2) = L(1, 2)\rho(1, 2) \tag{A.1}$$

The elements of the density matrix in the momentum representation for each particle may be noted

$$\rho_{k_1 k_2 k'_1 k'_2} = \rho_{1\,2\,1'2'} \tag{A.2}$$

where (12) means that a vector k_1 is associated with particle 1, k_2 with particle 2; similarly (1′2′) means that particle 1 has momentum k'_1, 2 momentum k'_2.

Then Eq. (A.1) takes the form

$$i\,\partial_t\,\rho_{1\,2\,1'2'} = \sum_{'''''} (L)_{1\,2\,1'2'1''2''1'''2'''}\rho_{1''2''1'''2'''} \tag{A.3}$$

with

$$(L_0)_{1\,2\,1'2'1''2''1'''2'''} = \delta_{1\,2\,1''2''}\,\delta_{1'2'1'''2'''}[(H_0)_{1\,2\,1\,2} - (H_0)_{1'2'1'2'}] \tag{A.4}$$

and

$$(\delta L)_{1\,2\,1'2'1''2''1'''2'''} = V_{1\,2\,1''2''}\,\delta_{1'2'1'''2'''} - V_{1'''2'''1'2'}\,\delta_{1\,2\,1''2''} \tag{A.5}$$

This notation may be considerably simplified if one notes by $(12)^\alpha$ the state $k_1^\alpha k_2^\alpha$. We then write

$$\rho_{\alpha\beta} = \rho_{(1\,2)^\alpha(1\,2)^\beta} = \rho_{k_1^\alpha k_2^\alpha k_1^\beta k_2^\beta} \tag{A.6}$$

Equation (A.3) becomes then

$$i\,\partial_t\,\rho_{\alpha\beta} = \sum_{\gamma\delta} L_{\alpha\beta\,\gamma\delta}\rho_{\gamma\delta} \tag{A.7}$$

with $L_{\alpha\beta\gamma\delta}$ given by Eq. (19). In particular [see Eqs. (20)],

$$(L_0)_{\alpha\beta\gamma\delta} = \delta_{\alpha\gamma}\,\delta_{\beta\delta}[(H_0)_{\alpha\alpha} - (H_0)_{\beta\beta}] \tag{A.8}$$

and

$$(\delta L)_{\alpha\beta\gamma\delta} = V_{\alpha\gamma}\,\delta_{\beta\delta} - V_{\delta\beta}\,\delta_{\alpha\gamma} \tag{A.9}$$

With central interactions, the possibilities of transitions induced by $V_{\alpha\beta}$ are restricted by the overall conservation of momentum to changes such that

$$k_1^\alpha + k_2^\alpha = k_1^\beta + k_2^\beta \tag{A.10}$$

The potential scattering problem corresponds to the study of the two-body system in the center of mass frame. Then

$$k_1 + k_2 = 0 \tag{A.11}$$

and both particles may be labeled with the same vector k,

$$k_1 = k, \qquad k_2 = -k \tag{A.12}$$

The notation $(12)^\alpha$ for the pair of particles may then be simplified using the single (relative) wave-vector k.

APPENDIX B. SCATTERING THEORY

Good reviews of formal scattering theory can be found in many textbooks.[11] Unfortunately often different and even conflicting notations are used. We therefore present a survey of scattering theory in the notations used in this chapter.

Asymptotic Operators and States

The formal solution of the Schrödinger equation [Eq. (2)] is

$$|\psi(t)\rangle = \mathrm{e}^{-iHt}|\psi(0)\rangle \tag{B.1}$$

Introducing the U evolution operator as

$$U(t_1, t_2) = e^{iH_0 t}\,\mathrm{e}^{-iH(t_1 - t_2)}\,\mathrm{e}^{-iH_0 t_2} \tag{B.2}$$

the state evolves in the Dirac interaction picture according to

$$|\tilde{\psi}(t)\rangle = \mathrm{e}^{iH_0 t}|\psi(t)\rangle = U(t, 0)|\psi(0)\rangle = U(t, 0)|\tilde{\psi}(0)\rangle \tag{B.3}$$

Asymptotic states are defined using a limit procedure in the Dirac picture

$$|\psi^{\pm}\rangle = \lim_{t \to \pm\infty} |\tilde{\psi}(t)\rangle = |\tilde{\psi}(\pm\infty)\rangle = U(\pm\infty, 0)|\psi(0)\rangle = W_{\pm}|\psi(0)\rangle \tag{B.4}$$

The so-defined $W_{\pm}$ operators may be shown to be unitary in the absence of bound states.[11]

The scattering operator S is defined through the relation between $|\psi^+\rangle$ and $|\psi^-\rangle$ originating from the same initial state:

$$|\psi^+\rangle = S|\psi^-\rangle \tag{B.5}$$

From the vanishing of the time derivative of $U(t, 0)$ in the asymptotic time limit, one gets the intertwining relations:

$$H_0 W_\pm = W_\pm H \tag{B.6}$$

Let us take as the initial state $|\psi(0)\rangle$ an eigenstate $|k\rangle$ of the unperturbed Hamiltonian H_0 [Eq. (4)]. In the course of time, the state $|\psi_k(t)\rangle$ evolves from $|k\rangle$ through Eq. (B.1) in the Schrödinger picture and $|\tilde{\psi}_k(t)\rangle$ through Eq. (B.3) in the Dirac picture. The corresponding asymptotic states are then defined through Eq. (B.4) as

$$|\psi_k^\pm\rangle = |\tilde{\psi}_k(\pm\infty)\rangle = W_\pm|k\rangle \tag{B.7}$$

They are unitary transforms of the unperturbed eigenstates, hence they are mutually orthogonal

$$\langle\psi_k^\pm|\psi_{k'}^\pm\rangle = 0 \quad \text{if} \quad k \neq k' \tag{B.8}$$

and form a complete set. The two sets of states are related through Eq. (B.5) by the scattering operator.

As the result of Eq. (B.7), we can write

$$W_\pm = \sum_k |\psi_k^\pm\rangle\langle k|; \qquad W_\pm^{-1} = \sum_k |k\rangle\langle\psi_k^\pm| \tag{B.9}$$

The scattering operator can thus be written as

$$S = W_+ W_-^{-1} = \sum_k |\psi_k^+\rangle\langle\psi_k^-| \tag{B.10}$$

and is obviously unitary

$$S^+ = S^{-1} \tag{B.11}$$

Möller Operators

We identify the space spanned by the asymptotic states $|\psi_k^\pm\rangle$ with the space spanned by the unperturbed eigenstates $|k\rangle$. The Möller operators $U_\pm$ are then defined in terms of the asymptotic operators $W_\pm$ or by taking the asymptotic time limits of the evolution operator [Eq. (B.2)]:

$$U_{\pm} = W_{\mp}^{-1} = U(0, \mp\infty) \tag{B.12}$$

Their unitarity follows from that of the $W_{\pm}$'s. The intertwining relation Eq. (B.6), now becomes

$$HU_{\pm} = U_{\pm}H_0 \tag{B.13}$$

Therefore the Möller operators can be interpreted as diagonalizing the Hamiltonian H in the $|k\rangle$ basis [compare Eq. (5)]. This means that the states defined by [see Eq. (6)]

$$|k^{\pm}\rangle = U_{\pm}|k\rangle \tag{B.14}$$

are eigenstates of the H operator. These states are orthogonal and complete

$$\langle k^{\pm}|k'^{\pm}\rangle = 0 \quad \text{if} \quad k \neq k' \tag{B.15}$$

$$\sum_k |k^{\pm}\rangle\langle k^{\pm}| = 1 \tag{B.16}$$

As the result,

$$U_{\pm} = \sum |k^{\pm}\rangle\langle k| \quad \text{and} \quad U_{\pm}^{-1} = \sum_k |k\rangle\langle k^{\pm}| \tag{B.17}$$

The scattering operator S [Eq. (B.10)] can then be written

$$S = U_{-}^{-1}U_{+} = \sum_{kk'} |k\rangle S_{kk'}\langle k'| \tag{B.18}$$

in terms of the elements of the S-matrix

$$S_{kk'} = \langle k|S|k'\rangle = \langle k^{-}|k'^{+}\rangle \tag{B.19}$$

Using Eq. (B.15), one verifies that the scattering operator [Eq. (B.18)] commutes with the unperturbed Hamiltonian

$$[H_0, S]_{-} = 0 \tag{B.20}$$

Another "scattering" operator S' may also be defined through

$$S' = \sum_k |k^{+}\rangle\langle k^{-}| = U_{+}U_{-}^{-1} \tag{B.21}$$

It commutes with the complete Hamiltonian

$$[H, S']_- = 0 \tag{B.22}$$

This operator S' thus defines a constant of motion of the total Hamiltonian.

Expansion Coefficients

At any time t, the state $|\psi(t)\rangle$ may be expressed in the unperturbed basis, with coefficients $c_k(t)$ [see Eq. (68)].

To work in the Dirac picture, we write

$$c_k(t) = a_k(t)\,\mathrm{e}^{-i\omega_k t} \quad \text{with} \quad c_k(0) = a_k(0) \tag{B.23}$$

In agreement with Eq. (B.4) we define

$$|\psi^\pm\rangle = \sum_k a_k^\pm |k\rangle \tag{B.24}$$

When we compare Eq. (B.24) with

$$|\psi^\pm\rangle = W_\pm \sum_k a_k(0)|k\rangle = \sum_k a_k(0)|\psi_k^\pm\rangle \tag{B.25}$$

we see that the asymptotic coefficients $a_k^\pm$ are expressed as linear combinations of the initial coefficients $a_k(0)$:

$$a_k^\pm = \sum_{k'} a_{k'}(0)\langle k|\psi_{k'}^\pm\rangle \tag{B.26}$$

where

$$\langle k|\psi_{k'}^\pm\rangle = \langle k|W_\pm|k'\rangle = (W_\pm)_{kk'} \tag{B.27}$$

We may also write

$$a^\pm = W_\pm a(0) \tag{B.28}$$

The a_k^+'s are also related to the a_k^-'s through the S-matrix, as

$$a^+ = W_+ W_-^{-1} a^- \tag{B.29}$$

or

$$a_k^+ = \sum_{k'} S_{kk'} a_{k'}^- \tag{B.30}$$

Since the Möller operators can be used to transform the total Hamiltonian H into the unperturbed one H_0 [Eq. (B.13)], the evolution [Eq. (B.1)] of the state vector $|\psi(t)\rangle$ may be written as

$$|\psi(t)\rangle = U_{\pm}\, e^{-iH_0 t}\, U_{\pm}^{-1}|\psi(0)\rangle \tag{B.31}$$

This way of writing the evolution of wave functions comes as close as possible to the decomposition in subdynamics and the transformation to the physical representation that we perform in the superspace formalism. We come back on this point in the next paragraph.

Using Eqs. (B.12), (B.4), and (B.24), one can write

$$U_{\pm}^{-1}|\psi(0)\rangle = W_{\mp}|\psi(0)\rangle = |\psi^{\mp}\rangle = \sum_k a_k^{\mp}|k\rangle \tag{B.32}$$

Inserting Eq. (B.32) into Eq. (B.31) and using Eq. (4) as well as Eq. (B.14) leads to the expansion of the state vector $|\psi(t)\rangle$ in the H-eigenbases $|k^{\pm}\rangle$:

$$\begin{aligned} |\psi(t)\rangle &= U_{\pm}\, e^{-iH_0 t} \sum_k a_k^{\mp}|k\rangle = U_{\pm} \sum_k a_k^{\mp}\, e^{-i\omega_k t}|k\rangle \\ &= \sum_k a_k^{\mp}\, e^{-i\omega_k t}|k^{\pm}\rangle = \sum_k c_k^{\mp}(t)|k^{\pm}\rangle \end{aligned} \tag{B.33}$$

(note the combination of + and − signs).

The coefficients of asymptotic expansion follow the unperturbed evolution

$$c_k^{\mp}(t) = c_k^{\mp}(0)\, e^{-i\omega_k t} \quad \text{with} \quad c_k^{\mp}(0) = a_k^{\mp} \tag{B.34}$$

Because

$$\sum_k c_k^{\pm}(0)|k^{\mp}\rangle = \sum_k c_k(0)|k\rangle \tag{B.35}$$

we also have

$$c_k(0) = \sum_{k'} c_{k'}^{\pm}(0)\langle k|k'^{\mp}\rangle \tag{B.36}$$

The set of elements

$$\langle k|k'^{\mp}\rangle = \langle k|U_{\mp}|k'\rangle = (U_{\mp})_{kk'} \tag{B.37}$$

form the (unitary) Möller matrices $U_{\mp}$. Then, from Eq. (B.36)

$$c(0) = U_{\mp}c^{\pm}(0) \quad \text{or} \quad c^{\pm}(0) = U_{\mp}^{-1}c(0) \tag{B.38}$$

in agreement with Eq. (B.28). Hence, also, from Eq. (B.30),

$$c^{+} = U_{-}^{-1} U_{+} c^{-} = S c^{-} \tag{B.39}$$

by definition of the S-matrix.

The In- and Out-Representations

Starting with Eq. (B.31), it is possible to define the transformed state $U_{\pm}^{-1}|\psi(t)\rangle$ as the state vector in one of the two following representations. The natural choice of notations is obviously, from Eq. (B.33),

$$U_{\pm}^{-1}|\psi(t)\rangle = \sum c_{k}^{\mp}(t)|k\rangle = |\psi^{\mp}(t)\rangle \tag{B.40}$$

These representations, which are not generally considered in the conventional treatments of scattering, in Hilbert space may be thought of as corresponding to the physical representation that is central in our presentation in superspace. This is why we consider them here.*

In these representations, the initial time state vector is identical with the asymptotic states in the Dirac picture, Eq. (B.6)

$$|\psi^{\mp}(0)\rangle = |\psi^{\mp}\rangle \tag{B.41}$$

In both representations, the state vector evolves in time according to the unperturbed Hamiltonian H_0:

$$|\psi^{\pm}(t)\rangle = \mathrm{e}^{-iH_0 t}\,|\psi^{\pm}\rangle \tag{B.42}$$

Equations (B.41) and (B.42) simply translate in terms of states the properties [Eq. (B.34)] of the coefficients.

It is clear that both evolutions to the future ($t > 0$) and to the past ($t < 0$) may be described in both representations, obtained by using U_{+} as well as U_{-}.

There is however one connection that looks more natural, as one sees by performing the decomposition [Eq. (B.31)] in the Dirac picture

$$|\tilde{\psi}(t)\rangle = \mathrm{e}^{iH_0 t}\, U_{\pm}\, \mathrm{e}^{-iH_0 t}\, U_{\pm}^{-1}|\psi(0)\rangle \tag{B.43}$$

This leads to the definition of the Möller operators in the Dirac picture,

$$\tilde{U}_{\pm}(t) = \mathrm{e}^{iH_0 t}\, U_{\pm}\, \mathrm{e}^{-iH_0 t} \tag{B.44}$$

* However, in superspace, the generator of motion in the physical representation is ϕ, here it is H_0. ϕ takes a block-diagonal form when H_0 is diagonal [see Eq. (44)].

Then, using Eq. (B.4),

$$|\tilde{\psi}(t)\rangle = \tilde{U}_{\pm}(t)|\psi^{\mp}\rangle \tag{B.45}$$

Taking the asymptotic time limits, one obtains from the definition [Eq. (B.4)]

$$|\psi^{(\pm)}\rangle = \tilde{U}_{[\pm]}(\pm\infty)\,|\psi^{[\mp]}\rangle \tag{B.46}$$

Therefore, by comparison with Eq. (B.5) and using Eq. (B.11), it comes

$$\tilde{U}_{\pm}(\mp\infty) = 1 \tag{B.47}$$

and

$$\tilde{U}_{\pm}(\pm\infty) = S^{\pm 1} \tag{B.48}$$

These results suggest the use of the U_+ transformation operator in Eq. (B.31) for positive times and of U_- for negative times.

In the evolution toward the future ($t > 0$), the natural representation will thus be conventionally $|\psi^-(t)\rangle$, with coefficients $c_k^-(t)$. This convention agrees with the chronological ordering convention that we use in the $i\varepsilon$-rule.

The explicit forms of the perturbed eigenstates of H can be obtained as the solutions of Lippman-Schwinger equations:

$$|k^{\pm}\rangle = |k\rangle + \sum_{k_1} \frac{t^{\pm}_{k_1 k}(\omega_k)}{\pm i\varepsilon + \omega_k - \omega_{k_1}} |k_1\rangle \tag{B.49}$$

Similarly, the coefficients of asymptotic expansion are also expressed in terms of the matrix elements of the transition operator

$$c_k^{\pm} = c_k + \sum_{k_1} \frac{t^{\pm}_{k k_1}(\omega_k)}{\pm i\varepsilon + \omega_k - \omega_{k_1}} c_{k_1} \tag{B.50}$$

The properties of the transition operator is studied in detail in Appendix C.

APPENDIX C. TRANSITION OPERATOR AND TRANSITION MATRIX

The general off-shell definition of the transition operator $t(z)$ whose matrix elements occur in the Eqs. (B.49) and (B.50) can be given in terms of the resolvents of the H and H_0 operators:

$$t(z)\frac{1}{z-H_0} = V\frac{1}{z-H} \tag{C.1}$$

or

$$\frac{1}{z-H_0}t(z) = \frac{1}{z-H}V \tag{C.2}$$

The resolvent of H, for which the following identity holds (e.g.)

$$\frac{1}{z-H} = \frac{1}{z-H_0} + \frac{1}{z-H_0}V\frac{1}{z-H} \tag{C.3}$$

can also be expressed in terms of the t-operator with the use of Eq. (C.1) or Eq. (C.2)

$$\frac{1}{z-H} = \frac{1}{z-H_0} + \frac{1}{z-H_0}t(z)\frac{1}{z-H_0} \tag{C.4}$$

Its matrix elements in the unperturbed basis are

$$\langle k|\frac{1}{z-H}|k'\rangle = \frac{1}{z-\omega_k}\delta_{kk'} + \frac{1}{z-\omega_k}t_{kk'}(z)\frac{1}{z-\omega_{k'}} \tag{C.5}$$

The t-matrix obeys the following integral equations:

$$t_{kk'}(z) = V_{kk'} + \sum_{k_1} V_{kk_1}\frac{1}{z-\omega_{k_1}}t_{k_1k'}(z) \tag{C.6a}$$

$$t_{kk'}(z) = V_{kk'} + \sum_{k_1} t_{kk_1}(z)\frac{1}{z-\omega_{k_1}}V_{k_1k'} \tag{C.6b}$$

The solution of these equations can be expressed as a perturbation series

$$t_{kk'}(z) = V_{kk'} + \sum V_{kk_1}\frac{1}{z-\omega_{k_1}}V_{k_1k'} + \sum V_{kk_1}\frac{1}{z-\omega_{k_1}}V_{k_1k_2}\frac{1}{z-\omega_{k_2}}V_{k_2k'} + \cdots \tag{C.7}$$

The t-matrix satisfies a set of relations, which are used repeatedly in this chapter. We therefore summarize them here.

(1)

$$t(z)^{+} = t(z^{cc}) \tag{C.8}$$

If one chooses $z = \omega \pm i\varepsilon$, this relation links the elements of the t-matrix on both sides of the real axis

$$[t^{+}(\omega)]^{+} = t^{-}(\omega) \tag{C.9}$$

or

$$(t^{+}_{k'k}(\omega))^{cc} = t^{-}_{kk'}(\omega) \tag{C.10}$$

(2) the expression of the completeness of the free states:

$$t_{kk'}(z) - t_{kk'}(z') = \sum_{k_1} t_{kk_1}(z) t_{k_1k'}(z') \left(\frac{1}{z - \omega_{k_1}} - \frac{1}{z' - \omega_{k_1}} \right) \tag{C.11}$$

If one chooses $z = \omega + i\varepsilon$ and $z' = \omega - i\varepsilon$, Eq. (C.11) gives the discontinuity of the t-matrix on the real axis:

$$t^{+}_{kk'}(\omega) - t^{-}_{kk'}(\omega) = \sum_{k_1} (-2\pi i)\, \delta(\omega - \omega_{k_1}) t^{+}_{kk_1}(\omega) t^{-}_{k_1k'}(\omega) \tag{C.12}$$

which, in turn, for $\omega = \omega_k$ and $k = k'$, gives the well-known optical theorem:

$$\text{Im } t^{+}_{kk}(\omega_k) = -\sum_{k_1} \pi\, \delta(\omega_k - \omega_{k_1}) |t^{+}_{kk_1}(\omega_k)|^2 \tag{C.13}$$

(3) the expression of the completeness of the scattering states (in the absence of bound states):

$$t^{+}_{kk'}(\omega) - t^{-}_{kk'}(\omega') = \sum_{k_1} t^{-}_{kk_1}(\omega_{k_1}) t^{+}_{k_1k'}(\omega_{k_1}) \times \left(\frac{1}{i\varepsilon + \omega_k - \omega_{k_1}} - \frac{1}{-i\varepsilon + \omega_{k'} - \omega_{k_1}} \right) \tag{C.14}$$

$$t^{+}_{kk'}(\omega') - t^{-}_{kk'}(\omega) = \sum_{k_1} t^{+}_{kk_1}(\omega_{k_1}) t^{-}_{k_1k'}(\omega_{k_1}) \times \left(\frac{1}{-i\varepsilon + \omega_k - \omega_{k_1}} - \frac{1}{i\varepsilon + \omega_{k'} - \omega_{k_1}} \right) \tag{C.15}$$

Two less familiar relations are also used in the text:

$$\frac{d}{dz} t_{k'k}(z) = -\sum_{k_1} t_{k'k_1}(z) \frac{1}{(z - \omega_{k_1})^2} t_{k_1 k}(z) \tag{C.16}$$

which can be easily proved using perturbation expansion, and

$$\frac{\mathrm{d}}{\mathrm{d}\omega_k} t^{-}_{kk'}(\omega_k) = \sum_{k_1} t^{+}_{kk_1}(\omega_{k_1}) \frac{1}{(i\varepsilon + \omega_{k_1} - \omega_k)^2} t^{-}_{k_1 k'}(\omega_{k_1}) \tag{C.17}$$

which is obtained by differentiation of Eq. (C.15).

APPENDIX D. ANALYTICAL CONTINUATION

The purpose of this appendix is to show that Eqs. (77) and (78), obtained by resumming the contributions from all subdynamics (0, k), are identical with the results derived for the evolution of the wave function using conventional techniques.

To solve Eq. (69) one first goes to the limit of continuous spectrum for H_0 and proceeds to the diagonalization of H in this basis. In this procedure (see Ref. 30, where it is explicitly performed in the exactly soluble case of a separable potential), the resolvent of H presents a cut on the positive real axis.

In contrast, as mentioned in the Introduction, we start with a discrete spectrum, in order to be able to organize the perturbation series, and we use the $i\varepsilon$-rule for the regularization of the propagators in the limit in which the spectrum becomes continuous.

The solution of Eq. (69) can be expressed as

$$c_{k_1}(t) = \frac{1}{2\pi i} \int_{\pm} \mathrm{d}z \ \mathrm{e}^{-izt} \sum_{k_2} \langle k_1 | \frac{1}{z - H} | k_2 \rangle c_{k_2}(0) \tag{D.1}$$

where the contour runs above the singularities of the resolvent of H. With the use of the expression Eq. (C.5) for this resolvent, Eq. (D.1) may be written as

$$c_{k_1}(t) = \frac{1}{2\pi i} \sum_{k_2} \int_{\pm} \mathrm{d}z \, \mathrm{e}^{-izt} \frac{1}{z - \omega_{k_1}} [\delta_{k_1 k_2} + t_{k_1 k_2}(z) \frac{1}{z - \omega_{k_2}}] c_{k_2}(0) \tag{D.2}$$

The singularities of the resolvent lie on the real axis: the poles at $z = \omega_{k_1}$ and $z = \omega_{k_2}$, and a cut on the positive real axis due to the t-matrix.

In order to give a meaning to the contributions arising from these singularities, let us displace the poles in the complex plane, the $z = \omega_{k_1}$ pole below the cut and $z = \omega_{k_2}$ above the cut, as shown in Fig. 1.

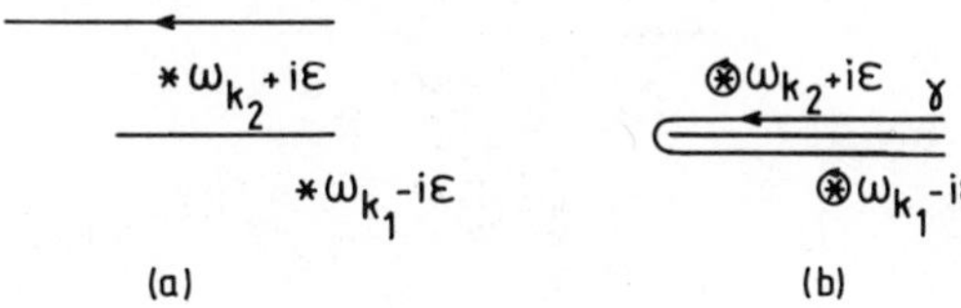

Fig. 1

The integration contour in Fig. 1*a* can then be replaced by the contour in Fig. 1*b*, and the residues at the poles evaluated:

$$c_{k_1}(t) = \mathrm{e}^{-i\omega_{k_1}t} \sum_{k_2} \left[\delta_{k_1k_2} + \frac{t^-_{k_1k_2}(\omega_{k_1})}{-i\varepsilon + \omega_{k_1} - \omega_{k_2}} \right] c_{k_2}(0)$$

$$+ \sum_{k_2} \mathrm{e}^{-i\omega_{k_2}t} \frac{t^+_{k_1k_2}(\omega_{k_2})}{i\varepsilon + \omega_{k_2} - \omega_{k_1}} c_{k_2}(0)$$

$$+ \frac{1}{2\pi i} \sum_{k_2} \int_\gamma \mathrm{d}z\, \mathrm{e}^{-izt} \frac{1}{z - \omega_{k_1} + i\varepsilon} t_{k_1k_2}(z) \frac{1}{z - \omega_{k_2} - i\varepsilon} c_{k_2}(0) \quad \text{(D.3)}$$

The integration on the contour around the cut can be transformed into an integration on the real axis

$$I_\gamma = -\frac{1}{2\pi i} \sum_{k_2} \int_\mu^\infty \mathrm{d}\omega\, \mathrm{e}^{-i\omega t} \frac{1}{\omega - \omega_{k_1} + i\varepsilon} [t^+_{k_1k_2}(\omega) - t^-_{k_1k_2}(\omega)]$$

$$\times \frac{1}{\omega - \omega_{k_2} - i\varepsilon} c_{k_2}(0) \quad \text{(D.4)}$$

which, with the use of the relation in Eq. (C.12), gives

$$I_\gamma = \sum_k \sum_{k_2} \mathrm{e}^{-i\omega_k t} \frac{t^+_{k_1k}(\omega_k)}{i\varepsilon + \omega_k - \omega_{k_1}} \cdot \frac{t^-_{kk_2}(\omega_k)}{-i\varepsilon + \omega_k - \omega_{k_2}} c_{k_2}(0) \quad \text{(D.5)}$$

Recollecting the various contributions, one finds for $c_{k_1}(t)$ the expression [compare with Eq. (83)]:

$$c_{k_1}(t) = \sum_{k_2} \sum_k \left[\delta_{k_1k} + \frac{t^+_{k_1k}(\omega_k)}{i\varepsilon + \omega_k - \omega_{k_1}} \right] \mathrm{e}^{-i\omega_k t}$$

$$\times \left[\delta_{kk_2} + \frac{t^-_{kk_2}(\omega_k)}{-i\varepsilon + \omega_k - \omega_{k_2}} \right] c_{k_2}(0) \quad \text{(D.6)}$$

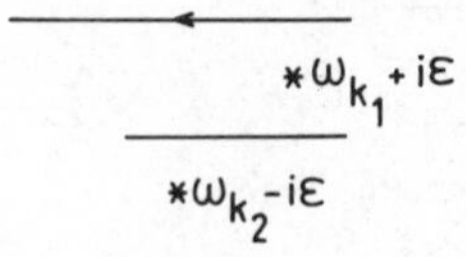

Fig. 2

The way the poles have been displaced from the cut is somewhat arbitrary. One could have moved them in the opposite way, as shown in Fig. 2, which would have given for $c_{k_1}(t)$ the following expression

$$c'_{k_1}(t) = \sum_{k_2} \sum_{k} \left(\delta_{k_1 k} + \frac{t^-_{k_1 k}(\omega_k)}{-i\varepsilon + \omega_k - \omega_{k_1}} \right) e^{-i\omega_k t} \times \left(\delta_{k k_2} + \frac{t^+_{k k_2}(\omega_k)}{i\varepsilon + \omega_k - \omega_{k_2}} \right) c_{k_2}(0) \tag{D.7}$$

This would correspond to the opposite choice of the chronological ordering for correlations of same degree in the $i\varepsilon$-rule, and would have led to the Λ'-transformation, which is applicable to observables (see text).

There exists also the possibility of displacing both poles on the same side of the cut. In that case, one still finds a superposition of oscillations, but one cannot cast the evolution into a form similar to Eq. (D.6) or Eq. (D.7). This does not lead to any decomposition in subdynamics with properties of Eqs. (24)–(26) and it is not possible to define a transformation leading to new states.

APPENDIX E. MORE ABOUT THE SECTORS (1, 0) AND (0, 1)

We shall first show that in the sector (1, 0), the properties of Eqs. (24)–(26) and (54) are satisfied.

The property that the $\Pi^{(k0)}$ form a complete set of subdynamics projectors is readily verified. The completeness of the $\Pi^{(k0)}$ [Eq. (86)] results immediately from Eq. (C.14):

$$\begin{aligned} \sum_k \overset{(k0)}{\underset{k_1 0 k_2 0}{\Pi}} &= \sum_k \left(\delta_{k_1 k} + \frac{t^+_{k_1 k}(\omega_k)}{i\varepsilon + \omega_k - \omega_{k_1}} \right) \left(\delta_{k k_2} + \frac{t^-_{k k_2}(\omega_k)}{-i\varepsilon + \omega_k - \omega_{k_2}} \right) \\ &= \delta_{k_1 k_2} + \frac{t^+_{k_1 k_2}(\omega_{k_2})}{i\varepsilon + \omega_{k_2} - \omega_{k_1}} + \frac{t^-_{k_1 k_2}(\omega_{k_1})}{-i\varepsilon + \omega_{k_1} - \omega_{k_2}} \\ &\quad + \sum_k \frac{t^+_{k_1 k}(\omega_k)}{i\varepsilon + \omega_k - \omega_{k_1}} \cdot \frac{t^-_{k k_2}(\omega_k)}{-i\varepsilon + \omega_k - \omega_{k_2}} = \delta_{k_1 k_2} \end{aligned} \tag{E.1}$$

For their orthogonality and idempotence [Eq. (87)], because one has

$$\left(\overset{(k0)}{\Pi}\ \overset{(k'0)}{\Pi}\right)_{k_1 0 k_2 0} = C_{k_1 0 k 0} \sum_{k_3} D_{k0k_30} C_{k_3 0 k' 0} D_{k'0k_20} \tag{E.2}$$

it is sufficient to prove that

$$\sum_{k_3} D_{k0k_30} C_{k_30k'0} = \delta_{kk'} \tag{E.3}$$

that is

$$\sum_{k_3} \left(\delta_{kk_3} + \frac{t^{-}_{kk_3}(\omega_k)}{-i\varepsilon + \omega_k - \omega_{k_3}} \right) \left(\delta_{k_3k'} + \frac{t^{+}_{k_3k'}(\omega_{k'})}{i\varepsilon + \omega_{k'} - \omega_{k_3}} \right) = \delta_{kk'} \tag{E.4}$$

This equality is the immediate result of Eq. (C.11).

The commutation of L and Π results from Eq. (89). Because

$$\left(\overset{(k0)}{\Pi} L\right)_{k_1 0 k_2 0} = C_{k_1 0 k 0} \sum_{k_3} D_{k0k_30} L_{k_3 0 k_2 0} \tag{E.5}$$

it is necessary to show that

$$\sum_{k_3} D_{k0k_30} L_{k_30k_20} = \omega_k D_{k0k_20} \tag{E.6}$$

Indeed, by virtue of Eq. (C.6b)

$$\begin{aligned}
&\sum_{k_3} \left(\delta_{kk_3} + \frac{t^{-}_{kk_3}(\omega_k)}{-i\varepsilon + \omega_k - \omega_{k_3}} \right) (\delta_{k_3k_2}\omega_{k_2} + V_{k_3k_2}) \\
&\quad = \omega_k\, \delta_{kk_2} + V_{kk_2} + \omega_{k_2} \frac{t^{-}_{kk_2}(\omega_k)}{-i\varepsilon + \omega_k - \omega_{k_2}} + \sum_{k_3} \frac{t^{-}_{kk_3}(\omega_k)}{-i\varepsilon + \omega_k - \omega_{k_3}} V_{k_3k_2} \\
&\quad = \omega_k \left(\delta_{kk_2} + \frac{t^{-}_{kk_2}(\omega_k)}{-i\varepsilon + \omega_k - \omega_{k_2}} \right)
\end{aligned} \tag{E.7}$$

One shows similarly that

$$\sum_{k_3} L_{k_10k_30} C_{k_30k0} = C_{k_10k0}\omega_k \tag{E.8}$$

As in this sector, the matrix elements of $\mathscr{H}$ are proportional to those of L, the property [Eq. (57)] of commutation of $\mathscr{H}$ and Π is also trivially satisfied.

We have emphasized the isomorphisms between ρ_{k0} and the wave function and between ρ_{0k} and the complex conjugate of the wave function. In this respect, it is quite easy to show the correspondence of our treatment with the usual procedure of diagonalization of the Hamiltonian. Indeed one can write the matrix elements of the different operators we have encountered in Section IV, under a factorized form.

In sector (1, 0), one has

$$L = H \times 1 \tag{E.9}$$

$$\Lambda = U_+ \times 1 \tag{E.10}$$

$$\Lambda^{-1} = U_+^{-1} \times 1 \tag{E.11}$$

with

$$H_{kk} = \omega_k; \qquad H_{kk'} = V_{kk'} \tag{E.12}$$

$$(U_+)_{kk'} = C_{k0k'0} \tag{E.13}$$

$$(U_+^{-1})_{kk'} = D_{k0k'0} \tag{E.14}$$

Equation (43) can be interpreted as

$$\Lambda^{-1} L \Lambda = U_+^{-1} H U_+ \times 1 = {}^d H \times 1 \tag{E.15}$$

with [compare Eq. (5)]

$${}^d H = H_0 \tag{E.16}$$

The unitarity of the superoperator Λ implying that of the operator U_+, U_+ is the unitary operator that diagonalizes H.

Similarly, in the sector (0, 1), the evolution is given by

$$\begin{aligned} i\,\partial_t\,\rho_{0k_1} &= L_{0k_1 0k_1}\rho_{0k_1} + \sum_{k_2} L_{0k_1 0k_2}\rho_{0k_2} \\ &= -\omega_{k_1}\rho_{0k_1} - \sum_{k_2} V_{k_2 k_1}\rho_{0k_2} \end{aligned} \tag{E.17}$$

Comparing with Eq. (67), it is clear that all the expressions are the same as those obtained previously for the sector (1, 0) if one changes ω_k into $-\omega_{k'}$ $V_{kk'}$ into $-V_{k'k}$.

Thus

$$\Lambda_{0k_1 0k} = C_{0k_1 0k} = \delta_{k_1 k} + \frac{t^-_{kk_1}(\omega_k)}{-i\varepsilon + \omega_k - \omega_{k_1}} = (U_+^{-1})_{kk_1} \tag{E.18}$$

$$(\Lambda^{-1})_{0k0k_2} = D_{0k0k_2} = \delta_{kk_2} + \frac{t^+_{k_2 k}(\omega_k)}{i\varepsilon + \omega_k - \omega_{k_2}} = (U_+)_{k_2 k} \tag{E.19}$$

In the sector (0, 1)

$$L = -1 \times H \tag{E.20}$$

$$\Lambda = 1 \times U_+^{-1} \tag{E.21}$$

$$\Lambda^{-1} = 1 \times U_+ \tag{E.22}$$

and we obtain the result:

$$\Lambda^{-1} L \Lambda = -1 \times U_+^{-1} H U_+ = -1 \times {}^d H \tag{E.23}$$

Collecting the results of Eqs. (E.15) and (E.23), we are able to define a unitary, factorizable superoperator which diagonalizes H.

Indeed as

$${}^d H = U_+^{-1} H U_+ = (U_+^{-1} \times U_+) H = \mathscr{U}_+^{-1} H \tag{E.24}$$

the superoperator $\mathscr{U}_+^{-1}$ has such properties. Its matrix elements are given by Eq. (11)

$$\begin{aligned}(\mathscr{U}_+^{-1})_{k_1 k_2 k_3 k_4} &= (U_+^{-1})_{k_1 k_3} (U_+)_{k_4 k_2} \\ &= \left(\delta_{k_1 k_3} + \frac{t^-_{k_1 k_3}(\omega_{k_1})}{-i\varepsilon + \omega_{k_1} - \omega_{k_3}} \right) \left(\delta_{k_2 k_4} + \frac{t^+_{k_4 k_2}(\omega_{k_2})}{i\varepsilon + \omega_{k_2} - \omega_{k_4}} \right)\end{aligned} \tag{E.25}$$

and those of its inverse by

$$\begin{aligned}(\mathscr{U}_+)_{k_1 k_2 k_3 k_4} &= (U_+)_{k_1 k_3} (U_+^{-1})_{k_4 k_2} \\ &= \left(\delta_{k_1 k_3} + \frac{t^+_{k_1 k_3}(\omega_{k_3})}{i\varepsilon + \omega_{k_3} - \omega_{k_1}} \right) \left(\delta_{k_2 k_4} + \frac{t^-_{k_4 k_2}(\omega_{k_4})}{-i\varepsilon + \omega_{k_4} - \omega_{k_2}} \right)\end{aligned} \tag{E.26}$$

However, as we have already shown in Section II, some matrix elements, such as $(\mathscr{U}_+)_{kkk'k'}$ are ill defined, due to the incompatibility of analytical continuations.

Using the $i\varepsilon$-rule with the choice of the antichronological ordering of the correlations of the same degree, one would have, in the sector (1, 0),

$$\Lambda_i = U_- \times 1; \qquad \Lambda_i^{-1} = U_-^{-1} \times 1 \tag{E.27}$$

and in the sector (0, 1),

$$\Lambda_i = 1 \times U_-^{-1}; \qquad \Lambda_i^{-1} = 1 \times U_- \tag{E.28}$$

The corresponding factorized superoperators,

$$\mathscr{U}_-^{-1} = U_-^{-1} \times U_-, \qquad \mathscr{U}_- = U_- \times U_-^{-1} \tag{E.29}$$

would present the same divergence difficulties as $\mathscr{U}_+$, $\mathscr{U}_+^{-1}$.

The mixing of $U_\pm$ with $U_\mp^{-1}$ into still another $\mathscr{U}$ would not eliminate the divergences entirely. In addition such an hybrid $\mathscr{U}^{-1}$ would no more diagonalize H.

Note also that we have used the Λ's obtained in sectors (1, 0) and (0, 1) only, to treat an operator H which does not belong to these sectors.

APPENDIX F. THE REGULARIZATION PROCEDURE IN PERTURBATION AND CONVOLUTION FORMS

1. The proof of Eq. (109a) proceeds in two steps. First we show that Eq. (109a) is true at the lowest order in the perturbation expansion

$$C^{(2)}_{k_1k_2kk} = \frac{1}{2\pi i}\int_{\gamma_0} \mathrm{d}z\, V_{k_1k}(-V_{kk_2})\frac{1}{z-\omega_{k_1}+\omega_{k_2}} \times \left(\frac{1}{z-\omega_k+\omega_{k_2}}+\frac{1}{z-\omega_{k_1}+\omega_k}\right)\frac{1}{z}$$

Taking the residue at $z = +i0$, one obtains

$$\begin{aligned} C^{(2)}_{k_1k_2kk} &= V_{k_1k}(-V_{kk_2})\frac{1}{i\varepsilon-\omega_{k_1}+\omega_{k_2}}\left(\frac{1}{i\varepsilon-\omega_k+\omega_{k_2}}+\frac{1}{i\varepsilon-\omega_{k_1}+\omega_k}\right) \\ &= \frac{V_{k_1k}}{i\varepsilon+\omega_k-\omega_{k_1}}\cdot\frac{V_{kk_2}}{-i\varepsilon+\omega_k-\omega_{k_2}} \end{aligned} \tag{F.1}$$

which is indeed Eq. (109a) at the lowest order.

Next Eq. (109a) is proved by recurrence: one shows that if it is valid at order n, it is also true at order $n + 1$. One may write

$$C^{(n+1)}_{k_1k_2kk} = \frac{1}{2\pi i}\int_{\gamma_0} dz \left[\sum_{k'_1} \frac{1}{z - \omega_{k_1} + \omega_{k_2}} V_{k_1k'_1}\mathscr{C}^{(n)}_{k'_1k_2kk}(z)\cdot\frac{1}{z}\right.$$

$$\left. + \sum_{k'_2} \frac{1}{z - \omega_{k_1} + \omega_{k_2}}(-V_{k'_2k_2})\mathscr{C}^{(n)}_{k_1k'_2kk}(z)\cdot\frac{1}{z}\right]$$

When the residue is taken at $z = +i0$, one obtains

$$C^{(n+1)}_{k_1k_2kk} = \frac{1}{i\varepsilon - \omega_{k_1} + \omega_{k_2}} \sum_{k'_1} V_{k_1k'_1} C^{(n)}_{k'_1k_2kk}$$

$$- \frac{1}{i\varepsilon - \omega_{k_1} + \omega_{k_2}} \sum_{k'_2} V_{k'_2k_2} C^{(n)}_{k_1k'_2kk} \tag{F.2}$$

Introducing in Eq. (F.2) the expressions for $C^{(n)}_{k'_1k_2kk}$ and $C^{(n)}_{k_1k'_2kk}$:

$$C^{(n)}_{k'_1k'_2kk} = \sum_{l:1}^{n-1} \frac{t^{+(l)}_{k'_1k}(\omega_k)}{i\varepsilon + \omega_k - \omega_{k'_1}} \cdot \frac{t^{-(n-l)\,l)}_{kk'_2}(\omega_k)}{-i\varepsilon + \omega_k - \omega_{k'_2}}$$

$$+ \delta_{k'_2k}\frac{t^{+(n)}_{kk'_1}(\omega_k)}{i\varepsilon + \omega_k - \omega_{k'_1}} + \delta_{k'_1k}\frac{t^{-(n)}_{kk'_2}(\omega_k)}{-i\varepsilon + \omega_k - \omega_{k'_2}} \tag{F.3}$$

the summation over k_1 and k'_2 can be performed with the use of Eq. (C.11), giving, after some relabeling and manipulation, the result

$$C^{(n+1)}_{k_1k_2kk} = \sum_{l:1}^{n}\left[\frac{t^{+(l)}_{k_1k}(\omega_k)}{i\varepsilon - \omega_{k_1} + \omega_{k_2}} \cdot \frac{t^{-(n+1-l)}_{kk_2}(\omega_k)}{-i\varepsilon + \omega_k - \omega_{k_2}}\right.$$

$$\left. - \frac{t^{+(l)}_{k_1k}(\omega_k)}{i\varepsilon + \omega_k - \omega_{k_1}} \cdot \frac{t^{-(n+1-l)}_{kk_2}(\omega_k)}{i\varepsilon - \omega_{k_1} + \omega_{k_2}}\right]$$

$$= \sum_{l:1}^{n} \frac{t^{+(l)}_{k_1k}(\omega_k)}{i\varepsilon + \omega_k - \omega_{k_1}} \cdot \frac{t^{-(n+1-l)}_{kk_2}(\omega_k)}{-i\varepsilon + \omega_k - \omega_{k_2}} \tag{F.4}$$

2. All matrix elements of the resolvent of the Liouvillian $(z - L)^{-1}$ can be expressed as a convolution of matrix elements of the resolvent of the

Hamiltonian $(z - H)^{-1}$ as

$$\langle\alpha\beta|\frac{1}{z-L}|\gamma\delta\rangle = \frac{1}{2\pi i}\int_{-\infty}^{+\infty} dx \langle\alpha|\frac{1}{x+\frac{z}{2}-H}|\gamma\rangle\langle\delta|\frac{1}{x-\frac{z}{2}-H}|\beta\rangle \tag{F.5}$$

Using Eq. (F.5), Eq. (23) becomes

$$\overset{\nu}{\Pi}_{\alpha\beta\gamma\delta} = \frac{1}{2\pi i}\int_{\gamma_\nu}' dz \frac{1}{2\pi i}\int_{-\infty}^{+\infty} dx \langle\alpha|\frac{1}{x+\frac{z}{2}-H}|\gamma\rangle\langle\delta|\frac{1}{x-\frac{z}{2}-H}|\beta\rangle \tag{F.6}$$

with the prime on the integration sign to recall that a regularization procedure has to be used.

More explicitly, let us consider the matrix elements of Π^0:

$$\overset{0}{\Pi}_{k_1k_2k_3k_4} := \frac{1}{2\pi i}\int_{\gamma_0}' dz \frac{1}{2\pi i}\int_{-\infty}^{+\infty} dx \langle k_1|\frac{1}{x+\frac{z}{2}-H}|k_3\rangle\langle k_4|\frac{1}{x-\frac{z}{2}-H}|k_2\rangle \tag{F.7}$$

In the perturbative form, the original $i\varepsilon$-rule tells us that, when the residue in $z = 0$ is taken, all propagators are to be treated $+i\varepsilon$.

This can be translated on the convolution form Eq. (F.7) by the following statement: after having isolated the singularity in $z = 0$ in the Eq. (F.7), all propagators in $x + z/2$ (including those contained in the t-matrix) have to be treated $+i\varepsilon$ and those in $(x - z/2)$ have to be treated $-i\varepsilon$.

Let us consider, for instance,

$$\overset{0}{\Pi}_{k_1k_2kk} = \frac{1}{2\pi i}\int_{\gamma_0} dz \frac{1}{2\pi i}\int_{-\infty}^{+\infty} dx \frac{1}{x+\frac{z}{2}-\omega_{k_1}} t_{k_1k}\left(x+\frac{z}{2}\right)$$
$$\times \frac{1}{x+\frac{z}{2}-\omega_k} \times \frac{1}{x-\frac{z}{2}-\omega_k} t_{kk_2}\left(x-\frac{z}{2}\right)\frac{1}{x-\frac{z}{2}-\omega_{k_2}} \tag{F.8}$$

in which Eq. (C.5) has been used to express the matrix elements of the resolvent of H. One isolates the pole in $z = 0$, which is due to the pinch of $x = \omega_k - z/2$ and $x = \omega_k + z/2$:

$$\overset{0}{\Pi}_{k_1k_2kk} = \frac{1}{2\pi i}\int_{\gamma_0} dz \frac{1}{2\pi i}\int_{-\infty}^{+\infty} dx \frac{1}{z}\left(\frac{1}{x - \frac{z}{2} - \omega_k} - \frac{1}{x + \frac{z}{2} - \omega_k}\right)$$

$$\times \frac{1}{x + \frac{z}{2} - \omega_{k_1}} t_{k_1k}\left(x + \frac{z}{2}\right) \frac{1}{x - \frac{z}{2} - \omega_{k_2}} t_{kk_2}\left(x - \frac{z}{2}\right) \quad \text{(F.9)}$$

and one calculates the residue in $z = 0$ treating the propagators in $(x + z/2)$, $+i\varepsilon$, and those in $(x - z/2)$, $-i\varepsilon$:

$$\overset{0}{\Pi}_{k_1k_2kk} = \frac{1}{2\pi i}\int_{-\infty}^{+\infty} dx \left(\frac{1}{x - i\varepsilon - \omega_k} - \frac{1}{x + i\varepsilon - \omega_k}\right)$$

$$\times \frac{1}{x + i\varepsilon - \omega_{k_1}} t^{+}_{k_1k}(x) \frac{1}{x - i\varepsilon - \omega_{k_2}} t^{-}_{kk_2}(x) \quad \text{(F.10)}$$

The last integration is easily performed and leads to Eq. (109).

For the element $A_{kkk'k'}$:

$$A_{kkk'k'} = \frac{1}{2\pi i}\int_{\gamma_0} dz \frac{1}{2\pi i}\int_{-\infty}^{+\infty} dx \frac{1}{x + \frac{z}{2} - \omega_k} t_{kk'}\left(x + \frac{z}{2}\right) \frac{1}{x + \frac{z}{2} - \omega_{k'}}$$

$$\times \frac{1}{x - \frac{z}{2} - \omega_{k'}} t_{k'k}\left(x - \frac{z}{2}\right) \frac{1}{x - \frac{z}{2} - \omega_k} \quad \text{(F.11)}$$

one has to treat a double pole in $z = 0$, due to the double pinch in $x = \omega_k - z/2$, $x = \omega_k + z/2$ and in $x = \omega_{k'} - z/2$, $x = \omega_{k'} + z/2$. One obtains

$$A_{kkk'k'} = \frac{1}{2\pi i}\int_{\gamma_0} dz \frac{1}{z^2}\frac{1}{2\pi i}\int_{-\infty}^{+\infty} dx \left(\frac{1}{x + \frac{z}{2} - \omega_k} - \frac{1}{x - \frac{z}{2} - \omega_k}\right)$$

$$\times \left(\frac{1}{x + \frac{z}{2} - \omega_{k'}} - \frac{1}{x - \frac{z}{2} - \omega_{k'}}\right) t_{kk'}\left(x + \frac{z}{2}\right) t_{k'k}\left(x - \frac{z}{2}\right)$$

(F.12)

and when taking the residue in $z = 0$, treating as stated above the $x + z/2$ with $+i\varepsilon$, and $x - z/2$ with $-i\varepsilon$, one has

$$\lim_{z \to +i0} \frac{1}{2\pi i} \frac{\mathrm{d}}{\mathrm{d}z} \int_{-\infty}^{+\infty} \mathrm{d}x \left(\frac{1}{x + \frac{z}{2} - \omega_k} - \frac{1}{x - \frac{z}{2} - \omega_k} \right)$$

$$\times \left(\frac{1}{x + \frac{z}{2} - \omega_{k'}} - \frac{1}{x - \frac{z}{2} - \omega_{k'}} \right) t_{kk'}\left(x + \frac{z}{2} \right) t_{k'k}\left(x - \frac{z}{2} \right) \tag{F.13}$$

Performing the derivation with respect to z, taking the limit $z = +i0$ and integrating over x,

$$\frac{1}{2} \int_{-\infty}^{+\infty} \mathrm{d}x \left[\frac{1}{(x + i\varepsilon - \omega_k)^2} + \frac{1}{(x - i\varepsilon - \omega_k)^2} \right] \delta(x - \omega_{k'}) |t_{kk'}^{+}(x)|^2$$

$$+ \frac{1}{2} \int_{-\infty}^{+\infty} \mathrm{d}x \left[\frac{1}{(x + i\varepsilon - \omega_{k'})^2} + \frac{1}{(x - i\varepsilon - \omega_{k'})^2} \right] \delta(x - \omega_k) |t_{kk'}^{+}(x)|^2$$

$$+ 2\pi i \int_{-\infty}^{+\infty} \mathrm{d}x \, \delta(x - \omega_k) \, \delta(x - \omega_{k'}) \frac{\mathrm{d}}{\mathrm{d}z} t_{kk'}\left(x + \frac{z}{2} \right) t_{k'k}\left(x - \frac{z}{2} \right) \Big|_{z:i0} \tag{F.14}$$

one immediately finds Eq. (111b).

For $\Pi^{(kk')}$, the $i\varepsilon$-rule is more complex, as one has to take differently into account monocorrelations appearing at the right or at the left of the distinguished propagator, and also the vacuum, that is uniformly treated $-i\varepsilon$.

For the convolutive form, this gives the following prescriptions:

For expressions appearing at the right of the monocorrelation (kk'), the result of the integration is given by:

$$t\left(x + \frac{z}{2} \right) \Rightarrow t^{-}(\omega_k)$$

$$t\left(x - \frac{z}{2} \right) \Rightarrow t^{+}(\omega_{k'})$$

$$\left(x + \frac{z}{2} - \omega_{k_1} \right)^{-1} \Rightarrow (-i\varepsilon + \omega_k - \omega_{k_1})^{-1}$$

$$\left(x - \frac{z}{2} - \omega_{k_2} \right)^{-1} \Rightarrow (i\varepsilon + \omega_{k'} - \omega_{k_2})^{-1}$$

$$\left(x + \frac{z}{2} - \omega_k\right)^{-1} \Rightarrow 2\pi i$$

$$\left(x - \frac{z}{2} - \omega_{k'}\right)^{-1} \Rightarrow -2\pi i \tag{F.15}$$

At the left of (kk'), one obtains:

$$t\left(x + \frac{z}{2}\right) \Rightarrow t^+(\omega_k)$$

$$t\left(x - \frac{z}{2}\right) \Rightarrow t^-(\omega_{k'})$$

$$\left(x + \frac{z}{2} - \omega_{k_1}\right)^{-1} \Rightarrow (i\varepsilon + \omega_k - \omega_{k_1})^{-1}$$

$$\left(x - \frac{z}{2} - \omega_{k_2}\right)^{-1} \Rightarrow (-i\varepsilon + \omega_{k'} - \omega_{k_2})^{-1}$$

$$\left(x + \frac{z}{2} - \omega_k\right)^{-1} \Rightarrow 2\pi i$$

$$\left(x - \frac{z}{2} - \omega_{k'}\right)^{-1} \Rightarrow 2\pi i \tag{F.16}$$

except if the same wave vector k_1 appears in propagators in $x + z/2$ and $x - z/2$. In this case, one has to write:

$$\frac{1}{\left(x + \frac{z}{2} - \omega_{k_1}\right)\left(x - \frac{z}{2} - \omega_{k_1}\right)} \Rightarrow -\frac{1}{-i\varepsilon + \omega_k - \omega_{k'}} \times \left(\frac{1}{i\varepsilon - \omega_{k'} + \omega_{k_1}} + \frac{1}{i\varepsilon + \omega_k - \omega_{k_1}}\right)$$

$$\frac{1}{\left(x + \frac{z}{2} - \omega_k\right)\left(x - \frac{z}{2} - \omega_k\right)} \Rightarrow 2\pi i \frac{1}{i\varepsilon + \omega_{k'} - \omega_k}$$

$$\frac{1}{\left(x + \frac{z}{2} - \omega_{k'}\right)\left(x - \frac{z}{2} - \omega_{k'}\right)} \Rightarrow 2\pi i \frac{1}{i\varepsilon + \omega_{k'} - \omega_k} \tag{F.17}$$

It can be checked on specific elements that these rules lead to Eqs. (115)–(117).

APPENDIX G. MORE ABOUT THE SECTOR (1, 1)

1. The property that the Π's form subdynamics is formulated in the relation in Eq. (26). As the $\Pi^{\nu}_{k_1k_2k_3k_4}$ are not continuous in their indices $k_1k_2k_3k_4$, as for instance

$$\overset{(kk')}{\Pi}_{k_1k_1kk'} \neq \lim_{k_2 \to k_1} \overset{(kk')}{\Pi}_{k_1k_2kk'}$$

one has to verify that the Eq. (26) is satisfied for all the different matrix elements. This is straightforward but quite lengthy and we shall not give all the proofs here. We shall simply limit ourselves to illustrate it on the simple element $[L, \Pi^0]_{kk_2kk}$.

Now,

$$\left(L\overset{0}{\Pi}\right)_{kk_2kk} = L_{kk_2kk_2}\overset{0}{\Pi}_{kk_2kk} + L_{kk_2kk}\overset{0}{\Pi}_{kkkk} + \sum_{k'} L_{kk_2kk'}\overset{0}{\Pi}_{kk'kk}$$

while $(\Pi^0 L)_{kk_2kk}$ vanishes at the same order in L^{-3}.

Introducing the explicit expressions for the elements of L and Π^0, one obtains

$$\left(L\overset{0}{\Pi}\right)_{kk_2kk} = (\omega_k - \omega_{k_2})\frac{t^-_{kk_2}(\omega_k)}{-i\varepsilon + \omega_k - \omega_{k_2}} - V_{kk_2}$$
$$- \sum_{k'} V_{k'k_2}\frac{t^-_{kk'}(\omega_k)}{-i\varepsilon + \omega_k - \omega_{k'}}$$

which vanishes, due to Eq. (C.6). The same procedure is used to show that $\mathscr{H}$ commutes with Π^0 [Eq. (57)].

2. As for the demonstration of Eq. (26), the proof that Λ^{-1} is the inverse of Λ is quite lengthy. One has to show that

$$(\Lambda^{-1}\Lambda)_{k_1k_2k_3k_4} = (\Lambda\Lambda^{-1})_{k_1k_2k_3k_4} = \delta_{k_1k_3}\,\delta_{k_2k_4}$$

for all $k_1k_2k_3k_4$ and, for instance, the element in which $k_1 = k_2$ in general requires a separate demonstration from the one in which $k_1 \neq k_2$.

We shall here only demonstrate that the properties of left and right inverse of Λ^{-1} invoke two different properties, the completeness of unperturbed states and that of the scattering states. For instance,

$$(\Lambda^{-1}\Lambda)_{kk'kk} = \Lambda^{-1}_{kk'kk} + \Lambda_{kk'kk} + \sum_{k_1} \Lambda^{-1}_{kk'kk_1}\Lambda_{kk_1kk}$$

Using the explicit forms of Λ and Λ^{-1} [Eqs. (120)–(125)], one obtains

$$(\Lambda^{-1}\Lambda)_{kk'kk} = \frac{t^{+}_{kk'}(\omega_{k'})}{i\varepsilon + \omega_{k'} - \omega_k} + \frac{t^{-}_{kk'}(\omega_k)}{-i\varepsilon + \omega_k - \omega_{k'}}$$
$$+ \sum_{k_1} \frac{t^{+}_{k_1k'}(\omega_{k'})}{i\varepsilon + \omega_{k'} - \omega_{k_1}} \cdot \frac{t^{-}_{kk_1}(\omega_k)}{-i\varepsilon + \omega_k - \omega_{k_1}}$$

which vanishes by virtue of Eq. (C.11).

For $(\Lambda\Lambda^{-1})$, one obtains, for example,

$$(\Lambda\Lambda^{-1})_{kk'kk} = \Lambda_{kk'kk} + \Lambda^{-1}_{kk'kk} + \sum_{k_1} \Lambda_{kk'kk_1}\Lambda^{-1}_{kk_1kk}$$
$$= \frac{t^{-}_{kk'}(\omega_k)}{-i\varepsilon + \omega_k - \omega_{k'}} + \frac{t^{+}_{kk'}(\omega_{k'})}{i\varepsilon + \omega_{k'} - \omega_k}$$
$$+ \sum_{k_1} \frac{t^{-}_{k_1k'}(\omega_{k_1})}{-i\varepsilon + \omega_{k_1} - \omega_{k'}} \cdot \frac{t^{+}_{kk_1}(\omega_{k_1})}{i\varepsilon + \omega_{k_1} - \omega_k}$$

which vanishes due to Eq. (C.15).

The relations Eqs. (C.16) and (C.17) are necessary to show that

$$(\Lambda^{-1}\Lambda)_{kkk'k'} = (\Lambda\Lambda^{-1})_{kkk'k'} = 0$$

Indeed these involve the $kkk'k'$ elements of Λ, in which derivatives of the t-matrix appear. For instance,

$$(\Lambda^{-1}\Lambda)_{kkk'k'} = \sum \Lambda^{-1}_{kkk_1k_2}\Lambda_{k_1k_2k'k'} + \Lambda^{-1}_{kkkk'}\Lambda_{kk'k'k'}$$
$$+ \Lambda^{-1}_{kkk'k'}\Lambda_{k'k'k'k'} + \Lambda^{-1}_{kkkk}\Lambda_{kkk'k'}$$

3. The explicit form of the $P\phi P$ operator as the cross section can be directly obtained from $P\Lambda^{-1}L\Lambda P$ [Eq. (43)]. The result can be reached more easily using already proven properties.

Indeed, from the fact that Π and L commute [Eq. (26)], and that Π is idempotent [Eq. (24)], one gets

$$L\Pi = \Pi L \Pi \tag{G.1}$$

Taking the $P - P$ component of this relation, one obtains

$$PLCA = ADLCA \tag{G.2}$$

Defining $\tilde{L}$ (which in previous works was also noted $\Omega\psi$ or θ), by

$$\tilde{L} = PLC \tag{G.3}$$

one finds that

$$\tilde{L} = ADLC \tag{G.4}$$

so that one now can write

$$P\phi P = P\Lambda^{-1}L\Lambda P = \chi^* DLC\chi = \chi^{-1}ADLC\chi = \chi^{-1}\tilde{L}\chi \tag{G.5}$$

Obviously, as one has also

$$[\Pi, \mathscr{H}]_- = 0 \tag{G.6}$$

one similarly obtains

$$P^p \mathscr{H} P = \chi^{-1}\tilde{\mathscr{H}}\chi \tag{G.7}$$

where $\tilde{\mathscr{H}}$ is defined by

$$\tilde{\mathscr{H}} = P\mathscr{H}C \tag{G.8}$$

For the determination of ϕ in the potential scattering problem, we have shown that among the diagonal elements of $\Lambda(\chi)$, only $\chi_{kk\,kk} = 1$ were playing a role, so that $P\phi P$ reduces to $\tilde{L}$.

From Eq. (G.3), one has

$$\tilde{L}_{kk\,kk} = \sum_{k_1} L_{kk\,k_1k} C_{k_1k\,kk} + \sum_{k_2} L_{kk\,kk_2} C_{kk_2\,kk} \tag{G.9}$$

and, using the expressions of the C's [Eq. (109)] as well as Eq. (C.6), one obtains Eq. (139):

$$\begin{aligned}\tilde{L}_{kk\,kk} &= \sum_{k_1} V_{kk_1} \frac{t^+_{k_1k}(\omega_k)}{i\varepsilon + \omega_k - \omega_{k_1}} - \sum_{k_2} V_{k_2k} \frac{t^-_{kk_2}(\omega_k)}{-i\varepsilon + \omega_k - \omega_{k_2}} \\ &= t^+_{kk}(\omega_k) - V_{kk} - t^-_{kk}(\omega_k) + V_{kk} = 2i \operatorname{Im} t^+_{kk}(\omega_k)\end{aligned} \tag{G.10}$$

Similarly, one recovers Eq. (138) from

$$\tilde{L}_{kkk'k'} = \sum_{k_1k_2} L_{kkk_1k_2} C_{k_1k_2k'k'}$$

$$= \sum_{k_1k_2} (V_{kk_1}\delta_{kk_2} - V_{k_2k}\delta_{kk_1}) \left[\delta_{k_1k'} + \frac{t^+_{k_1k'}(\omega_{k'})}{i\varepsilon + \omega_{k'} - \omega_{k_1}} \right]$$

$$\times \left[\delta_{k'k_2} + \frac{t^-_{k'k_2}(\omega_{k'})}{-i\varepsilon + \omega_{k'} - \omega_{k_2}} \right] \tag{G.11}$$

when the summations are performed using Eq. (C.6).

4. It has been insisted upon in the text the importance of the nonfactorizability of the Λ-transformation. It was already shown in Section V that when *all* its elements were taken into consideration, Λ could not be put in a factorized form. However, when only the physical representation is considered, this question is somewhat more delicate.

In terms of the Möller operators [Eq. (7)], these expressions can be written as

$$^p\rho_{kk} = \sum_{k_1k_2} (U_-^{-1})_{kk_1} \rho_{k_1k_2} (U_-)_{k_2k} \tag{G.12}$$

$$^p\rho_{kk'} = \sum_{k_1k_2} (U_+^{-1})_{kk_1} \rho_{k_1k_2} (U_+)_{k_2k'} \tag{G.13}$$

The form of Eq. (G.13) that $^p\rho_{kk'}$ takes is due to our chronological ordering convention, which was imposed on us by the requirements in the general formalism of the existence of a Lyapunov function.[27,28] Now, even if one had chosen the opposite ordering convention, which would have led to

$$^{p'}\rho_{kk'} = \sum_{k_1k_2} (U_-^{-1})_{kk_1} \rho_{k_1k_2} (U_-)_{k_2k'} \tag{G.14}$$

for the physical correlation, and although this form suggests the identification of our Λ with the factorizable unitary transformation

$$\mathscr{U}_{-kk'k_1k_2} = (U_-^{-1})_{kk_1} (U_-)_{k_2k'} \tag{G.15}$$

we have seen in Section V, that this could not be the case.

However the difference for reason of order of magnitude in L^{-3} is not manifested on $^p\rho$ itself when finite wave packets are considered.

These remarks are however not essential: the importance is in the time evolution, and, as emphasized in Section VI, $\partial_t {}^P\rho_{kk}$, the dominant order of which is $0(L^{-6})$, is not factorizable.

APPENDIX H. DIAGONALIZATION OF THE $\mathscr{H}$ SUPEROPERATOR

We have mentioned in Section III that the determination of the diagonal elements of the Λ-transformation requires, in general, as a supplementary condition, the diagonalization of the energy superoperator.

We shall now determine here the elements $\Lambda_{kkk'k'}$ (often noted $\chi_{kk'}$) and $\Lambda^{-1}_{kkk'k'}$ $(=\chi^*_{kk'})$ for which such condition has to be used.

The matrix elements of the energy superoperator $\mathscr{H}$ defined by Eq. (56) are given by

$$\mathscr{H}_{k_1k_2k_3k_4} = \tfrac{1}{2}(H_{k_1k_3}\delta_{k_4k_2} + H_{k_4k_2}\delta_{k_1k_3}) \tag{H.1}$$

in terms of the matrix elements of the Hamiltonian operator H itself.

The additional condition [Eq. (59)] is $(k' \neq k)$

$${}^P\mathscr{H}_{kkk'k'} = 0 \tag{H.2}$$

One proceeds in two steps, each involving the auxiliary superoperator $\tilde{\mathscr{H}}$ defined in Eq. (G.8). As $\tilde{\mathscr{H}}$ only possesses $P-P$ elements, we simplify the notation by writing only once the repeating index. Using Eqs. (H.1) and (109b), one finds that

$$\tilde{\mathscr{H}}_{kk} = \omega_k \tag{H.3}$$

and, with the help of Eq. (C.6), that

$$\tilde{\mathscr{H}}_{kk'} = \left(\frac{1}{\omega_{k'} - \omega_k}\right)_P |t^+_{kk'}(\omega_{k'})|^2 \tag{H.4a}$$

at the dominant order in the volume.

In Eq. (H.4a), the principal part and the $t^\pm$ elements may be explicitly written as limits for $z \to +i0$:

$$\tilde{\mathscr{H}}_{kk'} = \lim_{z\to i0} \frac{1}{2}\left(\frac{1}{z+\omega_{k'}-\omega_k} - \frac{1}{z+\omega_k-\omega_{k'}}\right) \times t_{kk'}(z+\omega_{k'})t_{k'k}(-z+\omega_{k'}) \tag{H.4b}$$

The operation of L-inversion has to be defined in such a way that it does not affect the superoperator $\mathscr{H}$, which is independent of L:

$$\mathscr{H}' = \mathscr{H} \tag{H.5}$$

Taking the $P - P$ elements of Eq. (57), expressing the commutativity of $\mathscr{H}$ and Π, one has

$$P\mathscr{H}CA = AD\mathscr{H}P \tag{H.6a}$$

that may be written as

$$\tilde{\mathscr{H}}A = A\tilde{\mathscr{H}}^* \tag{H.6b}$$

because $\mathscr{H}$ *and* P are starhermitian and $D = C^*$. One finds

$$\tilde{\mathscr{H}}^*_{kk} = \omega_k \tag{H.7}$$

and

$$\begin{aligned}(\tilde{\mathscr{H}}^*)_{kk'} &= \left(\frac{1}{\omega_k - \omega_{k'}}\right)_P |t^+_{kk'}(\omega_k)|^2 \\ &= \lim_{z \to i0} -\frac{1}{2}\left(\frac{1}{z + \omega_{k'} - \omega_k} - \frac{1}{z + \omega_k - \omega_{k'}}\right) \\ &\quad \times t_{kk'}(z + \omega_k)t_{k'k}(-z + \omega_k)\end{aligned} \tag{H.8}$$

Using Eqs. (106) and (G.9), one obtains from Eq. (H.6b)

$$(\omega_k - \omega_{k'})A_{kk'} = (\tilde{\mathscr{H}}^*)_{kk'} - \tilde{\mathscr{H}}_{kk'} \tag{H.9}$$

On the one hand, this equality [Eq. (H.9)] may be verified using Eqs. (H.4b) and (H.8), as well as the expression Eq. (111b) for $A_{kk'}$ that is derived in Appendix F. On the other hand, Eq. (H.9) may be considered as an equation for obtaining $A_{kk'}$ by means of an appropriate *division procedure* (see Ref. 31), which is necessary because the r.h.s. is known to contain a distribution (principal part) in $(\omega_k - \omega_{k'})$.

Defining, in the limit $z \to +i0$,

$$\frac{1}{\omega_k - \omega_{k'}}\left(\frac{1}{z + \omega_k - \omega_{k'}} - \frac{1}{z + \omega_{k'} - \omega_k}\right)$$
$$= \left[\frac{1}{(z + \omega_k - \omega_{k'})^2} + \frac{1}{(z + \omega_{k'} - \omega_k)^2}\right]$$
$$- \frac{1}{2}\left(\frac{1}{z + \omega_k - \omega_{k'}} + \frac{1}{z + \omega_{k'} - \omega_k}\right)\frac{\partial}{\partial z} \tag{H.10}$$

one obtains from Eq. (H.9)

$$A_{kk'} = \lim_{z \to i0} \frac{1}{2}\left\{\left[\frac{1}{(z + \omega_k - \omega_{k'})^2} + \frac{1}{(z + \omega_{k'} - \omega_k)^2}\right]\right.$$
$$\left. - \frac{1}{2}\left[\frac{1}{z + \omega_{k'} - \omega_k} + \frac{1}{z + \omega_k - \omega_{k'}}\right]\frac{\partial}{\partial z}\right\}$$
$$\times \left[t_{kk'}(z + \omega_k)t_{k'k}(-z + \omega_k) + t_{kk'}(z + \omega_{k'})t_{k'k}(-z + \omega_{k'})\right] \tag{H.11}$$

which is identical with Eq. (111b).

This division procedure Eq. [(H.10)] will now be used for determining χ. Indeed it was shown in Appendix G that the $P - P$ elements of ${}^P\mathcal{H}$, defined by Eq. (57), can be written in terms of $\tilde{\mathcal{H}}$ [Eq. (G.7)]. Then, using Eq. (118) and Eq. (H.3)

$${}^P\mathcal{H}_{kk} = \omega_k \tag{H.12a}$$

and

$${}^P\mathcal{H}_{kk'} = (\omega_k - \omega_{k'})\chi_{kk'} + \tilde{\mathcal{H}}_{kk'} \tag{H.12b}$$

The diagonality requirement [Eq. (H.2)] then imposes that

$$(\omega_k - \omega_{k'})\chi_{kk'} = -\tilde{\mathcal{H}}_{kk'} \tag{H.13}$$

This equation may be solved, using the division procedure [Eq. (H.10)] for determining $\chi_{kk'}$ from the known $\tilde{\mathcal{H}}_{kk'}$ [Eq. (H.4)].

One finds

$$\chi_{kk'} = \lim_{z \to i0} \frac{1}{2} \left\{ \left[\frac{1}{(z + \omega_k - \omega_{k'})^2} + \frac{1}{(z + \omega_{k'} - \omega_k)^2} \right] - \frac{1}{2} \left[\frac{1}{z + \omega_{k'} - \omega_k} + \frac{1}{z + \omega_k - \omega_{k'}} \right] \frac{\partial}{\partial z} \right\} \times t_{kk'}(z + \omega_{k'}) t_{k'k}(-z + \omega_{k'}) \tag{H.14a}$$

which may also be written as

$$\chi_{kk'} = \frac{1}{2} \left[\frac{1}{(i\varepsilon + \omega_k - \omega_{k'}} + \frac{1}{(i\varepsilon + \omega_{k'} - \omega_k)^2} \right] |t^+_{kk'}(\omega_{k'})|^2 + \frac{i\pi}{2} \delta(\omega_k - \omega_{k'}) \left[\left(\frac{\partial}{\partial \omega_k} t^+_{kk'}(\omega_k) \right) t^-_{k'k}(\omega_k) - t^+_{kk'}(\omega_k) \left(\frac{\partial}{\partial \omega_k} t^-_{k'k}(\omega_k) \right) \right] \tag{H.14b}$$

This then completely fixes the $P - P$ elements of the Λ-transformation. As a check, one has to verify that, as in the general theory,

$$A = \chi \chi^* \tag{H.15a}$$

or, using Eqs. (106) and (118),

$$A_{kk'} = \chi_{kk'} + \chi^*_{kk'} \tag{H.15b}$$

The elements $\chi^*_{kk'}$ are obtained from Eq. (H.14b) through L-inversion. As

$$\chi^*_{kk'} = (\chi'_{k'k})^{cc} \tag{H.16}$$

one finds

$$\chi^*_{kk'} = \frac{1}{2} \left[\frac{1}{(i\varepsilon - \omega_k + \omega_{k'})^2} + \frac{1}{(i\varepsilon - \omega_{k'} + \omega_k)^2} \right] |t^+_{kk'}(\omega_k)|^2 + \frac{i\pi}{2} \delta(\omega_k - \omega_{k'}) \left\{ \left[\frac{\partial}{\partial \omega_k} t^+_{kk'}(\omega_k) \right] \cdot t^-_{k'k}(\omega_k) - t^+_{kk'}(\omega_k) \left[\frac{\partial}{\partial \omega_k} t^-_{kk'}(\omega_k) \right] \right\} \tag{H.17}$$

When Eqs. (H.14a) and (H.17) are inserted into Eq. (H.15b), one indeed recovers $A_{kk'}$ as given by Eq. (111b).

As said in the text, for reasons of order of magnitude in the volume, the $P - P$ elements of Λ do not show up in the potential scattering problem. However they play an essential role in the problem of moderately dense gases as soon as one goes beyond the Boltzmann approximation.[32]

References

1. I. Prigogine and C. George, *Proc. Natl. Acad. Sci. USA* **80**, 4590 (1983).
2. I. Prigogine and B. Misra, Lecture at the "Workshop on Irreversible Processes in Quantum Mechanics and Quantum Optics," San Antonio, March 1982.
3. C. George, I. Prigogine, and L. Rosenfeld, *Det K. Dan. Vidensk. Selsk. Mat.-fys Medd.*, **38**, 12 (1972).
4. B. Misra, I. Prigogine, and M. Courbage, *Proc. Natl. Acad. Sci. USA* **76**, 3607, 4768 (1979).
5. M. Courbage and I. Prigogine, *Proc. Natl. Acad. Sci. USA* **80**, 2412 (1983).
6. I. Prigogine, *From Being to Becoming*, Freeman, San Francisco, 1980.
7. B. Misra and I. Prigogine, *Lett. Math. Phys.* **7**, 421 (1983).
8. F. Mayné and I. Prigogine, *Physica* **63** 1 (1973).
9. I. Prigogine and F. Mayné, in *Transport Phenomena*, Lecture Notes in Physics, Springer Verlag **31**, 34 (1974).
10. I. Prigogine and A. Grecos, in *Problems in the Foundations of Physics*, Soc. Italiana di Fisica, Bologna, Italy, Corso 72 308 (1979).
11. See any textbook on scattering theory, for example:
    M. L. Golberger and K. M. Watson, *Collision Theory*, Wiley, New York, 1964.
    R. G. Newton, *Scattering Theory of Waves and Particles*, McGraw Hill, New York, 1966.
    J. M. Jauch and F. Rohrlich, *The theory of Photons and Electrons*, Addison-Wesley, Cambridge, Mass., 1955.
    W. Heitler, *Quantum theory of Radiation*, Clarendon Press, Oxford, 1954.
12. L. Van Hove, *Physica* **21**, 517 (1955).
13. I. Prigogine, *Nonequilibrium Statistical Mechanics*, Wiley, New York, 1962.
14. This aspect goes beyond results which can be found in earlier papers, like
    S. Fujita, *Physica* **28**, 281 (1962).
    R. Résibois, *Physica* **29**, 721 (1963).
    I. Prigogine, *Adv. Chem. Phys.* **15**, 11 (1969).
15. C. George, *Physica* **65**, 277 (1973).
16. B. Misra and I. Prigogine, in *Proceedings of the Workshop on Long Time Predictions in Dynamical Systems*, V. Szebehely, C. Horton, and L. Reichl, eds., Wiley, New York, 1982.
    B. Misra, I. Prigogine, and C. Lockhart, *Phys. Rev.* **D25**, 921 (1982).
17. N. Bogoliubov and D. Shirkov, *Quantum Fields*, Benjamin, 1983.
18. S. Weinberg, *The First Three Minutes*, Basic Books, New York, 1977.
19. See, for example, G. Nicolis and I. Prigogine, *Self-organisation in non equilibrium systems*, Wiley-Interscience, New York, 1977.
20. F. Mayné (to appear).
21. I. Prigogine, C. George, F. Henin, and L. Rosenfeld, *Chem. Scripta* **4**, 5 (1973).

22. F. Henin, *Physica* **54**, 385 (1971; *Bull. Acad. R. Belg., Cl. Sc.* **56**, 1113 (1970).
23. M. de Haan, *Bull. Acad. R. Belg., Cl. Sc.* **63**, 69, 317, 605 (1977).
24. A. Grecos, T. Guo, and W. Guo, *Physica* **80A**, 421 (1975).
25. M. Courbage, *J. Math. Phys.* **23**, 646, 652 (1982).
26. C. George, F. Henin, F. Mayné, and I. Prigogine, *Hadronic J.* **1**, 520 (1978). F. Henin and F. Mayné, *Physica* **108A**, 281 (1981).
27. I. Prigogine, F. Mayné, C. George, and M. de Haan, *Proc. Natl. Acad. Sci. USA* **74**, 4152 (1977).
28. M. de Haan, C. George, and F. Mayné, *Physica* **92A**, 584 (1978). *Bull. Acad. R. Belg., Cl. Sc.* **63**, 586 (1977).
29. A. Grecos and I. Prigogine, *Physica* **59**, 77 (1972).
30. E. G. C. Sudarshan, in 1961 Brandeis Summer Institute Lectures in Theoretical Physics, Vol. 2, p. 201, Benjamin, New York, 1962.
31. F. Henin, *Mémoire, Acad. R. Belg., Cl. Sc.* (to appear).
32. F. Mayné (to appear).

STATISTICAL PHYSICS OF POLYMER SOLUTIONS: CONFORMATION-SPACE RENORMALIZATION-GROUP APPROACH

YOSHITSUGU OONO

Department of Physics and Materials Research Laboratory, University of Illinois at Urbana-Champaign, Urbana Illinois

CONTENTS

I. INTRODUCTION

A. General Introduction

The main aim of this chapter is twofold: to show that the renormalization-group approach is the most natural approach to solutions of flexible polymers, and to make it popular among physical chemists and polymer scientists.

There have been many misunderstandings about renormalization-group theories. For example, papers published in leading chemical physics journals contain such statements as "the scaling theory can be derived without any help of renormalization-group theory," "renormalization-group theory can give only exponents," in a certain formalism "there is no requirement that the polymerization index be large," and "to know the temperature dependence detailed knowledge of chain interaction potential functions is necessary."

The author hopes that such misunderstandings will disappear before long. To this end an intuitive background of the renormalization-group theory is explained, and simple but representative examples are actually calculated in great detail (even pedagogically).

Though the present chapter does not exhibit all the so-far-obtained results, it is designed to help the reader to participate in the new development of the statistical physics of polymer solutions. This chapter does cover, however, almost all the main topics discussed in Yamakawa's *Modern Theory of Polymer Solutions.*[1]

There have been several reviews related to the application of renormalization theories to polymers. Whittington's recent article[2] in this series summarizes rigorous results and exact enumeration results. No transport properties nor universal functions are discussed. McKenzie's review[3] is also almost solely devoted to exponents. There is a superb textbook by de Gennes[4] on scaling concepts. Those who want an overview of polymer physics should read it. That book, however, does not explain in detail how to actually calculate observable quantities theoretically. The author hopes that this chapter will be complimentary to the existing reviews and books. For later reference, the scaling argument is sketched in Appendix A.

Any prior knowledge of renormalization-group theory is not at all assumed; it will be given in due course. Moreover, the so-called magnet-polymer analogy or zero-component field theory approach[5,6] is completely avoided. Hence, no advanced knowledge (except a very rudimentary one of critical phenomena) is required. Our exposition of renormalization-group theory is not the standard one. Readers who want more standard knowledge should consult, for example, Wilson and Kogut,[7] Ma,[8] and Toulouse and Pfeuty.[9] The recent lecture notes by Fisher[10] are very nice. Since we are interested mainly in quantities other than exponents, we do not discuss real-space renormalization methods developed by Shapiro,[11] and others,[12–18] for which there is a review article by Stanley et al.[19a] and by Family.[19b]

The actual calculation of experimentally interesting quantities may be analytically rather demanding. For the representative examples full technical details are given; the author hopes that the chapter is not too hard to read. The glossary of symbols can be found at the end of the text.

We calculate experimentally observable quantities. We should pursue the logical and conceptual soundness of the theory before requiring the agreement of its consequences with experiments.

B. Problem

A polymer chain is a very long molecule consisting of many monomers. Since each monomer has a repulsive core, the chain cannot cross itself, or equivalently, no part of the chain can occupy the same space-point. This is due to the excluded-volume interaction. The importance of the excluded-volume was first pointed out by Flory.[20] The rumor that Debye did not accept its importance immediately indeed shows that the excluded-volume effect is both theoretically and intuitively very subtle. Thus Flory's deep insight cannot be overly praised.

Actually, there is not only repulsive interaction but attractive interactions, so the word "excluded-volume" may be misleading; the "Flory effect" may be a better name to describe the effect of intermolecular forces on polymer systems. More precisely, since the chain is immersed in solvent molecules, the interaction is an effective one with the solvent effect incorporated. Still the essence is the same; there is a short-range repulsive interaction with an attractive tail as the ordinary van der Waals force. In the study of the static properties of polymer solutions, the main problem is to study the effect of the quality of solvents (or the effect of temperature) on these properties.

For transport properties and dynamical properties there is an extra complication. Since the chain is immersed in a solvent, the motion of a part of the chain excites the motion of solvent molecules, which propagates to another part of the chain to exert frictional force. This is the (long-range) hydrodynamic interaction, which profoundly affects dynamical properties of a chain. The main problem of polymer solution dynamics, then, is to study the Flory effect and the hydrodynamic effect.

As will be explained later, the conventional description of the hydrodynamic effect must be carefully reconsidered. The chain is described as a semimicroscopic object, and the solvent is described hydrodynamically. This implies that there is a discrepancy in the levels of the descriptions of chains and solvents; the former is described on the kinetic level, but the latter on the hydrodynamic level. A renormalization-group idea is necessary to overcome this inconsistency in the existing theories.

C. Outline

In Section II, two main conventional approaches, perturbation and mean-field approaches, are reviewed. After specifying the model and the partition function in Section II.A, perturbation theory is considered in Section II.B, and the mean-field theory in Section II.C. In Section II.D, the concept of the renormalization of interactions is intuitively introduced. The k-tolerant walk is discussed briefly, and it is explained why the theta-temperature is such a nontrivial concept in many-body theory. In Section II.E we show, with a simple example, what types of properties of a model survive the renormalization transformation. The last two sections explain why renormalization is inevitable.

In Section III, the Gell-Mann-Low type renormalization-group theory is explained in a general setting. In Section III.A the concepts of universality and minimal models are introduced, and in Section III.B the renormalization-group equation and the theoretical foundation of scaling arguments are given. In Section III.C the renormalized perturbation theory and the necessity of the ε-expansion are explained. In Section III.D, the derivation of the asymptotic relation $\langle R^2 \rangle \sim N^{2\nu}$ is sketched. Abstract arguments in this section are illustrated with concrete examples in the next two sections.

In Section IV, static properties, that is, equilibrium properties of polymer systems, are studied. In Section IV.A, the simplest nontrivial universal function, the density distribution function of the end-to-end vector, is calculated. All the analytic details are given. In Section IV.B, the temperature (solvent-quality) dependence (i.e., the crossover behavior) of the end-to-end vector distribution function is calculated. The finite-order self-avoiding walk is also discussed briefly. In Section IV.C, the effect of the polymer concentration is discussed, and the osmotic pressure is calculated. Analytical details are given. To make the ε-expansion method more intuitive another expansion scheme, the homotopy expansion, is sketched in Section IV.D.

In Section V, transport properties are studied. In Section V.A, the intrinsic viscosity of dilute solutions is calculated with the aid of the Kirkwood-Riseman scheme. A detailed renormalization argument is given. In Section V.B, we reconsider the validity of the generally accepted fundamental equation (the so-called full-diffusion equation) and show that its validity beyond the order $\varepsilon = 4 - d$ is unclear, where d is the spatial dimensionality. In Section V.C, the initial decay rate of the dynamical scattering factor of dilute solutions is calculated. Unfortunately, there is so far no satisfactory renormalization-group theory of semidilute solution dynamics, though we briefly mention semidilute solutions in Section V.B.

In Section VI, universal results are summarized along with many figures. These results are noteworthy because they can be compared with experiments directly without any adjustable parameters. In Section VI.A, our theoretical results for dilute solutions are compared with experimental results. A new procedure to analyze experimental data is proposed. Section VI.B discusses semidilute solutions, and miscellaneous topics are briefly mentioned in Section VI.C.

Section VII is a summary.

Readers who are not interested in theoretical and calculational details may directly go to Section VI. Those who are not interested in calculational details can skip both Sections IV and V.

II. CRITICAL SURVEY OF REPRESENTATIVE THEORIES

In this section, two main approaches, the perturbative approach and the self-consistent or mean-field theoretic approach, are critically surveyed. An intuitive explanation for the necessity of the renormalization-group idea is given in the latter half of the section.

A. Partition Function

Main problems of polymer solution theory have been introduced in Section I.B. For simplicity the static dilute solution problem is studied; that is, the equilibrium statistical mechanics of a single chain is discussed.

The problem of equilibrium statistical mechanics is to calculate the partition function of the system. We assume that the microscopic state of a chain is designated by its conformation $C = \{\mathbf{r}_i\}_{i=1}^{N_0}$, where $\mathbf{r}_i$ is the position of the ith monomer, and N_0 the total number of monomers in a chain. The partition function $G(N_0, \mathbf{R})$ of a chain whose one end is at the origin and the other end at $\mathbf{R}$ can be expressed as

$$G(N_0, \mathbf{R}) = \sum_C W[C] \exp(-\beta V[C]) \tag{1}$$

where $\beta = 1/k_B T$ and $W[C]$ is the statistical weight of the conformation C when all the excluded-volume (or nonbonding) interactions among monomers (or elementary units of a chain) are switched off (i.e., $W[C]$ is the so-called unperturbed statistical weight factor for a chain conformation), and

$V[C]$ is the total interaction energy. Σ implies the sum (or integral) over all the conformations with $\mathbf{r}_1 = \mathbf{0}$ and $\mathbf{r}_{N_0} = \mathbf{R}$. Since the chain is immersed in solvent molecules, $V[C]$ is the effective potential energy after all the degrees of freedom for the solvent molecules have been averaged out.

When we normalize $G(N_0, \mathbf{R})$ so that

$$\int G(N_0, \mathbf{R})\, \mathrm{d}^d\mathbf{R} = 1$$

G is the density distribution function for the end-to-end vector $\mathbf{R}$, where $\mathrm{d}^d\mathbf{R}$ denotes the d-dimensional volume element. Its second moment is the mean-square end-to-end distance. The most famous problem is to calculate the following exponent

$$\lim_{N_0 \to \infty} \frac{\ln\langle R^2 \rangle}{\ln N_0} = 2\nu \tag{2}$$

Although polymer scientists believe in the existence of this ν, even the existence of this limit in d-space ($1 < d \leqslant 4$) is not established. This fact clearly shows how mathematically tough our excluded-volume problem is. Only recently, it was rigorously proved that $\nu = \frac{1}{2}$ if $d > 4$ by Brydges and Spencer.[21]

If the monomer–monomer effective interaction can be assumed to be approximately the superposition of two-body interactions (as is usually claimed) with the potential function given by $\phi(\mathbf{r})$, we can explicitly write $V[C]$ as

$$V[C] = \frac{1}{2} \sum_{\substack{i,j \\ i \neq j}} \phi(\mathbf{r}_i - \mathbf{r}_j) \tag{3}$$

where $i = j$ terms are omitted in the sum so as to eliminate the self-interaction of a monomer. Although ϕ is an effective potential, it has a hard core part and an attractive tail part as does the Lennard-Jones potential. This means that the two-body irreducible cluster integral [see Eq. (5)] may change its sign according to the temperature, but the three-body irreducible cluster integral does not change its sign.[22]

In this article we confine ourselves to nonelectrolyte polymer solutions, so that the effective two-body interaction potential ϕ is short-ranged. It is customary to assume the following form for ϕ:

$$\phi(\mathbf{r}) = k_B T v_0\, \delta(\mathbf{r}) \tag{4}$$

where v_0 is called the excluded-volume parameter and is conventionally identified with the (negative) two-body irreducible cluster integral as

$$v_0 = -\frac{1}{k_B T} \int [\exp(-\beta\phi(\mathbf{r})) - 1] \, d^d\mathbf{r} \tag{5}$$

In Section II.D we will see such an interpretation is far from reality, and we discard this naive interpretation.

As mentioned before, when the quality of the solution becomes poorer, the attractive part of the effective interaction potential becomes more important, so that v_0 decreases toward zero. There is a special temperature where v_0 vanishes, which corresponds to the Boyle temperature for imperfect gases; nothing dramatic happens for imperfect gases at this temperature, because higher-order irreducible cluster integrals are all positive. This temperature is conventionally identified as the theta temperature where ν defined by Eq. (2) becomes $\frac{1}{2}$, the value for a simple random walk. Later we will see that this is also far from reality; at the theta temperature v_0 identified as in Eq. (5) is definitely negative.

In any case, in this section we use the model with the interaction potential given by Eq. (4), which will be justified in Section III.A. We must be a bit more specific about our model. We choose the Gaussian unperturbed weight W, that is,

$$W[C] = \prod_{i=1}^{N_0-1} (2\pi)^{-d/2} \exp\{-\tfrac{1}{2}(\mathbf{r}_i - \mathbf{r}_{i+1})^2\} \tag{6}$$

where d is the spatial dimensionality. Since each component of the end-to-end vector is statistically independent, obeying the Gaussian distribution with the mean zero and the variance N_0, the unperturbed mean-square end-to-end distance $\langle R^2 \rangle_0$ is

$$\langle R^2 \rangle_0 = dN_0 \tag{7}$$

Note that our normalization of the length scale is different from that in most of the existing literature. For any polymer we can always choose the unit of length so that Eq. (7) is satisfied.

So far we have assumed a chain consisting of discrete monomers. However, we are not interested in things happening on the scale of monomers. Therefore, for the simplicity of analysis let us adopt a continuum version of the model first proposed by Edwards.[23] For this model, we have

$$G(N_0, \mathbf{R}) = \int_{\mathbf{c}(0)=\mathbf{0}}^{\mathbf{c}(N_0)=\mathbf{R}} \mathscr{D}[\mathbf{c}] \exp\left[-\frac{1}{2}\int_0^{N_0} \dot{\mathbf{c}}(\tau)^2 \, d\tau - \tfrac{1}{2}v_0 \times \int_0^{N_0} d\tau \int_{\substack{0 \\ |\tau-\sigma|>a}}^{N_0} d\sigma \, \delta(\mathbf{c}(\tau) - \mathbf{c}(\sigma)) \right] \tag{8}$$

where the conformation of a chain is specified by a continuous curve $\mathbf{c}(\tau)$ in d-space parametrized by the contour variable τ (Fig. 1) and $\dot{\mathbf{c}} = d\mathbf{c}/d\tau$. $\mathscr{D}[\mathbf{c}]$ is the uniform measure on the set of all the conformations connecting $\mathbf{0}$ and $\mathbf{R}$ and the integral is the functional integral over this set. The cut-off parameter a in Eq. (8) eliminates the self-interaction; $|\tau - \sigma| > a$ corresponds to the constraint $i \neq j$ in Eq. (3). Those who are not very familiar with path integrals may find some help in Appendix B. Detailed knowledge of path integrals is not necessary in this article except rudiments given in Appendix B. Standard references are Feynman[24] and Gel'fand and Yaglom.[25]

If we introduce the monomer density field ρ defined by (see Section IV.C.7 for more mathematical details)

$$\rho(\mathbf{r}) = \int_0^{N_0} d\tau \, \delta(\mathbf{r} - \mathbf{c}(\tau)) \tag{9}$$

the second term in the exponent of Eq. (8) can be rewritten as

$$\tfrac{1}{2}v_o \int d^d\mathbf{r} \rho(\mathbf{r})^2 \tag{10}$$

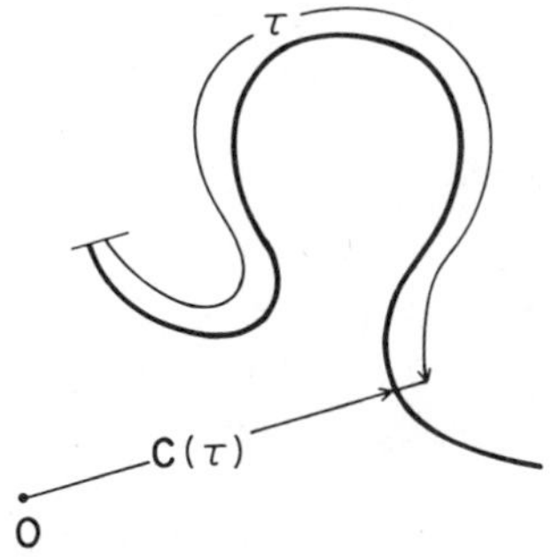

Fig. 1. Parametrization of a conformation $\{\mathscr{C}(\tau)\}_{\tau=0}^{N_0}$. $\mathscr{C}(\tau)$ is the position vector of the "monomer" unit at the contour variable τ.

provided the cutoff is ignored. This form clearly shows the physical meaning of the term.

B. (Bare) Perturbation Theory

According to the conventional interpretation of the theta temperature, $v_0 = 0$ in Eq. (8) at this temperature. Hence it is quite natural to perform perturbative calculations around this "ideal" state with respect to the excluded-volume parameter v_0. This (bare) perturbation approach has been devised by Teramoto,[30] Rubin,[31] Bueche,[32] Zimm et al.,[33] and others[34] and has culminated in the two-parameter theory compiled by Yamakawa.[1]

Ignoring the cutoff parameter a in Eq. (8), we have two parameters, v_0 and N_0, or equivalently v_0 and the mean-square end-to-end distance $\langle R^2 \rangle_0$ which is directly related to N_0 through Eq. (7). The Hamiltonian $\mathscr{H}_E$:

$$\mathscr{H}_E = \frac{1}{2}\int_0^{N_0} \dot{\mathbf{c}}(\tau)^2\, d\tau + \tfrac{1}{2} v_0 \underset{|\tau-\sigma|>a}{\int_0^{N_0} d\tau \int_0^{N_0}} d\sigma\, \delta(\mathbf{c}(\tau) - \mathbf{c}(\sigma)) \tag{11}$$

used to define the partition function [Eq. (8)] (see Appendix B) is invariant under the following reparametrization:

$$\begin{aligned} \mathbf{c} &\to C^{1/2}\mathbf{c} \\ N_0 &\to CN_0 \quad (\tau \to C\tau) \\ v_0 &\to C^{-\varepsilon/2} v_0 \end{aligned} \tag{12}$$

where $\varepsilon = 4 - d$.

Finding such a reparametrization which leaves the Hamiltonian invariant is in the literature often called the *dimensional analysis* of the model. This is because the reparametrization [Eq. (12)] can be reproduced by introducing the following "engineering dimension"

$$[\mathbf{c}] = C^{1/2}, \quad [N_0] = [\tau] = C, \quad [v_0] = C^{-\varepsilon/2} \tag{13}$$

where $[X]$ denotes the dimension of X. Note that each term in the Hamiltonian becomes dimensionless, that is, $[\mathscr{H}_E] = 1$, and so $\mathscr{H}_E$ is invariant under the reparametrization Eq. (12).

Since $[\langle R^2 \rangle] = C$, $\langle R^2 \rangle / N_0$ or $\langle R^2 \rangle / \langle R^2 \rangle_0$ is invariant under the reparametrization Eq. (12) (or is dimensionless). Hence $\langle R^2 \rangle / \langle R^2 \rangle_0$ must be a function of invariant (i.e., dimensionless) parameters. We have only one such parameter (if we ignore a), i.e., $z \propto v_0 N_0^{\varepsilon/2}$ with the proportionality constant of order 1. Hence, we have

$$\langle R^2 \rangle / \langle R^2 \rangle_0 = \alpha_R^2(z) \tag{14}$$

that is, the expansion factor of the end-to-end distance α_R^2 must be a function of z. Therefore, if we perform a perturbation calculation in terms of z, the result is a series in terms of z, which is the famous two-parameter-scheme perturbation series.[1] For the mean-square end-to-end distance $\langle R^2 \rangle$, several terms are known. Very recently Muthukumar and Nickel[35] have given the following result for the 3-space chain

$$\alpha_R^2(z) = 1 + \tfrac{4}{3}z - 2.075z^2 + 6.297z^3 - 25.057z^4 + 116.135z^5 - 594.717z^6 + \cdots \tag{15}$$

(They corrected the coefficient of the z^2-term given by Yamakawa and Tanaka[36] previously.)

The series Eq. (15) corresponds to the high-temperature expansion[37] in magnetic systems from which Domb and others extracted exponents and other universal quantities. Although this approach has not yet been performed actually in polymer physics, it should have been pursued seriously *before* renormalization approaches. However, it is very hard to apply this method to calculate universal functions depending on more than one variable, and there is a fundamental difficulty in the case of dynamical properties as is explained in Section V.B.

In contrast, the conventional use of perturbation series is much more naive. The series, such as Eq. (15), seems to have been considered as a convergent series until the mid-70s.[1,38] Serious doubt about its convergence has been cast by Suzuki et al.[39] numerically and by Edwards[40] and Oono[41,42] analytically (both stimulated by the work of Suzuki et al.). The expansion actually is one about a singular point (Appendix C).

If we are faithful to the conventional interpretation of parameters v_0 and N_0, then v_0 is of order 1, and N_0 is, say, 10^3 so that the parameter z is very big. This clearly shows that the naive microscopic interpretation of v_0 and N_0 is wrong. (About the difficulty of interpreting z microscopically, see Ref. 43.) Thus even as an asymptotic series, the series in terms of the z-parameter is, in practice, useless.

Sometimes the theory with the z-parameter is used to discuss the chain below the theta temperature where v_0 is negative.[1] However, it should be noted that the original model is unstable for $v_0 < 0$; it can be shown[42] rigorously that in the $N_0 \to \infty$ limit with $z \propto v_0 N_0^{1/2}$ being kept finite but negative the chain is almost surely (i.e., if we sample a chain from the equilibrium ensemble, then with probability 1) contained in a small sphere with the radius of order the bond length. Hence for the present model $\alpha_R^2(z) = 0$ for $z < 0$ and $\alpha_R^2(0) = 1$ (in the $N_0 \to \infty$ limit).

So far we have ignored the cutoff parameter a. Mathematically, this is not allowed. Rigorously speaking, Eq. (8) or Eq. (11) is not well defined in the $a \to 0$ limit. The correct interpretation of Eq. (8) suggested by Symanzik[44] (slightly modified from the original form) is

$$G(N_0, \mathbf{R}) = \lim_{a \to 0} \frac{G_a(N_0, \mathbf{R})}{\displaystyle\int G_a(N_0, \mathbf{R})\, \mathrm{d}^d\mathbf{R}} \tag{16}$$

where G_a is given by Eq. (8). This interpretation is justified for sufficiently small v_0 by Westwater[45] in 3-space. Readers may think that Eq. (16) is not different from Eq. (8), since the denominator is a number independent of **R**. However, this number is infinite in the $a \to 0$ limit and so is the numerator. Moreover, precisely speaking the limit must be interpreted as the weak limit (the limit does not exist in the ordinary sense); $G(N_0, \mathbf{R})$ makes sense only inside of the integral sign, or only in the following context

$$\int G(N_0, \mathbf{R}) f(\mathbf{R})\, \mathrm{d}^d\mathbf{R} \equiv \lim_{a \to 0} \frac{\displaystyle\int G_a(N_0, \mathbf{R}) f(\mathbf{R})\, \mathrm{d}^d\mathbf{R}}{\displaystyle\int G_a(N_0, \mathbf{R})\, \mathrm{d}^d R} \tag{17}$$

that is, any expectation value with respect to the density distribution function must be calculated as the $a \to 0$ limit of expectation values for the finite a. We cannot change the order of the limit and the integral in Eq. (17).

Very recently, Bovier et al.[46] have simplified the proof of the existence of the continuum limit discussed above, and have shown that the perturbation theory is asymptotic to all orders in v_0. Analogy to constructive field theoretical results[47,48] suggests that we may assume the Borel summability[49] of the resultant asymptotic series. Then, it is natural to expect that $\langle R^2 \rangle / \langle R^2 \rangle_0 = \alpha_R^2(z) \sim z^{2(2\nu - 1)/\varepsilon}$ asymptotically in the $z \to \infty$ limit.

If we assume, however, this to be true for any v_0, we will have a paradox. Since $z \propto N_0^{1/2} v_0$ in 3-space, we can make $z \to \infty$ by making v_0 very big with N_0 being kept finite (actually this limit is discussed in Bovier et al.[46]). If the cutoff a is kept finite and N_0 is also finite, then $\langle R^2 \rangle$ is finite even in the $v_0 \to \infty$ limit. Thus it is clear that the universal form $\langle R^2 \rangle / \langle R^2 \rangle_0 = \alpha_R^2(z)$ is correct only in the $N_0 \to \infty$, $v_0 \to 0$ limit with $z(\geqslant 0)$ being kept finite.

As we will see later, the renormalization-group theory result shows that $\langle R^2\rangle/\langle R^2\rangle_0$ is not a universal function of one parameter. However, if the excluded-volume parameter is sufficiently small, then $\langle R^2\rangle/\langle R^2\rangle_0$ is virtually universal. Thus the result is fully consistent with the consideration above.

It is customary[50] that the local conformation interaction is included in $\langle R^2\rangle_0$ and only global excluded-volume interactions are studied by the perturbation method. However, it is *not* clear what is local and what is global; there is no clear-cut distinction. This is closely related to the inadequacy of the conventional interpretation of the parameter of the model as we will see in Section II.D.

C. Mean-Field Theory

Historically speaking, the mean-field type appraoch to polymer static properties was devised first. The celebrated formula for ν by Flory[51]

$$\nu = \frac{3}{2+d} \qquad (d \leqslant 4) \tag{18}$$

was derived by this approach. Analogous but different approaches have been devised by Fixman,[52] Kurata et al.,[53] Ptitsyn[54] and others.

In 1965, Edwards modernized the whole approach by introducing the model discussed in Section II.A. His paper[23] was an epoch-making one, which catalyzed innovations in polymer physics. Reiss,[55] de Gennes,[56] Freed,[57] Kyselka,[58] and others have developed or improved this self-consistent field theory. The theory has been extended to more concentrated solutions,[59] to the theta point[60] and to the coil-globule transition.[61] We do not review the whole machinery here. A general reference is Freed.[57] A rather stochastic theoretical approach is given in the appendix of Ref. 60.

The essence of the argument is as follows. Consider the Hamiltonian Eq. (11) and assume that on the average the conformation $\mathbf{c}(\tau)$ has the following scaling property

$$\begin{aligned} \mathbf{c} &\to C^{\nu}\mathbf{c} \\ \tau &\to C\tau \end{aligned} \tag{19}$$

or

$$\mathbf{c}(C\tau) = C^{\nu}\mathbf{c}(\tau) \tag{19a}$$

under which the unperturbed and the interaction parts are scaled similarly.

Since $\dot{\mathbf{c}}(C\tau) \sim C^{\nu-1}\dot{\mathbf{c}}(\tau)$, we have

$$\int \dot{\mathbf{c}}(\tau)^2 \, \mathrm{d}\tau \to C^{2\nu-1} \int \dot{\mathbf{c}}(\tau)^2 \, \mathrm{d}\tau \tag{20}$$

$$v_0 \int \mathrm{d}\tau \int \mathrm{d}\sigma \, \delta(\mathbf{c}(\tau) - \mathbf{c}(\sigma)) \to C^{-\nu d+2} v_0 \int \mathrm{d}\tau \int \mathrm{d}\sigma \, \delta(\mathbf{c}(\tau) - \mathbf{c}(\sigma)) \tag{21}$$

Thus $C^{2\nu-1} = C^{-\nu d+2}$ gives Eq. (18). (If $d > 4$, $\nu = 3/(2+d) < \frac{1}{2}$, which is physically nonsense, so $\nu = \frac{1}{2}$; i.e., there is no contribution from the self-interactions). It is known that the Flory exponent Eq. (18) is very close to the exact values. In 1-space it is (obviously) exact (see Redner and Kang[62]). In 2-space it is believed that Eq. (18) gives the exact value;[63] however, it is rather an unfortunate agreement, which delayed the progress of polymer physics (but see Ref. 68).

To show that the mean-field theory does not work in general, let us consider the following model Hamiltonian

$$\mathscr{H}_2 = \frac{1}{2}\int_0^{N_0} \dot{\mathbf{c}}(\tau)^2 \, \mathrm{d}\tau + \frac{1}{3!} w_0 \underset{\substack{|\tau-\sigma_1|>a,\, |\tau-\sigma_2|>a \\ |\sigma_1-\sigma_2|>a}}{\int \mathrm{d}\tau \int \mathrm{d}\sigma_1 \int \mathrm{d}\sigma_2} \, \delta(\mathbf{c}(\tau) - \mathbf{c}(\sigma_1))\, \delta(\mathbf{c}(\tau) - \mathbf{c}(\sigma_2)) \tag{22}$$

The second term can be rewritten as

$$\frac{1}{3!} w_0 \int \mathrm{d}^d\mathbf{r} \rho(\mathbf{r})^3 \tag{23}$$

if the monomer density defined by Eq. (9) is used. This tells us that the second term is the three-body interaction term. Thus $\mathscr{H}_2$ describes the 2-tolerant walk. The *k-tolerant walk* has been introduced by Malakis[64] for which the walker may return to the same point up to k times but not more than k times. The 1-tolerant walk is the self-avoiding walk. It is clear that for the $k(\geqslant 2)$-tolerant walk, the 2-body irreducible cluster integral is zero. Thus according to the conventional interpretation of the parameter $v_0 = 0$, so that the exponent ν should be $\frac{1}{2}$. The perturbation series for $\langle R^2 \rangle / \langle R^2 \rangle_0$ is carried out in terms of w_0 since $[w_0] = 1$ in 3-space. Because the natural expansion parameter is independent of N_0 a naive argument analogous to that leading to the perturbation theory in Section II.B gives the N_0 independent ratio, $\langle R^2 \rangle / \langle R^2 \rangle_0$. Hence $\nu = \frac{1}{2}$ in 3-space.

Now, let us use the simplified version of the self-consistent field theory. Let us assume that for the most probable conformation **c** has the scaling property given by Eq. (19a). Then the second term of Eq. (22) is transformed to

$$w_0 \int d\tau \int d\sigma_1 \int d\sigma_2\, \delta(\mathbf{c}(\tau) - \mathbf{c}(\sigma_1))\, \delta(\mathbf{c}(\tau) - \mathbf{c}(\sigma_2))$$

$$\to C^{-2\nu d+3} w_0 \int d\tau \int d\sigma_1 \int d\sigma_2 \ldots \qquad (24)$$

Thus from Eqs. (20) and (24) we must have $C^{2\nu-1} = C^{-2\nu d+3}$, that is,

$$\nu = \frac{2}{1+d} \qquad (d \leqslant 3) \qquad (25)$$

This is correct for $d = 1$, and in agreement with the result of the conventional argument mentioned previously. More sophisticated mean-field theories give the same answer: $\nu = \frac{1}{2}$ in 3-space.[60,61]

However, the correct answer is that ν is identical to that for the self-avoiding walk $\nu_{SAW} \simeq \frac{3}{5}$ in 3-space. This conclusion has been reached with the aid of renormalization-group consideration,[65] and exact enumeration[66] supports it. A more formal justification has been given recently.[67] (No rigorous proof is yet known. An intuitive explanation is given in the next section.) Thus the mean-field theory is not correct even qualitatively (but see Ref. 68).

The reason for this incorrectness can be easily understood. In the mean-field theoretic approach ρ^2 is essentially replaced by $\rho\langle\rho\rangle$, where $\langle\rho\rangle$ is the mean monomer density distribution. However, the chain is not a droplet of a soup of monomers; monomers are highly localized and correlated because the chain is connected. In other words, the fluctuation of the monomer density field is so big that we cannot reliably replace it with its average. Thus, although mean-field theory can give suggestive results as in the cases of critical phenomena, conceptually it is completely wrong; it does not take into account the salient features of the system.

If, however, the polymer solution is concentrated, then the fluctuations of the monomer density field are not big. Therefore, it is expected that the mean-field theory can be used for a concentrated solution—the polymer melt. Actually, mean-field theory can show that $\nu = \frac{1}{2}$ for the melt.[59,60] This has been asserted by Flory[20] for a long time. However, there is no known theoretical reason for $\langle R^2\rangle_\Theta$ (the mean-square end-to-end distance at the theta temperature) and $\langle R^2\rangle$ in the melt to be identical. Moreover, as we will see in the next section, $\langle R^2\rangle_\Theta$ is not the same as the "unperturbed" mean-square end-to-end distance $\langle R^2\rangle_0$.

Incidentally, the Flory formula for the exponent Eq. (18) can also be obtained[68] by the interdimensional scaling method of Imry,[69] a non-mean-field argument.

D. Cumulative Effect of Interactions[65]

The purpose of this section is to give an intuitive explanation for the failure of the mean-field approach in more detail. This will be an introduction to the renormalization-group theory explained in the next section.

Since the actual polymer chain is fairly complicated, the true Hamiltonian cannot be given by Eq. (11). Thus what we are doing is not an honest microscopic statistical mechanical calculation; the model is not truly microscopic already at our starting point. Even if we insist that the theory must start from the "true Hamiltonian," there is no way to find it. We have to start from a model Hamiltonian, which is not unique for a given system because it depends on the scale of "resolution" or the level of description. In this section we will see to what extent the model Hamiltonian is not unique. This nonuniqueness is almost fatal to all the conventional arguments.

Let us first consider a chain model with the two-body intrachain interaction analogous to Eq. (4). Let us change the scale of description of the system. Figure 2 explains what we can expect. Even if the two-body interaction parameter is originally v_0 in the model, if we coarse-grain the resolving power of the description of the model, we must change its value, or renormalize it.

Now we can explain the result for the 2-tolerant walk mentioned in Section II.C. The original model assigns zero to the two-body interaction. However, as is shown in Fig. 3, if we coarse-grain the level of the description of the model, effective two-body interactions appear. Thus there is no essential difference between the 2-tolerant walk and the self-avoiding walk so

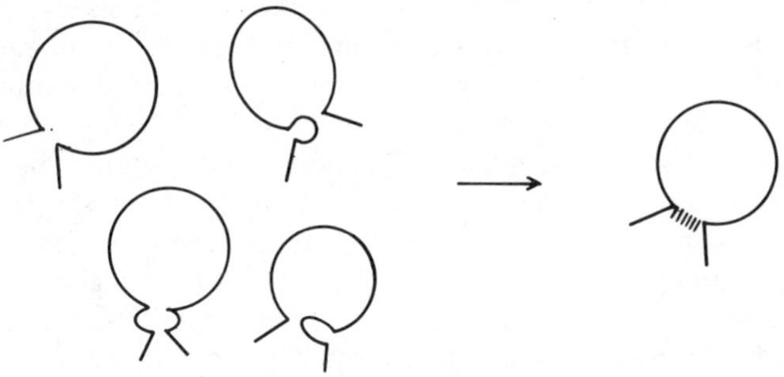

Fig. 2. Coarse-graining and renormalized two-body interactions. Two intrachain interactions closely located in space are regarded as a single effective two-body interaction if the level of the description of the system is coarse-grained.

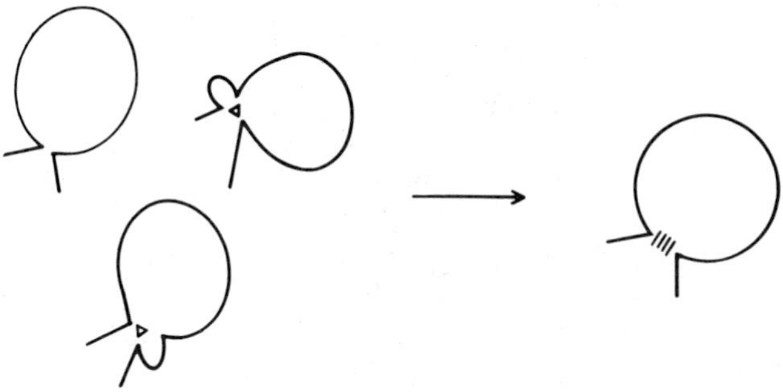

Fig. 3. Even the three-body interaction denoted by the small triangle looks like a two-body interaction if one of the two loops involved is sufficiently small.

long as we are concerned with the global behavior of the walks. We say that the 2-tolerant walk and the self-avoiding walk are in the same universality class. The argument can be extended to general k-tolerant walks; however big k may be, it belongs to the same universality class of the self-avoiding walk. If k is, say, 10^3, this conclusion may be intuitively unacceptable. In this case, to have significant effective two-body interaction we have to coarse-grain the system extensively. To have a long chain even after such extensive coarse-graining, we must have a very long chain at the beginning. For $k = 10^3$, we may need a walk with an unimaginably big number of steps. Thus theoretically the 10^3-tolerant walk is asymptotically indistinguishable from the self-avoiding walk, but for any practical purpose it is a mere random walk.

The preceding paragraph explains why the theta point cannot be identified with the vanishing point of the two-body irreducible cluster integral. To eliminate the global effect of the two-body interaction, the effective two-body interactions built up from many-body interactions must be neutralized by the original two-body interaction. Therefore the original two-body parameter v_0 must be negative at the theta-temperature.

Thus the statistical mechanics of the theta state of a chain is a much more difficult and subtle problem in many-body theory than the ordinary excluded-volume problem.[70–72] Duplantier's detailed calculation[73] shows that

$$\langle R^2 \rangle_\Theta = \langle R^2 \rangle_0 (A - B/\ln N)$$

where A and B are interaction-dependent positive parameters (see also Ref. 74). This casts serious doubt on the theoretical basis of the program of calculating the "unperturbed" end-to-end distance by the matrix manipulation method.[50]

Note that there is no clear-cut distinction between local interactions and global interactions; a local interaction in one description may become global in another description. Thus interactions between points that are separated from each other by any contour distance (the distance along the chain) all contribute to the excluded-volume effect more or less evenly. This also causes a conceptual difficulty for the matrix program mentioned previously.

It is now clear that the theta state cannot be the ideal standard state of a chain (or a solution). There is no theoretical reason for $\langle R^2 \rangle_\Theta$ to be independent of solvents at their theta-temperatures. Theoretically, the self-avoiding walk should be the (reference) unperturbed state of a chain exactly as the hard-core fluid is the reference state of liquids.[75]

As we see in Fig. 4, the contributions from, for example, the four-body

Fig. 4. Coarse-graining makes the four-body interaction denoted by the small square an effective 2- to 4-body interaction depending on how the interacting monomers are connected along the chain.

interactions depend on how the four contacting points are connected along the chain. The difference is reflected in the difference of the local monomer density. Thus the fluctuation of the monomer density field has a very drastic effect. Therefore we again conclude that mean-field theory cannot work successfully.

E. Naive Renormalization[65]

In the previous Section, II.D, we have discussed only the coarse-graining of the model, or the Kadanoff transformation (named after Kadanoff who introduced the block spin in the theory of critical phenomena).[76] Figure 5 illustrates that, if only the Kadanoff transformation is applied, the chain looks different from the original one. However, if the coarse-grained chain is appropriately shrunken, then at least statistically the resultant chain looks similar to a portion of the original chain. This scaling transformation combined with the Kadanoff transformation is called the renormalization-group transformation. We expect that an infinitely long chain is invariant under this transformation. This leads to the idea of renormalization-group theories following Wilson.[78,7] Since we will not use these schemes, we will not go into the details. We must note, however, that a very long chain can be described by highly renormalized Hamiltonian.

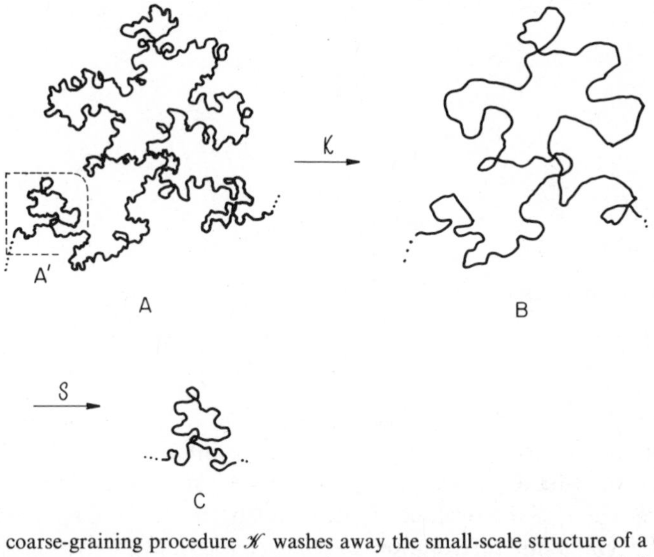

Fig. 5. A coarse-graining procedure $\mathscr{K}$ washes away the small-scale structure of a conformation. A shrinking (or scaling) of the resultant figure produces, at least in a statistical sense, a subconformation A′ in A. This shrinking treatment we denote by $\mathscr{S}$. The combinations of $\mathscr{K}$ and $\mathscr{S}$ is called the renormalization transformation. We regard the chain as a statistically fractal[77] object. The fractal dimension of the chain is given by $1/\nu$, where ν is defined by Eq. (2).

In this section, with the aid of a simple lattice random walk, we show how the renormalization transformation selectively accentuates the global (as opposed to local) structure and properties of a chain.

Consider a simple random walk on a d-dimensional cubic lattice.[79] Let $\mathbf{r}(n)$ be the position of a walker starting from the origin after n-steps, and $\mathbf{a}_n$ be the vector denoting the nth step. Then

$$\mathbf{r}(n) = \sum_{k=1}^{n} \mathbf{a}_k \tag{26}$$

We assume that $\mathbf{a}_k$ $(k = 1, \ldots, n)$ obey the same law of distribution with a finite second moment which we may normalize to be unity. Define a coarse-grained step, combining s steps as

$$\mathbf{b}_m \equiv \mathbf{a}_{(m-1)s+1} + \mathbf{a}_{(m-1)s+2} + \cdots + \mathbf{a}_{ms} \tag{27}$$

Next let us scale this step vector to define $\hat{\mathbf{a}}_m$:

$$\hat{\mathbf{a}}_m \equiv \mathbf{b}_m/\sqrt{s} \tag{28}$$

Define a renormalized random walk by

$$\mathbf{r}^{(1)}(n) = \sum_{k=1}^{n} \hat{\mathbf{a}}_k \tag{29}$$

We know

$$\mathbf{r}^{(1)}(n) = \mathbf{r}(ns)/\sqrt{s} \tag{30}$$

Performing this transformation l times, we get

$$\mathbf{r}^{(l)}(n) = s^{-l/2}\mathbf{r}(ns^l) \tag{31}$$

If the limit $l \to \infty$ exists in a certain sense, then we may say we have a fixed point for this renormalization-group transformation.

According to the central limit theorem[80] (we have assumed that the second moment of the original single step length is finite), we know $\mathbf{r}^{(\infty)}(n)$ obeys the Gaussian distribution. A remarkable fact is that for a simple random walk on the cubic lattice, if n is not very small (say, $n \gtrsim 10$), then the distribution of $\mathbf{r}(n)$ can be well described by that for $\mathbf{r}^{(\infty)}(n)$. For $n = 10$ a numerical comparison is given in Chandrasekhar's review article.[81] This suggests that the behavior obtained from the fixed-point Hamiltonian can well describe the actual system if the chain is not short; that is, n need not be 10^{10} but for $n \sim 10^2$ we may expect a reasonable agreement. There is no theoretical reason known for this happy consequence.

Next, let us study the structure of each trajectory (sample path). For this purpose, let us modify Eq. (31) a bit as

$$\mathbf{r}^{(n)}(t) = \frac{1}{\sqrt{n}}\mathbf{r}([nt]) \tag{32}$$

where [] is the Gauss symbol to denote the largest integer not exceeding the number in it. $\mathbf{r}^{(n)}(t)$ converges[82] (in probability[83]) to the (d-dimensional) Wiener process[84] [the (mathematical) Brownian motion]. From this construction we expect that local details of the lattice random walk are washed away in the Wiener process.

Let us study what properties of the lattice random walk survive the renormalization transformation. Table I summarizes salient features of the lattice random walk and the Wiener process. Being recursive or transitive must be a global property and it survives the transformation. So is $\langle R^2 \rangle \propto N$. However, the occurrence of multiple points is drastically affected by the renormalization. For the lattice random walk, no matter how big the spatial dimensionality is, there is finite expectation values for any multiple point.[85] In contrast to this, if the spatial dimensionality is 3, there is almost surely no triple points (triple collision of a single chain)[86a] for the Wiener process. This implies that all the higher-order interactions are essentially local ones, only binary collisions can affect the global properties of a chain.

TABLE I
Features of the Lattice Random Walk and the Wiener Process

Lattice random walk	Wiener process
$d = 2$ recursive $d > 2$ transitive	$d = 2$ recursive $d > 2$ transitive
The expectation value of the number of m-tuple points is proportional to the total number of steps (except logarithmic corrections for $d = 2$) for any $d \geqslant 2$.	For $d = 2$, with probability 1, there exist m-tuple points for any m (>0). For $d = 3$, with probability 1, there are double points, but again with probability 1, there are no higher-order multiple points. For $d \geqslant 4$, with probability 1, there exists no multiple point.
$\langle R^2 \rangle \propto N$	$\langle R^2 \rangle \propto N$

For 4-space, there is almost surely no multiple point for the Wiener process.[86b] Thus we may intuitively expect $\nu = \frac{1}{2}$ for $d \geqslant 4$; for $d > 4$, as has been mentioned previously, this is proved by Brydges and Spencer.[21]

The preceding paragraph may be a bit misleading and may apparently be contradictory to what is asserted in Section II.D. The precise implication is as follows. Since the Wiener process is the end product (=fixed point) of the renormalization transformation, the binary collision in it must be the completely renormalized binary collision. The true meaning of the statement "only binary collisions can affect the global properties" is that only through renormalized binary collisions are global properties affected by self-interactions.

III. OUTLINE OF RENORMALIZATION-GROUP APPROACH

We have seen in the previous section that our model Hamiltonians are not at all unique. Then, what can we unambiguously predict from such a model as defined by the Hamiltonian Eq. (11)? We can at best hope to achieve a phenomenological description of polymer solutions.

In this section, the renormalization-group approach is explained as a method to extract the large-scale phenomenology of our model. This section gives a rather "philosophical" argument. All the details will be given along with concrete examples in the following two sections.

A. Phenomenology, Universality, and Minimal Models

We are interested in macroscopic properties, that is, the thermodynamic and long-length and time-scale properties (e.g., transport coefficients) of polymer solutions. We want to derive the phenomenological description of solutions.

Since it is impossible to know the "true" microscopic Hamiltonian, we must abandon the calculation of the microscopic constants specific to the chemical species. Generally speaking, it is impossible to calculate such constants as the absolute value of $\langle R^2 \rangle$ truly microscopically. However, if the solution can be described phenomenologically by a few empirical parameters, there is hope to derive the description from the model Hamiltonian. This is what we can at most hope to achieve.

Generally, phenomenological theories contain a few parameters. For example, the Debye theory of the specific heat of solids contains the Debye temperature as a material constant, and the Navier-Stokes equation contains, for example, bulk viscosity. These phenomenological parameters depend on the microscopic details of the systems. However, *the structure of the phenomenological theory or equation is universal; it is independent of the microscopic details of a class of systems.* We call this class a *universality class.*

For example, with only a change of parameters the Navier-Stokes equation can be used for liquid water as well as for air.

To derive the structure of a phenomenological theory, any model in a (universality) class can be used. But there must be the most convenient or conceptually simplest microscopic model, which we call the *minimal model.* We have only to study this model in order to find the phenomenological description of systems in the universality class. (Here we pretend that the universality class is given as a well-defined set. Actually, the boundary of the set is fuzzy and we often do not know precisely what characterizes the universality class. See the explanation later.)

If we can successfully absorb the microscopic details of our microscopic model into a few phenomenological parameters, we say the model is renormalizable. Thus the renormalizability of the model implies the existence of a well-behaved phenomenological description of the model. (The usage of *renormalizability* here is much more vague and generalized than the standard usage.)

Generally, we may proceed as follows. First, we expect the existence of a phenomenological description of the systems (in our case, polymer systems) from experimental results or from theoretical argument. Anyway, non-existence of any phenomenological description implies that we are forced to use a special description for each system, in which case there will be not much theoretical interest in studying such systems.

Once we expect the existence of a certain phenomenological description, all the systems in the class under consideration should give the same phenomenology. Therefore we try to find the minimal model for the class. This class (the universality class) is not well defined at the beginning of the study, and is made more well-defined later through the feedback process between theory and experiments (as has been the case in the critical phenomena).

In the case of a single polymer chain, the most important property is that the chain is a connected object. The simplest way to incorporate the connectivity is to use a random walk as the unperturbed chain. Since we are interested in the large-scale behavior, we may replace it by the Wiener process as we have seen in Section II.E. Another important feature of the chain is the short-range intrachain interaction. The formally simplest way to model this is by use of the δ-function. What we then have obtained as our minimal model is the Edwards model.[23] Although we adopt a continuum model, we must still respect the discrete atomic nature of the chain. Therefore, we need a cutoff a to eliminate an unphysical self-interaction of the chain unit. Thus our model Hamiltonian is Eq. (11). The parameter a represents the microscopic scale of description of the chain. If we change a, other parameters should be readjusted accordingly as we have discussed in Section II.D.

For nondilute polymer solutions, many chains must be considered simultaneously, but the extension of the Edwards model to these cases is straightforward. This is discussed in Section IV.C.

For the dynamics of dilute solutions, not only the chain connectivity and the intrachain interaction, but the hydrodynamic effect are important, since the chain is immersed in a solvent. The minimal model for this case is discussed in Section V.B.

B. Renormalization-Group Equation

Let us consider the calculation of an (equilibrium) quantity Q. We assume that a statistical mechanical procedure for the calculation is known. We denote the result of a statistical mechanical calculation for this quantity by Q_B; Q_B depends on parameters in the model Hamiltonian (we call them *bare parameters*), especially on the cutoff parameter a. Varying a corresponds to changing the model Hamiltonian.

In the resultant phenomenological description of the chain, we expect to find at least two parameters: one is the parameter N, describing the total length of the chain, and the other is v, describing the quality of the solvent. The parameter N and its microscopic counterpart N_0 must be proportional to M, the molecular weight, so that N must be proportional to N_0. The parameter v must be a function of v_0. From the dimension-analytic argument, there must be one more parameter, the phenomenological length scale L, which is the phenomenological counterpart of the cutoff a. Very crudely, we may regard a as the size of the chain unit. Thus this a is closely related to the proportionality factor between the molecular weight M and the bare parameter N_0. We need an analog related to the proportionality factor between M and the phenomenological parameter N; this is the length scale L. The molecular weight is experimentally observable, but the number of units N (or N_0) is not observable; we do not know what the unit is; the unit cannot be defined unambiguously. (Even the statistical unit, the Kuhn unit in conventional theories, is different from the monomer unit in the synthetic- or structural-chemistry sense.) This implies that the choice of N (equivalently L) is arbitrary; we can change L with N_0, v_0, and a being kept constant. Equivalently, we have the following equation:

$$L \frac{\partial}{\partial L} Q_B|_{N_0, v_0, a} = 0 \tag{33}$$

For convenience, let us introduce dimensionless parameters u_0 and its phenomenological counterpart u by [see Eq. (13)]

$$u_0 = v_0 L^{\varepsilon/2} \tag{34}$$

$$u = vL^{\varepsilon/2} \tag{35}$$

The relation between N and N_0 can be written as

$$N = Z_N(u_0, a/L)N_0 \tag{36}$$

where we have used the proportionality between N and N_0. Z_N must be dimensionless, so that it can depend only on dimensionless parameters, u_0, a/L and other dimensionless parameters if they exist (e.g., as in the case of the solution dynamics; see Section V.A). The parameter u must be a function of u_0, and other dimensionless parameters,

$$u = u(u_0, a/L) \tag{37}$$

If Q is a directly observable quantity without any prefactor such as the mean-square end-to-end distance $\langle R^2 \rangle$, then numerically Q_B must be equal to Q. If Q is the distribution function (not normalized), we may have a prefactor such that

$$Q = Z_Q^{-1} Q_B \tag{38}$$

The multiplicative constants Z_Q, Z_N are called the *renormalization constants.* The relation Eqs. (36) and (37) are independent of quantities we calculate, if the model and the scheme of calculation are fixed. Thus, for example, if we get Z for one quantity, we need not calculate it in the calculation of any other quantities. Generally, we have

$$Q = Z_Q^{-1} Q_B(u_0(u, a/L), Z_N^{-1}N, a) \tag{39}$$

We are not interested in what happens at the atomic scale level. This implies that we are interested in the limit $a \to 0$. In this phenomenological limit, we expect

$$Q = Q(N, u, L) \tag{40}$$

that is, Q can be described by the phenomenological parameters only. Thus Q_B has the following expression

$$Q_B = Z_Q Q(Z_N N_0, u_0(u), L) \tag{41}$$

where $u_0(u)$ is the relation between u and u_0 determined by Eq. (37).

Combining Eqs. (41) and (33), we get

$$L\frac{\partial}{\partial L} Z_Q Q(Z_N N_0, u_0(u), L)|_{N_0, u_0, a} = 0 \tag{42}$$

Even in the $a \to 0$ limit Z_Q, Z_N, and u_0 (or u) are L-dependent, since they depend on u_0 which is defined by Eq. (34). With the aid of the chain rule of differentiation, we have

$$\left[\left(L\frac{\partial}{\partial L} Z_Q\right) + Z_Q L\frac{\partial}{\partial L}\right] Q = Z_Q\left\{L\left(\frac{\partial \ln Z_Q}{\partial L}\right) + L\frac{\partial}{\partial L} + L\left(\frac{\partial u}{\partial L}\right)\frac{\partial}{\partial u} + \left(\frac{\partial Z_N N_0}{\partial L}\right)\frac{\partial}{\partial N}\right\} Q = 0 \tag{43}$$

Note that

$$\left.\frac{\partial Z_N N_0}{\partial L}\right|_{N_0} = \frac{\partial Z_N}{\partial L} N_0 = N\frac{\partial \ln Z_N}{\partial L}$$

Hence we get from Eq. (43) the following partial differential equation which dictates the functional form of the phenomenological description of Q:

$$\left[L\frac{\partial}{\partial L} + \beta(u)\frac{\partial}{\partial u} + \gamma_N(u) N\frac{\partial}{\partial N} + \gamma_Q(u)\right] Q = 0 \tag{44}$$

where

$$\beta(u) = L(\partial u/\partial L)_{N_0, v_0, a \to 0} \tag{45}$$

$$\gamma_N(u) = L(\partial \ln Z_N/\partial L)_{N_0, v_0, a \to 0} \tag{46}$$

$$\gamma_Q(u) = L(\partial \ln Z_Q/\partial L)_{N_0, v_0, a \to 0} \tag{47}$$

Equation (44) is called the *renormalization-group equation*[87] for Q. For each macroscopically observable quantity, we can construct such an equation, provided a well-defined phenomenological functional form like Eq. (40) can be expected. The functions β and γ_N are independent of Q, so that the renormalization-group equations for various quantities look pretty much the same.

In the phenomenological limit $a \to 0$, Q_B may not be finite. This implies that this quantity Q is quite sensitive to the microscopic details of the model. Even in this case, with appropriate choices of the renormalization constants $Z_Q, Z_N, \ldots$, we may get a finite well-defined functional form $Q(N, u, L)$. In this case, inevitably $Z_Q, Z_N, \ldots$, become unbounded. The physical implication of this is as follows. The quantity Q is quite sensitive to the microscopic details of the model; nevertheless, the microscopic details can be absorbed into the phenomenological parameters, so that the functional form itself is insensitive to the microscopic details.

In the conventional two-parameter theory,[1] two-parameter phenomenology is expected so that they adopted the microscopic description with two parameters. Hence it is obvious that the two-parameter theory cannot explain why there are only two important phenomenological parameters. We must stress that our consideration is diametrically different from such a short-circuit one. As has been explained in Sections II.D and II.E, there is a convincing reason to believe that even if a microscopic model contains more than two parameters, we need eventually two parameters to describe the renormalized (i.e., phenomenological) result. This is the reason why we choose the model that contains only two parameters. Note that the argument in Sections II.D and II.E *explains* the two-parameter phenomenology. Without a renormalization-group argument, this is impossible.

C. Renormalized Perturbation Theory and ε-Expansion

As has been explained in Section II.B, the natural expansion parameter for the Edwards Hamiltonian Eq. (11) is $v_0 N_0^{\varepsilon/2}$, where $\varepsilon = 4 - d$. Hence in 3-space we have trouble in the $N_0 \to \infty$ limit. In 4-space, $\varepsilon = 0$, so that we do not have any difficulty. However, we are in 3-space. Thus if we want to perform a perturbation calculation of the excluded-volume interaction, we must expand not only in v_0 but in ε also. Therefore we have the following double expansion for Q_B

$$Q_B = \sum_m Q_m^0(\varepsilon) v_0^m \tag{48}$$

$$Q_m^0(\varepsilon) = \sum_n Q_{mn}^0 \varepsilon^n \tag{49}$$

where the coefficient Q_{mn}^0 is calculated in 4-space. As we will see in concrete examples, $Q_m^0(\varepsilon)$ is singular (in the cutoff $a \to 0$ limit) at $\varepsilon = 0$. The physical origin of this singularity is the sensitivity of the quantity Q_m^0 to the microscopic details (see Section IV.A for more details).

We expect the phenomenological functional form of Q to be expandable in an asymptotic expansion,

$$Q = \sum_m Q_m(\varepsilon) v^m \tag{50}$$

$$Q_m(\varepsilon) = \sum_n Q_{mn} \varepsilon^n \tag{51}$$

where Q_m depends on other renormalized parameters also. In this expansion, the coefficients are believed to be well defined, since the origin of the divergence of $Q_m^0(\varepsilon)$ is in the sensitivity to the microscopic details of the model which should have been successfully eliminated by the introduction of renormalization constants Z_Q, Z_N and the $v_0 - v$ relation.

The preceding paragraph suggests how to determine Z_Q, Z_N and v (or u): absorb order by order in v (or in u) the principal part (i.e., $n < 0$ terms) of the Laurent expansion [Eq. (49)]. The resultant expansion [Eq. (50)] is called the *renormalized perturbation expansion.*[88] The most important feature of this is that v is of order ε, so even in 3-space v is of order 1 (see the next section). Although it is believed that Eq. (50) is divergent, we may use this as an asymptotic expansion. Since v is of order unity, the expansion can be useful in practice. This is in marked contrast to the bare perturbation expansion explained in Section II.C.

What we will do to estimate the phenomenological functional form is to compare the renormalized perturbation result and the functional form dictated by the renormalization-group equation [Eq. (44)].

D. Fixed Points and Scaling Concepts

For the mean-square end-to-end distance $\langle R^2 \rangle$, the renormalization-group equation [Eq. (44)] reads

$$\left[L\frac{\partial}{\partial L} + \beta(u)\frac{\partial}{\partial u} + \gamma_N(u) N \frac{\partial}{\partial N} \right] \langle R^2 \rangle = 0 \tag{52}$$

where $\gamma_{\langle R^2 \rangle} = 0$ because, as has been explained in Section III.B, $\langle R^2 \rangle$ is directly observable (in contrast to N). As we will see in Section IV.A, we can calculate β and γ_N as follows to order ε,

$$\beta(u) = \pi^{-2} u(u^* - u) \tag{53}$$

$$\gamma_N(u) = \frac{u}{(2\pi)^2} \tag{54}$$

where $u^* = \varepsilon\pi^2/2$.

To solve Eq. (52), let us introduce the characteristic equations following the standard method to solve quasilinear partial differential equations,[89]

$$\frac{\mathrm{d}L}{L} = \frac{\mathrm{d}u}{\beta(u)} = \frac{\mathrm{d}N}{N\gamma_N(u)} = \frac{\mathrm{d}t}{t} \tag{55}$$

where t, the dilation factor, is introduced to parametrize the solution. Equation (55) is equivalent to the following set of equations:

$$\frac{\mathrm{d}L}{\mathrm{d}t} = \frac{L}{t} \tag{56}$$

$$\frac{\mathrm{d}u}{\mathrm{d}t} = \frac{\beta(u)}{t} \tag{57}$$

$$\frac{\mathrm{d}N}{\mathrm{d}t} = \frac{N\gamma_N(u)}{t} \tag{58}$$

From Eq. (57) we get

$$u = u^* \frac{At^{\varepsilon/2}}{1 + At^{\varepsilon/2}} \tag{59}$$

where A is a constant (which must be positive). Equations (54), (59), and (58) tell us that N is an increasing function of t. Thus the $N \to \infty$ limit is realized in the $t \to \infty$ limit. In this limit, Eq. (59) gives $u \simeq u^*$. This implies that the phenomenology of a very long chain is governed by the nontrivial zero of $\beta(u)$. Since u^* is of order ε, for a very long chain the dimensionless interaction parameter is almost of order ε. Hence, the expansion Eq. (50) becomes a series in ε(the ε-expansion).[88] Note that if A is very small, even for a fairly long chain, u can be far smaller than u^*. This happens if the solvent is not good (if the solution is close to the theta state). This observation opens up the possibility of describing the crossover behavior as we will see in Section IV.B.

In this subsection, let us confine ourselves to the good-solvent limit. Then, for a sufficiently long chain we may assume $u \simeq u^*$. Hence we may put $\beta \simeq 0$ and $\gamma_N = \gamma^* = \varepsilon/8$ in Eq. (52) to obtain

$$\left(L\frac{\partial}{\partial L} + \gamma^* N \frac{\partial}{\partial N}\right)\langle R^2\rangle = 0 \tag{60}$$

The general solution is

$$\langle R^2 \rangle = f(LN^{-1/\gamma^*}) \tag{61}$$

where f is an as yet undetermined well-behaved function.

From the reparametrization invariance of the Edwards Hamiltonian discussed in Section II.B, we have

$$C\langle R^2 \rangle = f(C^{1-1/\gamma^*}LN^{-1/\gamma^*}) \tag{62}$$

where C is an arbitrary positive number. Let us choose C so that

$$C^{1-1/\gamma^*}N^{-1/\gamma^*}L = 1 \quad \text{or} \quad C = N^{1/(\gamma^*-1)}L^{-\gamma^*/(\gamma^*-1)} \tag{63}$$

Thus we get

$$\langle R^2 \rangle = f(1)L^{\gamma^*/(\gamma^*-1)}N^{1/(1-\gamma^*)} \tag{64}$$

Hence we can identify

$$(1-\gamma^*)^{-1} = 2\nu \quad \text{or} \quad \gamma^* = 1 - \frac{1}{2\nu} \tag{65}$$

The argument in the previous paragraph is a typical one to determine asymptotic relations. We will encounter more complicated examples in Sections IV and V. Since we have used the results of the renormalized perturbation [Eqs. (53) and (54)], our argument seems to be contaminated with the perturbative calculation. However, there are good reasons[90] to believe that $\beta(u)$ and $\gamma_N(u)$ qualitatively behave just as Eqs. (53) and (54). Thus our argument is free from our primitive technology. The renormalization-group equation is the *only* way to logically and naturally derive the asymptotic power laws. Once this functional form is derived, there may be methods other than renormalization-group theories to estimate actual values of exponents. But these other methods are not logically self-contained.

IV. STATIC PROPERTIES

In this section, the application of the renormalization-group theory to static properties (thermodynamic and static correlations) is explained with the aid of two representative examples, the end-to-end vector distribution function for a single chain in a dilute solution, and the osmotic pressure of the semidilute polymer solution.

A. Density Distribution Function for the End-to-End Vector of the Self-Avoiding Chain[91a]

Let $G_B(N_0, \mathbf{R})$ be the (bare) partition function for a chain of total contour length N_0 whose end-to-end vector is $\mathbf{R}$. If this is normalized so that with the normalization constant $\mathcal{N}_0$

$$\mathcal{N}_0^{-1} \int G_B(N_0, \mathbf{R})\, \mathrm{d}^d\mathbf{R} = 1 \tag{66}$$

$\mathcal{N}_0^{-1} G_B(N_0, \mathbf{R})$ is the density distribution function for the end-to-end vector of the self-avoiding chain.

We use, as the minimal model, the Edwards Hamiltonian $\mathscr{H}_E$ [Eq. (11)]. Then G_B can be written as (see Appendix B)

$$G_B(N_0, \mathbf{R}) = \int_{\mathbf{c}(0)=\mathbf{0}}^{\mathbf{c}(N_0)=\mathbf{R}} \mathscr{D}[\mathbf{c}] \exp(-\mathscr{H}_E) \tag{67}$$

1. *Bare Perturbation*

From Eq. (67) we can straightforwardly expand the interaction as

$$\begin{aligned} G_B(N_0, \mathbf{R}) = \int_{\mathbf{c}(0)=\mathbf{0}}^{\mathbf{c}(N_0)=\mathbf{R}} \mathscr{D}[\mathbf{c}] \exp\left(-\frac{1}{2}\int_0^{N_0} \dot{\mathbf{c}}^2\, \mathrm{d}\tau\right)\Bigg[1 - \frac{1}{2} v_0 \int \mathrm{d}\tau \int \mathrm{d}\sigma \\ \times \delta(\mathbf{c}(\tau) - \mathbf{c}(\sigma)) + \frac{1}{2}\left(\frac{v_0}{2}\right)^2 \int \mathrm{d}\tau \int \mathrm{d}\sigma \int \mathrm{d}\tau' \int \mathrm{d}\sigma' \\ \times \delta(\mathbf{c}(\tau) - \mathbf{c}(\sigma))\, \delta(\mathbf{c}(\tau') - \mathbf{c}(\sigma')) + \cdots \Bigg] \end{aligned} \tag{68}$$

Using the Markovian property of the Wiener process [or essentially Eq. (B.5) in Appendix B], Eq. (68) can be written as

$$\begin{aligned} G_B(N_0, \mathbf{R}) = G_0(N_0, \mathbf{R}) - v_0 \int_0^{N_0} \mathrm{d}\tau \int_0^{N_0} \mathrm{d}\sigma \int \mathrm{d}^d\mathbf{r}\, G_0(N_0 - \tau, \mathbf{R} - \mathbf{r}) \\ \times G_0(\tau - \sigma, \mathbf{0}) G_0(\sigma, \mathbf{r}) + O(v_0^2), \end{aligned} \tag{69}$$

where G_0 is the Gaussian distribution function defined by

$$G_0(N_0, \mathbf{R}) = (2\pi N_0)^{-d/2} \exp(-\tfrac{1}{2}R^2/N_0) \tag{70}$$

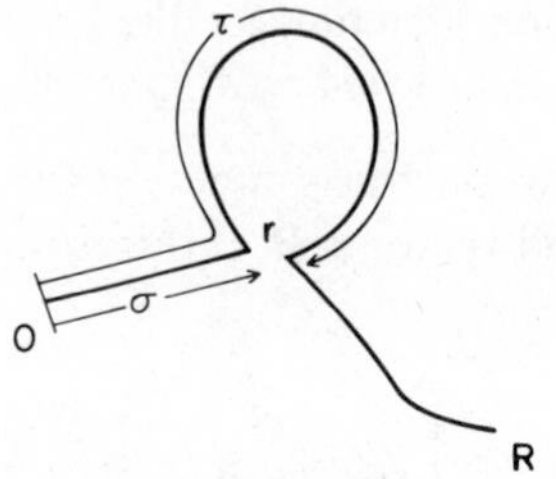

Fig. 6. Explanation of the variables in the second term of Eq. (69). **r** denotes the point of the excluded-volume interaction.

The parametrization of the second term of Eq. (69) is explained in Fig. 6. As has been explained, after renormalization, the interaction parameter becomes of order ε, so that for the calculation to order ε, $O(v_0)$ is sufficient.

Integrating over **r** and reparametrizing the loop circumference as s, we can rewrite Eq. (69) as

$$G_B(N_0, \mathbf{R}) = G_0(N_0, \mathbf{R}) - u_0 L^{-\varepsilon/2} I \tag{71}$$

with

$$I = \int_0^{N_0} \mathrm{d}s(N_0 - s) G_0(N_0 - s, \mathbf{R}) G_0(s, \mathbf{0}) \tag{71a}$$

where we have used the dimensionless coupling constant introduced in Eq. (34). If we introduce the variables

$$t = \frac{s}{N_0 - s} \tag{72}$$

and

$$\alpha = \frac{R^2}{2N_0} \tag{73}$$

then Eq. (71a) can be rewritten as

$$I = (2\pi)^{-d} N_0^{2-d} \mathrm{e}^{-\alpha} \mathrm{G}(\alpha), \tag{74}$$

where

$$G(\alpha) = \int_0^\infty dt(1+t)^{1-\varepsilon} t^{-2+\varepsilon/2} e^{-\alpha t} \tag{75}$$

We must calculate $G(\alpha)$ to order ε^0, since $G(\alpha)$ appears with u_0, which after renormalization becomes of order ε.

Let us calculate $G(\alpha)$ following Baldwin[92] whose method simplifies the calculation given in Ref. 91a.

$$\begin{aligned} G(\alpha) &= \int_0^\infty dt(1+t)^{1-\varepsilon} e^{-\alpha t} \frac{2}{\varepsilon} \frac{1}{(\varepsilon/2)-1} \frac{d^2}{dt^2} t^{\varepsilon/2} \\ &= (1+t)^{1-\varepsilon} e^{-\alpha t} \frac{2}{\varepsilon} \frac{1}{(\varepsilon/2)-1} \frac{d}{dt} t^{\varepsilon/2} \Big|_0^\infty - \int_0^\infty dt \frac{d}{dt} \\ &\quad \times [(1+t)^{1-\varepsilon} e^{-\alpha t}] \frac{2}{\varepsilon} \frac{1}{(\varepsilon/2)-1} \frac{d}{dt} t^{\varepsilon/2} \end{aligned} \tag{76}$$

Note that the first term due to the integration by parts is infinitely big. However, this divergence comes from the $t \to 0$ limit, so that it is independent of α, that is, a pure constant. This can be eliminated by changing the origin of the free energy (this corresponds to the mass subtraction in the field theoretic approaches[88]). Ignoring the harmless first term, we get

$$G(\alpha) = -\int_0^\infty dt \frac{d}{dt} [(1+t)^{1-\varepsilon} e^{-\alpha t}] \frac{2}{\varepsilon} \frac{1}{(\varepsilon/2)-1} \frac{d}{dt} t^{\varepsilon/2} \tag{77}$$

Again integrating this by parts, we get

$$G(\alpha) = \int_0^\infty dt \left\{ \frac{d^2}{dt^2} [(1+t)^{1-\varepsilon} e^{-\alpha t}] \right\} t^{\varepsilon/2} \frac{2}{\varepsilon} \left(\frac{\varepsilon}{2} - 1 \right)^{-1} \tag{78}$$

Since the integral is regular even at $\varepsilon = 0$, we may expand this in terms of ε before integration to get

$$\begin{aligned} \frac{\varepsilon}{2}\left(\frac{\varepsilon}{2} - 1\right) G(\alpha) &= \int_0^\infty dt \frac{d^2}{dt^2} [(1+t) e^{-\alpha t}] - \varepsilon \int_0^\infty dt \frac{d^2}{dt^2} [(1+t) e^{-\alpha t} \\ &\quad \times \ln(1+t)] + \frac{\varepsilon}{2} \int_0^\infty dt \frac{d^2}{dt^2} [(1+t) e^{-\alpha t}] \ln t + O(\varepsilon^2) \end{aligned} \tag{79}$$

The remaining calculation is straightforward; we get

$$G(\alpha) = \frac{2}{\varepsilon}(1 - \alpha) - 2 - \alpha + (\alpha - 1)(\hat{\gamma} + \ln \alpha) + O(\varepsilon) \tag{80}$$

where $\hat{\gamma}$ is Euler's constant, $\hat{\gamma} = 0.577\ldots$. Note that Baldwin's scheme gives the order ε^{-1} term very simply. Thus, combining Eq. (80) with Eqs. (74) and (71), we get the bare perturbation result for the (yet unnormalized) density distribution function for the end-to-end vector **R** of the self-avoiding walk;

$$G_B(N_0, \mathbf{R}) = G_0(N_0, \mathbf{R})\left\{1 - u_0 L^{-\varepsilon/2}(2\pi)^{-d/2} N_0^{\varepsilon/2} \times \left[\frac{2}{\varepsilon}(1 - \alpha) - 2 - \alpha + (\alpha - 1)(\hat{\gamma} + \ln \alpha)\right]\right\} + O(\varepsilon) \tag{81}$$

where we have used $G_0 = (2\pi N_0)^{-d/2} \exp(-\alpha)$.

As we see in Eq. (80), $G(\alpha)$ has a first order pole in ε at $\varepsilon = 0$, which is the manifestation of the sensitivity of the perturbation terms to the microscopic details of the model as has been discussed in Section III.B. Following the prescription for the renormalized perturbation given in Section III.C, these singularities must be absorbed into the renormalization constants.

2. *Renormalization*

We introduce the phenomenological counterpart N, u of model parameters N_0, u_0. N must be proportional to N_0, and u must be a function of u_0 as

$$N = Z_N N_0 \equiv (1 + u\hat{B} + \cdots)N_0 \tag{82}$$

$$u = u(u_0) \equiv u_0(1 - \hat{D}u_0 + \cdots) \tag{83}$$

where $\hat{B}$ and $\hat{D}$ are yet unspecified numerical coefficients. Since G is not directly observable (because it is not normalized), we may have a prefactor such that

$$G(N, \mathbf{R}) = Z_G^{-1} G_B(N_0, \mathbf{R}) \equiv (1 + \hat{A}u + \cdots)^{-1} G_B \tag{84}$$

Note that **R** needs no renormalization, since it is a directly measurable quantity (at least in principle).

We have only to regularize G to order u (or u_0). Hence we may simply replace N_0 by N in the terms multiplied with u_0 in Eq. (81), but N_0 in G_0 must be carefully calculated as

$$G_0(Z_2^{-1}N, \mathbf{R}) = G_0(N, \mathbf{R})\left[1 - \hat{B}u\left(\frac{\varepsilon - 4}{2} + \alpha\right) + \cdots\right] \tag{85}$$

where we have replaced u_0 by u. Thus Eq. (84) can be written explicitly as

$$G(N, \mathbf{R}) = (1 - \hat{A}u + \cdots)G_0(N, \mathbf{R})\left\{1 - \hat{B}u\left(\frac{\varepsilon - 4}{2} + \alpha\right) - uL^{-\varepsilon/2}N^{\varepsilon/2}(2\pi)^{-2+\varepsilon/2} \times \left[\frac{2}{\varepsilon}(1 - \alpha) - 2 - \alpha + (\alpha - 1)(\hat{\gamma} + \ln \alpha)\right]\right\} \tag{86}$$

$\hat{A}$ and $\hat{B}$ are so chosen to eliminate the singularities ε^{-1}. Comparing the order u terms, we get the following requirement

$$-\hat{A} - \hat{B}\left(\frac{\varepsilon - 4}{2} + \alpha\right) - L^{-\varepsilon/2}N^{\varepsilon/2}(2\pi)^{-2+\varepsilon/2}\left[\frac{2}{\varepsilon}(1 - \alpha) + \cdots\right] = O(\varepsilon^0) \tag{87}$$

that is, ignoring $O(\varepsilon^0)$ terms, we obtain

$$\hat{A} - 2\hat{B} + \alpha\hat{B} + \frac{1}{(2\pi)^2}\frac{2}{\varepsilon}(1 - \alpha) = 0 \tag{88}$$

This must be an identity in α, so we get

$$\hat{A} = \hat{B} = (2\pi^2\varepsilon)^{-1} \tag{89}$$

Putting this into Eq. (86), we finally obtain the renormalized perturbation expansion of the (unnormalized) density distribution function for the end-to-end vector $\mathbf{R}$ as

$$G(N, \mathbf{R}) = \left(\frac{1}{2\pi N}\right)^{2-\varepsilon/2} e^{-\alpha}\left\{1 + \frac{u}{(2\pi)^2}\left[1 + \alpha + (1 - \alpha)(\hat{\gamma} + \ln \alpha) - (1 - \alpha)\ln\frac{2\pi N}{L}\right]\right\} \tag{90}$$

3. *Renormalization of the Interaction Parameter*

To calculate $\beta(u)$ in the renormalization-group Eq. (44) we need $\hat{D}$ in Eq. (83). This is the first-order correction to u_0. To get Eq. (83), let us calculate the effective binary interaction u_1. The intuitive idea has already been explained in Section II.D. We have only to actually calculate the contribution from the effective two-body diagrams. The following formula may be obvious from the parametrization given in Fig. 7.

$$u_1 = u_0 - 4u_0^2 \int d^d\boldsymbol{\rho} \int_0^1 dx \int_0^1 dy G_0(x, \boldsymbol{\rho}) G_0(y, \boldsymbol{\rho}) \tag{91}$$

we need only the singularity, so that we may arbitrarily choose the upper limit of the integration range. The integral is singular at $\varepsilon = 0$. This singularity must be absorbed into the definition of $\hat{D}$. Introducing Eq. (83) into Eq. (91), we get

$$u_1 = u + \hat{D}u^2 - 4u^2 \int d^p\boldsymbol{\rho} \int_0^1 dx \int_0^1 dy G_0(x, \boldsymbol{\rho}) G_0(y, \boldsymbol{\rho}) \tag{92}$$

Comparing the order u^2 terms, we can fix $\hat{D}$ as

$$\hat{D} = 4\left[\int d^d\boldsymbol{\rho} \int_0^1 dx \int_0^1 dy G_0(x, \boldsymbol{\rho}) G_0(y, \boldsymbol{\rho})\right]_s \tag{93}$$

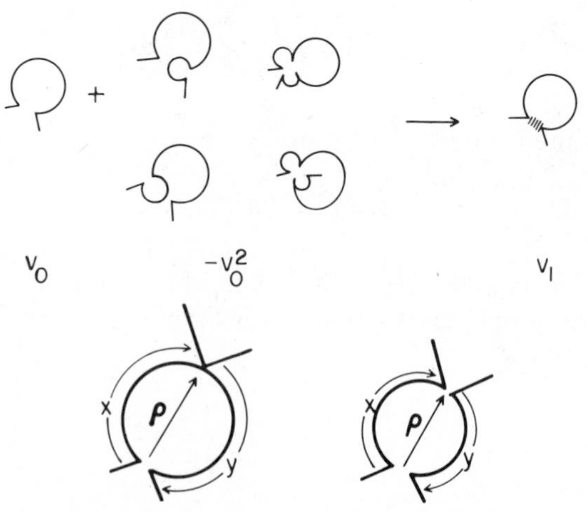

Fig. 7. Calculation of the effective two-body interaction parameter. To calculate the contribution from small loops, the conformation is parametrized as is shown in diagrams.

where $[\,]_s$ denotes the singular part. Hence we have

$$\hat{D} = \frac{2}{\varepsilon\pi^2} \tag{94}$$

or

$$u_0 = u + \frac{2}{\varepsilon\pi^2} u^2 + \cdots \tag{95}$$

A more physically satisfactory scheme to get this relation is given in Section V.C.4.

4. *The Renormalization-Group Equation for $G(N, \mathbf{R})$*

Since the parameter L is completely independent of the microscopic detail, as we have seen in Section III.B, we have

$$L \frac{\partial}{\partial L} G_B(N_0, \mathbf{R}) = 0 \tag{96}$$

which corresponds to Eq. (33). Rewriting G_B in terms of renormalized quantities as Eq. (41), we get the renormalization-group equation

$$\left[L \frac{\partial}{\partial L} + \beta(u) \frac{\partial}{\partial N} + \gamma_N(u) N \frac{\partial}{\partial N} + \gamma_G(u) \right] G(N, \mathbf{R}) = 0 \tag{97}$$

which corresponds to Eq. (44) and β, γ_N, and γ_G are defined in Eqs. (45)–(47). We have calculated Z_G, Z_N, and u in Sections IV.A.2 and IV.A.3, so that now we can calculate β, γ_N, and γ_G.

From the definition of $\beta(u)$, we get

$$\beta(u) = L \left(\frac{\partial u}{\partial L} \right)_{v_0} = L \frac{\partial u}{\partial u_0} \left(\frac{\partial u_0}{\partial L} \right)_{v_0} \tag{98}$$

Since $u_0 = v_0 L^{\varepsilon/2}$, we can calculate as

$$L \left(\frac{\partial u_0}{\partial L} \right)_{v_0} = \frac{\varepsilon}{2} u_0 \tag{99}$$

Combining Eqs. (98) and (99), we get a convenient formula for β:

$$\beta(u) = \frac{\varepsilon}{2}\left(\frac{\partial \ln u_0}{\partial u}\right)^{-1} \tag{100}$$

Using the relation of Eq. (95),

$$\begin{aligned}\beta(u) &= \frac{\varepsilon}{2}\left\{\frac{\partial\left[\ln u + \ln\left(1 + \frac{2}{\varepsilon\pi^2}u\right)\right]}{\partial u}\right\}^{-1} \\ &= \tfrac{1}{2}\varepsilon u\left(1 + \frac{u}{\frac{\varepsilon}{2}\pi^2 + u}\right)^{-1} \simeq \tfrac{1}{2}\varepsilon u\left(1 - \frac{2}{\varepsilon\pi^2}u\right) + O(u^3)\end{aligned} \tag{101}$$

This is the result already mentioned in Eq. (53).

Since Z_G and Z_N are identical to order u, so must be γ_N and γ_G. From the definitions of γ_N and β and the value of $\hat{B}$ given in Eq. (89), we get

$$\begin{aligned}\gamma_N(u) &= L\left(\frac{\partial \ln Z_N}{\partial L}\right) = \frac{\partial \ln Z_N}{\partial u}\beta(u) \\ &= \frac{1}{2\pi^2\varepsilon}\beta(u) = \frac{u}{4\pi^2} + O(u^2)\end{aligned} \tag{102}$$

Following the argument in Section III.D, to get the asymptotic form of the distribution function, we need $\gamma_N^* = \gamma_N(u^*)$ and $\gamma_G^* = \gamma_G(u^*)$, where $\beta(u^*) = 0$ $(u^* \neq 0)$. From Eqs. (101) and (102), we get

$$\gamma_N^* = \gamma_G^* = \frac{\varepsilon}{8} \tag{103}$$

For a sufficiently long chain in a good solvent, we have seen in Section III.D that we may put $u \simeq u^*$. Thus the renormalized perturbation result reads

$$\begin{aligned}G(N, \mathbf{R}) = \left(\frac{1}{2\pi N}\right)^{2-\varepsilon/2} e^{-\alpha}\bigg\{1 + \frac{\varepsilon}{8}\bigg[&1 + \alpha + (1-\alpha)(\hat{\gamma} + \ln\alpha) \\ &- (1-\alpha)\ln\frac{2\pi N}{L}\bigg]\bigg\}\end{aligned} \tag{104}$$

If α becomes big, this can be negative, which is physically nonsense. To remedy this defect, we exponentiate the order ε term to get

$$G(N, \mathbf{R}) = \left(\frac{1}{2\pi N}\right)^{2-\varepsilon/2} \exp\left\{-\alpha + \frac{\varepsilon}{8}\left[1 + \alpha + (1-\alpha)(\hat{\gamma} + \ln \alpha) - (1-\alpha)\ln\frac{2\pi N}{L}\right]\right\} \tag{105}$$

This may seem artificial and cannot be fully justified rigorously. However, the term containing $\ln(2\pi N/L)$ needs to be exponentiated to be consistent with the functional form of G explained in the next subsection. Moreover, we could have applied our perturbative calculation to $\ln G$ to impose the positivity of G; this naturally gives Eq. (105). In any case, there is always inevitable ambiguity to guess the correct interpolation formula from the perturbative calculation.

5. *Functional Form of the Density Distribution Function*

At the fixed point $u = u^*$, the renormalization-group Eq. (97) simplifies to

$$\left[L\frac{\partial}{\partial L} + \gamma_N^* N \frac{\partial}{\partial N} + \gamma_G^*\right] G(N, \mathbf{R}) = 0 \tag{106}$$

Hence, $G(N, \mathbf{R})$ must have the following general functional form

$$G(N, \mathbf{R}) = L^{-\gamma_G^*} f(LN^{-1/\gamma_N^*}, \mathbf{R}) \tag{107}$$

where f is an as yet unspecified well-behaved function. The argument above corresponds to that given up to Eq. (61). Here $\mathbf{R}$ is considered as a mere parameter.

From the reparametrization invariance of the Edwards Hamiltonian discussed in Section II.B, and the fact that $G\,\mathrm{d}^d\mathbf{R}$ must be invariant (hence $[G] = C^{-d/2}$), we get for the arbitrary positive constant C

$$C^{-d/2} G(N, \mathbf{R}) = (CL)^{-\gamma_G^*} f(C^{1-1/\gamma_N^*} L N^{-1/\gamma_N^*}, C^{1/2}\mathbf{R}) \tag{108}$$

which corresponds to Eq. (62). We fix C exactly as Eq. (63) to get

$$G(N, \mathbf{R}) = L^{-\gamma_G^*(d/2-\gamma_G^*)/(\gamma_N^*-1)} N^{-(d/2-\gamma_G^*)/(\gamma_N^*-1)} \times f(\mathbf{R}/N^{1/2(1-\gamma_N^*)} L^{\gamma_N^*/2(1-\gamma_N^*)}) \tag{109}$$

or, if we absorb L into the definition of f, we get

$$G(N, \mathbf{R}) = N^{-(d/2-\gamma_G^*)/(\gamma_N^*-1)} f(\mathbf{R}/N^{1/2(1-\gamma_N^*)}) \tag{110}$$

where f denotes the appropriate function in each context.

We already know the relation between γ_N^* and ν given by Eq. (65). On the other hand, γ_G^* is related to the exponent γ defined by the following asymptotic relation,

$$C_N \sim N^{\gamma-1}\mu^N \tag{111}$$

where C_N is the total number of conformations for the walk with the number of steps N and μ is a numerical constant (connectivity constant).[2] Renormalization results cannot give the value of μ; $\ln \mu$ is essentially the free energy per chain unit, but we have arbitrarily shifted the origin of the free energy to eliminate the divergence in Eq. (76). What we can get from Eq. (110) is the term proportional to $N^{\gamma-1}$:

$$N^{\gamma-1} \sim \int G(N, \mathbf{R})\, \mathrm{d}^d\mathbf{R} = N^{-(d/2-\gamma_G^*)/(\gamma_N^*-1)+d/2(1-\gamma_N^*)} \times \int f(\mathbf{x})\, \mathrm{d}^d\mathbf{x} \tag{112}$$

Hence

$$\gamma - 1 = \frac{\gamma_G^*}{\gamma_N^* - 1} \quad \text{or} \quad \gamma_G^* = \frac{\gamma - 1}{2\nu} \tag{113}$$

Using these results, we can rewrite Eq. (107) as follows

$$G(N, \mathbf{R}) = N^{\gamma-\nu d-1} f(\mathbf{R}/N^{\nu}) \tag{114}$$

This is the form derived by Fisher[93] long ago, and justified field theoretically by des Cloizeaux.[94]

This form suggests that the natural variable is not α but the variable proportional to $R^2/N^{2\nu}$. Indeed, if we introduce the following variable

$$x = \frac{\alpha}{(2\pi N/L)^{\varepsilon/8}} \tag{115}$$

Eq. (105) can be rewritten as

$$G(N, \mathbf{R}) = (2\pi N)^{-2(1-3\varepsilon/16)} L^{\varepsilon/8} \exp\left[-x + \frac{\varepsilon}{8}x(1-\hat{\gamma}) + \frac{\varepsilon}{8}\ln x - \frac{\varepsilon}{8}x\ln x + \frac{\varepsilon}{8}(1+\hat{\gamma})\right] \tag{116}$$

where we have used the following approximation

$$x = \alpha\left[1 - \frac{\varepsilon}{8}\ln\left(\frac{2\pi N}{L}\right)\right] \tag{117}$$

Thus, besides the multiplicative constant, the function depends on L only through x. Unfortunately, N and L are not observable quantities. Therefore, there is no way to actually calculate the value of x even if $\mathbf{R}$ is specified. As we will see later this difficulty can be overcome if the distance is scaled with the mean-square end-to-end distance $\langle R^2\rangle^{1/2}$. We must therefore calculate $\langle R^2\rangle$.

6. *The Mean-square End-to-End Distance*

We can calculate $\langle R^2\rangle$ from Eq. (105), but the most efficient way is to expand the Fourier transform of $G_B(N_0, \mathbf{R})$:

$$\tilde{G}_B(N_0, \mathbf{k}) \equiv \int G_B(N_0, \mathbf{R})\, \mathrm{e}^{i\mathbf{k}\cdot\mathbf{R}} \mathrm{d}^d\mathbf{R}, \tag{118}$$

$$= \int G_B(N_0, \mathbf{R})\, \mathrm{d}^d\mathbf{R} - \tfrac{1}{2}\int G_B(N_0, \mathbf{R})(\mathbf{k}\cdot\mathbf{R})^2\, \mathrm{d}^d\mathbf{R} + \cdots \tag{119}$$

where we have used $\langle \mathbf{R}\rangle = \mathbf{0}$. If we define $\mathcal{N}_0$ (as before) as

$$\mathcal{N}_0 = \int G_B(N_0, \mathbf{R})\, \mathrm{d}^d\mathbf{R} \tag{120}$$

then Eq. (119) becomes

$$\tilde{G}_B(N_0, \mathbf{R}) = \mathcal{N}_0 - \frac{1}{2d}\mathcal{N}_0\langle R^2\rangle_B k^2 + \cdots \tag{121}$$

where we have used the following relation

$$\langle(\mathbf{k}\cdot\mathbf{R})^2\rangle_B = \sum_i\sum_j \langle k_i k_j R_i R_j\rangle_B = d^{-1}\langle R^2\rangle_B k^2 \tag{122}$$

with subscript B denoting bare quantities.

From Eq. (69) we get

$$\tilde{G}_B(N_0, \mathbf{k}) = \exp(-\tfrac{1}{2}k^2 N_0) - \frac{u_0 L^{-\varepsilon/2}}{(2\pi)^{d/2}} \int_0^{N_0} \mathrm{d}\tau (N_0 - \tau)\tau^{-d/2} \times \exp(-\tfrac{1}{2}k^2(N_0 - \tau)) \tag{123}$$

$$= 1 - \tfrac{1}{2}k^2 N_0 + \cdots - u_0 \frac{L^{-\varepsilon/2}}{(2\pi)^{d/2}} \int_0^{N_0} \mathrm{d}\tau (N_0 - \tau)\tau^{-d/2} + u_0 \times \frac{L^{-\varepsilon/2}}{(2\pi)^{d/2}} \int_0^{N_0} \mathrm{d}\tau \tfrac{1}{2}(N_0 - \tau)^2 \tau^{-d/2} k^2 + \cdots \tag{124}$$

Thus we get

$$\mathcal{N}_0 = 1 - \frac{u_0 L^{-\varepsilon/2}}{(2\pi)^{d/2}} \int_0^{N_0} \mathrm{d}\tau (N_0 - \tau)\tau^{-d/2} + \cdots \tag{125}$$

and

$$\frac{1}{2d} \mathcal{N}_0 \langle R^2 \rangle_B = \tfrac{1}{2}N_0 - \tfrac{1}{2}u_0 \frac{L^{-\varepsilon/2}}{(2\pi)^{d/2}} \int_0^{N_0} \mathrm{d}\tau (N_0 - \tau)^2 \tau^{-d/2} + \cdots \tag{126}$$

which give

$$\langle R^2 \rangle_B = d \left[N_0 + u_0 \frac{L^{-\varepsilon/2}}{(2\pi)^{d/2}} \int_0^{N_0} \mathrm{d}\tau (N_0 - \tau)\tau^{-1+(\varepsilon/2)} + \cdots \right] \tag{127}$$

$$= dN_0 \left[1 + \frac{u}{(2\pi)^2} \left(\frac{2}{\varepsilon} + \ln \frac{2\pi N_0}{L} - 1 \right) \right] \tag{128}$$

The singular term can be removed by the introduction of Z_N defined by Eq. (36). Then we get the renormalized result

$$\langle R^2 \rangle = 4(1 - \tfrac{3}{8}\varepsilon)(2\pi)^{\varepsilon/8} N (N/L)^{\varepsilon/8} \tag{129}$$

Here already the order ε term containing N has been exponentiated to be in conformity with Eq. (64), the result of the renormalization-group equation. The result to order ε^2 can be found in Ref. 91b.

A more directly observable quantity is the mean-square radius of gyration $\langle R_G^2 \rangle$, which is defined by

$$\langle R_G^2 \rangle_B = \frac{1}{N_0} \int_0^{N_0} d\tau \langle (\mathbf{c}(\tau) - \mathbf{c}_{CM})^2 \rangle \tag{130}$$

where $\mathbf{c}_{CM}$ is the center-of-mass position of the chain. It is known that $\langle R_G^2 \rangle / \langle R^2 \rangle$ is a universal ratio, a ratio independent of any chemical details, and to order ε it is shown by Witten and Schäfer[95,96] that

$$\langle R_G^2 \rangle / \langle R^2 \rangle = \frac{1}{6}\left(1 - \frac{\varepsilon}{96}\right) + O(\varepsilon^2) \tag{131}$$

7. *Normalized End-to-End Distance Distribution Function*

Since $\langle R^2 \rangle$ is at least in principle a measurable quantity, let us introduce $\mathbf{r}$

$$\mathbf{r} \equiv \frac{\mathbf{R}}{\langle R^2 \rangle^{1/2}} \tag{132}$$

Then we have

$$x = 2(1 - \tfrac{3}{8}\varepsilon) r^2 \tag{133}$$

Equation (116) can be streamlined a bit

$$G(N, \mathbf{R}) = x^{\varepsilon/8} \exp\left[-x^{1+\varepsilon/8} + \frac{\varepsilon}{8} x(1 - \hat{\gamma}) \right] \tag{134}$$

where all the multiplicative numerical constants are discarded. Introducing Eq. (133) into Eq. (134), we get

$$G(N, \mathbf{R}) = |\mathbf{r}|^{\varepsilon/4} \exp\left\{ -\left[2\left(1 - \frac{3}{8}\varepsilon\right) \mathbf{r}^2 \right]^{1+\varepsilon/8} + \frac{\varepsilon}{4}(1 - \hat{\gamma})\mathbf{r}^2 \right\} \tag{135}$$

Here we have used the following calculation:

$$\frac{\varepsilon}{8}(1 - \hat{\gamma})x = \frac{\varepsilon}{8}(1 - \hat{\gamma})2\left(1 - \frac{3}{8}\varepsilon\right)\mathbf{r}^2 \simeq \frac{\varepsilon}{4}(1 - \hat{\gamma})\mathbf{r}^2 \tag{136}$$

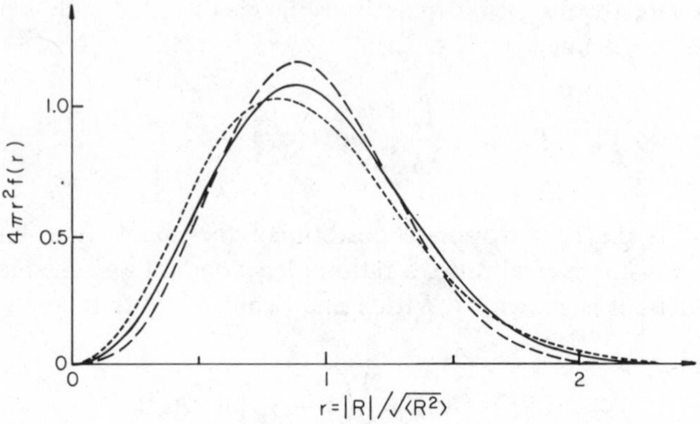

Fig. 8. The end-to-end distance distribution function for the self-avoiding walk. Dotted line represents the Gaussian distribution function [Eq. (139)]; solid line represents our result to order ε [Eq. (137)]; dashed line represents the exact enumeration estimate by Domb et al. [Eq. (140)]. The distance scale is normalized so that the mean-square end-to-end distance is unity.

When we normalize Eq. (135) as a function of **r**, we get[91a]

$$f(\mathbf{r}) = 0.3|\mathbf{r}|^{0.25} \exp(-1.4|\mathbf{r}|^{2.25} + 0.1|\mathbf{r}|^2) \tag{137}$$

where we have used

$$\left[2\left(1 - \frac{3}{8}\varepsilon\right)\mathbf{r}^2\right]^{1+\varepsilon/8} \equiv 2\left[1 - \frac{\varepsilon}{8}(3 - \ln 2)\right]|\mathbf{r}|^{2+\varepsilon/4} \simeq 1.4|\mathbf{r}|^{2.25} \tag{138}$$

In Fig. 8, the function we have just obtained is compared with the Gaussian function

$$f_G(\mathbf{r}) = 0.33 \exp(-1.5|\mathbf{r}|^2) \tag{139}$$

and the exact enumeration estimate by Domb et al.[97]

$$f_E(\mathbf{r}) = 0.271|\mathbf{r}|^{0.5} \exp(-1.31|\mathbf{r}|^{2.5}) \tag{140}$$

As is seen from the figure, our lowest nontrivial-order calculation captures the salient features of the self-avoiding walk. If the best estimate of the exponents is used, our result and Eq. (140) become very close.

B. Temperature Dependence

In the last section we discussed the end vector distribution function in the long-chain self-avoiding limit. However, the renormalized perturbation series contains much more information. As mentioned in Section III.D, we can study the Gaussian self-avoiding crossover behavior.

1. *General Consideration*

According to the interpretation given in Section III.B, the renormalized parameter v is the macroscopic counterpart of the excluded-volume parameter v_0. Therefore, u defined by Eq. (35) is the parameter describing the quality of the solvent. We have seen that the value of u at the nontrivial fixed point u^* of the renormalization-group transformation corresponds to the self-avoiding (good-solvent) limit. At $u = 0$, which corresponds to $u_0 = 0$, the chain is Gaussian. Thus values of u between u^* and 0 should correspond to various qualities of solvents.

Actually, as has been stressed in Section II.D, the true theta state is the consequence of a subtle balance of repulsive and attractive interactions. Therefore, the minimal model for the theta state is not given by the Edwards Hamiltonian but by the following Hamiltonian[70] (see also Ref. 42)

$$\mathscr{H} = \frac{1}{2}\int \dot{\mathbf{c}}^2\, d\tau + \frac{1}{2}v_0 \int d\tau \int d\sigma\, \delta(\mathbf{c}(\tau) - \mathbf{c}(\sigma)) + \frac{1}{3!}w_0 \int d\tau \int d\sigma_1 \int d\sigma_2\, \delta(\mathbf{c}(\tau) - \mathbf{c}(\sigma_1))\, \delta(\mathbf{c}(\tau) - \mathbf{c}(\sigma_2)) \tag{141}$$

where w_0 is the three-body interaction parameter, and the formula should be understood with an appropriate cutoff. In d-space, following the dimensional analysis in Section II.B, we get $[v_0] = C^{-(4-d)/2}$ and $[w_0] = C^{-(3-d)}$. Therefore, if we make the second term in Eq. (141) tractable by choosing $d = 4 - \varepsilon$, then we have $[w_0] = C^{1-\varepsilon}$, so that the dimensionless combination $w_0 N_0^{\varepsilon-1}$ becomes the natural parameter. This implies that in the $N_0 \to \infty$ limit, we cannot effectively study the effect of the third term which is reasonably important in 3-space where $\varepsilon = 1$. In addition if we work in 3-space we cannot treat the second term in a systematic way, as we have seen in Section III.C. Thus the ε-expansion techniques are powerless to study the self-avoiding theta chain crossover behavior. According to Duplantier[73]

$$\langle R^2 \rangle = dN(A - B/\ln N + vCN^{1/2}(\ln N)^{-4/11}) \tag{142}$$

where A, B, and C are positive constants. This equation contains the parameter v, but the equation can cover the immediate vicinity of the theta state only. There have been no closed form equation which describes the crossover between the self-avoiding walk and the theta state.

Therefore, as an approximation, we approximate the theta state by the Gaussian unperturbed state; this is the same identification as in the conventional theories. Then what we have to do in order to understand the crossover behavior is to find the u dependence of various quantities. For the density distribution function for the end-to-end vector we already know the renormalized perturbation result [Eq. (90)]. As in the case of the self-avoiding limit, we need a functional form like Eq. (110). From these two results can we then construct an interpolation formula.

2. *Functional Form Required by the Renormalization-Group Equation*

Since we are interested in u equal neither to 0 nor to u^*, we have to solve Eq. (97) with β and $\gamma_N = \gamma_G$ given by Eq. (101) and Eq. (102), respectively.

First, we must solve the characteristic equations corresponding to Eq. (55) (here we do not introduce the dilation parameter t), or

$$L\frac{\mathrm{d}u}{\mathrm{d}L} = \beta(u) \tag{143}$$

$$N\frac{\mathrm{d}u}{\mathrm{d}N} = \frac{\beta(u)}{\gamma_N(u)} \tag{144}$$

$$\frac{\mathrm{d}G}{\mathrm{d}u} = -\frac{G\gamma_G(u)}{\beta(u)} \tag{145}$$

From these equations we get

$$\ln L = \int_{u_1}^{u} \frac{\mathrm{d}u}{\beta(u)} + C_1 \tag{146}$$

$$\ln N = \int_{u_1}^{u} \frac{\gamma_N(u)}{\beta(u)}\,\mathrm{d}u + C_2 \tag{147}$$

$$\ln G = -\int_{u_1}^{u} \frac{\gamma_G(u)}{\beta(u)}\,\mathrm{d}u + C_3 \tag{148}$$

where u_1 and C_1–C_3 are constants. The variable $\mathbf{R}$ in G is considered a parameter. Thus, following the standard theory of quasilinear partial differential equations,[89] we get the following general solution to Eq. (97)

$$G(N, \mathbf{R}) = \exp\left[-\int_{u_1}^{u} \frac{\gamma_G}{\beta}\, du\right] F\left(L \exp\left[-\int_{u_1}^{u} \frac{du}{\beta}\right],\right.$$
$$\left. N \exp\left[-\int_{u_1}^{u} \frac{\gamma_N}{\beta}\, du\right], \mathbf{R}\right) \tag{149}$$

where F is a well-behaved yet unknown function. For convenience, let us choose $u_1 = u^*/2$ (if we choose another value of u_1, the function F must be modified. But since F is not fixed anyway, we may still choose u_1 arbitrarily in $(0, u^*)$). Then we have

$$-\int_{u_1}^{u} \frac{du}{\beta(u)} = -\frac{\pi^2}{u^*} \ln \frac{u}{u^* - u} \tag{150}$$

$$-\int_{u_1}^{u} \frac{\gamma_i}{\beta(u)}\, du = \tfrac{1}{4} \ln\left[2\left(\frac{u^* - u}{u^*}\right)\right], \qquad (i = N \text{ or } G) \tag{151}$$

Thus Eq (144) becomes

$$G(N, \mathbf{R}) = \left[2\left(\frac{u^* - u}{u^*}\right)\right]^{1/4} F\left(L\left(\frac{2}{u^* - u}\right)^{-2/\varepsilon}, N\left[2\left(\frac{u^* - u}{u^*}\right)\right]^{1/4}, \mathbf{R}\right) \tag{152}$$

From the reparametrization invariance of the Edwards Hamiltonian, we get the following relation corresponding to Eq. (108)

$$G(N, \mathbf{R}) = C^{-d/2}\left[2\left(\frac{u^* - u}{u^*}\right)\right]^{1/4} F\left(C^{-1}L\left(\frac{2}{u^* - u}\right)^{-2/\varepsilon},\right.$$
$$\left. C^{-1}N\left[2\left(\frac{u^* - u}{u^*}\right)\right]^{1/4}, C^{1/2}\mathbf{R}\right) \tag{153}$$

Since C is an arbitrary positive constant, we may choose C so that

$$C^{-1}N\left[2\left(\frac{u^* - u}{u^*}\right)\right]^{1/4} = 1 \tag{154}$$

Then Eq. (153) can be rewritten as

$$G(N, \mathbf{R}) = N^{-d/2}\left(\frac{2}{1+w}\right)^{-d/8+1/4} F\left(N^{-1}\left(\frac{2}{1+w}\right)^{-1/4} Lw^{-2/\varepsilon},\right.$$
$$\left.\mathbf{R}N^{-1/2}\left(\frac{2}{1+w}\right)^{-1/8}\right) \tag{155}$$

where F is again a well-behaved but as yet unknown function and

$$w = \frac{u}{u^* - u} \quad \text{or} \quad u = \frac{w}{1+w}u^* \tag{156}$$

The functional form Eq. (155) implies that the following variables are the natural ones:

$$\zeta = s\left(\frac{N}{L}\right)^{\varepsilon/2} w\left(\frac{2}{1+w}\right)^{\varepsilon/8} \tag{157}$$

$$X = R^2\,\frac{(1+w)^{1/4}}{2N} \tag{158}$$

where s is a constant that will be chosen so that our final equation becomes simple. We see that the normalized density distribution function for the end-vector is a function of ζ and X.

3. *Universal Form of the End-to-End Vector Distribution*

To construct an interpolation formula from the renormalized perturbation result Eq. (90), we rewrite it in terms of ζ and X. To this end, let us find α in terms of ζ and X. Inspection of Eq. (90) suggests that $2^{\varepsilon/8}s$ should be $(2\pi)^{\varepsilon/2}$, since N appears always with 2π. From Eq. (157), assuming that ε is small, we get

$$w = s^{-1}\left(\frac{N}{L}\right)^{-\varepsilon/2} \zeta\left(\frac{1+\zeta}{2}\right)^{\varepsilon/8} = \left(\frac{2\pi N}{L}\right)^{-\varepsilon/2} \zeta(1+\zeta)^{\varepsilon/8} \tag{159}$$

Hence

$$(1+w)^{1/4} = \left\{1 + \zeta(1+\zeta)^{\varepsilon/8}\left[1 - \frac{\varepsilon}{2}\ln\left(\frac{2\pi N}{L}\right)\right]\right\}^{1/4} + o(\varepsilon),$$
$$= [1 + \zeta(1+\zeta)^{\varepsilon/8}]^{1/4}\left(\frac{2\pi N}{L}\right)^{-\varepsilon\zeta/8(1+\zeta)} + o(\varepsilon) \tag{160}$$

Thus we get from Eq. (158)

$$\alpha = X\left(\frac{2\pi N}{L}\right)^{\varepsilon\zeta/8(1+\zeta)}[1 + \zeta(1 + \zeta)^{\varepsilon/8}]^{-1/4} \tag{161}$$

Replacing α in Eq. (90) with Eq. (161) and u with Eq. (156), we get

$$\begin{aligned} G(N, \mathbf{R}) = {} & (2\pi N)^{-2+\varepsilon/2-\varepsilon\zeta/8(1+\zeta)}[X(1+\zeta)^{-1/4}]^{\varepsilon\zeta/8(1+\zeta)} \\ & \times \exp\left\{-[X(1+\zeta(1+\zeta)^{\varepsilon/8})^{-1/4}]^{1+\varepsilon\zeta/8(1+\zeta)} + (1-\hat{\gamma})\frac{\varepsilon}{8}\right. \\ & \left.\times \frac{\zeta}{1+\zeta}X(1+\zeta)^{-1/4} + \frac{\varepsilon}{8}\frac{\zeta}{1+\zeta}(1+\hat{\gamma})\right\} \end{aligned} \tag{162}$$

See Appendix D for details. This has the form of Eq. (155) required by the renormalization-group equation.

However, now we have the same difficulty that we have encountered in Section IV.A.4, that is, even if we know $\mathbf{R}$, there is no way to calculate X. This can again be overcome by normalizing the distance by $\langle R^2 \rangle^{1/2}$.

Since we can get the renormalized perturbation results for $\langle R^2 \rangle$ from Eq. (128), we have only to rewrite it in terms of ζ. This is much simpler than in the case of the end-vector distribution function; we have

$$\langle R^2 \rangle = 4N\left(\frac{2\pi N}{L}\right)^{\varepsilon\zeta/8(1+\zeta)}\left(1 - \frac{\varepsilon}{4} - \frac{\varepsilon}{8}\frac{\zeta}{1+\zeta}\right) \tag{163}$$

In terms of $\mathbf{r} = \mathbf{R}/\langle R^2 \rangle^{1/2}$, we have the normalized density distribution function for the end-vector as[98]

$$\begin{aligned} f(\mathbf{r}, \zeta) = {} & \left(\frac{2}{\pi}\right)^{2-\varepsilon/2} 2^{\varepsilon\zeta/8(1+\zeta)}\left\{\exp\left[\frac{\hat{\gamma}}{8}\frac{\varepsilon\zeta}{1+\zeta} - \left(\frac{1}{4}\frac{\zeta}{1+\zeta} + \frac{1}{2}\right)\varepsilon\right]\right\}|\mathbf{r}|^{\varepsilon\zeta/4(1+\zeta)} \\ & \times \exp\left[-(2\mathbf{r}^2)^{1+(1/8)\varepsilon\zeta/(1+\zeta)}\left(1 - \frac{\varepsilon}{4} - \frac{\varepsilon\zeta}{8(1+\zeta)}\right)\right. \\ & \left. + \frac{\varepsilon}{4}\frac{\zeta}{1+\zeta}(1-\hat{\gamma})|\mathbf{r}|^2\right] \end{aligned} \tag{164}$$

Thus the distribution contains only one parameter ζ, which describes the quality of the solvent: $\zeta = 0$ corresponds to the Gaussian limit and $\zeta = \infty$ the self-avoiding limit. Equation (164) smoothly interpolates the Gaussian function and the function for the self-avoiding walk [Eq. (137)].

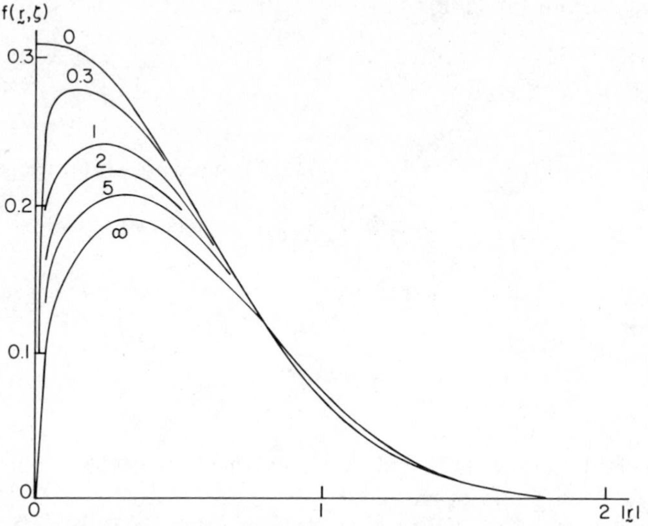

Fig. 9. The temperature dependence of the end-to-end vector distribution function given by Eq. (164). The numbers in the figure denote the values of the parameter ζ, which corresponds to the conventional z parameter. (However, see Section IV.B.4 for an essential difference between ζ and z). $\zeta = 0$ is the Gaussian limit. As soon as ζ becomes positive, $f = 0$ at the origin because of the excluded-volume effect. The distance scale is normalized so that the mean-square end-to-end distance is unity.

Equation (164) is shown in Fig. 9 for several values of ζ. The function always vanishes at the origin so long as $\zeta > 0$, which manifests the self-avoiding properties.

4. *Weak Nonuniversality of Expansion Factor*

Equation (163) implies that

$$\alpha_R^2 \equiv \langle R^2 \rangle / \langle R^2 \rangle_0 = \left(\frac{2\pi N}{L} \right)^{\varepsilon\zeta/8(1+\zeta)} \left(1 - \frac{\varepsilon}{8} \frac{\zeta}{1+\zeta} \right) \tag{165}$$

or

$$\alpha_R^2 = \left[\frac{1 + \zeta(1+\zeta)^{\varepsilon/8}}{1+w} \right]^{1/4} \left(1 - \frac{\varepsilon}{8} \frac{\zeta}{1+\zeta} \right) \tag{166}$$

Since α_R^2 contains not only ζ but w, which depends on the details of the system, α_R^2 is less universal than Eq. (164). However, if the renormalized excluded-volume parameter u is sufficiently small that is, $u/u^* \ll 1$, then w is small, so that, if we introduce the Z parameter by

$$Z = \left(\frac{2\pi N}{L}\right)^{\varepsilon/2} w \simeq \left(\frac{2\pi N}{L}\right)^{\varepsilon/2} \frac{u}{u^*} \tag{167}$$

then Eq. (166) can be approximated as

$$\alpha_R^2 = (1 + Z)^{1/4}\left(1 - \frac{\varepsilon}{8}\frac{Z}{1 + Z}\right) \tag{168}$$

This is exact in the $N \to \infty$, $u \to 0$ limit with Z being kept constant. Thus if the renormalized interaction is small, we get an approximate universal law Eq. (168). Here the word "universal" is used in the sense that the microscopic details can be absorbed in a single parameter Z.

The parameter Z may be a counterpart of the conventional z parameter discussed in Section II.B, since both quantities are proportional to the square root of the molecular weight (in 3-space). However, there is a fundamental difference; the excluded-volume parameter appearing in z is a bare parameter v_0, whereas that in Z is a renormalized one v. Therefore, the parameter obtained from the comparison of experimental results with the theoretical results cannot be interpreted microscopically. (Even the value of z experimentally obtained must not be interpreted microscopically.)

In contrast to the approximate universal ratio Eq. (168), $\langle R_G^2\rangle/\langle R^2\rangle$ is a true universal ratio, where $\langle R_G^2\rangle$ is the mean-square radius of gyration, as we see[98]

$$\langle R_G^2\rangle/\langle R^2\rangle = \frac{1}{6}\left(1 - \frac{\varepsilon}{96}\frac{\zeta}{1 + \zeta}\right) \tag{169}$$

which contains only ζ; no w. Unfortunately, this is only weakly dependent on the solvent quality, so it is very hard to use the formula to experimentally determine ζ.

5. *Another Route Connecting Gaussian and Self-Avoiding Limits*[99]

So far we have discussed the excluded-volume parameter dependence of a finite-length (but long) chain. There is another path that connects the Gaussian chain and the self-avoiding walk (SAW): the path via the finite-order self-avoiding walk (FSAW).[100-102]

The FSAW was originally considered by Domb and Hioe[100] and by Wall[101] as an intermediate model to study the SAW. In this model, a monomer is assumed to interact only with other monomers within a certain prescribed contour distance M from it. If $m = 0$, then the model is nothing but a simple random-walk-type chain; if $m = N$, the total length of the chain, then the model is the SAW.

The FSAW is closely related to the blob argument[4,103] employed to explain the crossover between dilute and semidilute solutions. In this argument, due to the screening, the excluded-volume interaction is effective only within a blob of size ξ, the monomer density correlation length. That is, a monomer can effectively interact with monomers within the contour distance $\sim\xi^{1/\nu}$. Outside of this range, the excluded-volume interactions are screened. Thus, in the blob argument, a single polymer chain is regarded as a FSAW of order $\sim\xi^{1/\nu}$.

A model Hamiltonian for the FSAW of (bare) order m_0 is

$$\mathscr{H}_{\mathrm{FSAW}} = \frac{1}{2}\int_0^{N_0} \dot{\mathbf{c}}^2\, d\tau + \frac{1}{2} v_0 \int_0^{N_0} d\tau \int_{\substack{0 \\ m_0 > |\tau-\sigma| > a}}^{N_0} d\sigma\, \delta(\mathbf{c}(\tau) - \mathbf{c}(\sigma)) \tag{170}$$

Within our conformation scheme, the difference between Eq. (170) and the Edwards Hamiltonian Eq. (11) causes only a slight technical complication. (We must assume that m_0 is much larger than a.) Therefore we quote the results below for the mean-square end-to-end distance $\langle R^2 \rangle$ and for the density distribution function for $\mathbf{r} = \mathbf{R}/\langle R^2 \rangle^{1/2}$. (Note, however, in the field-theoretic formalism, the FSAW is much more complicated than the SAW, because there is no simple way to describe the Hamiltonian Eq. (170) with the aid of the zero-component field.)

The end-to-end distance is

$$\langle R^2 \rangle = \langle R^2 \rangle_{\mathrm{SAW}} \left(\frac{m}{N}\right)^{\varepsilon/8} \exp\left[\frac{\varepsilon}{8}\left(1 - \frac{m}{N}\right)\right] \tag{171}$$

where $\langle R^2 \rangle_{\mathrm{SAW}}$ is given by Eq. (129). The density distribution function for $\mathbf{r}$ is given by

$$\begin{aligned} f(\mathbf{r}, \theta) = {} & (d/2\pi)^{d/2} (2r^2)^{\varepsilon[1-\exp(-2r^2\theta/(1-\theta))]/8} \\ & \times \exp\left[-\frac{\varepsilon}{4}\theta - (2r^2)^{1+\varepsilon[1-\exp(-2r^2\theta/(1-\theta))]/8} \left(1 - \frac{\varepsilon}{4} - \frac{\varepsilon}{8}\theta\right) \right. \\ & + \frac{\varepsilon}{8}\left(\frac{1-\theta}{\theta}(\mathrm{e}^{-2r^2\theta/(1-\theta)} - 1) + 2r^2\right) + \frac{\varepsilon}{8}(2r^2 - 1) \\ & \left. \times \{Ei(-2r^2\theta/(1-\theta)) - \hat{\gamma} - \ln\theta - \mathrm{e}^{-2r^2\theta/(1-\theta)} \ln(2r^2)\} \right] \end{aligned} \tag{172}$$

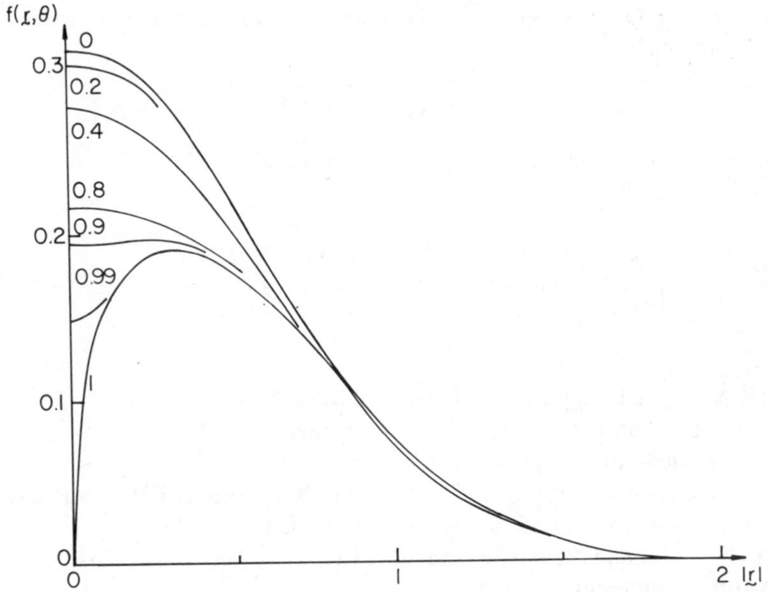

Fig. 10. The end-to-end vector distribution function for the finite order self-avoiding walk given by Eq. (172). The numbers in the figure represent the relative order $\theta = m/N$. $\theta = 1$ corresponds to the self-avoiding chain, and $\theta = 0$ to the Gaussian chain. As soon as θ becomes less than 1, the chain end can return to the origin, so $f > 0$ at the origin. The distance scale is normalized so that the mean-square end-to-end distance is unity.

where $\theta = m/N$, $\hat{\gamma} \simeq 0.577$, and $Ei(-x) = -\int_x^\infty dt\, e^{-t}/t$. For several ratios θ, Eq. (172) is shown in Fig. 10. Note that, in contrast to the crossover described in Section IV.B.3, where the excluded-volume parameter is varied but is effective between any pair of monomers, the distribution function is nonzero at the origin as soon as θ becomes less than 1. This is what we intuitively expect.

One more different route from the random walk to the self-avoiding walk will be discussed in Section IV.D.

C. Concentration Dependence

When the concentration of macromolecules is increased, chains begin to overlap each other. Therefore the interactions within a chain and between two chains are interferred or mediated by the existence of other chains. The main effect is the screening of the excluded-volume effect, as was first pointed out by Edwards.[59] In this section a renormalized version of Edwards theory due mainly to Ohta[104] is explained (for another nonfield theoretic scheme see des Cloiseaux[105]).

1. *Partition Function*

Our minimal model Hamiltonian is a straightforward extension of the Edwards Hamiltonian Eq. (11):

$$\mathscr{H} = \frac{1}{2}\sum_{\alpha=1}^{n}\int_0^{N_0^\alpha} \dot{\mathbf{c}}_\alpha(\tau)^2\, d\tau + \frac{1}{2}v_0 \sum_{\alpha,\beta}\int_0^{N_0^\alpha} d\tau \int_0^{N_0^\beta} d\sigma\, \delta(\mathbf{c}_\alpha(\tau) - \mathbf{c}_\beta(\sigma)) \tag{173}$$

where N_0^α is the bare polymer "length" of the αth chain, $\mathbf{c}_\alpha(\tau)$ $(0 \leqslant \tau \leqslant N_0^\alpha)$ is the conformation of the αth chain with τ being the contour variable, v_0 is the bare excluded-volume parameter, and n is the total number of polymer chains. We assume that the system is in a box V of volume V. It is understood that there is a cutoff to eliminate the self-interaction of the unit of the chain [as in Eq. (11)] *and* the interaction more than once between one pair of chain elements on different chains.

The partition function of the system is

$$Z = \prod_\alpha \int_{c_\alpha \in V} \mathscr{D}[\mathbf{c}_\alpha] \exp[-\mathscr{H}] \tag{174}$$

where $\mathscr{D}$ is the uniform measure (see Appendix B) on the set of conformations in the box V.

Following Kac,[106] we introduce a normal random field ψ such that

$$\langle \psi(\mathbf{r})\psi(\mathbf{r}')\rangle_\psi = v_0\, \delta(\mathbf{r} - \mathbf{r}') \tag{175}$$

$$\langle \psi(\mathbf{r})\rangle_\psi = 0 \tag{176}$$

where $\langle\,\rangle_\psi$ is the average over the random field. Since ψ is Gaussian, we have

$$\langle \exp(i\psi(\mathbf{r}) + i\psi(\mathbf{r}'))\rangle_\psi \propto \exp[-v_0\, \delta(\mathbf{r} - \mathbf{r}')] \tag{177}$$

where we ignored the infinitely big constant $v_0\, \delta(\mathbf{0})$; in any case the model is well defined only if the delta-function is mollified (see Section II.B). Thus we interpret delta-functions in the theory as a short-hand notation of a well-defined short-range interactions. Then Eq. (177) is correct with the neglect of a constant, which corresponds to a shift of the origin of the energy.

Using Kac's method, Eq. (174) can be rewritten as

$$Z = \left\langle \prod_{\alpha} G(N_0^{\alpha}, \psi) \right\rangle_{\psi} \tag{178}$$

where

$$G(N_0^{\alpha}, \psi) = \int_{c_\alpha \in V} \mathscr{D}[\mathbf{c}_\alpha] \exp\left[-\frac{1}{2} \int_0^{N_0^{\alpha}} \dot{\mathbf{c}}(\tau)^2 \, d\tau - i \int_0^{N_0^{\alpha}} \psi(\mathbf{c}(\tau)) \, d\tau \right] \tag{179}$$

The shift of the origin of the energy axis has already been performed. Equation (179) is the partition function of a single chain of bare total contour length N_0^{α} confined in the box V with the imaginary random field $i\psi(\mathbf{r})$. There is no self-interaction. Equation (178) can be rewritten as follows:

$$Z = \left\langle \exp\left[n \int_0^{\infty} P_0(N_0) \ln G(N_0, \psi) \, dN_0 \right] \right\rangle_{\psi} \tag{180}$$

where P_0 is the (bare) probability density distribution function for the chain length; since the bare length must be proportional to the molecular weight (see Section III.B), P_0 is essentially the molecular weight distribution function. Another method to introduce the field ψ is discussed in Section IV.C.7.

We can explicitly describe the average $\langle \rangle_{\psi}$ with the aid of the functional integral (this may be understood naively as the limit of an infinite multiple integral)

$$Z = \int \mathscr{D}[\psi] \exp\left[n \int_0^{\infty} P_0(N_0) \ln G(N_0, \psi) \, dN_0 - \frac{1}{2v_0} \int \psi(\mathbf{r})^2 \, d^d\mathbf{r} \right] \tag{181}$$

where the normalization constant is absorbed in the "measure" $\mathscr{D}[\psi]$.

For convenience, following Edwards,[59] we divide the random field ψ into two parts, ψ_0 and ψ_1,

$$\psi_0 \equiv \frac{1}{V} \int_V \psi(\mathbf{r}) \, d^d\mathbf{r} \tag{182}$$

$$\psi_1 \equiv \psi - \psi_0 \tag{183}$$

where ψ_0 is the zero wave-number component of the random field. Note that

$$\int \psi_1(\mathbf{r})\, \mathrm{d}^d\mathbf{r} = 0 \tag{184}$$

and

$$\int (\psi_0 + \psi_1)^2\, \mathrm{d}^d\mathbf{r} = V\psi_0^2 + \int \psi_1^2\, \mathrm{d}^d\mathbf{r} \tag{185}$$

Equations (181) and (185) imply that ψ_0 and ψ_1 are statistically independent Gaussian stochastic fields.

Using this decomposition, we have from Eq. (179)

$$G(N_0^\alpha, \psi) = \mathrm{e}^{-iN_0^\alpha \psi_0} G(N_0^\alpha, \psi_1) \tag{186}$$

Hence Eq. (181) can be rewritten as

$$Z = \int \mathrm{d}\psi_0 \left(\frac{V}{2\pi v_0}\right)^{1/2} \int \mathscr{D}[\psi_1] \exp\Bigg[-in \int_0^\infty P_0(N_0) N_0\, \mathrm{d}N_0 \psi_0 + n \int_0^\infty P_0(N_0) \ln G(N_0, \psi_1)\, \mathrm{d}N_0 - \frac{1}{2v_0} \int \psi_1^2\, \mathrm{d}^d r - \frac{V}{2v_0} \psi_0^2\Bigg] \tag{187}$$

where we have used the fact that ψ_0 is a Gaussian variable with

$$\langle \psi_0 \rangle_\psi = 0, \qquad \langle \psi_0^2 \rangle_\psi = \frac{v_0}{V} \tag{188}$$

The integration with respect to ψ_0 can be performed easily to yield

$$Z = \mathrm{e}^{-v_0 n^2 \langle\langle N_0 \rangle\rangle^2 V} \int \mathscr{D}[\psi_1] \exp\Bigg[n \int_0^\infty P_0(N_0) \ln G(N_0, \psi_1)\, \mathrm{d}N_0 - \frac{1}{2v_0} \int \psi_1^2\, \mathrm{d}^d\mathbf{r}\Bigg] \tag{189}$$

where $\langle\langle N_0 \rangle\rangle$ is the number-average length of the chain (or the average with respect to P_0). This is the most convenient starting point for the study of semidilute solutions.

2. *Calculation of the Partition Function*

We cannot calculate Eq. (189) exactly; therefore we have to consider a perturbative calculation. To this end, we perform the dimensional analysis (in the sense explained in Section II.B) of the model.

Since our model is essentially a direct product of single chain models, the analysis performed in Section II.B is still valid. Again we must work in (4-ε)-space. However, there is also a new element. Equation (188) implies that $[\psi]^2 = [v_0/V]$ or $=[v_0 c]$, where c is the polymer number density. On the other hand, Eq. (186) tells us $[N_0\psi] = 1$ or $[\psi] = [N_0]^{-1}$. Hence $[v_0 c] = [N_0]^{-2}$ or $[v_0 c N_0^2] = 1$. This implies that our model has two natural parameters in the vanishing cutoff limit, one is $v_0 N_0^{\varepsilon/2}$ and the other is $v_0 c N_0^2$. If the concentration of macromolecules is nonzero, the latter is not small. Thus our perturbative calculation should be in terms of v_0 [in $(4-\varepsilon)$-space] with $v_0 c N_0^2$ being kept constant. Such a perturbation theory can be devised if we expand G in terms of the field ψ_1.

We have to calculate ln G perturbatively. The formula for G given by Eq. (179) can be written as

$$G(N_0, \psi_1) = \int d^d\mathbf{r} \int d^d\mathbf{r}' G(N_0, \mathbf{r}, \mathbf{r}'; \psi_1) \tag{190}$$

where

$$G(N_0, \mathbf{r}, \mathbf{r}'; \psi_1) = \int_{\mathbf{c}(0)=\mathbf{r}}^{\mathbf{c}(N_0)=\mathbf{r}'} \mathscr{D}[\mathbf{c}] \exp\left[-\frac{1}{2}\int \dot{\mathbf{c}}^2\, d\tau - i\int \psi_1(\mathbf{c}(\tau))\, d\tau\right] \tag{191}$$

The path integral [Eq. (191)] is performed on the set of paths connecting $\mathbf{r}$ and $\mathbf{r}'$. We have the following partial differential equation according to the fundamental property of the Feynman-Kac formula (see Appendix B):

$$\left(\frac{\partial}{\partial N_0} - \frac{1}{2}\Delta\right) G(N_0, \mathbf{r}, \mathbf{r}'; \psi_1) = -i\psi_1(\mathbf{r}) G(N_0, \mathbf{r}, \mathbf{r}'; \psi_1) + \delta(N_0)\,\delta(\mathbf{r} - \mathbf{r}') \tag{192}$$

Note that because of the field ψ_1, the system is not translationally symmetric. Equation (192) can be transformed into an integral equation:

$$G(N_0, \mathbf{r}, \mathbf{r}'; \psi_1) = G_0(N_0, \mathbf{r} - \mathbf{r}') - i\int_0^{N_0} d\tau \int_V d^d\boldsymbol{\rho}\, G_0(\tau, \mathbf{r}' - \boldsymbol{\rho})\psi(\boldsymbol{\rho}) \times G(N_0 - \tau, \mathbf{r}, \boldsymbol{\rho}; \psi_1) \tag{193}$$

or into its Fourier-transformed version:

$$\tilde{G}(N_0, \mathbf{k}, \mathbf{k}'; \psi_1) = G_0(N_0, \mathbf{k})\, \delta(\mathbf{k} - \mathbf{k}') - i \int_0^{N_0} \mathrm{d}\tau \int \mathrm{d}^d\mathbf{q} (2\pi)^{-d} G_0(\tau, \mathbf{k}') \psi(q)$$
$$\times \tilde{G}(N_0 - \tau, \mathbf{k} - \mathbf{q}, \mathbf{k}'; \psi_1) \tag{194}$$

where G_0 is the Gaussian function given in Eq. (70), and

$$\tilde{G}(N_0, \mathbf{k}, \mathbf{k}'; \psi_1) = \int \mathrm{d}^d\mathbf{r} \int \mathrm{d}^d\mathbf{r}'\, \mathrm{e}^{-i\mathbf{k}\cdot\mathbf{r} - i\mathbf{k}'\cdot\mathbf{r}'} G(N_0, \mathbf{r}, \mathbf{r}'; \psi_1) \tag{195}$$

We solve Eq. (194) iteratively, and the result can be shown diagrammatically as

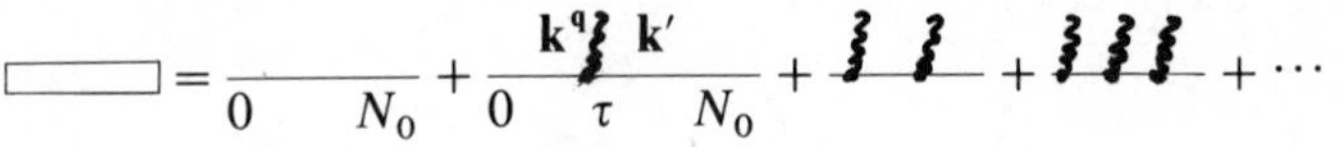

(196)

where

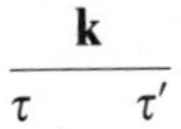

$= \mathrm{e}^{-(1/2)[k^2(\tau' - \tau)]}$

$= -i\psi_1(\mathbf{q})$

$= \int \mathrm{d}\tau (2\pi)^{-d} \int \mathrm{d}^d\mathbf{q}\, \delta(\mathbf{k} + \mathbf{q} - \mathbf{k}')$

Thus combining this result and Eq. (190) yields

$$G(N_0, \psi) = V - \int_0^{N_0} \mathrm{d}x (N_0 - x)(2\pi)^{-d} \int \mathrm{d}^d\mathbf{q} \tilde{G}_0(x, \mathbf{q}) \psi_1(\mathbf{q}) \psi_1(-\mathbf{q}) + \cdots \tag{197}$$

Introducing this into Eq. (189), the lowest nontrivial order result is obtained as

$$Z = \mathrm{e}^{-v_0 n^2 \langle\langle N_0 \rangle\rangle^2 / 2V} V^n Z_1 \tag{198}$$

with

$$Z_1 = \int \mathscr{D}[\psi_1] \exp\left\{\left[n \int_0^\infty dN_0 P_0(N_0) \ln\left(1 - \frac{1}{V}\int_0^{N_0} dx(N_0 - x)\right.\right.\right.$$
$$\left.\left.\times (2\pi)^{-d} \int d^d\mathbf{q}\hat{G}_0(x, \mathbf{q})\psi_1(\mathbf{q})\psi_1(-\mathbf{q}) + \cdots\right)\right]$$
$$\left. - \frac{1}{2v_0}(2\pi)^{-d} \int d^d\mathbf{q}\psi_1(\mathbf{q})\psi_1(-\mathbf{q})\right\} \tag{198a}$$

where the V in Eq. (197) is already separated out as V^n in Eq. (198). The quantity $S_0(\mathbf{q})$:

$$S_0(\mathbf{q}) = \int_0^{N_0} d\tau(N_0 - \tau)\tilde{G}_0(\tau, \mathbf{q}) \tag{199}$$

appearing in Eq. (198a) is the unperturbed scattering function itself. Hence, if we denote the average over the molecular weight distribution by $\langle\langle\ \rangle\rangle$, then Eq. (198a) becomes

$$Z_1 = \int \mathscr{D}[\psi_1] \exp\left[-(2\pi)^{-d} \int d^d\mathbf{q}\left[\frac{n}{V}\langle\langle S_0(\mathbf{q})\rangle\rangle\psi_1(\mathbf{q})\psi_1(-\mathbf{q})\right.\right.$$
$$\left. + \frac{1}{2v_0}\psi_1(\mathbf{q})\psi_1(-\mathbf{q})]\right\} \tag{200}$$

If the polymer number density $c = n/V$ is zero, then the first term in the square bracket disappears; we have only the $|\psi_1|^2/2v_0$ term. This implies that the inverse of the coefficient of $|\psi_1|^2$ is the effective excluded-volume parameter. Thus if $c \neq 0$

$$\left[\frac{1}{v_0} + 2c\langle\langle S_0(\mathbf{q})\rangle\rangle\right]^{-1} = \frac{v_0}{1 + 2cv_0\langle\langle S_0(\mathbf{q})\rangle\rangle} \tag{201}$$

is the effective excluded-volume parameter (*to the order* we are working). Thus the effective excluded-volume becomes smaller: the screening effect first pointed out by Edwards.[59]

To this order, the integration over ψ_1 is again Gaussian. Thus the problem is essentially to calculate the Fredholm determinant.[107] We use the following equation:

$$\int \mathscr{D}[\psi_1] \exp\left[-\frac{1}{2V}\sum_{\mathbf{q}} \psi_1(\mathbf{q})F(\mathbf{q})^{-1}\psi_1(-\mathbf{q})\right] = \exp\left[\frac{1}{2} V \frac{1}{(2\pi)^d} \int \mathrm{d}^{dq} \ln\left(\frac{F(\mathbf{q})}{V}\right)\right] \tag{202}$$

Now we can evaluate Eq. (198) as

$$Z = V^n \exp\left[-\frac{v_0 n^2}{2V} \langle\langle N_0 \rangle\rangle^2 - \frac{V}{2}(2\pi)^{-d} \int \mathrm{d}^d q \ln\left(1 + \frac{2nv_0}{V} \langle\langle S_0(\mathbf{q})\rangle\rangle\right)\right] \tag{203}$$

Hence, the (bare) free energy is given by

$$\begin{aligned} F_B &= -k_B T \ln Z \\ &= -k_B T n \ln V + \frac{k_B T v_0^2}{2V} n^2 \langle\langle N_0 \rangle\rangle^2 + \frac{k_B T}{2} \frac{V}{(2\pi)^d} \\ &\quad \times \int \mathrm{d}^d \mathbf{q} \ln\left(1 + \frac{2nv_0}{V} \langle\langle S_0(\mathbf{q})\rangle\rangle\right) \end{aligned} \tag{204}$$

3. *Bare Perturbation Result for Osmotic Pressure*

By definition, the (bare) osmotic pressure π is calculated as

$$\begin{aligned} \pi_B &= -\frac{\partial F_B}{\partial V} \\ &= k_B T \left\{\frac{n}{V} + \frac{\langle\langle N_0\rangle\rangle^2}{2V^2} n^2 v_0 - \frac{1}{2(2\pi)^d} \int \mathrm{d}^d \mathbf{k} \left[\ln\left(1 + \frac{2nv_0}{V}\langle\langle S_0(\mathbf{k})\rangle\rangle\right)\right.\right. \\ &\quad \left.\left. - \frac{2nv_0 \langle\langle S_0(\mathbf{k})\rangle\rangle / V}{1 + \frac{2nv_0}{V}\langle\langle S_0(\mathbf{k})\rangle\rangle}\right]\right\} \end{aligned} \tag{205}$$

We know

$$\langle\langle S_0(\mathbf{k})\rangle\rangle = 2\frac{\langle\langle N_0\rangle\rangle}{k^2} - \frac{4}{k^4}(1 - \langle\langle \mathrm{e}^{-N_0 k^2/2}\rangle\rangle) \tag{206}$$

We can calculate the osmotic pressure using this formula. Here, however, we use a simpler but sufficiently accurate approximation (within 3% error) to have a closed analytic form for the osmotic pressure. Note the following asymptotic behaviors of $\langle\langle S_0(\mathbf{k})\rangle\rangle$:

$$\begin{aligned}\langle\langle S_0(\mathbf{k})\rangle\rangle &\to \tfrac{1}{2}\langle\langle N_0^2\rangle\rangle \qquad (\mathbf{k}\to \mathbf{0})\\ &\to \frac{2}{k^2}\langle\langle N_0\rangle\rangle \qquad (\mathbf{k}\to\infty)\end{aligned} \tag{207}$$

This suggests that the following approximation is correct in both asymptotic regions

$$\langle\langle S_0(\mathbf{k})\rangle\rangle \simeq \frac{\langle\langle N_0\rangle\rangle^2}{\frac{1}{2}\langle\langle N_0\rangle\rangle k^2 + 2\mu} \tag{208}$$

where

$$\mu = \frac{\langle\langle N_0\rangle\rangle^2}{\langle\langle N_0^2\rangle\rangle} = \frac{M_n}{M_w} \tag{209}$$

with M_n being the number-averaged molecular weight and M_w the weight-averaged molecular weight. The approximation is *exact* when the molecular weight distribution function is exponential, that is, if $P_0(N_0) = \langle\langle N_0\rangle\rangle^{-1} e^{-N_0/\langle\langle N_0\rangle\rangle}$, then Eq. (208) is exact with $\mu = \frac{1}{2}$. Therefore we adopt Eq. (208) in the following calculation. Equation (205) can be written explicitly as

$$\begin{aligned}\pi_B = k_BT\Bigg\{c + \frac{A}{4}c + \frac{1}{2(2\pi)^d}\left(\frac{2}{N_0}\right)^{d/2}\\ \times \int d^dz\left[\ln\left(1+\frac{A}{z^2+2\mu}\right) - \frac{A}{z^2+2\mu+A}\right]\Bigg\}\end{aligned} \tag{210}$$

where $A = 2v_0cN_0^2$. Note that the term in square brackets contains order A^2 terms. Since $A^2/N_0^2 \sim cv_0(cv_0N_0^2)$, the second term can give the order ε term. The integral can be analytically performed to order ε as is given in Appendix E. The result is

$$\begin{aligned}\frac{\pi_B}{ck_BT} = 1 + \tfrac{1}{2}X_0 - \frac{u_0}{4\pi^2}X_0\Bigg\{\left[\frac{2}{\varepsilon} + \frac{3}{2} - \hat{\gamma} + \ln\left(\frac{2\pi N_0}{L}\right) - \ln 2\right]\\ - \frac{\mu^2}{X_0^2}\ln\mu + \left(\frac{\mu^2}{X_0^2} - 1\right)\ln(\mu + X_0) - \left(1 + \frac{\mu}{X_0}\right)\Bigg\}\end{aligned} \tag{211}$$

where $X_0 = v_0cN_0^2$, $\hat{\gamma}$ is Euler's constant and μ is given by Eq. (209).

4. *Renormalization of the Osmotic Pressure*

Since the polymer number density c and the polydispersity parameter μ are observable, it should not be renormalized (see Section III.B). Also the osmotic pressure is directly observable. Thus we can renormalize only the excluded-volume parameter v_0 and the bare chain length N_0. Note that the parameter μ is a given quantity even if N_0 is microscopically not uniquely defined. Let us use Eqs. (82) and (83) to renormalize Eq. (211). Since we have already fixed $\hat{B}$ and $\hat{D}$, we may use them to renormalize Eq. (211). Here, however, we determine the coefficient $\hat{D}$ independently from our former results, assuming that $\hat{B}$ is known. This is a more natural way to determine $\hat{D}$ than is given in Section IV.A.3.

Let us introduce a renormalized quantity $\hat{X}$ as

$$\begin{aligned}\hat{X} &= cuN^2 \\ &= X_0[1 + (-\hat{D} + 2\hat{B})u + \cdots]\end{aligned} \tag{212}$$

Replacing X_0 in Eq. (211) with $\hat{X}$, we get

$$\begin{aligned}\frac{\pi}{ck_BT} = 1 + \frac{\hat{X}}{2}[1 - (-\hat{D} + 2\hat{B})u + \cdots] - \frac{u}{4\pi^2}\Bigg\{\hat{X}\Bigg[\frac{2}{\varepsilon} + \frac{3}{2} - \hat{\gamma} \\ + \ln\frac{2\pi N_0}{L} - 1 + \frac{\mu}{\hat{X}} - \ln 2\Bigg] - \frac{\mu^2}{\hat{X}}\ln\mu + \left(\frac{\mu^2}{\hat{X}} - \hat{X}\right)\ln(\mu + \hat{X})\Bigg\}\end{aligned} \tag{213}$$

Comparing the singular terms to order u, we obtain the following identity:

$$-\frac{\hat{X}}{2}(-\hat{D} + 2\hat{B}) - \frac{1}{4\pi^2}\hat{X}\frac{2}{\varepsilon} = 0 \tag{214}$$

so that

$$\hat{D} = 2\hat{B} + \frac{1}{\pi^2\varepsilon} = \frac{2}{\pi^2\varepsilon} \tag{215}$$

where we have used Eq. (89). This is in agreement with Eq. (94).

Thus we get the osmotic pressure in terms of $\hat{X}$ as

$$\begin{aligned}\frac{\pi}{ck_BT} = 1 + \frac{\hat{X}}{2} - \frac{\varepsilon}{8}\Bigg\{\hat{X}\left(\frac{3}{2} - \hat{\gamma} + \ln\frac{2\pi N}{L} - \ln 2\right) - \frac{\mu^2}{\hat{X}}\ln\mu - \hat{X} + \mu \\ + \left(\frac{\mu^2}{\hat{X}} - \hat{X}\right)\ln(\mu + \hat{X})\Bigg\}\end{aligned} \tag{216}$$

5. *Renormalization-Group Equation for the Osmotic Pressure*

The osmotic pressure π depends on the polymer number density c and the distribution function of their degree of polymerization P

$$\pi = \pi(c, P) \tag{217}$$

Since the bare length N_0 and the renormalized length N are proportional as

$$N = Z_N N_0 \tag{218}$$

the bare distribution function P_0 and observable distribution function P must be related as

$$P(\cdot) = Z_N^{-1} P_0(Z_N^{-1} \cdot) \quad \text{or} \quad P_0(\cdot) = Z_N P(Z_N \cdot) \tag{219}$$

Therefore the renormalization-group equation for the osmotic pressure is given by

$$\left[L \frac{\partial}{\partial L} + \beta(u) \frac{\partial}{\partial u} - \gamma_N(u) \int_0^\infty dt P(t) \frac{\delta}{\delta P(t)} \right] \pi = 0 \tag{220}$$

where β and γ_N are given as before by Eqs. (45) and (46). The operator $\delta/\delta P$ is the functional derivative. The general solution to Eq. (220) at the nontrivial fixed point $u = u^*$ is

$$\pi(c, P) = f(c, L^{-(1-2\nu)/2\nu} P(\cdot)) \tag{221}$$

where we have already used the relation between ν and $\gamma_1(u^*)$ given by Eq. (65).

Next, we perform the dimensional analysis in the sense explained in Section II.B. Let $[N] = C$. Then

$$\begin{aligned} [\pi V] = 1 \Rightarrow [\pi] &= [V]^{-1} = C^{-d/2} \\ [c] &= [V]^{-1} = C^{-d/2} \\ [P] &= [N]^{-1} = C^{-1} \end{aligned} \tag{222}$$

Therefore, we have for an arbitrary positive constant C,

$$\pi(c, P(\cdot), L) = C^{d/2} \pi(C^{-d/2} c, C^{-1} P(\cdot), LC) \tag{223}$$

Combining this and Eq. (221) yields

$$\pi(c, P(\cdot), L) = C^{d/2} f(C^{-d/2} c, L^{-(1-2\nu)/\nu} P(\cdot) C^{-1/2\nu}) \tag{224}$$

Let us choose $C = c^{2/d}$, then generally we have

$$\pi(c, P(\cdot)) = cf(c^{-1/d\nu} P(\cdot)) = cf(P(c^{1/d\nu} \cdot)) \tag{225}$$

where f is an appropriate function. This implies that π/c is parametrized by $c^{1/\nu d}\langle\langle N \rangle\rangle$ and higher moments of P. If we consider only up to the second moment, we must have

$$\pi = cf(c\langle\langle N \rangle\rangle^{\nu d}, \mu) \tag{226}$$

that is, π/c is a function of μ and $c\langle\langle N \rangle\rangle^{\nu d}$.

6. *The Universal Form of the Osmotic Pressure*

The comparison of Eqs. (226) and (216) suggests that $\hat{X}$ should be rewritten in terms of a parameter proportional to $ucN^{d\nu}$ (see Appendix A also). Let us introduce

$$X = \hat{X}\left(\frac{2\pi N}{L}\right)^{-\varepsilon/4} (\ln 2 + \hat{\gamma} - \tfrac{1}{2})^{\varepsilon/4}$$

or

$$X = cuN^2\left(\frac{2\pi N}{L}\right)^{-\varepsilon/4} (\ln 2 + \hat{\gamma} - \tfrac{1}{2})^{\varepsilon/4} \tag{227}$$

Then we can rewrite Eq. (216) as

$$\frac{\pi}{ck_BT} = 1 + \frac{X}{2} \exp\frac{\varepsilon}{4}\left[\frac{\mu^2}{X^2} \ln \mu - \left(\frac{\mu^2}{X^2} - 1\right) \ln(\mu + X) + \frac{\mu}{X}\right] \tag{228}$$

X may be written as c/c^* by introducing some concentration scale c^*, but as it is, Eq. (228) cannot be compared with experiments without using adjustable parameter (e.g., c^*). However, as has been done by Wiltzius et al.,[108] we can eliminate the arbitrary constant, if we know the second virial coefficient experimentally. From Eq. (228) in the $X \to 0$ limit we obtain

$$\frac{\pi}{ck_BT} = 1 + \left[\frac{1}{2}\left(1 + \frac{\varepsilon}{8}\right) + \frac{\varepsilon}{8} \ln \mu\right] X + \cdots = 1 + A_2 c + \cdots \tag{229}$$

where A_2 is the second virial coefficient. Thus in 3-space

$$A_2 c = X\left(\frac{9}{16} - \frac{\varepsilon}{8}\ln\frac{M_w}{M_n}\right) \tag{230}$$

Since A_2 and c are measurable, Eq. (230) can be used to eliminate all the ambiguity in the parameter. A comparison with experiments will be given in Section VI.

Next, let us consider the asymptotic behavior of Eq. (228) in the $X \to \infty$ limit. We have the des Cloizeaux law:[6]

$$\frac{\pi}{ck_BT} \sim X^{1+\varepsilon/4} \sim c^{1+\varepsilon/4} \tag{231}$$

This is in conformity with the scaling argument $\pi \sim c^{d\nu/(d\nu-1)}$ [see Appendix A, Eq. (A.5)] since $d\nu - 1 = 1 - \varepsilon/4$.

The parameter $X \propto cN^{d\nu}$ contains c and N, so that there are two typical methods to make X very large; one is by making the polymer concentration higher and the other is the limit $N \to \infty$. Anyway, to apply the renormalization-group argument, N must be sufficiently large. However, at the same time the correlation length of the monomer density fluctuation must be much larger than the (cutoff)$^{1/2}$. Hence, the concentration cannot be very big. Thus, although, the resultant Eq. (228) looks valid even in the high concentration limit, we cannot justify the derivation of the formula in this limit.[177]

7. *Calculation of the Correlation Length*

To calculate the correlation length of the fluctuation of the monomer concentration, a convenient way is to introduce the instantaneous monomer density field

$$\rho(\mathbf{r}) = \sum_{\alpha=1}^{n} \int \delta(\mathbf{c}_\alpha(\tau) - \mathbf{r})\, d\tau \tag{232}$$

Mathematically, $\delta(\mathbf{c}_\alpha(\tau) - \mathbf{r})$ is the so-called Donsker's δ-function,[109] whose justification can be seen in Ref. 110. For the true meaning of the Eq. (232) see Kubo.[111] Ignoring the cutoff implicit in Eq. (173), it can be rewritten as

$$\mathscr{H} = \frac{1}{2}\sum_{\alpha=1}^{n} \int_0^{N_0^\alpha} \dot{\mathbf{c}}_\alpha(\tau)^2\, d\tau + \tfrac{1}{2}v_0 \int \rho(\mathbf{r})^2\, d^d\mathbf{r} \tag{233}$$

Then Eq. (174) can be written as

$$Z = \prod_{\alpha} \int \mathscr{D}[\mathbf{c}_\alpha] \int \mathscr{D}[\rho] \prod_{\mathbf{r}} \delta\left(\rho(\mathbf{r}) - \sum_{\alpha=1}^{n} \int \delta(\mathbf{c}_\alpha(\tau) - \mathbf{r})\,\mathrm{d}\tau\right)$$
$$\times \exp\left[-\frac{1}{2}\sum_{\alpha=1}^{n}\int_0^{N_0^\alpha} \dot{\mathbf{c}}_\alpha(\tau)^2\,\mathrm{d}\tau - \tfrac{1}{2}v_0 \int \rho(\mathbf{r})^2\,\mathrm{d}^d\mathbf{r}\right] \tag{234}$$

where the relation Eq. (232) is incorporated into Eq. (233) with the aid of (a "continuous" product of) δ-functions. Next, we rewrite the δ-function in terms of its Fourier transform as (see Ref. 110),

$$Z = \prod_{\alpha} \int \mathscr{D}[\mathbf{c}_\alpha] \int \mathscr{D}[\rho] \int \mathscr{D}[\psi] \exp\left\{i \int \mathrm{d}^d\mathbf{r}\psi(\mathbf{r})\left[\rho(\mathbf{r}) - \sum_{\alpha=1}^{n}\right.\right.$$
$$\left.\left.\times \int \delta(\mathbf{c}_\alpha(\tau) - \mathbf{r})\,\mathrm{d}\tau\right] - \frac{1}{2}\sum_{\alpha=1}^{n}\int_0^{N_0^\alpha} \dot{c}_\alpha(\tau)^2\,\mathrm{d}\tau - \tfrac{1}{2}v_0\int \rho(\mathbf{r})^2\,\mathrm{d}^d\mathbf{r}\right\}$$
$$= \prod_{\alpha} \int \mathscr{D}[\mathbf{c}_\alpha] \int \mathscr{D}[\rho] \int \mathscr{D}[\psi] \exp\left[i \int \mathrm{d}^d\mathbf{r}\psi(\mathbf{r})\rho(\mathbf{r}) - \tfrac{1}{2}v_0 \int \rho(\mathbf{r})^2\,\mathrm{d}^d r\right]$$
$$\times \prod_{\alpha=1}^{n} G(N_0^\alpha, \psi) \tag{235}$$

where we have used Eq. (179). If we integrate out the ρ variables, we recover Eq. (178); that is, the transformation of Eq. (234) into (235) is a method to formally introduce the stochastic field used in the Kac method.[106]

Let $\rho_{\mathbf{k}}$ be the Fourier transform of Eq. (232),

$$\rho_{\mathbf{k}} = \sum_{\alpha=1}^{n} \int \mathrm{d}\tau\, e^{i\mathbf{c}_\alpha(\tau)\cdot\mathbf{k}} \tag{236}$$

Then the static structure factor $S(\mathbf{k})$ is given by

$$S(\mathbf{k}) = \langle \rho_{\mathbf{k}}\rho_{-\mathbf{k}} \rangle \tag{237}$$

where the average is with respect to the partition function Eq. (235). Note that, since ρ is not uniquely defined (because we do not know the elementary unit of the chain), $S(\mathbf{k})$ is not directly observable; we must multiply it by a renormalization constant (see Section III.B).

For convenience, we rewrite Eq. (235) in terms of the Fourier transforms of ρ and ψ as

$$Z = \prod_{\alpha} \int \mathscr{D}[\mathbf{c}_\alpha] \int \mathscr{D}[\rho] \int \mathscr{D}[\psi] \exp\left[i \int_{\mathbf{k}} \psi_{\mathbf{k}} \rho_{-\mathbf{k}} - \tfrac{1}{2} v_0 \int_{\mathbf{k}} \rho_{\mathbf{k}} \rho_{-\mathbf{k}} \right] \times \prod_{\alpha=1}^{n} G(N_0^\alpha, \psi) \tag{238}$$

where $\int_{\mathbf{k}} = (2\pi)^{-d} \int d^d\mathbf{k}$. We may freely choose or change multiplicative constants. We have

$$\langle \rho_k \rho_{-k} \rangle = Z^{-1} \prod_{\alpha} \int \mathscr{D}[\mathbf{c}_\alpha] \int \mathscr{D}[\rho] \int \mathscr{D}[\psi] \rho_{\mathbf{k}} \rho_{-\mathbf{k}} \times \exp\left[i \int_{\mathbf{k}} \psi_{\mathbf{k}} \rho_{-\mathbf{k}} - \tfrac{1}{2} v_0 \int_{\mathbf{k}} \rho_{\mathbf{k}} \rho_{-\mathbf{k}} \right] \prod_{\alpha=1}^{n} G(N_0^\alpha, \psi) \tag{239}$$

We perform the integration with respect to ρ first. We need

$$\int d\rho_{\mathbf{k}}\, d\rho_{-\mathbf{k}} \rho_{\mathbf{k}} \rho_{-\mathbf{k}} \exp\left[i \int_{\mathbf{k}} \psi_{\mathbf{k}} \rho_{-\mathbf{k}} - \tfrac{1}{2} v_0 \int_{\mathbf{k}} \rho_{\mathbf{k}} \rho_{-\mathbf{k}} \right] = \left(\frac{1}{v_0} - \frac{1}{v_0^2} \psi_{\mathbf{k}} \psi_{-\mathbf{k}} \right) \exp\left(-\frac{1}{2v_0} \int_{\mathbf{k}} \psi_{\mathbf{k}} \psi_{-\mathbf{k}} \right) \tag{240}$$

where $d\rho_{\mathbf{k}}\, d\rho_{-\mathbf{k}} = d\rho_{\mathbf{k}}\, d\bar{\rho}_{\mathbf{k}}$ is interpreted in the standard way. Hence Eq. (239) can be written as

$$S(\mathbf{k}) = \frac{1}{v_0} - \frac{1}{v_0^2} \langle \psi_{\mathbf{k}} \psi_{-\mathbf{k}} \rangle \tag{241}$$

where $\langle\, \rangle$ is with respect to Eq. (189). This is the starting point to calculate the scattering function. To calculate $S(\mathbf{k})$ to order ε, the approximation Eq. (197) is not enough;[112] we need up to $O(\psi^4)$ terms. Therefore, the calculation becomes complicated; we will quote important results only in Section VI.

Although there are many technical complications, we can obtain any and all static phenomenological relations required by experiments.

D. Homotopy Expansion

As we have seen, the ε-expansion method allows us to systematically calculate almost any phenomenological consequence of the model. Moreover, as we will see in Section VI, the result is semiquantitatively in agreement with experimental results. However, it is not very easy to intuitively understand what is going on in $(4 - \varepsilon)$-space.

Here, though only briefly, another systematic method for studying the self-avoiding walk is introduced.[113] The method sheds some light on the nature of the ε-expansion calculation. We proceed by introducing another path connecting the Gaussian random walk (RW) and the self-avoiding walk (SAW). If the interaction between monomers close to each other along the chain is assigned a comparatively stronger excluded-volume effect than the interaction between monomers further apart along the chain, we may achieve a transition from SAW to RW. A naive model is given by the Hamiltonian

$$\mathscr{H}_\theta = \frac{1}{2}\int_0^{N_0} \dot{\mathbf{c}}(\tau)^2\, \mathrm{d}\tau + \tfrac{1}{2}v_0 \int \mathrm{d}\tau \int_{|\tau-\sigma|>a} \mathrm{d}\sigma |\tau - \sigma|^{-1/2+\theta}\, \delta(\mathbf{c}(\tau) - \mathbf{c}(\sigma)) \tag{242}$$

where θ is the parameter parametrizing a continuous path (i.e., homotopy) between the SAW and RW (it is a homotopy parameter). We consider this model in *3-space*. If $\theta = \frac{1}{2}$, we have the Edwards Hamiltonian Eq. (11), and if $\theta = 0$, we may essentially apply a mean-field theory argument, as can be easily seen from the dimensional analysis; $[v_0] = C^{-\theta}$ [compare this with $[v_0] = C^{-\varepsilon/2}$ given in Eq. (13)].

Thus the parameter θ *smoothly* changes the model between SAW and RW. This is quite different from the two paths we have already considered in Section IV.B. There, v_0 or the range parameter m_0 describes the crossover, but they *cannot* smoothly change the dimension of the excluded-volume parameter. Therefore in Section IV.B we needed a parameter which smoothly changes the dimension of the excluded-volume parameter; we were forced to work in $(4 - \varepsilon)$-space.

Actually, the naive model defined by the Hamiltonian Eq. (242) is not very faithful to the original idea. Consider Fig. 11. In these conformations, there are more than one self-contacts. For the interaction at p, the original Hamiltonian assigns the weight $|\tau - \sigma|^{-1/2+\theta}$, but if the chain has a contact at q already, then we may regard the small loop made by these two contacts as a single chain. Hence, the weight should be $l^{-1/2+\theta}$, where l is the contour length of the shortest loop allowing shortcuts through other contacting points (see Fig. 11*b*). This model is still a bit incomplete, because if the end-to-end vector $\mathbf{R}$ becomes very small, then there is another virtual contact between two parts (the head and the tail) of the chain.

Thus our model Hamiltonian is

$$\mathscr{H}_0 = \frac{1}{2}\int \dot{\mathbf{c}}(\tau)^2\, \mathrm{d}\tau + \tfrac{1}{2}v_0 \int \mathrm{d}\tau \int_{|\tau-\sigma|>a} \mathrm{d}\sigma\, l(\tau, \sigma)^{-1/2+\theta}\, \delta(\mathbf{c}(\tau) - \mathbf{c}(\sigma)) \tag{243}$$

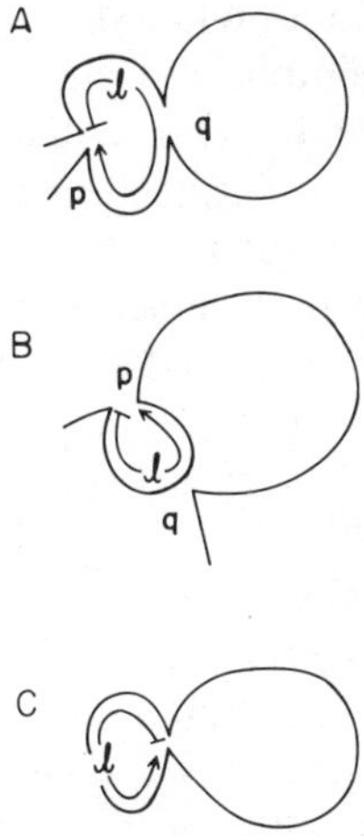

Fig. 11. Explanation of the contour length l. For the lowest nontrivial-order calculation, the rule given here is sufficient.

where $l(\tau, \sigma)$ is the shortest contour distance between $\mathbf{c}(\tau)$ and $\mathbf{c}(\sigma)$ along the chain allowing shortcuts through contacting points, pretending that both ends are connected (see Fig. 11*c*). This rule is sufficient to the lowest nontrivial order calculation.

We can study the system defined by Eq. (243) perturbatively in *3-space*. The necessary calculation for, for example, the end-to-end vector distribution function is quite parallel to that we have shown in detail in Section IV.A. For example, we can calculate the mean-square end-to-end distance as

$$\langle R^2 \rangle = 3N\left[1 - \frac{\theta}{4}\left(0.913 - \ln\frac{N}{L}\right)\right] \simeq 3\left(1 - \frac{\theta}{4}0.913\right)N\left(\frac{N}{L}\right)^{\theta/4} \tag{244}$$

where L is the phenomenological length scale as usual. This is comparable to the results [Eq. (129)] due to the standard ε-expansion. We obtain from Eq. (244)

$$\nu = \frac{1}{2}\left(1 + \frac{\theta}{4} + \cdots\right) \tag{245}$$

At $\theta = \frac{1}{2}$, which is the model we want to study, this is exactly the same as the $O(\varepsilon)$ result with $\varepsilon = 1$. Actually, $\theta = \varepsilon/2$ is true in the calculation of the exponent ν. The density distribution function $f(\mathbf{r})$ for the end-to-end vector in $\mathbf{r} = \mathbf{R}/(\langle R^2 \rangle)^{1/2}$ is more complicated analytically than Eq. (137), but numerically, if we put $\theta = \varepsilon/2$, indistinguishable from it.

Thus our expansion scheme with respect to θ, the homotopy parameter, can give at least comparable results to the ε-expansion method. The model Hamiltonian Eq. (243) has a built-in screening term. If $\theta < \frac{1}{2}$ the screening factor $l^{-1/2+\theta}$ screens global excluded-volume interactions. At $\theta = 0$, this screening is complete to make the polymer a Markovian chain; $\nu = \frac{1}{2}$. The naive identification, $\theta = \varepsilon/2$, suggests that changing the dimension is essentially screening the excluded-volume effect. This intuitive interpretation of the consequence of the change of the spatial dimensionality is quite natural: if the dimension is larger, it is more difficult for the chain to return to the starting point, and thus global interactions become rarer.

In the semidilute solutions, as we have seen in Section IV.C.2, the excluded-volume parameter is intrinsically screened by the extensive overlap of many chains. Therefore, introduction of an extra screening effect may be less drastic than in the case of dilute solutions. This suggests that the ε-expansion is more reliable for semidilute solutions than for dilute solutions. Theoretical results so far obtained are not contradictory to this suggestion (see Section VI).

V. TRANSPORT PHENOMENA

So far we have shown how the renormalization-group theory can be used to calculate static quantities. One of the merits of the conformation-space approach is the unified treatment of both equilibrium and nonequilibrium problems. Since the field-theoretic approach requires the existence of partition functions, it cannot be applied to nonequilibrium problems.

Unfortunately, nonequilibrium problems are almost always much more difficult than equilibrium ones, so that the theory of nonequilibrium properties of solutions is still not well developed in contrast to the equilibrium situations we have seen in Section IV. For example, there is no reliable and conceptually sound theory of semidilute solution dynamics, though the construction of a tractable theory is under way.

A. Transport Coefficients from Kirkwood-Riseman Theory, Especially $[\eta]$[114,115]

In this section, after a general discussion on the Kirkwood-Riseman scheme,[116] a detailed calculation of the intrinsic viscosity using this same scheme but *without* preaveraging the Oseen tensor is given.

1. *Kirkwood-Riseman Scheme*

As we have seen in Section III.A, any minimal model which can describe the dilute solution dynamics must describe the hydrodynamic interaction between monomers in a chain. As we will clearly see in Section V.B, this interaction is due to the fluctuation of the solvent velocity field excited by the movement of the chain.

In conventional approaches[1,117] the solvent velocity field is described hydrodynamically. Though this is an unsatisfactory description of the system as we will see in Section V.B, we follow the conventional theory in this section and describe the solvent hydrodynamically.

Let $\mathbf{F}(\tau)$ be the force exerted on the solvent by the unit of the chain at the contour variable τ, $\mathbf{u}(\tau)$ be the velocity of the unit at τ, and $\mathbf{v}(\mathbf{r})$ be the solvent velocity field without the chain. If a chain is immersed in the solvent the original velocity field $\mathbf{v}$ is perturbed by the forces $\mathbf{F}$ exerted on the solvent by the chain. The effect can be described by the Oseen tensor,

$$T_{\alpha\beta}[\mathbf{r}, \mathbf{r}'] = (2\pi)^{-d} \int d^d\mathbf{k} \frac{1}{\eta_0 k^2} \left(\delta_{\alpha\beta} - \frac{k_\alpha k_\beta}{k^2} \right) e^{i\mathbf{k}\cdot(\mathbf{r}-\mathbf{r}')} \tag{246}$$

where η_0 is the solvent viscosity. This is nothing but the hydrodynamic linear response kernel, so the perturbed velocity field $\tilde{\mathbf{v}}$ becomes

$$\tilde{\mathbf{v}}(\mathbf{r}) = \mathbf{v}(\mathbf{r}) + \int_0^{N_0} T[\mathbf{r}, \mathbf{c}(\tau)]\mathbf{F}(\tau)\, d\tau \tag{247}$$

On the other hand, if the translational friction coefficient of the chain unit is ζ_0 (again, this is a bare quantity, because we have no definite description of the chain unit and any quantity whose definition is microscopically arbitrary must be a bare quantity),

$$\mathbf{F}(\tau) = -\zeta_0(\tilde{\mathbf{v}}(\mathbf{c}(\tau)) - \mathbf{u}(\tau)) \tag{248}$$

Combining Eqs. (247) and (248), we get the Kirkwood-Riseman equation[116] for $\mathbf{F}$ as

$$\mathbf{F}(\tau) = \zeta_0[\mathbf{u}(\tau) - \mathbf{v}(\mathbf{c}(\tau))] - \zeta_0 \int_0^{N_0} d\tau' T(\tau, \tau')\mathbf{F}(\tau') \tag{249}$$

where

$$T(\tau, \tau') = T[\mathbf{c}(\tau), \mathbf{c}(\tau')] \tag{250}$$

The second term contains an implicit cutoff to eliminate the self-hydrodynamic interaction.

Once **F** is obtained, we can calculate the intrinsic viscosity $[\eta]$, translational and rotatory diffusion constants, and so on. For example, according to Kramers[118]

$$[\eta] = -\frac{N_A}{M\eta_0 g}\int_0^{N_0} \langle F_x(\tau)\tilde{c}_y(\tau)\rangle \,\mathrm{d}\tau \tag{251}$$

where $\tilde{\mathbf{c}}$ is the position vector of the unit at τ from the center of mass of the chain, and **F** is the solution to Eq. (249) with

$$v(\mathbf{c}(\tau)) = g\mathbf{e}_x\tilde{\mathbf{c}}(\tau)\cdot\mathbf{e}_y \tag{252}$$

$$\mathbf{u}(\tau) = \tfrac{1}{2}g\tilde{\mathbf{c}}(\tau)\times\mathbf{e}_z \tag{253}$$

where g is the shear strength, and $\mathbf{e}_x$ is the unit vector in the x-direction, and so forth. Equation (252) is the shear flow and Eq. (253) describes the rigid rotation. Thus in the present formalism, the relaxation time for the conformation is assumed to be much longer than the relaxation time of the solvent velocity fluctuation. Therefore this is consistent with the Kirkwood-Riseman equation which ignores the solvent velocity fluctuation.[119] Note that Eqs. (252) and (253) are meaningful even if the space has more than 3 dimensions, since we are considering a 2-dimensional flow.

The translational friction coefficient f also has an expression in terms of $\mathbf{F}(\tau)$ as

$$f\mathbf{u} = \int_0^{N_0} \mathrm{d}\tau\langle\mathbf{F}(\tau)\rangle \tag{254}$$

where **F** is the solution to Eq. (249) with $\mathbf{v} = \mathbf{0}$, and **u** is a uniform field.

Since the Kirkwood-Riseman scheme is an approximation, there have been attempts to make it more precise.[120] However, as we will see, the use of the Oseen tensor is a rather crude approximation to the description of the solvent dynamics, so such attempts are not of much intrinsic value.

2. *Dimensional Analysis*

To apply the Kirkwood-Riseman scheme, we must solve Eq. (249). This is a linear integral equation, but its kernel $T(\tau, \tau')$ explicitly depends on the instantaneous conformation of the chain. Hence, we must solve **F** for each conformation, but this is almost impossible. Thus in conventional theories a (bare) perturbation approach[121–123] or preaveraging approach is used which replaces T with its average $\langle T \rangle$ over the equilibrium conformation ensemble.

As we will see, this perturbative approach faces the same difficulty as in the excluded-volume effect explained in Section II.B. The preaveraging approach is not a very bad method if the problem of solving Eq. (249) is a mathematical one. However, as we have seen, the model is indeed a minimal model in the sense explained in Section III.B, so it presupposes the renormalization. We can combine the preaveraging approximation with the renormalized perturbation scheme.[114] The results suggest that preaveraging causes a minor inaccuracy (within 10 or 15%) with the renormalization-group approach. However, we do not discuss this approximation since we do not need it.

To discuss the inadequacy of the bare perturbational approach, we first study the reparametrization symmetry of our minimal model, that is, the Edwards Hamiltonian Eq. (11) and the Kirkwood-Riseman Eq. (249). According to the result given in Section II.B, the symmetry of the Edwards Hamiltonian is reproducible from the following dimensional analysis:

$$[\mathbf{c}] = C^{1/2},\ [N_0] = [\tau] = C,\ [v_0] = C^{-\varepsilon/2} \qquad (255)\ [=(13)]$$

Since $[\mathbf{r}] = [\mathbf{c}] = C^{1/2}$, $[\mathbf{k}] = C^{-1/2}$, so that the dimension of the Oseen tensor can be obtained from Eq. (246) as

$$[T] = [k]^{d-2} = C^{(2-d)/2} \tag{256}$$

Equation (249) tells us that

$$[\mathbf{F}] = [\zeta_0]\left[\int \mathrm{d}\tau\right][T][\mathbf{F}]$$

or

$$[\zeta_0]\left[\int \mathrm{d}\tau\right][T] = 1 \tag{257}$$

Therefore

$$[\zeta_0] = C^{-\varepsilon/2} \tag{258}$$

Since in the hydrodynamic perturbation theory the perturbation parameter is proportional to ζ_0, Eq. (258) implies that $\zeta_0 N_0^{\varepsilon/2}$ is the natural perturbation parameter. Thus exactly the same argument given in Section II.B invalidates this approach also. However, the analogy with the excluded-volume problem explained in Section III.3 suggests that the ε-expansion method can be applied in our present case as well. This observation was first exploited by Jasnow and Moore[124] (see also Ref. 125).

3. *Bare Perturbation Calculation of* $[\eta]$

Anticipating the renormalization of the friction coefficient ζ_0, we may solve Eq. (249) with Eqs. (252) and (253) iteratively. Our starting equation is

$$\mathbf{F}(\tau) = -\tfrac{1}{2}\zeta_0 g(\tilde{c}_y(\tau)\mathbf{e}_x + \tilde{c}_x(\tau)\mathbf{e}_y) - \zeta_0 \int_0^{N_0} d\tau' T(\tau, \tau')\mathbf{F}(\tau') \tag{259}$$

Hence to $O(\zeta_0^2)$, we get,

$$\mathbf{F}(\tau) = -\tfrac{1}{2}\zeta_0 g(\tilde{c}_y(\tau)\mathbf{e}_x + \tilde{c}_x(\tau)\mathbf{e}_y) + \tfrac{1}{2}\zeta_0^2 g \int_0^{N_0} d\tau' T(\tau, \tau')(\tilde{c}_y(\tau')\mathbf{e}_x + \tilde{c}_x(\tau')\mathbf{e}_y) \tag{260}$$

Note that we have solved the Kirkwood-Riseman equation for each instantaneous conformation. Introducing this solution to Kramers' formula Eq. (251), we get

$$[\eta]_B = \frac{N_A}{M\eta_0}\frac{\zeta_0}{2}\left[\int_0^{N_0} \langle \tilde{c}_y(\tau)^2 \rangle \, d\tau - \zeta_0 \int_0^{N_0} d\tau \int_0^{N_0} d\tau' \langle T_{xx}(\tau, \tau')\tilde{c}_y(\tau')\tilde{c}_y(\tau) + T_{xy}(\tau, \tau')\tilde{c}_x(\tau')\tilde{c}_y(\tau)\rangle\right] \tag{261}$$

where $\langle\rangle$ is the equilibrium average. Our task is to calculate this in $(4-\varepsilon)$-space. We need to calculate $\langle \tilde{c}_y(\tau)^2\rangle$ to $O(v_0)$, but the second term is only needed to $O(v_0^0)$ since there is an extra ζ_0 which is eventually of order ε.

Let us calculate the first term. Using the spherical symmetry of the equilibrium ensemble

$$\langle \tilde{c}_y(\tau)^2 \rangle = d^{-1}\langle \tilde{\mathbf{c}}(\tau)^2\rangle = \frac{1}{dN_0^2}\int_0^{N_0} d\sigma \int_0^{N_0} d\sigma' \langle (\mathbf{c}(\tau) - \mathbf{c}(\sigma))\cdot(\mathbf{c}(\tau) - \mathbf{c}(\sigma'))\rangle \tag{262}$$

where we have used

$$\tilde{\mathbf{c}}(\tau) = \mathbf{c}(\tau) - \mathbf{c}_{CM} = \frac{1}{N_0}\int_0^{N_0} d\sigma[\mathbf{c}(\tau) - \mathbf{c}(\sigma)] \tag{263}$$

To calculate Eq. (262) we utilize

$$\langle (\mathbf{c}(\tau) - \mathbf{c}(\sigma)) \cdot (\mathbf{c}(\tau) - \mathbf{c}(\sigma')) \rangle$$
$$= \tfrac{1}{2}\{\langle (\mathbf{c}(\tau) - \mathbf{c}(\sigma))^2 \rangle + \langle (\mathbf{c}(\tau) - \mathbf{c}(\sigma'))^2 \rangle - \langle (\mathbf{c}(\sigma) - \mathbf{c}(\sigma'))^2 \rangle\} \tag{264}$$

Thus the problem is reduced to the calculation of each term on the r.h.s. of Eq. (264). The first term in Eq. (261) is thus proportional to

$$I = \frac{1}{dN_0^2} \int_0^{N_0} \mathrm{d}\tau \int_0^{N_0} \mathrm{d}\sigma \int_0^{\sigma} \mathrm{d}\sigma' \{h(\tau, \sigma) + h(\tau, \sigma') - h(\sigma, \sigma')\} \tag{265}$$

where

$$h(\tau, \sigma) = \langle (\mathbf{c}(\tau) - \mathbf{c}(\sigma))^2 \rangle \tag{266}$$

Using the symmetry of h, we can simplify Eq. (265) to

$$I = \frac{1}{2\,dN_0} \int_0^{N_0} \mathrm{d}x \int_0^{N_0} \mathrm{d}y h(x, y) \tag{267}$$

To calculate Eq. (267) the best way is to calculate its generating function $G(\mathbf{k}, x, y)$:

$$G(\mathbf{k}, x, y) = \langle \mathrm{e}^{i\mathbf{k}\cdot(\tilde{\mathbf{c}}(x) - \tilde{\mathbf{c}}(y))} \rangle \tag{268}$$

$$= \mathcal{N}_0 - \frac{\mathcal{N}_0}{2d} k^2 h(x, y) + \cdots \tag{269}$$

This is parallel to what we have done in Section IV.A.1. The calculation, which is straightforward but tedious and which might discourage readers to use renormalization-group theories, is given in Appendix F, since, even in the original reference,[114] no detail was given. The final result is

$$h(x, y) = d(x - y)\left\{1 - \frac{u_0}{(2\pi)^2}\left[-\frac{2}{\varepsilon} + 1 - \ln\frac{2\pi(x - y)}{L} - \frac{y}{x - y}\ln\frac{x}{y} - \frac{N_0 - x}{x - y}\ln\frac{N_0 - y}{N_0 - x} + \frac{1}{2}\left(1 - \frac{x - y}{N_0}\right)\right]\right\}, \ (x > y) \tag{270}$$

Thus Eq. (267) can be calculated as (again this is a straightforward calculation)

$$I = \frac{N_0^2}{6}\left\{1 - \frac{u_0}{(2\pi)^2}\left[-\frac{2}{\varepsilon} - \ln\frac{2\pi N_0}{L} + \frac{13}{12}\right]\right\} + O(\varepsilon) \tag{271}$$

Next, let us calculate the second term in Eq. (261). More explicitly it reads

$$\begin{aligned} J = \int_0^{N_0} \mathrm{d}\tau \int_0^{N_0} \mathrm{d}\tau' \int \frac{\mathrm{d}^d\mathbf{k}}{(2\pi)^d}\frac{1}{k^2}\Bigg\{&\left(1 - \frac{k_x^2}{k^2}\right)\langle \tilde{c}_y(\tau)\tilde{c}_y(\tau') \\ &\times \exp[i\mathbf{k}\cdot(\tilde{\mathbf{c}}(\tau) - \tilde{\mathbf{c}}(\tau'))]\rangle + \left(-\frac{k_x k_y}{k^2}\right)\langle \tilde{c}_y(\tau)\tilde{c}_x(\tau') \\ &\times \exp[i\mathbf{k}\cdot(\tilde{\mathbf{c}}(\tau) - \tilde{\mathbf{c}}(\tau'))]\rangle\Bigg\} \end{aligned} \tag{272}$$

(In the original references,[114,127] the positions of { were wrong due to transcription mistakes; the results are correct.) We have only to calculate this to $O(\varepsilon^0)$, so that the average can be taken over the unperturbed ensemble. The best method to perform the calculation is to introduce the following generating function

$$J(\mathbf{k}, \mathbf{l}) = \langle \exp i[\mathbf{k}\cdot\tilde{\mathbf{c}}(\tau) - \mathbf{l}\cdot\tilde{\mathbf{c}}(\sigma)]\rangle \tag{273}$$

$$= \exp[-\tfrac{1}{2}(Ak^2 + Bl^2 - 2C\mathbf{k}\cdot\mathbf{l})] \tag{274}$$

where

$$A = \tfrac{1}{3}N_0\left(1 - \frac{3\tau(N_0 - \tau)}{N_0^2}\right), \qquad B = \tfrac{1}{3}N_0\left(1 - \frac{3\sigma(N_0 - \sigma)}{N_0^2}\right),$$

$$C = \frac{(\tau + \sigma) - |\tau - \sigma|}{2} + \frac{\tau^2 + \sigma^2}{2N_0} - (\tau + \sigma) + \frac{N_0}{3} \tag{275}$$

From Eq. (274) we get by appropriate differentiation

$$\langle \tilde{c}_y(\tau)\tilde{c}_y(\sigma)\,\mathrm{e}^{i\mathbf{k}\cdot(\mathbf{c}(\tau) - \mathbf{c}(\sigma))}\rangle = [(C - A)(C - B)k_y^2 + C]\,\mathrm{e}^{-(1/2)k^2(A+B-2C)} \tag{276}$$

$$\langle \tilde{c}_y(\tau)\tilde{c}_x(\sigma)\,\mathrm{e}^{i\mathbf{k}\cdot(\mathbf{c}(\tau) - \mathbf{c}(\sigma))}\rangle = (C - A)(C - B)k_x k_y\,\mathrm{e}^{-(1/2)k^2(A+B-2C)} \tag{277}$$

Introducing these formulas into Eq. (272) we have

$$J = \int_0^{N_0} \mathrm{d}\tau \int_0^{N_0} \mathrm{d}\sigma \frac{1}{(2\pi)^d} \int \mathrm{d}^d\mathbf{k} \frac{1}{k^2} \left\{ (C - A)(C - B) \left[k_y^2 - 2 \frac{k_x^2 k_y^2}{k^2} \right] + C \left[1 - \frac{k_x^2}{k^2} \right] \right\} \tag{278}$$

Performing the **k**-integration (Appendix G) and other integrations, we get

$$J = 2N^{2+\varepsilon/2}(2\pi)^{-2+\varepsilon/2} \left[\frac{5}{48(d+2)} + \left(1 - \frac{1}{d}\right)\left(\frac{1}{3\varepsilon} - \frac{4}{9}\right) \right] \tag{279}$$

Combining Eqs. (271) and (279) with Eq. (261) we finally get the bare perturbation result as follows

$$\frac{M}{N_A} [\eta]_B = \tfrac{1}{2}\xi_0 L^{-\varepsilon/2} \left\{ \frac{N_0^2}{6} \left[1 - \frac{u_0}{(2\pi)^2} \left(-\frac{2}{\varepsilon} + \frac{13}{12} - \ln \frac{2\pi N_0}{L} \right) \right] - \frac{1}{2\pi^2} \xi_0 N_0^2 \left[\frac{5}{48(d+2)} + \left(1 - \frac{1}{d}\right) \frac{1}{d-2} \times \left(\frac{1}{3\varepsilon} - \frac{4}{9} + \frac{1}{6} \ln \frac{2\pi N}{L} \right) \right] \right\} \tag{280}$$

where we have introduced ξ_0 defined by

$$\xi_0 \equiv \left(\frac{\zeta_0}{\eta_0} \right) L^{\varepsilon/2} \tag{281}$$

4. *Renormalization*

Now, the most tedious part of calculation is over. As we see explicitly in Eq. (280), there are $1/\varepsilon$-terms, which must be eliminated. The parameters that we may renormalize, or for which we may introduce phenomenological counterparts, are u_0, N_0 as in static problems and ξ_0 or ζ_0. All of these parameters crucially depend on the choice of the chain unit, which is almost arbitrary.

Since our renormalization scheme is the same as the one we have used in Section IV, we need not repeat the renormalization of u_0 and N_0 as is noted in Section III.B. We introduce one more renormalization constant Z_ζ as

$$\hat{\zeta} = Z_\zeta \zeta_0 \tag{282}$$

where $\hat{\zeta}$ is the phenomenological counterpart of ζ_0, or equivalently

$$\xi = Z_\zeta \xi_0 \tag{283}$$

where ξ is again the phenomenological counterpart of ξ_0. This relation must be such that ξ vanishes when ξ_0 vanishes, so that we assume

$$\xi_0 = \xi(1 + E_1\xi + E_2 u + \cdots) \tag{284}$$

Introducing Eqs. (82) and (83) with $\hat{B}$, $\hat{D}$ given by Eqs. (89) and (94), respectively, and Eq. (284) into Eq. (280), we get

$$\frac{M}{N_A}[\eta] = \frac{\xi}{2} L^{-\varepsilon/2}(1 + E_1\xi + E_2 u + \cdots)\left\{\frac{1}{6}N^2\left(1 - \frac{1}{\pi^2\varepsilon}u\right)\right.$$
$$\left.\times\left[1 - \frac{u}{(2\pi)^2}\left(-\frac{2}{\varepsilon} + \cdots\right)\right] - \frac{1}{2\pi^2}\xi N^2\frac{1}{8\varepsilon} + \cdots\right\} \tag{285}$$

where we have ignored regular terms. Thus we must have the identity

$$\frac{1}{6}\left[E_1\xi + E_2 u - \frac{1}{\pi^2\varepsilon}u + \frac{1}{2\pi^2\varepsilon}u\right] - \frac{\xi}{16\pi^2\varepsilon} = 0 \tag{286}$$

that is,

$$E_1 = \frac{3}{8\pi^2\varepsilon} \quad \text{and} \quad E_2 = \frac{1}{2\pi^2\varepsilon} \tag{287}$$

Note that $[\eta]$ itself should not be renormalized, because it is directly observable. The renormalized perturbation result is

$$\frac{M}{N_A}[\eta] = \frac{\xi}{2}L^{-\varepsilon/2}\left\{\frac{N^2}{6}\left[1 - \frac{u}{(2\pi)^2}\left(\frac{13}{12} - \ln\frac{2\pi N}{L}\right)\right] - \frac{1}{(2\pi)^2}\xi N^2\right.$$
$$\left.\times\left[-\frac{7}{72} + \frac{1}{16}\ln\frac{2\pi N}{L}\right]\right\} \tag{288}$$

5. *The Renormalization-Group Equation for* $[\eta]$

This section corresponds to Section IV.A.4 for the static quantity. Since we have an extra parameter ξ, we need one more β-function. The starting point is always

$$L\frac{\partial}{\partial L}[\eta]_B = 0 \tag{289}$$

Using the chain rule as we have done in Section III.B, we have

$$\left[L\frac{\partial}{\partial L} + \beta(u)\frac{\partial}{\partial u} + \beta_\xi(u, \xi)\frac{\partial}{\partial \xi} + \gamma_N(u)N\frac{\partial}{\partial N}\right][\eta] = 0 \tag{290}$$

where β and γ_N are given by Eqs. (101) and (102), respectively, and

$$\beta_\xi = L\left(\frac{\partial \xi}{\partial L}\right)_{v_0,\zeta_0} = \frac{\varepsilon}{2}\left(\frac{\partial \xi}{\partial \ln u_0} + \frac{\partial \xi}{\partial \ln \xi_0}\right) = \xi\left(\frac{\varepsilon}{2} - \frac{3}{16\pi^2}\xi - \frac{1}{4\pi^2}u\right) \tag{291}$$

Exactly as has been discussed in Section III.D, the hydrodynamic effect for a sufficiently long chain can be described by the value ξ^* which satisfies $\beta_\xi(u^*, \xi^*) = 0$, where $\beta(u^*) = 0$. There are four possibilities:

$$\left.\begin{array}{lll} \text{(a)} & u^* = 0,\ \xi^* = 0 & \text{free-draining Gaussian limit} \\ \text{(b)} & u^* = \pi^2\varepsilon/2,\ \xi^* = 0 & \text{free-draining self-avoiding limit} \\ \text{(c)} & u^* = 0,\ \xi^* = 8\pi^2\varepsilon/3 & \text{nondraining Gaussian limit} \\ \text{(d)} & u^* = \pi^2\varepsilon/2,\ \xi^* = 2\pi^2\varepsilon & \text{nondraining self-avoiding limit} \end{array}\right\} \tag{292}$$

For a sufficiently long chain, (d) is realized if the bare friction coefficient is nonzero. (c) is an approximate model of a long chain at the theta point as we have discussed in Section III.D. The fixed points in the static cases (a) and (b) are not at all realistic; they imply that the translational friction coefficient is zero, which is total nonsense. In the free-draining theory, the friction coefficient in the unperturbed term is assumed to be nonzero, and in higher-order terms is assumed to be zero. This is an inconsistency, since both friction coefficients must be identical. Thus, in the theory of solutions, free-draining theories are meaningless. (In concentrated systems, due to extensive hydrodynamic screening effect,[128,119] free-draining models may be useful.)

For any choice of (u, ξ) from (a)–(d), Eq. (290) becomes

$$\left(L\frac{\partial}{\partial L} + \gamma_N(u^*)N\frac{\partial}{\partial N}\right)[\eta] = 0 \tag{293}$$

Hence $[\eta]$ must have the following functional form

$$\frac{M\eta_0}{N_A}[\eta] = H(LN^{-1/\gamma_N^*}) \tag{294}$$

where we have used the fact that M, η_0, and N_A are given constants.

Following a line of thought exactly parallel to Section IV.A.5, we perform a dimensional analysis (again in the sense of Section II.B) of $[\eta]$. Note that the dimensions of M and N are different; M is a *given* constant. Thus $[M] = 1$ but $[N] = C$, where C is the dimension defined in Section II.B.

From the Kirkwood-Riseman Eq. (260)

$$[\mathbf{F}] = [\zeta_0][\tilde{\mathbf{c}}] = C^{(1-\varepsilon)/2} \tag{295}$$

Next, from the Kramers formula Eq. (251)

$$[[\eta]] = [\mathbf{F}][\mathbf{c}][\tau] = C^{d/2} \tag{296}$$

This can easily be seen from Eq. (261) also, since $[[\eta]] = [\zeta_0][c_y]^2[\tau]$. Thus

$$\frac{M[\eta]}{N_A}\eta_0 = C^{-d/2}H(C^{1-1/\gamma_N^*}LN^{-1/\gamma_N^*}) \tag{297}$$

for any positive number C. Hence we may choose the arbitrary positive constant C so that

$$C^{1-1/\gamma_N^*}LN^{-1/\gamma_N^*} = 1 \tag{298}$$

which is exactly the same as Eq. (63). We finally get

$$\frac{M[\eta]}{N_A}\eta_0 = H(1)N^{d\nu}L^{-d(2\nu-1)/2} \tag{299}$$

where the relation Eq. (65) between ν and γ_N^* has been used. Thus asymptotically we have [see Appendix A, Eq. (A.4)],

$$[\eta] \sim M^{d\nu-1} \tag{300}$$

since $N \propto M$. Note that for the free-draining cases $H(1) = 0$, that is, $[\eta] = 0$.

If we exponentiate the first-order terms in Eq. (288) we have the form of

Eq. (299) as

$$\frac{M}{N_A}[\eta] = \frac{1}{12}\frac{1}{(2\pi)^2} L^{d/2}\xi^*\left(\frac{2\pi N}{L}\right)^{d\nu} \exp\left[\frac{7}{6}\frac{\xi^*}{(2\pi)^2} - \frac{13}{12}\frac{u^*}{(2\pi)^2}\right] \tag{301}$$

In particular, at the nondraining and self-avoiding fixed point (d), Eq. (301) reduces to

$$\frac{M}{N_A}[\eta] = \frac{\varepsilon}{24}\exp\left(\frac{43}{96}\varepsilon\right)L^{d/2}\left(\frac{2\pi N}{L}\right)^{d\nu} \tag{302}$$

with

$$\nu = \frac{1}{2}\left(1 + \frac{\varepsilon}{8}\right) \tag{303}$$

At the nondraining Gaussian fixed point (c), our result reads

$$\frac{M}{N_A}[\eta] = \frac{\varepsilon}{18}\exp\left(\frac{7}{9}\varepsilon\right)(2\pi N)^{d\nu} \tag{304}$$

with

$$\nu = \tfrac{1}{2} \tag{305}$$

6. *Crossover Behavior of the Intrinsic Viscosity*

As has been discussed in Section IV.B, we need the full solution of the renormalization-group equation [Eq. (290)]. Using the standard theory of the quasilinear partial differential equations,[89] we have the general solution to this equation as (Appendix H)

$$\frac{M[\eta]}{N_A}\eta_0 = F(Lw^{-2/\varepsilon}, N(1+w)^{-1/4}, (1+w)^{3/4}(1-\hat{z})/\hat{z}) \tag{306}$$

where w is defined by Eq. (156) and

$$\hat{z} = \frac{u^*\xi}{u\xi^*} \tag{307}$$

with (u^*, ξ^*) being given by the value at (d) in Eq. (292). The invariance of $[\eta]$

under the reparametrization requires

$$\frac{M\eta_0}{N_A}[\eta] = C^{-d/2}F\left(CLw^{-2/\varepsilon}, CN(1+w)^{-1/4}, (1+w)^{3/4}\frac{1-\hat{z}}{\hat{z}}\right) \tag{308}$$

which is a generalization of Eq. (297). We may choose C to simplify the formula:

$$CN(1+w)^{-1/4} = 1 \tag{309}$$

which is essentially Eq. (154). Then Eq. (308) becomes

$$\frac{M\eta_0}{N_A}[\eta] = N^{d/2}(1+w)^{-d/8}G(\zeta, \delta) \tag{310}$$

where G is an as yet undetermined function, ζ is given by Eq. (157) and

$$\delta = (1+w)^{3/4}\frac{1-\hat{z}}{\hat{z}} \tag{311}$$

The parameter δ is a new one which describes the partial-draining effect. Note that if u is very small, then $\hat{z} \sim +\infty$ so that $\delta \simeq -1$; δ is exactly -1 for the nondraining Gaussian limit.

The remaining task is to find the function G from the renormalized perturbation result Eq. (288). We can follow what we have done in Section IV.B.3. We know the relation [Eq. (160)] between w and ζ. We must find the equation for ξ in terms of ζ and δ. We have (Appendix I).

$$\xi = \xi^*\{1 + \delta[1 + \zeta(1+\zeta)^{\varepsilon/8}]^{-3/4}\}^{-1}\zeta(1+\zeta)^{\varepsilon/8} \times \left(\frac{2\pi N}{L}\right)^{-\varepsilon/2 + \varepsilon\zeta\{1 - 3[\delta(1+\zeta)^{-3/4}/(1+\delta(1+\zeta)^{-3/4})]/4\}/2(1+\zeta)} \Big/ [1 + \zeta(1+\zeta)^{\varepsilon/8}] \tag{312}$$

Using this and Eq. (160), Eq. (288) can be rewritten in the form Eq. (310) required by the renormalization-group equation as

$$\frac{[\eta]}{N^{d/2}(1+w)^{-d/8}} = \frac{N_A}{12M}(2\pi)^{-\varepsilon/2}2\pi^2\varepsilon\frac{\zeta(1+\zeta)^{\varepsilon/8}}{[1+\zeta(1+\zeta)^{\varepsilon/8}]^{1/2+\varepsilon/8}} \times \{1 + \delta[1 + \zeta(1+\zeta)^{\varepsilon/8}]^{-3/4}\}^{-1} \times \exp\left[\frac{43}{96}\varepsilon\frac{\zeta}{1+\zeta} - \frac{7}{12}\varepsilon\frac{\zeta}{1+\zeta}\frac{\delta(1+\zeta)^{-3/4}}{1+\delta(1+\zeta)^{-3/4}}\right]. \tag{313}$$

Of course, $[\eta]$ itself cannot be universal, so that, as we expect, we cannot eliminate N and w from Eq. (313).

However, as is clear from Eq. (310) or Eq. (313), we need two variables to describe the universal behavior of the transport and dynamical properties of dilute solutions. This apparently contradicts the consensus[1] that the partial draining effect is unimportant. One of the successes of our renormalization approach is that we can explain why the partial-draining effect is not very important in the actual situations. As we have discussed in Section IV.B.4, the condition for the universal behavior of the expansion factors such as α_R^2 is the limit $N \to \infty$, $u \to 0$ with $Z \propto N^{\varepsilon/2}u$ being kept constant. Experimentally (as we will see in Section VI), α^2 seems universal. Thus we may think that the renormalized excluded-volume parameter is fairly small in the realistic situations. Then, as we have already discussed, $\hat{z}$ is very big, so Eq. (311) tells us that $\delta \simeq -1$. Thus one variable virtually disappears in the realistic situations. (See Section VII for a more careful statement.)

B. Critical Consideration on the Description of Solution Dynamics

In Section V.A we have adopted the Kirkwood-Riseman scheme, in which the Oseen tensor is used to describe the effect of the solvent velocity field [see Eq. (247). The scheme can be approximately justified from the so-called (Kirkwood) full-diffusion equation[129,1] description of polymer solutions,

$$\frac{\mathrm{d}}{\mathrm{d}t} P = \mathscr{L}_F^* P \tag{314}$$

where P is the distribution function of the conformation $\{\mathbf{c}\}$ and the operator $\mathscr{L}_F^*$ is given by

$$\mathscr{L}_F^* = \int \mathrm{d}\tau \int \mathrm{d}\tau' \frac{\delta}{\delta \mathbf{c}(\tau)} \left[\frac{1}{\zeta_0} \delta(\tau - \tau') + T(\tau, \tau') \right] \left(\frac{\delta}{\delta \mathbf{c}(\tau')} + \frac{\delta \mathscr{H}_E}{\delta \mathbf{c}(\tau')} \right) \tag{315}$$

Here, again, we use the Oseen tensor, and $k_B T$ is assumed to be unity. This operator implies that while the chain is described at the kinetic level, the solvent is described at the hydrodynamic level. That is, the fluctuation of the chain is taken into account explicitly (thus we have the Fokker-Planck type equation), but the solvent velocity field is described deterministically [see Eq. (247); if $\mathbf{F}$ is given, $\tilde{\mathbf{v}}$ is determined uniquely]. Thus, there is an inconsistency in the description of the system. One might argue as follows. A polymer chain is very long; thus the solvent molecules are relatively tiny. Therefore the solvent can be described hydrodynamically. This argument is, however, completely wrong. Since the Brownian motion of the chain is taken into account and since the fluctuating driving force is independent at each point on the chain, we should also take into account the Brownian motion of the solvent elements. Hence the inconsistency in describing the chain and solvent at different levels must be taken seriously.

An internally consistent description of a polymer solution has already been proposed by the present author,[130] which employs the kinetic-level description of both the chain and the solvent:

$$\frac{\partial}{\partial t}\mathbf{c}(\tau, t) = \mathbf{u}(\mathbf{c}(\tau, t), t) - \frac{1}{\zeta_0}\frac{\delta \mathscr{H}_E}{\delta \mathbf{c}(\tau, t)} + \boldsymbol{\theta}(\tau, t), \tag{316}$$

$$\rho_0 \frac{\partial}{\partial t}\mathbf{u}(\mathbf{r}, t) = -\int_0^{N_0} \mathrm{d}\tau \frac{\delta \mathscr{H}_E}{\delta \mathbf{c}(\tau, t)}\,\delta(\mathbf{r} - \mathbf{c}(\tau, t)) + \eta_0\,\Delta\mathbf{u}(\mathbf{r}, t) - \nabla p(\mathbf{r}, t) + \mathbf{f}(\mathbf{r}, t) \tag{317}$$

where $\mathbf{u}(\mathbf{r}, t)$ is the solvent velocity, $\{\mathbf{c}(\tau, t)\}_{\tau=0}^{N_0}$ is the conformation at time t, ρ_0 is the solvent density, p is the pressure and $\boldsymbol{\theta}$ and $\mathbf{f}$ are independent Gaussian white noises with zero means and

$$\langle \boldsymbol{\theta}(\tau, t)\boldsymbol{\theta}(\tau', t')\rangle = 2\zeta_0^{-1}\mathbf{1}\,\delta(t - t')\,\delta(\tau - \tau') \tag{318}$$

$$\langle \mathbf{f}(\mathbf{r}, t)\mathbf{f}(\mathbf{r}', t')\rangle = -2\eta_0\,\Delta\delta(\mathbf{r} - \mathbf{r}')\,\delta(t - t')\mathbf{1}, \tag{319}$$

$\mathbf{1}$ being the d × d unit tensor, and Δ the Laplacian. These noises satisfy the fluctuation-dissipation theorem. We have chosen $k_B T = 1$. The simultaneous equations Eq. (316) and Eq. (317) are very similar to that for the binary fluid critical dynamics.[131,132] The minimal model described by these equations has not yet been studied, but as we will see in Appendix J, to order ε, it can be reduced to the full-diffusion equation model Eq. (314). In other words, the full-diffusion equation is justifiable to order ε. The order ε^2 correction is quite complicated and practically intractable. The effects of these corrections have not yet been studied. If there is a flow field (as in the case of studying non-Newtonian effects), then the order ε^2 term becomes even non-Markovian; the memory effect of the solvent velocity field must be taken into account.[132]

Our conclusion is that any result so far obtained concerning polymer solutions is at most correct to order ε, and that the calculation to order ε^2 is prohibitively difficult. We have *no* theoretical method which is reliable to order higher than ε.

The simultaneous equations Eq. (316) and Eq. (317) have been extended by Shiwa to more concentrated solutions, and dynamical scaling results are reproduced.[133]

C. Full Dynamics Calculations

Our calculation in Section V.A is not a full dynamics calculation. So far there have been only a few quantities calculated from the full-diffusion Eq. (314) with the aid of the renormalization-group methods. As an example, here we show how to calculate the longest relaxation time of a dilute solution and the initial decay rate of the dynamical scattering factor. Since we start from the full-diffusion equation we should not expect any result reliable further than to order ε.

1. *The Longest Relaxation Time for Dilute Solutions*[134]

Let $\langle\rangle$ denote the average over $P(t)$. We have

$$\frac{\mathrm{d}}{\mathrm{d}t}\langle f\rangle(t) = \int f\frac{\mathrm{d}P}{\mathrm{d}t}\mathscr{D}[\mathbf{c}] = (f, \mathscr{L}_F^* P) \tag{320}$$

where the scalar product (,) is defined as

$$(f, g) = \int fg\mathscr{D}[\mathbf{c}] \tag{321}$$

Hence the adjoint operator $\mathscr{L}_F$ of $\mathscr{L}_F^*$ with respect to this scalar product can be used to define the time-dependent (i.e., Heisenberg picture) operators as

$$f(t) = \mathrm{e}^{t\mathscr{L}_F} f \tag{322}$$

Then Eq. (320) can be solved as

$$\langle f\rangle(t) = (\mathrm{e}^{t\mathscr{L}_F} f, P_0) \tag{323}$$

where P_0 is the initial distribution function, which we usually assume to be the equilibrium distribution function. The operator $\mathscr{L}_F$ (which is called the generator of a Markov process) is given by

$$\mathscr{L}_F = \int \mathrm{d}\tau \int \mathrm{d}\sigma \left[\frac{\delta}{\delta\mathbf{c}(\tau)} - \frac{\delta\mathscr{H}_E}{\delta\mathbf{c}(\tau)}\right] D(\tau, \sigma)\frac{\delta}{\delta\mathbf{c}(\sigma)} \tag{324}$$

where $\mathscr{H}_E$ is the Edwards Hamiltonian Eq. (11), and D is defined as

$$D(\tau, \sigma) = \frac{1}{\zeta_0}\delta(\tau - \sigma)\mathbf{1} + T(\tau, \sigma) \tag{325}$$

Equation (323) tells us that the smallest (nonzero) eigenvalue of $-\mathscr{L}_F$ is the reciprocal of the longest relaxation time of the system.

First, we decompose $\mathscr{L}_F$ as follows:

$$\mathscr{L}_F = \mathscr{L}_0 + \mathscr{L}_1^V + \mathscr{L}_1^H \tag{326}$$

where

$$\mathscr{L}_0 = \frac{1}{\zeta_0}\int d\tau\left[\frac{\delta}{\delta\mathbf{c}(\tau)} - \frac{\delta\mathscr{H}_0}{\delta\mathbf{c}(\tau)}\right]\frac{\delta}{\delta\mathbf{c}(\tau)} \tag{327}$$

$$\mathscr{L}_1^V = -\frac{1}{\zeta_0}\int d\tau\left(\frac{\delta\mathscr{H}_1}{\delta\mathbf{c}(\tau)}\right)\frac{\delta}{\delta\mathbf{c}(\tau)} \tag{328}$$

$$\mathscr{L}_1^H = \int d\tau\int d\sigma\left[\frac{\delta}{\delta\mathbf{c}(\tau)} - \left(\frac{\delta\mathscr{H}_0}{\delta\mathbf{c}(\tau)}\right)\right]T(\tau,\sigma)\frac{\delta}{\delta\mathbf{c}(\sigma)} \tag{329}$$

with $\mathscr{H}_1 = \mathscr{H}_E - \mathscr{H}_0$ and

$$\mathscr{H}_0 = \frac{1}{2}\int_0^{N_0} \dot{\mathbf{c}}(\tau)^2\, d\tau \tag{330}$$

We regard $\mathscr{L}_1^V$ and $\mathscr{L}_1^H$ as operators of order ε.

As is well known, the unperturbed operator Eq. (327) can be diagonalized with the aid of the free-draining Rouse coordinate $\boldsymbol{\xi}(p)$ $(p = 0, 1, 2, \ldots)$:[1]

$$\boldsymbol{\xi}(p) = \sqrt{\frac{2}{N_0}}\int_0^{N_0} d\tau\mathbf{c}(\tau)\cos\frac{\pi p\tau}{N_0} \tag{331}$$

as

$$-\mathscr{L}_0 = \sum_p\left[\frac{\partial}{\partial\boldsymbol{\xi}(p)} - \left(\frac{\pi p}{N_0}\right)^2\boldsymbol{\xi}(p)\right]\frac{\partial}{\partial\boldsymbol{\xi}(p)} \tag{332}$$

Therefore the smallest eigenvalue and its eigenfunction for $-\mathscr{L}_0$ is obtained as

$$\phi_\alpha = \left(\frac{\pi}{N_0}\right)\xi_\alpha(1),\ (\alpha = 1, \ldots, d) \tag{333}$$

$$\lambda_0 = \zeta_0^{-1}\left(\frac{\pi}{N_0}\right)^2$$

that is, d functions are degenerated, where ξ_α is the αth component of $\boldsymbol{\xi}$. We need the first order corrections due to $\mathscr{L}_1^V$ and $\mathscr{L}_1^H$. Since these operators are spherically symmetric, the degeneracy among $\{\phi_\alpha\}$ is not lifted. Hence we may simply use the first-order perturbation theory familiar in the elementary quantum mechanics. Let us introduce the scalar product $\langle\ ,\ \rangle$ by

$$\langle f, g\rangle = (f, gP_e) \tag{334}$$

where P_e is the equilibrium distribution function. Note that ϕ_α is normalized with respect to this scalar product, and that $\mathscr{L}_F$ and $\mathscr{L}_0$ are self-adjoint with respect to it. The perturbative corrections can be obtained as

$$\delta\lambda^H = -\langle\phi_x, \mathscr{L}_1^H\phi_x\rangle \tag{335}$$

$$\delta\lambda^v = -\langle\phi_x, \mathscr{L}_1^V\phi_x\rangle \tag{336}$$

In these formulas we may ignore the excluded-volume effect in P_e. Thus we have[134]

$$\delta\lambda^H = \frac{1}{\eta_0}\frac{2}{N_0}\left(\frac{\pi}{N_0}\right)^2 \int_0^{N_0} d\tau \int_{\substack{0\\|\tau-\sigma|>a}}^{N_0} d\sigma\left(1-\frac{1}{d}\right)\frac{1}{(2\pi)^d}\int d^d\mathbf{k}k^{-2}\cos\frac{\pi\tau}{N_0}\cos\frac{\pi\sigma}{N_0}$$
$$\times \exp(-\tfrac{1}{2}k^2|\tau-\sigma|) \tag{337}$$

and

$$\delta\lambda^V = \frac{1}{\zeta_0}\frac{v_0}{N_0}\int_0^{N_0} d\tau \int_{\substack{0\\|\tau-\sigma|>a}}^{N_0} d\sigma \frac{1}{(2\pi)^d}\int d^d\mathbf{k}k_x^2\left(\cos\frac{\pi\tau}{N_0} - \cos\frac{\pi\sigma}{N_0}\right)^2$$
$$\times \exp(-\tfrac{1}{2}k^2|\tau-\sigma|) \tag{338}$$

Thus, after an elementary calculation, we get ($d = 4$)

$$\lambda = \lambda_0 + \delta\lambda^V + \delta\lambda^H$$
$$= \frac{1}{\zeta_0}\left(\frac{\pi}{N_0}\right)^2\left\{1 - \frac{u_0}{2\pi^2}\left[\frac{1}{2}\ln\frac{2\pi N_0}{a} + I\right] + \frac{3\xi_0}{16\pi^2}\left[\ln\frac{2\pi N_0}{a} + J\right]\right\} \tag{339}$$

with

$$I = \frac{1}{2}\left[1 + \frac{4}{\pi^2} + \mathrm{ci}(\pi) - \hat{\gamma} - \ln\pi - \frac{3}{\pi}\mathrm{si}(\pi) + \frac{2}{\pi}\mathrm{si}(2\pi)\right] \tag{340}$$

$$J = \mathrm{ci}(\pi) - \hat{\gamma} - \ln\pi - \frac{1}{2} - \frac{\mathrm{si}(\pi)}{\pi} \tag{341}$$

where $\hat{\gamma}$ is Euler's constant, $\text{si}(x) = -\int_x^\infty dt \sin t/t$ and $\text{ci}(x) = -\int_x^\infty dt \cos t/t$. Equation (339) can easily be renormalized exactly as we have seen in Section V.A to yield

$$\lambda = \frac{1}{\zeta}\left(\frac{\pi}{N}\right)^2 \left\{1 - \frac{u}{2\pi^2}\left[\frac{1}{2}\ln\frac{2\pi N}{L} + I\right] + \frac{3\xi}{16\pi^2}\left[\ln\frac{2\pi N}{L} + J\right]\right\} \tag{342}$$

The renormalization-group equation for λ is identical to the one for the intrinsic viscosity [Eq. (306)]. Thus the functional form required by the renormalization-group theory is

$$\lambda = F\left(Lw^{-2/\varepsilon},\ N(1+w)^{-1/4},\ (1+w)^{3/4}\,\frac{1-\hat{z}}{\hat{z}}\right) \tag{343}$$

where w and $\hat{z}$ are as in Section V.A, and F is an as yet unspecified function.

The dimension of λ can be found as follows. $[\lambda]$ must be identical to $[\mathscr{L}_0]$. Since $[\delta/\delta\mathbf{c}] = C^{-3/2}$, $[\mathscr{L}_0] = [\zeta]^{-1}[\tau]C^{-3} = C^{-d/2}$. Hence $[\lambda] = C^{-d/2}$. Thus Eq. (343) must have the following scaling property:

$$\lambda = C^{d/2}F\left(CLw^{-2/\varepsilon},\ CN(1+w)^{-1/4},\ (1+w)^{3/4}\,\frac{1-\hat{z}}{\hat{z}}\right) \tag{344}$$

Thus, using the variables ζ and δ introduced in Section V.A.6, we have

$$\lambda = N^{-d/2}(1+w)^{d/8}\Lambda(\zeta, \delta) \tag{345}$$

where Λ is a function we will determine by comparison with Eq. (342). The final result reads

$$\begin{aligned}\lambda\eta_0[(1+w)^{-1/4}N]^{d/2} &= \frac{1}{2}(2\pi)^{\varepsilon/2}\,\frac{1-(1+Z)^{-3/4}}{Z}(1+Z)^{(1/2)+(\varepsilon/8)}\\ &\quad\times\exp\left[-\frac{\varepsilon}{8}J\frac{Z}{1+Z} - \frac{3}{8}\varepsilon I\right.\\ &\quad\left.\times\frac{Z}{1+Z}\,\frac{(1+Z)^{-3/4}}{1-(1+Z)^{-3/4}}\right]\end{aligned} \tag{346}$$

2. *The Initial Decay Rate of the Dynamical Scattering Factor*

The dynamical scattering factor is defined by

$$S(\mathbf{k}, t) = \langle \rho_{\mathbf{k}}(t)\rho_{-\mathbf{k}}(0)\rangle \tag{347}$$

where the average is over the initial equilibrium ensemble, and

$$\rho_{\mathbf{k}}(t) = \int d\tau \exp[i\mathbf{k} \cdot \mathbf{c}(\tau, t)] \tag{348}$$

which is the Fourier transform of the time-dependent monomer density field.

Ackasu pointed out that the initial decay rate $\Omega(\mathbf{k})$ of $S(\mathbf{k}, t)$ can be calculated very easily.[135] From Eq. (323), we can rewrite Eq. (347) as

$$S(\mathbf{k}, t) = \langle \rho_{-\mathbf{k}}(0)\, e^{t\mathscr{L}_F} \rho_{\mathbf{k}}(0) \rangle \tag{349}$$

Therefore

$$\Omega(\mathbf{k}) = -\frac{d}{dt} \ln S(\mathbf{k}, t)\bigg|_{t=0} = -S(\mathbf{k}, 0)^{-1} \langle \rho_{-\mathbf{k}} \mathscr{L}_F \rho_{\mathbf{k}} \rangle \tag{350}$$

where $S(\mathbf{k}, 0)$ is the static coherent scattering function, and $\mathscr{L}_F$ is explicitly given by Eq. (324). Although Eq. (350) is calculable nonperturbatively if there is no excluded-volume effect,[135] since our model is a minimal model which does not have any direct microscopic meaning, we *must* perform the renormalization-group type calculation for the logical consistency. Moreover, our starting point, the full-diffusion equation [Eq. (314)], is justifiable only up to order ε as we have seen in Section V.C. Thus if we require strict consistency of the theory the only theoretically possible way we already have is the renormalized perturbation approach.

The calculation to order $\varepsilon = 4 - d$ has been performed by Lee et al.;[136] the calculation is much easier than the intrinsic viscosity; we quote our numerical results in Section VI.A.7.

It is conceivable to use the form of the Oseen tensor in 3-space in the $(4 - \varepsilon)$-space calculation. As we have seen in Section V.B, however, the true minimal model is Eq. (316) and Eq. (317) and we have to work in $(4 - \varepsilon)$-space, so that the ad hoc use of the 3-space Oseen tensor spoils the theoretical consistency. Furthermore, the 3-space Oseen tensor is *not* renormalizable, so that the final result explicitly depends on the "microscopic (not uniquely given, or bare)" friction constant of the chain unit; that is, we cannot obtain a well-defined phenomenological description of the system. Hence, we do not pursue this line of approach.

VI. SUMMARY OF OBTAINED RESULTS

So far we have focused mainly on conceptual and technical details of one application of renormalization-group theory to static and dynamical properties of polymer solutions. We have shown, at least in principle, that any desired static quantity can be calculated by the ε-expansion method. It is only a matter of time before we will have a complete renormalization theory for semidilute solution dynamics.

In this section, results already obtained by the renormalization-group theories are summarized and are compared with experimental results when possible. As we will see, general semiquantitative agreement with experiments is encouraging, but the accuracy of experiments are still insufficient, especially in the case of dynamical quantities, to prove or to disprove the main qualitative conclusion of the renormalization-group theories, that is, the existence of universality.

In Section A, we will first discuss our principle in comparing theoretical and experimental results, and compare our results for dilute solutions with experiments. A new class of plots which have not previously been utilized in the analysis of experimental results will be introduced. These plots are free from any adjustable parameter and also from the concept of the theta-point, which should *not* be regarded as the standard universal state of dilute polymer solutions.

In Section B, we compare our results for semidilute solutions with experiments. Our comparison is confined only to static quantities.

In Section C, other interesting results not directly relevant to real experiments are discussed briefly.

A. Comparison With Experiments—Dilute Solutions

1. *What Are the Predictions of Renormalization-Group Theory?*

As we have discussed in Section III, the renormalization-group theory we are using (the Gell-Mann-Low type) can predict phenomenological consequences of a universality class. The functional form of a macroscopic observable in terms of phenomenological parameters is then a prediction of our theory. For example, as is given in Eq. (288) or Eq. (313), the intrinsic viscosity is given in terms of N, L and w or u, which are phenomenological parameters we cannot calculate.

Therefore the true prediction of the theory are the consequences free from any such unobtainable parameters. Let us consider an example of static predictions. As we mentioned in Sect. 4.B, the ratio of the mean-square radius of gyration $\langle R_G^2 \rangle$ and the mean-square end-to-end distance $\langle R^2 \rangle$ depends on only one parameter ζ as

$$\frac{\langle R_G^2 \rangle}{\langle R^2 \rangle} = 6^{-1}\left(1 - \frac{\varepsilon}{96}\frac{\zeta}{1+\zeta}\right) \tag{351}$$

Also the following ratio is universal[95]

$$U_{A/S} = \frac{A_2 M^2}{\langle R_G \rangle^{d/2} N_A} \tag{352}$$

where A_2 is the osmotic second virial coefficient, M the molecular weight of the polymer and N_A Avogadro's constant; $U_{A/S}$ is a function of ζ only, that is, [see Eqs. (355) and (356) later][98,115]

$$U_{A/S} = 1.808 \frac{\zeta(1+\zeta)^{1/8}}{1 + \zeta(1+\zeta)^{1/8}} \times \exp\left\{\left[\frac{13}{48} - \frac{1}{8}(1 - 4\ln 2)\right]\frac{\zeta}{1+\zeta} + \frac{1}{2}\right\} \tag{353}$$

From Eqs. (351) and (353), we can construct a plot, for example, $\langle R_G^2 \rangle / \langle R^2 \rangle$ vs. $U_{A/S}$. The resultant plot, or the functional relation between $\langle R_G^2 \rangle / \langle R^2 \rangle$ and $U_{A/S}$ is the true prediction of our renormalization theory. Note that these quantities are directly observable (at least, in principle), and the comparison of experiments and theoretical results has *no room* for adjustable parameters.

For transport properties, as we have seen in Section V.A, we need not only ζ but a hydrodynamic parameter δ, so that the quantities corresponding to Eqs. (351) and (353) are functions of ζ and δ. We may say that to find the functional forms of quantities depending only on ζ and δ is the task of the renormalization-group approach. We call these quantities universal ratios or universal quantities.

2. *Universal Ratios*

Although we have not given the result for the translational friction coefficient, an analogous (but simpler) calculation to that given in Section V.A gives us[115,127]

$$\frac{f}{[N(1+w)^{-1/4}]^{-1+d/2}} = 2\pi^2(2\pi)^{-\varepsilon/2}\eta_0\{1 + \delta[1 + \zeta(1+\zeta)^{\varepsilon/8}]^{-3/4}\}^{-1} \times [1 + \zeta(1+\zeta)^{\varepsilon/8}]^{(1/4)(1-\varepsilon/2)-1}\zeta(1+\zeta)^{\varepsilon/8} \times \exp\left[\frac{\varepsilon}{16}\frac{\zeta}{1+\zeta}\frac{1}{1+\delta(1+\zeta)^{-3/4}}\right] \tag{354}$$

Other necessary quantities are [114, 98]

$$\frac{A_2 M^2}{N_A N^{2-(\varepsilon/2)}(1+w)^{-d/8}} = \frac{\varepsilon}{16}(2\pi)^{2-(\varepsilon/2)}\zeta[1+\zeta(1+\zeta)^{\varepsilon/8}]^{-1/2} \times \exp\left[-\frac{\varepsilon}{8}\frac{\zeta}{1+\zeta}(1-4\ln 2)\right] \tag{355}$$

and

$$\frac{\langle R_G^2\rangle}{N(1+w)^{-1/4}} = \frac{2}{3}[1+\zeta(1+\zeta)^{\varepsilon/8}]^{1/4}\exp\left[-\frac{13}{96}\varepsilon\frac{\zeta}{1+\zeta}-\frac{\varepsilon}{4}\right] \tag{356}$$

From these equations and the equation for the intrinsic viscosity [Eq. (313)] we can construct the following universal ratios:

$$U_{\eta/S} = \frac{M[\eta]}{N_A\langle R_G^2\rangle^{d/2}} \tag{357}$$

$$U_{A/\eta} = \frac{A_2 M}{[\eta]} \tag{358}$$

$$U_{f/S} = \frac{(f/\eta_0)^{1/(d-2)}}{\langle R_G^2\rangle^{1/2}} \tag{359}$$

$$U_{\eta/f} = \eta_0 \frac{(M[\eta]/N_A)^{(d-2)/d}}{f} \tag{360}$$

Here the symbol $U_{X/Y}$ suggests that $U_{X/Y}$ is essentially the ratio of X and Y. $U_{A/\eta}$ was considered by Yamakawa.[1] Although we can calculate these ratios separately, we have

$$U_{\eta/f}\,U_{f/S} = U_{\eta/S}^{1/d} \tag{361}$$

This may be used to check the internal consistency of our calculation. [In the good solvent limit Eq. (361) becomes 1.565 = 1.597 and in the Gaussian limit 1.877 = 1.793 in 3-space]. The four ratios above are related to conventional universal ratios as follows (in 3-space):

$$\text{Intrinsic viscosity constant:}^{[20,1]}\ \Phi = N_A U_{\eta/S}/6^{1.5} \tag{362}$$

$$\text{Flory-Scheraga-Mandelkern parameter:}^{[137]}\ \beta = (N_A/100)^{1/3} U_{\eta/f} \tag{363}$$

$$\text{Flory's } P \text{ factor:}^{20,1} \quad P = U_{f/S}/\sqrt{6} \tag{364}$$

$$\text{Penetration function: } \Psi = U_{A/S}/4\pi^{3/2} = 0.0449 U_{A/S} \tag{365}$$

The prefactors in these formulas have definite geometrical meaning in the Gaussian limit. However, we are interested in the entire crossover behavior, and then the meaning of prefactors is not clear. Therefore, we only use ratios $U_{X/Y}$ from now on.

We can construct a more complicated universal ratio

$$U_{\eta D/S} = \frac{[\eta] D M \eta_0}{N_A k_B T \langle R_G^2 \rangle} \tag{366}$$

where D is the translational diffusion constant; $D = k_B T/f$. This universal ratio was suggested by des Cloizeaux.[138] It is identical to $U_{\eta/S}/U_{f/S}^{d-2}$.

Explicit formulas in terms of ζ and δ are as follows

$$U_{\eta/S} = \left(\frac{3}{2}\right)^{3/2} \frac{1}{12} (2\pi)^{-1/2} 2\pi^2 \frac{\zeta(1+\zeta)^{1/8}}{1+\zeta(1+\zeta)^{1/8}} \{1 + \delta[1 + \zeta(1+\zeta)^{1/8}]^{-3/4}\}^{-1}$$
$$\times \exp\left\{\frac{23}{32}\frac{\zeta}{1+\zeta} + \frac{1}{2} - \frac{7}{12}\frac{\zeta}{1+\zeta}\frac{\delta(1+\zeta)^{-3/4}}{1+\delta(1+\zeta)^{-3/4}}\right\} \tag{367}$$

$$U_{A/\eta} = \tfrac{3}{2}\{1 + \delta[1 + \zeta(1+\zeta)^{1/8}]^{-3/4}\}$$
$$\times \exp\left\{-\left[\frac{43}{96} + \frac{1}{8}(1 - 4\ln 2)\right]\frac{\zeta}{1+\zeta} + \frac{7}{12}\frac{\zeta}{1+\zeta}\frac{\delta(1+\zeta)^{-3/4}}{1+\delta(1+\zeta)^{-3/4}}\right\} \tag{368}$$

$$U_{f/S} = \sqrt{3\pi}\,\pi \frac{\zeta(1+\zeta)^{1/8}}{1+\zeta(1+\zeta)^{1/8}} \{1 + \delta[1 + \zeta(1+\zeta)^{1/8}]^{-3/4}\}^{-1}$$
$$\times \exp\left\{\frac{13}{192}\frac{\zeta}{1+\zeta} + \frac{1}{8} + \frac{1}{32}\frac{\zeta}{1+\zeta}\frac{1}{1+\delta(1+\zeta)^{-3/4}}\right\} \tag{369}$$

$$U_{\eta/f} = \left(\frac{1}{24}\right)^{1/3} \pi^{-1}\left(\frac{1+\zeta(1+\zeta)^{1/8}}{\zeta(1+\zeta)^{1/8}}\{1 + \delta[1 + \zeta(1+\zeta)^{1/8}]^{-3/4}\}\right)^{2/3}$$
$$\times \exp\left\{\frac{\zeta}{1+\zeta}\frac{[-(7/24)\delta(1+\zeta)^{-3/4} - (1/16)]}{1+\delta(1+\zeta)^{-3/4}} + \frac{43}{192}\frac{\zeta}{1+\zeta}\right\} \tag{370}$$

$$U_{\eta D/S} = \frac{1}{8} e^{1/4} \exp\left\{\frac{\zeta}{1+\zeta}\left[\frac{7}{12} + \frac{7}{12}\frac{(1+\zeta)^{-3/4}}{1+\delta(1+\zeta)^{-3/4}} - \frac{1}{16}\frac{1}{1+\delta(1+\zeta)^{-3/4}}\right]\right\} \quad (371)$$

3. *Asymptotic Values of Universal Ratios*

Although most of the ratios explicitly given above depend on two parameters ζ and δ, if we consider the Gaussian (i.e., $\zeta \to 0$) and the good solvent (i.e., $\zeta \to \infty$) limits, they are independent of δ. In the former case $\delta = -1$, and in the latter, since δ always appears with the factor $[1 + \zeta(1+\zeta)^{1/8}]^{-3/4}$, the equation becomes independent of δ.

The asymptotic results are summarized in Tables II and III. As we see from them, overall agreement is quite encouraging. Unfortunately, however,

TABLE II
Universal Ratios Containing Transport Coefficients in Asymptotic Regimes

Universal ratios (asymptotic values)	Nondraining self-avoiding chain		Nondraining Gaussian chain	
	Result of the expansion with $\varepsilon = 1$	Experimental results	Result of the expansion with $\varepsilon = 1$	Experimental results[k]
$U_{\eta/S}$ [Eq. (357)]	4.078	~4.5[a] ~3.5[b]	5.768	~6.22[e] [~6.12][f]
$U_{A/\eta}$ [Eq. (358)]	1.196	1.3[c,b]	0	0
$U_{f/S}$ [Eq. (359)]	12.067	12.1[d]	15.189	[~14.7][f] [~17.0][g] ~16.3[h] ~14.8[i] ~13.5[d]
$U_{\eta/f}$ [Eq. (360)]	0.1297		0.1236	[~0.126][f] [~0.12][g]
$U_{A/S}$ [Eq. (352)]	4.88	4.9[a,b]	0	0

[a] Miyaki et al. (Ref. 162) for poly(D,L-β-methyl-β-propiolactone) in tetrahydrofuran.
[b] Hirosye et al. (Ref. 163) for poly(D-β-hydroxybutyrate) in trifluoroethanol.
[c] Miyaki et al. (Ref. 165) for polystyrene in benzene.
[d] ter Meer et al. (Ref. 166) for polymethylmethacrylate in acetone.
[e] Miyaki et al. (Ref. 160) for polystyrene in cyclohexane.
[f] Zimm (Ref. 167), asymptotic estimate.
[g] de la Torre et al. (Ref. 168).
[h] ter Meer et al. (Ref. 166) for polymethylmethacrylate in butylchloride.
[i] Schmidt and Burchard (Ref. 169) for polystyrene.
[j] Guttman et al. (Ref. 170).
[k] The experimental results in [ ] are by the Monte Carlo method.

TABLE III
Universal Ratios Customarily Defined[l]

Conventional universal ratios (asymptotic values)	Nondraining self-avoiding chain		Nondraining Gaussian chain	
	Present theory	Experimental results	Present theory	Experimental results[k]
Intrinsic viscosity constant Φ Eq. (362)	1.67×10^{23}	$\sim 1.8 \times 10^{23}$[a] $\sim 1.4 \times 10^{23}$[b]	$\sim 2.36 \times 10^{23}$	2.55×10^{23}[e] [$\sim 2.51 \times 10^{23}$][f]
Flory's P factor Eq. (364)	4.926	~ 4.9[d]	6.201	[~ 6][f] [~ 7][g] 6.65[g] ~ 6.0[i] ~ 5.5[j]
Flory-Scheraga-Mandelkern parameter β Eq. (363)	2.360×10^6		2.249×10^6	[$\sim 2.29 \times 10^6$][f] [$\sim 2.2 \times 10^6$][g]
Penetration function Ψ Eq. (365)	0.219	~ 0.22[a, b]	0	0

[l] See Table II for footnotes.

we cannot extend our theory to be reliable to, say, $O(\varepsilon^2)$. As we have discussed in Section V.B, our analytically tractable model is only correct to order ε.

As we will see in Section VI.A.5, the good solvent asymptotic limit is very difficult to achieve in practice. Thus we need the full crossover behavior.

4. *Approximate Universal Relations*

As we see in Eqs. (367)–(371), except for the purely static ratio $U_{A/S}$ given by Eq. (353) these ratios depend on ζ and δ. Therefore, strictly speaking, there is no unique functional relation between any pair of universal ratios. Moreover, we have seen in Section IV.B.4 that the expansion factor α_R^2 is not a universal function of ζ. We have, however, pointed out that if the renormalized excluded-volume parameter is sufficiently small, α_R^2 can be regarded as a universal function of Z. In this limit $U_{A/S}$ becomes

$$U_{A/S} = 1.808 \frac{Z}{1+Z} \exp\left[\frac{13}{48}\frac{Z}{1+Z} + \frac{1}{2} - \frac{1}{8}\frac{Z}{1+Z}(1 - 4\ln 2)\right] \quad (372)$$

Thus we may construct the α_R^2 vs $U_{A/S}$ approximate universal plot.

Experimentally α_R^2 is not easy to measure, so we replace α_R^2 with α_G^2, the expansion factor for the radius of gyration:

$$\alpha_G^2 = \frac{\langle R_G^2 \rangle}{\langle R_G^2 \rangle_0} = (1+Z)^{1/4}\left(1 - \frac{13}{96}\frac{Z}{1+Z}\right) \tag{373}$$

The result is correct to order ε, and we have set $\varepsilon = 1$. We can replace 1/4 with a better estimate of the exponent $(4\nu - 2 \simeq 0.35)$ to partially take into account the contribution from higher order corrections in ε. In *all* cases this procedure makes theoretical results to order ε agree better with experiments.

Figure 12 shows the comparison of the theoretically obtained $\alpha_G^2 - U_{A/S}$ relation and experiments. We have used only the experimental results for

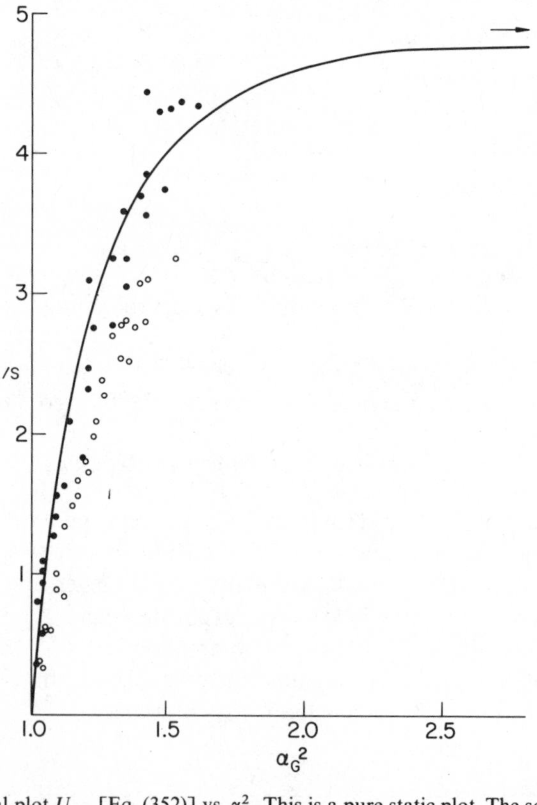

Fig. 12. Universal plot $U_{A/S}$ [Eq. (352)] vs. α_G^2. This is a pure static plot. The solid curve is the renormalization-group result [Eqs. (372) and (373)]. The arrow shows the asymptotic value of $U_{A/S}$ in the self-avoiding limit ($\simeq 4.88$). • is for polychloroprene in *t*-decalin by Fujita's group (Ref. 158), and ○ is for poly-*p*-methylstyrene in diethyl succinate by Tanaka et al.[159]

which $\langle R_G^2 \rangle$ and $\langle R_G^2 \rangle_0$ are measured in the *same* solvent. The reason for this has already been explained in Section II.D; there is no theoretical reason to believe the independence of $\langle R_G^2 \rangle_0$ from the choice of the solvent.

As we have discussed in the end of Section V.A, in the same limit as we have discussed above, $\delta = -1$ and $\zeta = Z$, so that universal ratios Eqs. (367)–(371) become universal functions of Z only; that is,

$$U_{\eta/S} = 1.206 \frac{Z}{1+Z}[1-(1+Z)^{-3/4}]^{-1} \times \exp\left\{\frac{23}{32}\frac{Z}{1+Z} + \frac{1}{2} + \frac{7}{12}\frac{Z}{1+Z}\frac{(1+Z)^{-3/4}}{1-(1+Z)^{-3/4}}\right\} \tag{374}$$

$$U_{A/\eta} = 1.5[1-(1+Z)^{-3/4}] \times \exp\left\{-\left[\frac{43}{96} + \frac{1}{8}(1-4\ln 2)\right]\frac{Z}{1+Z} - \frac{7}{12}\frac{Z}{1+Z}\frac{(1+Z)^{-3/4}}{1-(1+Z)^{-3/4}}\right\} \tag{375}$$

$$U_{f/S} = 9.64[1-(1+Z)^{-3/4}]^{-1}\frac{Z}{1+Z}\exp\left\{\frac{13}{192}\frac{Z}{1+Z} + \frac{1}{8} + \frac{1}{32}\frac{Z}{1+Z}\frac{1}{1-(1+Z)^{-3/4}}\right\} \tag{376}$$

$$U_{\eta/f} = 0.110\left\{\frac{1+Z}{Z}[1-(1+Z)^{-3/4}]\right\}^{2/3}\exp\left[\frac{7}{24}\frac{Z}{1+Z}\frac{(1+Z)^{-3/4}}{1-(1+Z)^{-3/4}} - \frac{1}{16}\frac{Z}{1+Z}\frac{1}{1-(1+Z)^{-3/4}} + \frac{43}{192}\frac{Z}{1+Z}\right] \tag{377}$$

$$U_{\eta D/S} = 0.1605\exp\left\{\frac{Z}{1+Z}\left[\frac{7}{12} + \frac{7}{12}\frac{(1+Z)^{-3/4}}{1-(1+Z)^{-3/4}} - \frac{1}{16}\frac{1}{1-(1+Z)^{-3/4}}\right]\right\} \tag{378}$$

(The equation given in Ref. 115 for $U_{\eta D/S}$ is wrong.) Thus we may construct universal plots of α_G^2 vs. $U_{X/Y}$ as are given in Figs. 13 and 14. Figure 14 shows that the accuracy of experiments is insufficient to prove or to disprove the

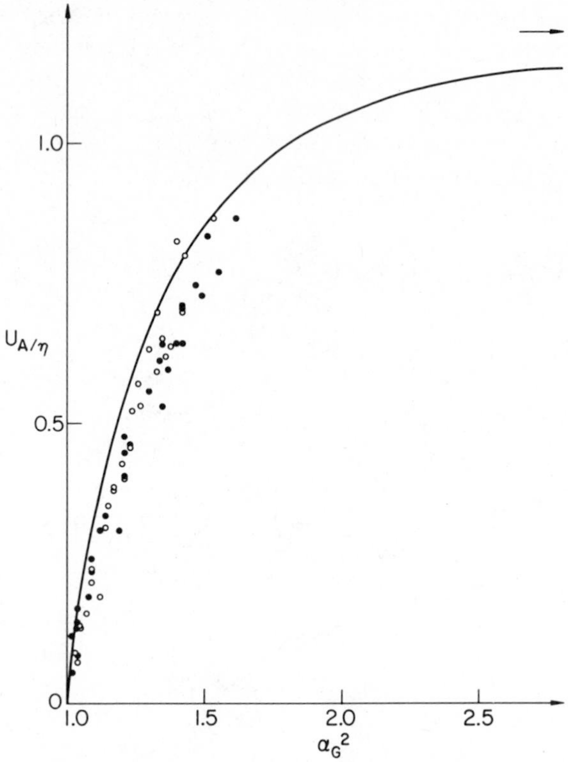

Fig. 13. Universal plot $U_{A/\eta}$ [Eq. (358)] vs. α_G^2. The curve is the result of the renormalization-group theory [Eqs. (375) vs. (373)]. • and ○ are the same as in Fig. 12. The arrow indicates the asymptotic value (~ 1.2).

universality hypothesis. However, Figs. 12 and 13 suggest that the renormalized excluded-volume is small. $U_{f/S}$ vs. α_G^2 is given in Fig. 15.

If we require that α_G^2 be calculated from the results using the same solvent, then there is no way to obtain the values for the good-solvent limit. (The only possible way is to do experiments under pressure. This is a very interesting possibility, but the author has never heard of such an experiment.) Thus we must have the theta-point-free methods for the analysis of experimental data. Fortunately, the plot of a universal ratio against another, that is, the $U_{X/Y}$ vs. $U_{X'/Y'}$ plot, is an *ideal* method satisfying the requirement.

Figures 16 and 17 are examples with experimental points. Other examples are given in Figs. 18–20. As is stressed in Section VI.A.2, $U_{X/Y}$ can be directly calculated from observed data, so that these plots are compared with experiments *without* any ambiguity. Figure 20 replaces Fig. 12 in Ref. 115. Thus the assertion in Ref. 115 that $U_{\eta D/S}$ is virtually constant is false.

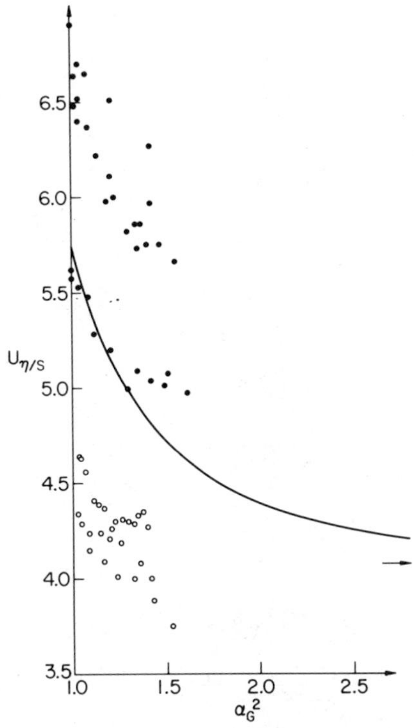

Fig. 14. Universal plot $U_{\eta/S}$ [Eq. (357)] vs. α_G^2. The curve is the result of the renormalization-group theory [Eq. (374) vs. Eq. (373)]. • and ○ are the same as in Fig. 12. The arrow indicates the asymptotic value (~4.1). Recent experimental results at the theta point ($\alpha_G^2 = 1$) suggest that $U_{\eta/S}$ is about 6.1–6.2 (Ref. 160). Although the decreasing tendency of the curve seems to be supported by experiments, the scattering of data is so wild that we cannot say anything certain from experimental results.

5. *Slowness of Dynamical Crossover Behavior*

It has been stressed by Weill and des Cloizeaux,[138] Akcasu et al.,[139] and very long ago by Stockmayer and Albrecht[140] that the asymptotic good-solvent regime is harder to realize for dynamical properties than for static properties.

The best way to observe the difference between the static and dynamical crossover rate is to calculate the effective indices defined by

$$2\nu_G = \frac{\partial \ln \langle R_G^2 \rangle}{\partial \ln N} \tag{379}$$

$$(d-2)\nu_H = \frac{\partial \ln f}{\partial \ln N} \tag{380}$$

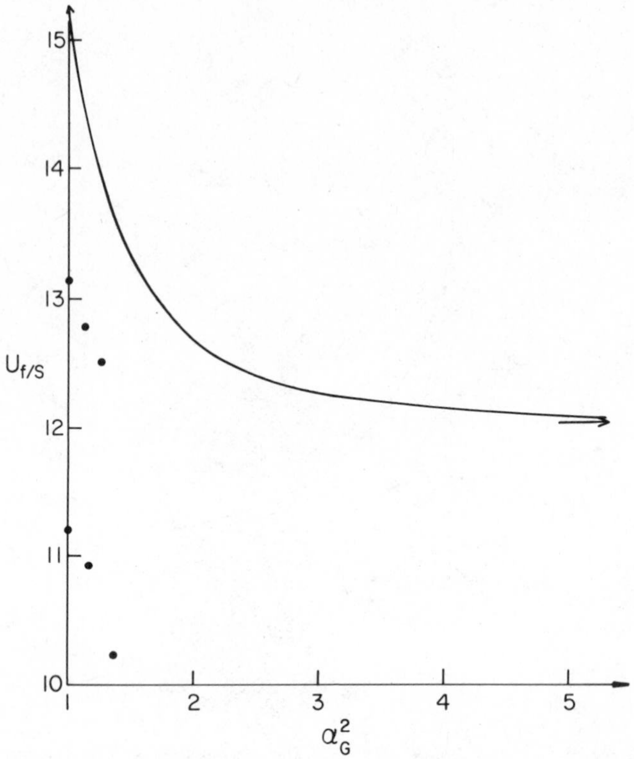

Fig. 15. Universal plot $U_{f/S}$ [Eq. (359)] vs. α_G^2. The curve is the result of the renormalization-group theory [Eq. (359) vs. Eq. (373)]. Experimental points are taken from Nose and Chu.[161] The arrow indicates the asymptotic value (~ 12). This asymptotic value agrees with experimental results of ter Meer et al.[166] sited in Table II.

Fig. 16. Universal vs. universal plot $U_{A/\eta}$ [Eq. (358)] vs. $U_{A/S}$ [Eq. (352)]. $U_{A/S}$ is bounded from above ($\lesssim 4.9$). Recent experimental results[162,163] suggest that the curve ends in the hatched region in the self-avoiding limit. • (with or without pips) are due to Fijita's group for polychloroprene (Ref. 158), ○ (with or without pips) are due to Tanaka et al.[159] poly-*p*-methylstyrene, and × are due to Kato et al.[164] polystyrene in toluene. Solvents are as follows: •, *t*-decalin; •-, 1-chlorobutane; •̍, carbon tetrachloride; ○, diethyl succinate; ○-, toluene; ○̍, dichloroethane; -○, cyclohexane; ○̩, methylethylketone. This plot suggests a universal relation between $U_{A/\eta}$ and $U_{A/S}$.

Fig. 17. Universal vs. universal plot $U_{\eta/S}$ [Eq. (357)] vs. $U_{A/S}$ [Eq. (352)]. The notations are exactly the same as in Fig. 16. Like Fig. 14, this plot magnifies the experimental error. According to the recent experimental results,[162,163] the universal curve ends in the hatched region.

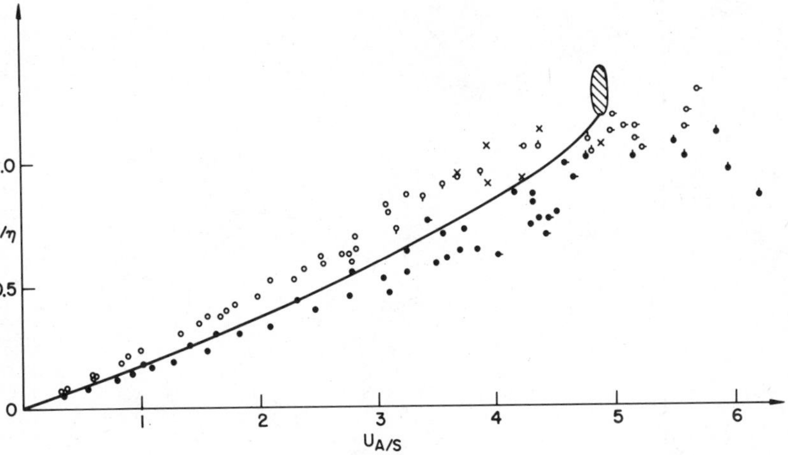

Fig. 16

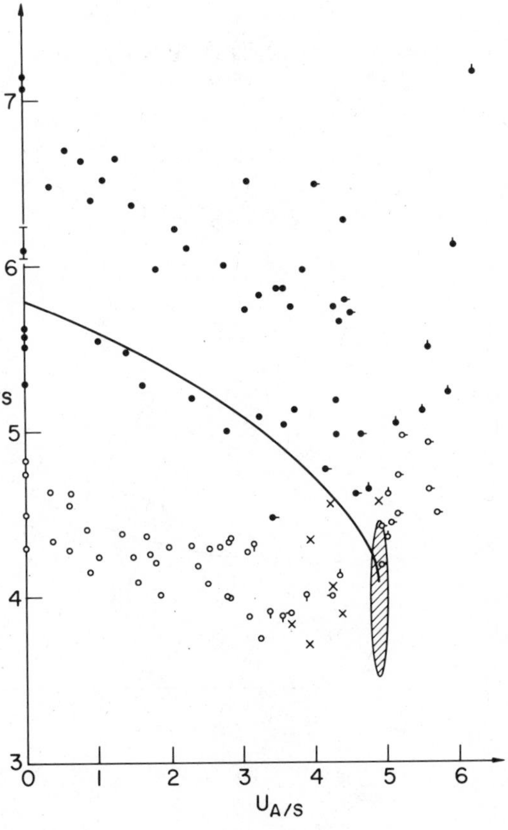

Fig. 17

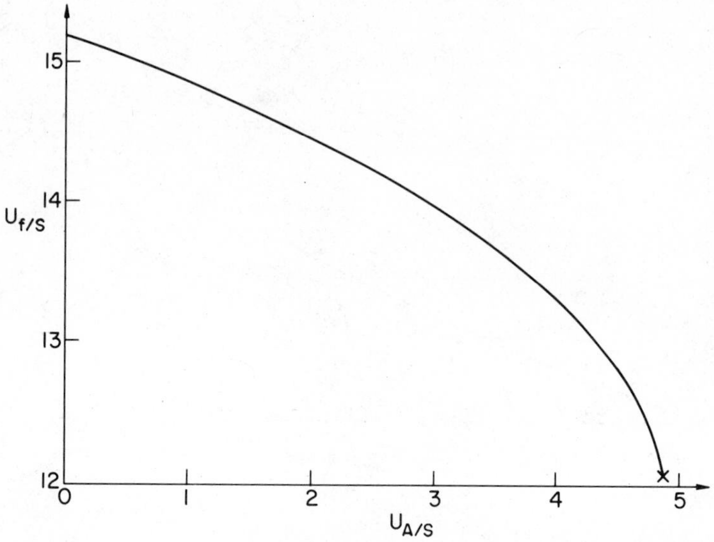

Fig. 18. Universal vs. universal plot $U_{f/S}$ [Eq. (359)] vs. $U_{A/S}$ [Eq. (353)]. The calculated curve ends at ×.

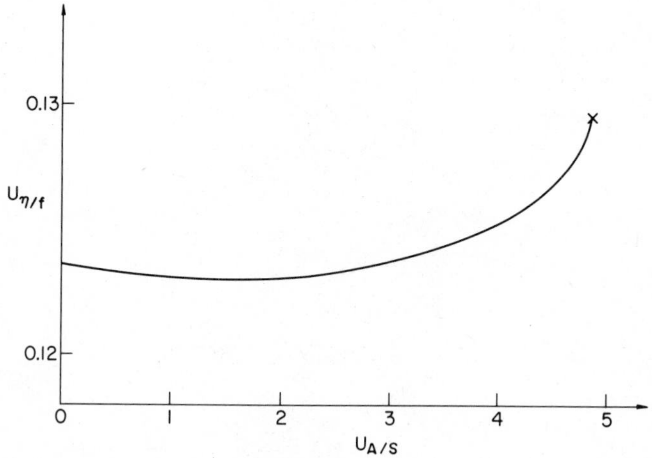

Fig. 19. Universal vs. universal plot $U_{\eta/f}$ [Eq. (360)] vs. $U_{A/S}$ [Eq. (353)]. The calculated curve ends at ×.

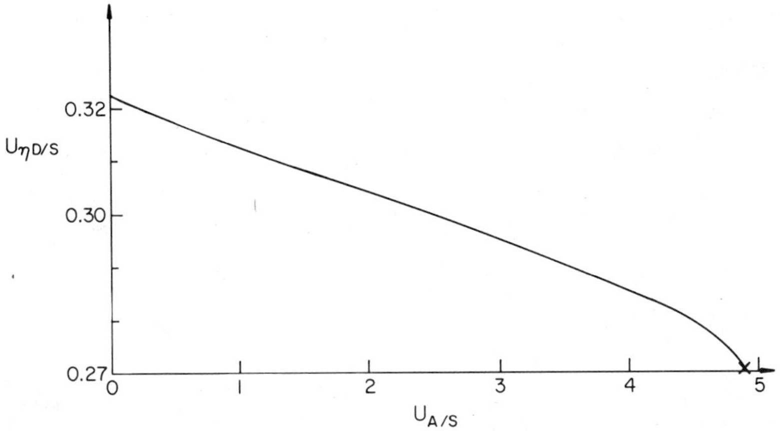

Fig. 20. Universal vs. universal plot $U_{\eta D/S}$ [Eq. (366)] vs. $U_{A/S}$ [Eq. (353)].

where ν_G is the exponent for the static radius, and ν_H is the exponent for the hydrodynamic radius. Asymptotically they must coincide if we start from the full-diffusion equation model [Eq. (314)], since there is only one length-scale; this justifies the scaling theory. [However, there is no reason to believe the reliability of Eq. (314) to higher order in ε (Section V.B); there is still a theoretical possibility that ν_H is definitely larger than ν_G.] These two indices can be explicitly calculated from Eqs. (354) and (356) as

$$\nu_G = \frac{1}{2} + \frac{1}{4}\left\{ \frac{-\dfrac{13}{96}\dfrac{Z}{1+Z}}{1 + \dfrac{83}{96}Z} + \frac{1}{4}\frac{Z}{1+Z} \right\} \tag{381}$$

$$\begin{aligned}\nu_H = \frac{1}{2} &+ \frac{1}{64}\frac{Z}{(1+Z)^2}\left\{\frac{1}{1-(1+Z)^{-3/4}} - \frac{3}{4}\frac{Z(1+Z)^{-3/4}}{[1-(1+Z)^{-3/4}]^2}\right\} \\ &+ \frac{1}{4}\left\{1 - \frac{3}{4}\frac{Z}{1+Z} - \frac{3}{4}\frac{Z}{1+Z}\frac{(1+Z)^{-3/4}}{1-(1+Z)^{-3/4}}\right\}\end{aligned} \tag{382}$$

where we have already assumed that the renormalized excluded-volume parameter is sufficiently small. In Fig. 21 the universal plot of ν_H and ν_G vs. $U_{A/S}$ is given, which clearly shows the slowness of a dynamical quantity in approaching its asymptotic value.

Note that our result is free from any arbitrariness[141] inherent in the blob argument[138,139,142] or modified blob arguments.[143]

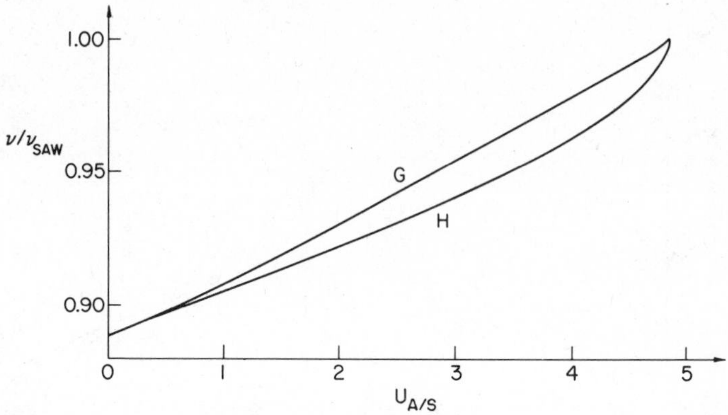

Fig. 21. Universal plot of v_H or v_G divided by $v(Z \to \infty) \equiv v_{SAW}$. G denotes the static index and H the hydrodynamic index. The plot clearly shows the slowness of the dynamical crossover.

6. *Static Scattering Factor* $S(\mathbf{k})$

The static (coherent) scattering function for a single chain has been obtained by Witten and Schäfer[95,96] by the renormalization-group method in a series expansion in k^2. (Witten[144] later considered the large k asymptotic form also.) A closed form to order ε can be found in Ohta et al.[145] In terms of the $I(\mathbf{k}) \equiv S(\mathbf{k})/S(0)$ function, comparison with a recent experimental data[146] is shown in Fig. 22. In Ref. 145 it is suggested that if we consider the higher order corrections, the value of $k^2\langle R_G^2\rangle$ of the crossing point between the Debye function and the result for the good-solvent limit[145] become smaller. It is stressed by Noda et al.[146] that the blob model cannot improve the agreement between theory and experiments. If we adopt the value $\nu \simeq 0.588$,[174] then we can improve the agreement of our results with recent experimental results. We may expect that the calculation to $O(\varepsilon^2)$ will give a satisfactory result.

7. *Initial Decay Rate of the Dynamical Scattering Factor*

An outline of the method to calculate the initial decay rate $\Omega(\mathbf{k})$ of the dynamical scattering factor was discussed in Section V.C.2. It can be shown that $\eta_0\Omega(\mathbf{k})/k^d k_B T$, where η_0 is the solvent viscosity and k_B is the Boltzmann constant, is a universal function of $k\langle R_G^2\rangle^{1/2}$. Both at the theta condition and in the good-solvent limit, our theoretical results[136] to order ε are compared with experimental results by Han and Akcasu[142] and by Kurata et al.[175] in Figs. 23 and 24. The agreement between our results and experimental results are not very good especially for the theta state. If we shift curves as is also suggested in Han and Akcasu, we can achieve excellent agreement, but we do not use this ad hoc procedure.

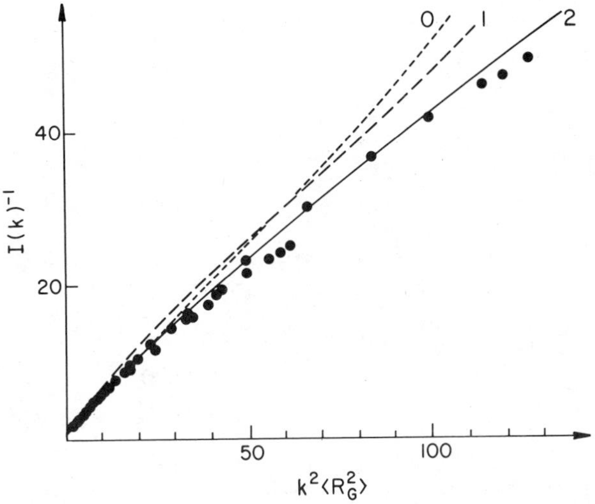

Fig. 22. The universal plot $I(\mathbf{k}) = S(\mathbf{k})/S(0)$ vs. $k^2\langle R_G^2\rangle$. 0: the Debye function, the result for the Gaussian chain; 1: the result to order ε by Ohta et al.,[145] 2: the order ε result with ν being replaced by the best estimate $\nu \simeq 0.588$.[176] Thus the last result almost quantitatively agrees with the recent accurate experimental results by Noda et al.[146]

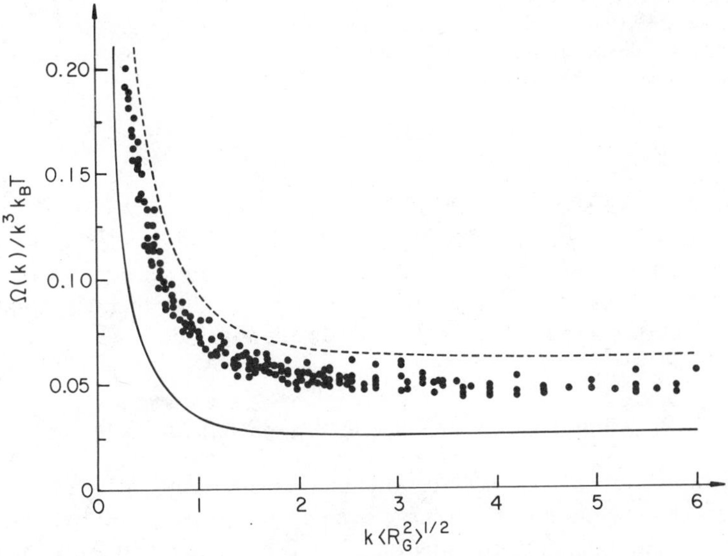

Fig. 23. The universal plot $\Omega(\mathscr{K})k^3k_BT$ vs. $k\langle R_G^2\rangle^{1/2}$ at the theta state, where the solvent viscosity is normalized to unity. The solid curve is the renormalization-group theoretical result to order ε ($\varepsilon = 1$), and the broken curve is the nonrenormalization-group theoretical result (without preaveraging approximation) by Akcasu et al.[135] Experimental points are taken from Han and Akcasu.[142] (A much better agreement is obtained by Baldwin and Lee recently.)

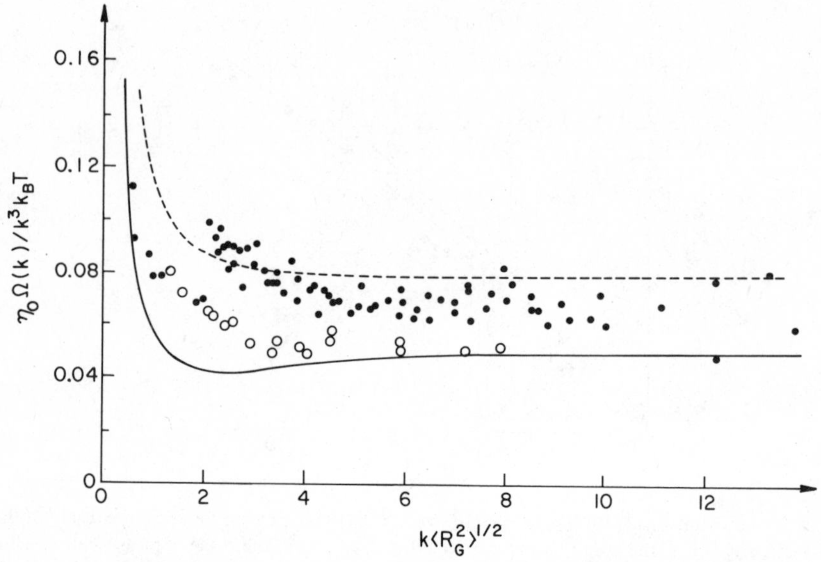

Fig. 24. The universal plot $\eta_0\Omega(\mathscr{K})/k^3k_BT$ vs. $k\langle R_G^2\rangle^{1/2}$ in the good-solvent limit. The solid curve is the renormalization-group theoretical result to order ε, and the broken curve is the result of the approximate treatment of the excluded-volume effect by Akcasu et al.[135] Experimental points are: •, Han and Akcasu,[142] and 0, Kurata et al.[175b] The renormalization-group result is not monotonically decreasing. Our calculation is, however, too crude to assert the existence of a minimum point on the curve.

We can also show that $\Omega(\mathbf{k})/k^2D_0$ is a universal function of $k\langle R_G^2\rangle^{1/2}$, where $D_0 = \lim_{k\to 0}\Omega(k)/k^2$. In Fig. 25, our theoretical result[136] and experimental results by Akcasu and Han[142] at the theta state are compared. In this case the agreement is satisfactory. This suggests that there is significant theoretical or experimental error in the absolute value of $\Omega(\mathbf{k})$.

B. Comparison With Experiments—Semidilute Solutions

Unfortunately, we cannot have any theoretical result for the dynamical properties of semidilute solutions; our comparison will be severely limited.

The osmotic pressure given by Eq. (228) is compared with experiments in Fig. 26. From Eq. (228), we can calculate the osmotic compressibility as

$$\frac{1}{{}_Ak_BT}\left(\frac{\partial\pi}{\partial c}\right)_T = 1 + \frac{1}{8}\left[9X - 2 + \frac{2\ln(1+X)}{X}\right]$$

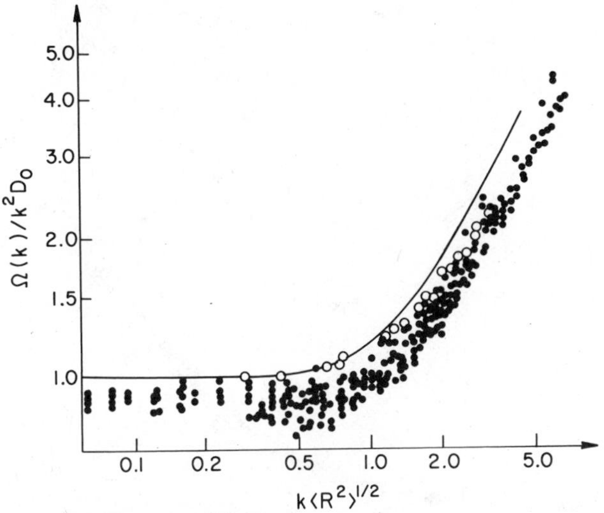

Fig. 25. The universal plot $\Omega(\mathscr{K})/k^2D_0$ vs. $k\langle R_G^2\rangle^{1/2}$ at the theta state. The solid curve is the renormalization-group theoretical result to order ε. By definition $\Omega(\mathscr{K})/k^2D_0$ must be unity at the $k\langle R_G^2\rangle^{1/2} \to 0$ limit. The experimental points denoted by • due to Han and Akcasu[142] do not satisfy this because of inappropriate choice of D_0. If we multiply a constant to ensure the correct limiting behavior of experimental results, we have an excellent agreement of our renormalization-group result with experimental data. There is a discrepancy between our results and the experimental results by Kurata et al.[175a] denoted by o.

$$\times \exp\left\{\frac{1}{4}\left[\frac{1}{X} + \left(1 - \frac{1}{X^2}\right)\ln(1 + X)\right]\right\} \tag{383}$$

where c is the polymer number density.

As we discussed in Section IV.C.6, the variable X can be related to the second-virial coefficient by Eq. (230). We can compare Eq. (383) with experimental results without any adjustable parameter. This has been done by Wiltzius et al.[108] as is shown in Fig. 27.

A comparison of the correlation length ξ is given in Fig. 28.[147]

Schäffer[148] recently published an extensive comparison of the ε-expansion result and the experimental results.

C. Other Results by Renormalization-Group Theory

1. *Test Chain in Semidilute Solutions*

The mean square end-to-end distance of a test chain of length N immersed in a semidilute solution of polymers with different molecular weights can be calculated along the line given in Section IV.C. The result reads to order ε in

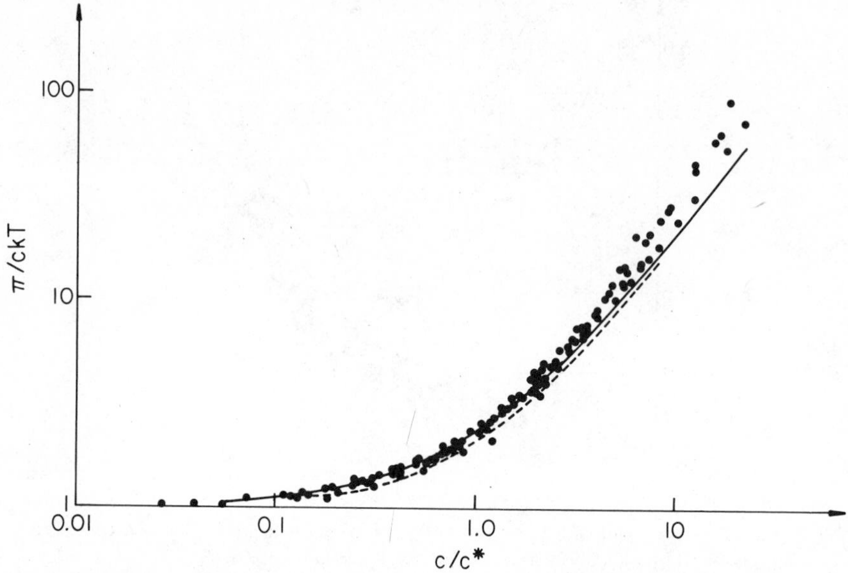

Fig. 26. The osmotic pressure vs. c/c^*, where c is the polymer number density, and c^* is an adjustable parameter mentioned in Section IV.C.6. The experimental points for poly-(α-methylstyrene) in toluene are taken from Noda et al.[176] The solid curve shows the monodisperse case $M_w/M_n = \mu^{-1} = 1$, and the broken line denotes a polydisperse case with $M_w/M_n = 100$. The polydispersity effect is very small. The discrepancy between theory and experiment for larger values of c is due to the first-order approximation. We can augment our theoretical results with the best estimate of the exponent ν; if we replace $\varepsilon/4$ in Eq. (228) with $(2 - d\nu)/(d\nu - 1)$ [compare with Eq. (A.5) in Appendix A] with $\nu \simeq 0.588$, we get an excellent agreement between our result and experimental results. (See Figs. 27 and 28.)

the good solvent limit as

$$\langle R^2 \rangle = \langle R^2 \rangle_{\mathrm{SAW}} \exp\left[\frac{\varepsilon M_n}{4M} XG\left(2\frac{M}{M_n}\left(\frac{M_n}{M_w} + X\right)\right)\right] \tag{384}$$

where M_n and M_w are, respectively, the number and weight average molecular weight of the solute polymers, M is the molecular weight of the test chain, $\langle \mathbf{R}^2 \rangle_{\mathrm{SAW}}$ is its mean-square end-to-end distance in the dilute limit, and X is given by Eq. (227), that is, essentially proportional to $cM_n^{\nu d}$ with c being the polymer number density. The function G is given by

$$G(x) = \frac{3}{x} + \frac{1}{2} e^x \mathrm{Ei}(-x)\left(1 - \frac{4}{x} + \frac{6}{x^2}\right) - \left(\frac{1}{x} + \frac{3}{x^2}\right)(\hat{\gamma} + \ln x) \tag{385}$$

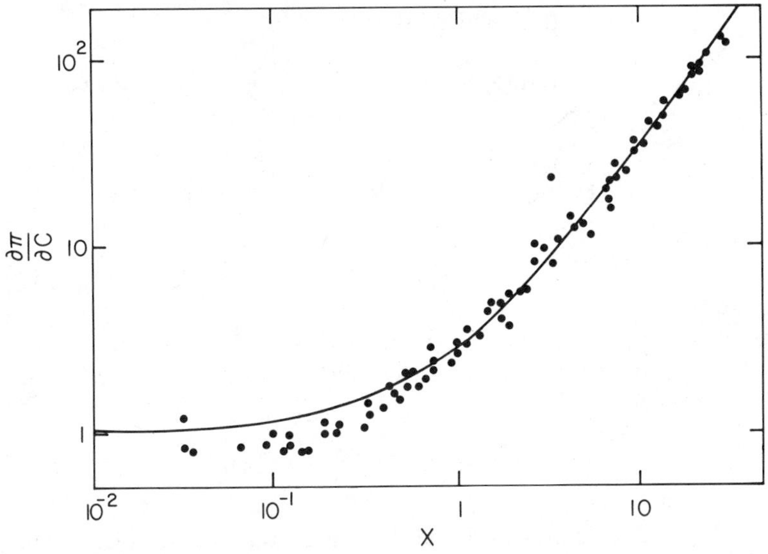

Fig. 27. The universal plot of the osmotic compressibility $\partial\pi/\partial c$ vs. X, which can be calculated from the second virial coefficient as is discussed in Section IV.C.6. Experimental points for polystyrene in toluene and methylethylketone (MEK) are taken from Wiltzius et al.[108] In this figure our theoretical result is augmented as is explained in Fig. 26. Even without this augmentation, as is shown by Wiltzius et al.[108] our result gives a very good fit to their experimental result *without* any adjustable parameter. (Courtesy of Professor T. Ohta.)

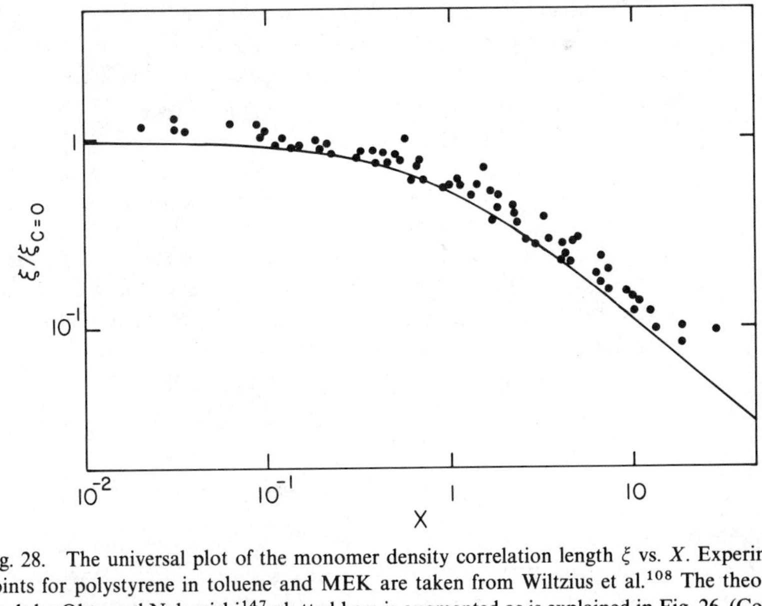

Fig. 28. The universal plot of the monomer density correlation length ξ vs. X. Experimental points for polystyrene in toluene and MEK are taken from Wiltzius et al.[108] The theoretical result by Ohta and Nakanishi[147] plotted here is augmented as is explained in Fig. 26. (Courtesy of Professor T. Ohta.)

For sufficiently overlapped ($X \to \infty$) solutions, $MXG/M_n \sim \frac{1}{2} \ln (M_w X/M_n)$. Hence Eq. (384) becomes

$$\langle R^2 \rangle \sim \langle R^2 \rangle_{\mathrm{SAW}} \left(\frac{M_n}{MX} \right)^{\varepsilon/8} \sim M M_n^{\varepsilon/8} X^{-\varepsilon/8} \sim M \rho^{-\varepsilon/8} \tag{386}$$

where ρ is the monomer number density. This is in conformity with Eq. (A.7) in Appendix A.

Next, let us consider the situation in which $M/M_n \to \infty$ with X being fixed, that is the situation with a very long chain in a semidilute solution. In this limit, Eq. (384) becomes

$$\langle R^2 \rangle \sim M^{1+[1-X/(\mu+X)]\varepsilon/8} M_n^{\varepsilon X/8(\mu+X)} \tag{387}$$

where $\mu = M_n/M_w$. This implies that if the concentration is sufficiently high (X very big), then $\langle R^2 \rangle \sim M$; that is, even if the chain is very long, it behaves like a random walk. An interesting feature is that if the polydispersity of the solution is very large ($\mu \searrow 0$), $\langle R^2 \rangle \sim M$ even if X is not very big, that is, in a solution with a large polydispersity, the screening effect is very efficient. The result [Eq. (387)] seems to contradict with the scaling results.[149,150]

In the scaling theories Eq. (386) is derived by the blob argument (see Appendix A). Comparing this with Eq. (171) we see the effective order of the corresponding FSAW for the test chain is $m \sim M_n/X$. This is consistent with the identification $m \sim \xi^{1/\nu}$ to order ε^0 (since m appears in the $O(\varepsilon)$ term, we cannot correctly find the $O(\varepsilon)$ term for m. We expect $m \sim (M_n/X)^{1/(\nu d-1)}$ from $m \sim \xi^{1/\nu}$ [see Appendix A, (A.6)]).

2. *Rings, Stars, Polymers Interacting with Walls, etc.*

Given a configuration of a chain, say a branched polymer, can we do what we have done for linear polymers? Such a problem is an exercise without any conceptual difficulty, though a considerable number of messy calculations are required except for rings.[151,152] Here we only mention references.[153,154]

However, the effect of branching on transport properties is important in practice. Thus, for example, it would be useful to calculate the intrinsic viscosity.

3. *Effect of Flow Fields*

Under a given flow field, a chain is deformed. The study of the flow problem is very important as the foundation for the study of, for example, the effect of polymer additives on turbulent flows.[155] However, these problems are highly nonequilibrium ones, so generally very hard to attack.

For potential flows, however, the full diffusion equation can be used to obtain the stationary distribution function. For a steady 2-dimensional elongational flow such that

$$\mathbf{v} = S(x, -y, 0, \ldots) \tag{388}$$

where S is a constant, the end-to-end vector distribution function and the isotropic scattering function have been calculated by Ohta and Yamazaki.[156] Calculation is quite demanding.

Since this calculation is based on the Edwards Hamiltonian Eq. (11), if the flow strength becomes very big (or $\zeta_0 S N_0^2 > \pi^2$), the model becomes physically nonsense; there is no force to keep the total chain length finite, so that $\langle R^2 \rangle > N^2$, an absurdity. Thus we must incorporate the effect of nonharmonicity into the model; our minimal model is insufficient for strong flow problems, so that we must modify our model. However, so far no elegant way has been proposed which can be used with the renormalization-group theory.

VII. SUMMARY

We have shown how to use renormalization-group theory to calculate directly observable quantities by representative examples, and have exhibited our results. As we have seen in Section VI, the overall agreement of our lowest-nontrivial-order calculation with experimental results is very encouraging.

Here we summarize conceptual conclusions, followed by a summary of more practical conclusions.

1. Renormalization-group (RG) theories are the only possible theories which justify scaling arguments and various asymptotic functional forms. Only with the help of RG theories can the phenomenological description of polymer solutions with a minimal number of phenomenological parameters be elucidated theoretically. Thus RG theories are the only conceptually and logically sound basic theories of polymer solutions.

2. Unfortunately, however, our developing (i.e., still primitive) theoretical tools impose severe limitations on the actual implementation of RG theories. The only possible method so far devised which can give universal functions in 3-space is the renormalized perturbation theory, or the ε-expansion method. For the unified treatment of static and dynamic properties, the conformation-space RG approach explained in this article seems to be the best. Though sometimes less flexible as has been mentioned in Section IV.B.5, the field theoretic approach (which is equivalent to the conformation-space approach

for static problems) is generally more efficient for static problems, as has been demonstrated by Schäfer.[148]

3. (1) and (2) imply that the ε-expansion method is almost the only way to theoretically study the macroscopic description of polymer solutions. (The homotopy method is a variant of the ε-expansion method.) Comparison of various ε-expansion results and experimental results in Section VI suggests that even the lowest nontrivial order calculations can give reasonable results. Rigorously speaking, there is no strong a priori reason to believe the reliability of the ε-expansion results. However, the fact that the lower critical dimension for polymers is zero, that is, far from $d = 3$, may be relevant.

It must be stressed, however, that there has been *no* argument about the reliability of conventional approaches. As we have seen in Section II, all the non-RG theories are conceptually wrong and/or logically not self-contained (i.e., without the help of RG theories, they cannot produce any reliable results). Hence, even if the results of these theories happen to agree with experiments, we can hardly regard this agreement as a theoretical success.

4. As we have seen in Section V, all the starting points of conventional dynamical theories of polymer solutions are justifiable only to order ε.

5. Thus the ε-expansion, or the renormalized perturbation method is the only logically self-contained and practically feasible theoretical method so far devised to study polymer solutions. Hence we must accept the reasonable agreement exhibited in Section VI with gratitude to Nature. Even if there were no good agreement between the ε-expansion results and experiments, we must stick to logically sound theories. Indeed, theorists should put the first priority on conceptual and logical soundness. Only when this credo can lead us to agreement with experiments, can we say that our theory is successful.

6. May we say then that our theory is successful? We may say yes with respect to qualitative aspects. Semiquantitatively it is a bit too early to say yes or no. However, truly quantitative agreement of our results with experiments should *not* be expected because of our primitive analytic tools.

7. The renormalized excluded-volume parameter v seems small in actual polymer solutions. Under this condition the universal functions are virtually functions of only one variable $Z \propto vN^{1/2}$. Then the partial draining effect does not affect any universal functions. If experimentally the draining effect is observed definitely, what we should do first is to assume v to be not small. In this case we need two parameters ζ and δ to describe the crossover behavior as we have seen in Eqs. (367)–(371).

If, as Fixman claims,[173] even at the theta state no scaling is found, we must reconsider our minimal model, that is, the Edwards Hamiltonian Eq. (11) + the Langevin equations [Eqs. (316) and (317)]. What we have pursued in this review is the consequence of *this* minimal model. If the results disagree with real or numerical experiments, we *must* discard our minimal model; the minimal model must evolve through the feedback process between theory and experiments as is discussed in Section II.A. We should, however, avoid ad hoc, purely empirical modifications of our model.

8. The crossover behavior of dilute solutions can be studied without any adjustable parameter. Experimental results should be analyzed by the "theta-point-free methods" as we have done in Section VI.A., since the "theta state" cannot be regarded the standard state of a polymer chain. Thus the use of expansion factors α, α_η, and soon, should be discouraged.

9. As we have seen in Section VI, existing experiments are generally not sufficiently accurate to prove or disprove the existence of universality for transport properties.

The study of polymer solutions has not yet been completed. We can summarize outstanding problems unsolved as follows.

10. Semidilute solution dynamics, especially the concentration dependence of transport properties, must be finished. Actual calculations are under way.

11. Transport properties of dilute solutions must be studied without the Kirkwood-Riseman scheme. This is not a fundamental problem, because there is no need of extra theoretical machinery. Time dependent correlation functions must be calculated as well. These studies are under way.

12. According to Baumgärtner's simulation,[157] the effect of entanglement is relevant even to the dynamics of a single chain in poor solutions. Thus the effect of entanglement in poor solutions must be studied as well.

13. The boundary between the semidilute region and the concentrated region is not well understood.

14. The effect of strong flows on the properties of solutions has not yet been studied.

Among these remaining problems (12)–(14) are conceptually more difficult than (10) or (11).

Note added in proof (1985, January)

The original manuscript reflects the status of the solution theory at the end of 1983. Since then, the problem (11) mentioned above has been virtually finished, and considerable progress has been made in the static part of problem (13). As to problem (10), the cooperative diffusion constant has been calculated which agrees well with experimental results.

Acknowledgments

The author is grateful to C. Peter Flynn, Director of the Materials Research Laboratory at the University of Illinois, for his generous support; and Ralph O. Simmons, Head of the Department of Physics of the University of Illinois, for the opportunity to present an earlier version of this work as an advanced graduate course.

The author thanks Phillip H. Geil, the leader of Polymer Group at the University of Illinois, and Philip R. Baldwin for carefully reading the manuscript. He received detailed comments and corrections from Phil Geil, Biman Bagchi, Yasuhiro Shiwa, and Phil Baldwin. They have been incorporated in the final version of the text. The author wishes to thank them sincerely for their help.

The author wishes to express his gratitude to Takao Ohta for fruitful collaboration since 1980 and for unpublished results of his group; to Kyozi Kawasaki for his encouragement.

A special thanks is due to Karl F. Freed for giving the author the opportunity to continue his own theoretical program: this was not allowed until he came to the United States.

Last, but not least, the author thanks Letitia L. Watts for her excellent typing of the manuscript and for her patience.

The work is, in part, supported by NSF-MRL Grant DMR-80-20250, and by the Research Board of the University of Illinois.

APPENDIX A: A SKETCH OF SCALING ARGUMENTS

For later convenience, we sketch the scaling arguments for polymer solutions. The theoretical basis of these arguments are provided by renormalization-group theories as we see in Sections III–V.

In the scaling argument for polymer solutions we assume that there is one representative length scale in the system. For dilute solutions, this is the root-mean-square end-to-end distance $\langle R^2 \rangle^{1/2}$ or the root-mean-square radius of gyration $\langle R_G^2 \rangle^{1/2}$, and for semidilute solutions the correlation length ξ of the monomer density field. As we see in Section V.B, this scaling hypothesis is correct to order ε, but, there is *no* theoretical reason to believe this to be exact for transport properties to higher orders.

For dilute solutions, $\langle R^2 \rangle^{1/2} \propto \langle R_G^2 \rangle^{1/2} = R_G \sim N^{\nu}$ is *assumed*, where N is the degree of polymerization. This relation is *derived* in Section III.D. The static scattering function $S(\mathbf{k})$ has the following form:

$$S(\mathbf{k}) = N^2 f(R_G \mathbf{k}) \tag{A.1}$$

which is a consequence of dimensional analysis. If k is sufficiently large, then $S(\mathbf{k})/N$ should not depend on N because $S(\mathbf{k})$ probes only local configurations. Hence $f(x) \sim x^{-1/\nu}$ in the $x \to \infty$ limit, that is,

$$S(\mathbf{k}) \sim |\mathbf{k}|^{-1/\nu} \tag{A.2}$$

For dilute solution dynamics, we are interested in the single chain dynamics. Assuming the non-draining limit, the translational friction coefficient f must be given by the Stokes formula. Hence we have

$$f \sim R_G^{d-2} \sim N^{\nu(d-2)} \tag{A.3}$$

This can be derived also by a dimensional analysis. The dynamical exponent is defined by $f \sim N^{\nu(z-2)}$, so that we get $z = d$ (but see Section V.B). The intrinsic viscosity $[\eta]$ is proportional to the hydrodynamic volume, so that $M[\eta] \sim N^{d\nu}$ or

$$[\eta] \sim N^{d\nu - 1} \tag{A.4}$$

where M is the molecular weight.

The osmotic pressure π can be expanded as

$$\frac{\pi}{k_B T} = \frac{\rho}{N} + \rho\left(\frac{\rho}{N}\right)^2 R_G^d + \cdots$$

where ρ is the monomer number density, and the second term assumes the hard-core like mutual exclusion of two polymer chains. Generally, this implies

$$\frac{\pi}{k_B T} = \frac{\rho}{N} f\left(\frac{\rho R_G^d}{N}\right)$$

where $f(x)$ must be constant in the $x \to 0$ limit. In the $x \to \infty$ limit, due to the extensive overlap of chains π should be independent of N. Hence we have $f(x) \sim x^{1/(d\nu - 1)}$ or we have obtained the des Cloizeaux law[6]

$$\pi \sim \rho^{d\nu/(d\nu - 1)} \tag{A.5}$$

The correlation length must have the following functional form

$$\xi = R_G f(\rho/\rho^*)$$

where ρ^* is the overlap threshold $\rho^* \sim N/R_G^d \sim N^{1-\nu d}$. If $\rho > \rho^*$, ξ should be independent of N, so that $f \sim x^{\nu/(1-d\nu)}$. Hence

$$\xi \sim \rho^{\nu/(1-d\nu)} \tag{A.6}$$

The size of the test chain $\langle R^2 \rangle$ must be

$$\langle R^2 \rangle = R_G^2 f(\rho/\rho^*)$$

In the melt, $\langle R^2 \rangle \sim N$, so that $f(x) \sim x^{(2\nu-1)/(d\nu-1)}$, or

$$\langle R^2 \rangle \sim \rho^{(2\nu-1)/(d\nu-1)} N \tag{A.7}$$

APPENDIX B: FEYNMAN-KAC FORMULA

Let us consider the Green's function G_0 in d-space (i.e., $\mathbf{R}^d$) for the diffusion equation

$$\left(\frac{\partial}{\partial t} - \frac{1}{2}\Delta\right) G_0(t, \mathbf{r}|\mathbf{r}_0) = \delta(t)\,\delta(\mathbf{r} - \mathbf{r}_0) \tag{B.1}$$

with the boundary condition

$$G_0(t, \mathbf{r}|\mathbf{r}_0) \to 0 \quad \text{as} \quad |\mathbf{r} - \mathbf{r}_0| \to \infty \tag{B.2}$$

We know

$$G_0(t, \mathbf{r}|\mathbf{r}_0) = \left(\frac{1}{2\pi}\right)^d \int \mathrm{d}^d\mathbf{k} \exp[-\tfrac{1}{2}k^2 t - i\mathbf{k}\cdot(\mathbf{r} - \mathbf{r}_0)] \tag{B.3}$$

$$= (2\pi t)^{-d/2} \exp[-(\mathbf{r} - \mathbf{r}_0)^2/2t] \tag{B.4}$$

We can easily show that $G_0(t, \mathbf{r}|\mathbf{r}_0)$ has the following *Markovian property*

$$G_0(t, \mathbf{r}|\mathbf{r}_0) = \int \mathrm{d}^d\mathbf{r}_1 G(t - t_1, \mathbf{r}|\mathbf{r}_1) G(t_1, \mathbf{r}_1|\mathbf{r}_0) \tag{B.5}$$

Hence, more generally, we get

$$G_0(t, \mathbf{r}|\mathbf{r}_0) = \int \cdots \int \left(\prod_{i=1}^{N-1} \mathrm{d}^d\mathbf{r}_i\right) \prod_{i=1}^{N} G(t_i - t_{i-1}, \mathbf{r}_i|\mathbf{r}_{i-1}) \tag{B.6}$$

where $(t_N, \mathbf{r}_N) = (t, \mathbf{r})$, and $t_i = i\,\Delta t + t_0$ with $\Delta t = (t - t_0)/N$. Let $\Delta\mathbf{r}_i = \mathbf{r}_i - \mathbf{r}_{i+1}$. Then Eq. (B.6) can be rewritten as

$$G_0(t, \mathbf{r}|\mathbf{r}_0) = \prod_{i=1}^{N-1} \int \mathrm{d}^d\mathbf{r}_i (2\pi\,\Delta t)^{-d/2} \exp\left[-\sum_{i=1}^{N} \frac{(\Delta\mathbf{r}_i)^2}{2\,\Delta t}\right] \tag{B.7}$$

If Δt is sufficiently small, then we may formally write

$$\sum \frac{(\Delta\mathbf{r}_i)^2}{\Delta t} \to \int_0^t \left(\frac{\mathrm{d}\mathbf{r}}{\mathrm{d}t}\right)^2 \mathrm{d}t \tag{B.8}$$

Thus we end up with the *path integral* representation of the Green's function for Eq. (B.1) with Eq. (B.2) as

$$G_0(t, \mathbf{r}|\mathbf{r}_0) = \int_{\mathbf{r}(0)=\mathbf{r}_0}^{\mathbf{r}(t)=\mathbf{r}} \mathscr{D}[\mathbf{r}] \exp\left[-\frac{1}{2} \int_0^t \left(\frac{d\mathbf{r}}{dt}\right)^2 dt \right] \tag{B.9}$$

where the integration symbol is used to denote the summation over all the paths $\{\mathbf{r}(s): s \in [0, t]\}$ such that $\mathbf{r}(0) = \mathbf{r}_0$ and $\mathbf{r}(t) = \mathbf{r}$. The measure ("volume element") $\mathscr{D}$ denotes the uniform measure on this set; that is, in the language of statistical mechanics, the measure $\mathscr{D}$ assigns the same weight to all conformations of a chain connecting $\mathbf{r}_0$ and $\mathbf{r}$. A mathematical justification of this measure is discussed by Itô.[26]

Next, let us consider the problem with a potential function

$$\left(\frac{\partial}{\partial t} - \frac{1}{2}\Delta + V\right) G(t, \mathbf{r}|\mathbf{r}_0) = \delta(t)\,\delta(\mathbf{r} - \mathbf{r}_0) \tag{B.10}$$

with the boundary condition Eq. (B.2). If V is a constant, we get

$$G(t, \mathbf{r}|\mathbf{r}_0) = G_0(t, \mathbf{r}|\mathbf{r}_0)\, e^{-tV} \tag{B.11}$$

Intuitively, we may expect that for a very short span Δt of time $V(\mathbf{r}(t))$ is virtually constant. Thus we may use Eq. (B.11) for small t, so that we get

$$G(t, \mathbf{r}|\mathbf{r}_0) = \prod_{i=1}^{N} \int d^d\mathbf{r}_i (2\pi\,\Delta t)^{-d/2} \exp\left[-\sum_{i=1}^{N} \frac{(\Delta \mathbf{r}_i)^2}{2\,\Delta t} - \sum_{i=1}^{N} V(\mathbf{r}_i)\,\Delta t \right] \tag{B.12}$$

The procedure which transformed Eq. (B.8) into Eq. (B.9) gives

$$G(N, \mathbf{r}|\mathbf{r}_0) = \int_{\mathbf{c}(0)=\mathbf{r}_0}^{\mathbf{c}(N)=\mathbf{r}} \mathscr{D}[\mathbf{c}] \exp\left[-\frac{1}{2} \int_0^N \left(\frac{d\mathbf{c}}{d\tau}\right)^2 d\tau - \int_0^N V(\mathbf{c}(\tau))\, d\tau \right] \tag{B.13}$$

This is the *Feynman-Kac formula* for the Green's function.[27,28] The original Kac formula[28] is

$$G(N, \mathbf{r}|\mathbf{r}_o) = \int_{\mathbf{c}(0)=\mathbf{r}_0}^{\mathbf{c}(N)=\mathbf{r}} \mathscr{D}_W[\mathbf{c}] \exp\left[-\int_0^N V(\mathbf{c}(\tau))\, d\tau \right] \tag{B.14}$$

where $\mathscr{D}_W$ is the measure for the Wiener process.[29]

Since the trajectory (i.e., the sample path) of the Wiener process is, as is explained in Section II.E, the continuous version of the simple random walk, Eq. (B.14) can be interpreted as the conformational partition function; $\mathscr{D}_W[\mathbf{c}]$ gives the unperturbed statistical weight for the conformation and $\exp[-\int V\,d\tau]$ is the Boltzmann factor for each conformation. Compare Eq. (B.14) with Eq. (1).

We may have a much more general Boltzmann factor in Eq. (B.14): if $F(\{\mathbf{c}\})$ is a map from the set of conformations to the energy, $\exp(-F)$ can be used in Eq. (B.14). An example is

$$F(\{\mathbf{c}\}) = \frac{1}{2}\int_0^N d\tau \int_0^N d\sigma \phi(\mathbf{c}(\tau) - \mathbf{c}(\sigma)) \tag{B.15}$$

where ϕ is a pair potential. If we want to include the connectedness of the chain in the energy factor, then the formula corresponding to Eq. (B.13)

$$G(N, \mathbf{r}|\mathbf{r}_0) = \int_{\mathbf{c}(0)=\mathbf{r}_0}^{\mathbf{c}(N)=\mathbf{r}} \mathscr{D}[\mathbf{c}] \exp\left[-\frac{1}{2}\int_0^N \left(\frac{d\mathbf{c}}{d\tau}\right)^2 d\tau - F(\{\mathbf{c}\})\right] \tag{B.16}$$

is the natural expression. Hence we may regard

$$\mathscr{H}(\{\mathbf{c}\}) = \frac{1}{2}\int_0^N \left(\frac{d\mathbf{c}}{d\tau}\right)^2 d\tau + F(\{\mathbf{c}\}) \tag{B.17}$$

as the total Hamiltonian of the chain.

APPENDIX C: THE SINGULARITY AT $z = 0$[42]

If v_0 is negative, the model, Eq. (11) is meaningless. It is easy to show that this is still true even if v_0 is infinitesimally small.

For simplicity, let us consider the following lattice polymer model on the simple cubic lattice with the unit lattice spacing:

1. The polymer consists of N elements connected successively by bonds of length unity. These elements are on the lattice points.
2. To any pair of elements on the same lattice point, the interaction energy $k_B T v_0$ is assigned.

Let Z_N be the partition function for the lattice polymer, whose one end is at the origin. If $v_0 < 0$, the following inequality is obvious

$$\exp\left(\frac{-N^2 v_0}{2}\right) < Z_N < 6^N \exp\left(\frac{-N^2 v_0}{2}\right) \tag{C.1}$$

where 6^N is the total number of conformations on the 3-space simple cubic lattice. If $v_0 > 0$, then obviously

$$Z_N < 6^N \tag{C.2}$$

since all the Boltzmann factors are less than unity. Hence we have

$$\lim_{N \to \infty} N^{-3/2} \ln Z_N = \begin{cases} 0 & \text{for} \quad z_{1/2} > 0 \\ \dfrac{-z_{1/2}}{2} & \text{for} \quad z_{1/2} < 0 \end{cases} \tag{C.3}$$

where the limit is taken with $z_{1/2} = N^{1/2} v_0$ being kept finite. Thus even if v_0 is infinitesimally small, the model system exhibits a "phase transition." Even if $z_\mu = N^\mu v_0$ $(\frac{1}{2} < \mu < 1)$ is kept finite, there is a "phase transition" at $z_\mu = 0$.

APPENDIX D: DERIVATION OF EQ. (162)

Since u is of order ε, in the square bracket of Eq. (90), we may simply replace α with $X(1 + \zeta)^{1/4}$ [see Eq. (161)]. From Eqs. (159) and we get

$$u = \frac{\zeta u^*}{1 + \zeta} \tag{D.1}$$

where we have discarded $O(\varepsilon^2)$ terms, and u^* is $\varepsilon\pi^2/2$ [see Eq. (53)].

Now we can rewrite Eq. (90) as follows

$$G(N, \mathbf{R}) = (2\pi N)^{-2+\varepsilon/2} \exp\left\{-X\left(\frac{2\pi N}{L}\right)^{\varepsilon\zeta/8(1+\zeta)} [1 + \zeta(1 + \zeta)^{\varepsilon/8}]^{-1/4} \right.$$
$$+ \frac{\varepsilon\zeta}{8(1 + \zeta)}\Bigg[1 + X(1 + \zeta)^{-1/4} + (1 - X(1 + \zeta)^{1/4})$$
$$\left.\times (\hat{\gamma} + \ln(X(1 + \zeta)^{-1/4}) - (1 - X(1 + \zeta)^{-1/4}) \ln \frac{2\pi N}{L}\Bigg]\right\} \tag{D.2}$$

If the first term in the curly bracket is written as

$$-X[1 + \zeta(1 + \zeta)^{\varepsilon/8}]^{-1/4}\left[1 + \frac{\varepsilon}{8}\frac{\zeta}{1 + \zeta} \ln \frac{2\pi N}{L} + \cdots\right] \tag{D.3}$$

the second term in the square bracket in Eq. (D.3) cancels the term in Eq.

(D.2) to order ε. The remaining term in Eq. (D.3) and the $-X(1+\zeta)^{1/4} \times \ln[X(1+\zeta)^{-1/4}]$ term in Eq. (D.2) can be combined to yield the first term in the curly bracket of Eq. (162). Equations (155) and (160) tell us that if $(2\pi N)^{-2+\varepsilon/2-\varepsilon\zeta/8(1+\zeta)}$ is factored out from G, G must be a function of ζ and X. Thus there should not appear $X \ln (2\pi N/L)$ in Eq. (D.2).

APPENDIX E: INTEGRAL IN EQ. (210)

We need

$$J = \int d^d\mathbf{z}\left[\ln\left(1+\frac{A}{z^2+2\mu}\right)-\frac{A}{z^2+2\mu+A}\right] \tag{E.1}$$

$$= S_{d-1}\int_0^\infty dz\, z^{d-1}\left[\ln\left(1+\frac{A}{z^2+2\mu}\right)-\frac{A}{z^2+2\mu+A}\right] \tag{E.2}$$

where S_{d-1} is the volume of the unit $(d-1)$-sphere given by

$$S_{d-1} = \frac{2\pi^{d/2}}{\Gamma(d/2)} \tag{E.2a}$$

Introducing $y = z^2$, and integrating once by parts, we get

$$J = \frac{S_{d-1}}{d} A^2 K \tag{E.3}$$

where

$$K = \int_0^\infty dy\, y^{2-\varepsilon/2}(y+2\mu+A)^{-2}(y+2\mu)^{-1} \tag{E.3a}$$

This integral is divided into two pieces to get

$$K = \int_0^\infty dy\, y^{-1-\varepsilon/2}(1-e^{-y}) + \int_0^\infty dy\left(\frac{y^2}{(y+2\mu+A)^2(y+2\mu)} - \frac{1-e^{-y}}{y}\right)\left(1-\frac{\varepsilon}{2}\ln y+\cdots\right) = \frac{2}{\varepsilon}\Gamma\left(1-\frac{\varepsilon}{2}\right) - \ln 2\mu + \frac{2B}{A} \times \ln\frac{2\mu}{B} - \frac{B}{A} - \frac{B^2}{A^2}\ln\frac{2\mu}{B} - \hat{\gamma} + O(\varepsilon) \tag{E.4}$$

where $B = 2\mu + A$.

APPENDIX F: CALCULATION OF *G*(k, *x*, *y*)

The Fourier transform of Eq. (268) has already been given in Oono and Ohta,[126] but here we directly calculate

$$G(\mathbf{k}, x, y) = e^{-k^2(x-y)/2} - v_0 \sum_{i=1}^{6} B_i \tag{F.1}$$

where B_i are diagrammatically given in Fig. 29. They are given as $(x > y)$

$$B_1 = e^{-k^2(x-y)/2} \int_0^y dt(y-t)(2\pi t)^{-d/2} \tag{F.2}$$

$$B_2 = e^{-k^2(x-y)/2} \left[(x-y)^{2-d/2} \int_0^1 dt(1-t)(2\pi)^{-d/2} t^{-d/2} \right. \\ \left. \times \exp[(x-y)tk^2] \right] \tag{F.3}$$

$$B_3 = e^{-k^2(x-y)/2} \int_0^x dt(N-x-t)(2\pi t)^{-d/2} \tag{F.4}$$

$$B_4 = \int_0^y ds \int_0^{x-y} dt \int_{\mathbf{q}} \exp[-\tfrac{1}{2}q^2(y-s) - \tfrac{1}{2}(\mathbf{k}+\mathbf{q})^2 t - \tfrac{1}{2}k^2(x-y-t)] \tag{F.5}$$

$$B_6 = \int_0^y ds \int_0^{N-x} dt \int_{\mathbf{q}} \exp[-\tfrac{1}{2}q^2(y-s) - \tfrac{1}{2}(\mathbf{k}+\mathbf{q})^2(x-y) - \tfrac{1}{2}q^2t^2] \tag{F.6}$$

B_5 is analogous to B_4. Note that B_2 is exactly the first-order term of the Fourier transform of the end-vector distribution function for a chain of length $x - y$.

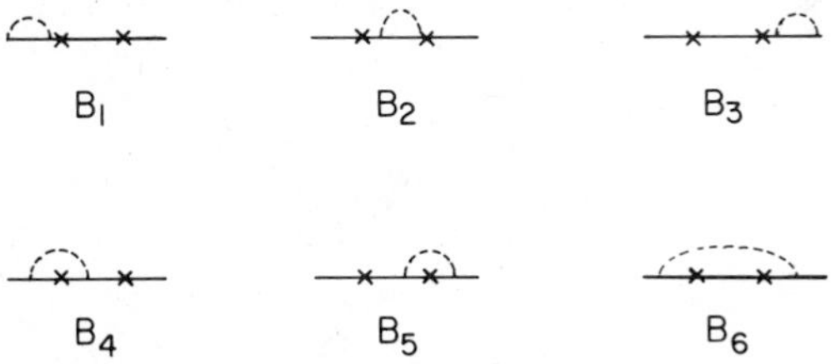

Fig. 29. Diagrammatic expressions of B_1–B_6.

Evaluation of B_1–B_6 is straightforward; setting $M = x - y$, we get to $O(\varepsilon)$

$$B_1 = (2\pi)^{-2+\varepsilon/2}\, \mathrm{e}^{-k^2M/2}\left[-\frac{2}{\varepsilon} - 1 - \ln y\right] \tag{F.7}$$

$$B_2 = (2\pi)^{-2+\varepsilon/2}\, \mathrm{e}^{-k^2M/2}\left\{-1 - \frac{2}{\varepsilon} - \ln M + \frac{1}{2}Mk^2\left[-1 + \ln M + \frac{2}{\varepsilon}\right] + \cdots\right\} \tag{F.8}$$

$$B_3 = (2\pi)^{-2+\varepsilon/2}\, \mathrm{e}^{-k^2M/2}\left[-\frac{2}{\varepsilon} - 1 - \ln(N - x)\right] \tag{F.9}$$

$$B_4 = (2\pi)^{-2+\varepsilon/2}\, \mathrm{e}^{-k^2M/2}\left[\frac{2}{\varepsilon} + 1 - \ln\frac{y+M}{My} - \frac{1}{2}k^2M \times \left(\frac{1}{2}\frac{y}{y+M} - \frac{y}{M}\ln\frac{M+y}{y}\right)\right] \tag{F.10}$$

$$B_5 = (2\pi)^{-2+\varepsilon/2}\, \mathrm{e}^{-k^2M/2}\left[\frac{2}{\varepsilon} + 1 - \ln\frac{N-y}{(N-x)M} - \frac{1}{2}k^2M\left(\frac{N-x}{N-y} - \frac{N-x}{M}\ln\frac{N-y}{N-x}\right)\right] \tag{F.11}$$

$$B_6 = (2\pi)^{-2+\varepsilon/2}\, \mathrm{e}^{-k^2M/2}\left[\ln\frac{y}{x-y}\frac{N-y}{N} - \frac{1}{4}k^2M \times \left(\frac{M}{N-y} - 1 + \frac{M}{x} - \frac{M}{N}\right)\right] \tag{F.12}$$

APPENDIX G: NECESSARY EQUATIONS FOR k-INTEGRATION

Necessary equations to get Eq. (279) are summarized. Let us introduce the following (standard) abbreviation

$$\frac{1}{(2\pi)^d}\int \mathrm{d}^d\mathbf{k} \equiv \int_{\mathbf{k}} \tag{G.1}$$

$$U = \int_{\mathbf{k}} \mathrm{e}^{-k^2 f/2} = (2\pi)^{-d/2} f^{-d/2} \tag{G.2}$$

By appropriate integration of Eq. (G.2) with respect to f, we get

$$V = \int_{\mathbf{k}} k^{-2}\,\mathrm{e}^{-k^2 f/2} = (2\pi)^{-d/2}\frac{1}{d-2}f^{-1+\varepsilon/2} \tag{G.3}$$

From the spherical symmetry, we have

$$\int_{\mathbf{k}} \frac{k_\alpha^2}{k^2}\,\mathrm{e}^{-k^2 f/2} = \frac{U}{d}, \qquad (\alpha = x, y, \ldots) \tag{G.4}$$

Analogously, we get

$$\int_{\mathbf{k}} \frac{k_\alpha^2}{k^4}\,\mathrm{e}^{-k^2 f/2} = \frac{V}{d}, \qquad (\alpha = x, y, \ldots) \tag{G.5}$$

Again using the spherical symmetry and elementary properties of Gaussian distribution

$$\int_{\mathbf{k}} \frac{k_\alpha^2 k_\beta^2}{k^4}\,\mathrm{e}^{-k^2 f/2} = \frac{U}{d(d+2)}, \qquad (\alpha \neq \beta) \tag{G.6}$$

APPENDIX H: DERIVATION OF EQ. (306)

The characteristic equations for Eq. (290) read

$$\frac{\mathrm{d}L}{L} = \frac{\mathrm{d}u}{\beta} = \frac{\mathrm{d}\xi}{\beta_\xi} = \frac{\mathrm{d}N}{N\gamma_N} = \frac{\mathrm{d}[\eta]}{0} \tag{H.1}$$

Therefore, we can easily obtain the following partial relations for the characteristic curve

$$L\exp\left(-\int_{u_1}^{u}\frac{\mathrm{d}u}{\beta}\right) = \text{const.} \tag{H.2}$$

$$N\exp\left(-\int_{u_1}^{u}\frac{\gamma_N}{\beta}\,\mathrm{d}u\right) = \text{const.} \tag{H.3}$$

and $[\eta] = \text{const.}$ See Eqs. (146) and (147). We have to solve one more equation:

$$\frac{\mathrm{d}\xi}{\mathrm{d}u} = \frac{\beta_\xi}{\beta} = \frac{3}{16}\frac{\xi}{u(u^*-u)}\left[\xi^* - \xi + \frac{4}{3}(u^*-u)\right] \tag{H.4}$$

Introducing $x = u/u^*$ and $y = \xi/\xi^*$, we can rewrite this as

$$\frac{x}{y}\frac{dy}{dx} = \frac{3}{4}\frac{1-y}{1-x} + \frac{1}{4} \tag{H.5}$$

If we further introduce

$$z = \frac{(1-x)^{3/4}}{x}\, y \tag{H.6}$$

Eq. (H.5) becomes separable as

$$\frac{dz}{dx} = -\frac{3}{4}(1-x)^{-7/4} z^2 \tag{H.7}$$

Hence, the solution we need is

$$(1-x)^{-(3/4)}\left(1 - \frac{x}{y}\right) = (1+w)^{3/4}(1 - \hat{z}^{-1}) = \text{const.} \tag{H.8}$$

where w is defined in Eq. (156) and $\hat{z} = y/x = u^*\xi/\xi^* u$ [Eq. (307)]. Hence the standard theory of quasilinear partial differential equation gives Eq. (306).

APPENDIX I: DERIVATION OF EQ. (312)

From Eq. (311) we have

$$1 + \delta(1+w)^{-3/4} = \frac{1}{\hat{z}} \tag{I.1}$$

From Eq. (307),

$$\xi = \xi^* u \hat{z}/u^* \tag{I.2}$$

$$= \xi^* \hat{z} \frac{\zeta(1+\zeta)^{\varepsilon/8}}{1 + \zeta(1+\zeta)^{\varepsilon/8}} \left(\frac{2\pi N}{L}\right)^{-\varepsilon/2 + \varepsilon\zeta/2(1+\zeta)} \tag{I.3}$$

where Eq. (159) and $u = u^*w/(w+1)$ have been used. Combining Eq. (I.1) and Eq. (160), we obtain

$$z = [1 + \delta(1 + \zeta(1 + \zeta)^{\varepsilon/8})^{-3/4}]^{-1} \times \left(\frac{2\pi N}{L}\right)^{-(3/8)\varepsilon[\zeta/(1+\zeta)]\{\delta(1+\zeta)^{-3/4}/[1+\delta(1+\zeta)^{-3/4}]\}} \tag{I.4}$$

Thus Eqs. (I.3) and (I.4) yield the desired result, Eq. (312).

APPENDIX J: DERIVATION OF THE FULL-DIFFUSION EQUATION

Our minimal model is defined by Eq. (316) and Eq. (317) with the Gaussian driving forces whose correlation functions are given by Eqs. (318) and (319). We further assume that the solvent is incompressible. Then we can eliminate the pressure from Eq. (317) to get

$$\frac{\partial}{\partial t}\mathbf{u}(\mathbf{r}, t) = -\left[\int_0^{N_0} d\tau \frac{\delta \mathscr{H}}{\delta \mathbf{c}(\tau, t)} \delta(\mathbf{r} - \mathbf{c}(\tau, t))\right]_\perp + \eta_0 \Delta \mathbf{u} + \mathbf{f}_\perp \tag{J.1}$$

where $\perp$ denotes the transversal components. For simplicity we set $\rho_0 = 1$.

Although we can proceed within the stochastic differential equation, we use the equivalent Fokker-Planck formalism. Let $P = P(\{\mathbf{c}\}, \{\mathbf{u}\}, t)$ be the joint time-dependent distribution function for the conformation and the solvent velocity field. We have the following Fokker-Planck equation for P:

$$\frac{d}{dt} P = \mathscr{L} P \tag{J.2}$$

where

$$\mathscr{L} = \mathscr{L}_0 + \mathscr{L}_{\text{int}} \tag{J.3}$$

with

$$\mathscr{L}_0 = \int d\tau \frac{\delta}{\delta \mathbf{c}(\tau)} \frac{1}{\zeta_0}\left[\frac{\delta}{\delta \mathbf{c}(\tau)} + \frac{\delta \mathscr{H}_E}{\delta \mathbf{c}(\tau)}\right] + \int_\mathbf{k} \frac{\delta}{\delta \mathbf{u}_\mathbf{k}} k^2 \eta_0 \left[\frac{\delta}{\delta \mathbf{u}_{-\mathbf{k}}} + \mathbf{u}_\mathbf{k}\right] \tag{J.4}$$

and

$$\mathscr{L}_{\text{int}} = -\lambda_0 \int_\mathbf{k} \int d\tau \left\{\frac{\delta}{\delta \mathbf{c}(\tau)} [\mathbf{u}_\mathbf{k}\, e^{-i\mathbf{k}\cdot\mathbf{c}(\tau)}] + \frac{\delta}{\delta \mathbf{u}_\mathbf{k}} \left[\frac{\delta \mathscr{H}_E}{\delta \mathbf{c}(\tau)}\right]_\perp e^{i\mathbf{k}\cdot\mathbf{c}(\tau)}\right\} \tag{J.5}$$

$\mathscr{L}_0$ is the unperturbed Fokker-Planck operator, and $\mathscr{L}_{\text{int}}$ is the mode-coupling term. λ_0 is the mode-coupling constant, which is introduced for the sake of indicating the order of perturbation. After renormalization in $(4 - \varepsilon)$-space, λ_0 becomes of order $\varepsilon^{1/2}$.

Let $W(\{\mathbf{c}\}, \{\mathbf{u}\})$ be the (normalized) equilibrium distribution function:

$$\mathscr{L}W(\{\mathbf{c}\}, \{\mathbf{u}\}) = 0 \tag{J.6}$$

If there is no systematic macroscopic flow, W is factorizable as

$$W(\{\mathbf{c}\}, \{\mathbf{u}\}) = W_c(\{\mathbf{c}\})W_u(\{\mathbf{u}\}) \tag{J.7}$$

and therefore

$$\mathscr{L}_0 W = 0 \tag{J.8}$$

which implies

$$\mathscr{L}_{\text{int}} W = 0 \tag{J.9}$$

Let us introduce the projection operator $\mathscr{P}$

$$\mathscr{P} \equiv W_u(\{\mathbf{u}\}) \int_{\mathbf{u}} \tag{J.10}$$

where

$$\int_{\mathbf{u}} \equiv \int \mathrm{d}\{\mathbf{u}\} \tag{J.11}$$

is the functional integral over the velocity field. By a straightforward calculation we can easily show that

$$P(\{\mathbf{c}\}, t) = \int_{\mathbf{u}} P(\{\mathbf{c}\}, \{\mathbf{u}\}, t) \tag{J.12}$$

obeys the following equation

$$\frac{\mathrm{d}}{\mathrm{d}t} P\{\mathbf{c}\}, t) = \mathscr{L}_{\text{eff}} P(\{\mathbf{c}\}, t) \tag{J.13}$$

where

$$\mathscr{L}_{\text{eff}} = \langle \mathscr{L} \rangle_{\mathbf{u}} + \langle \mathscr{L}(1 - \mathscr{P})(\hat{\delta} - \mathscr{L})^{-1}(1 - \mathscr{P})\mathscr{L} \rangle_{\mathbf{u}} \tag{J.14}$$

with $\hat{\delta}$ being a positive infinitesimal number and $\langle \rangle_u$ is

$$\langle \cdots \rangle_u = \int_{\mathbf{u}} \ldots W_u(\{\mathbf{u}\}) \tag{J.15}$$

Our task is to write $\mathscr{L}_{\text{eff}}$ explicitly to order ε.

Obviously,

$$\langle \mathscr{L}_0 \rangle_u = \int d\tau \frac{\delta}{\delta \mathbf{c}(\tau)} \frac{1}{\zeta_0} \left(\frac{\delta}{\delta \mathbf{c}(\tau)} + \frac{\delta \mathscr{H}_E}{\delta \mathbf{c}(\tau)} \right) \tag{J.16}$$

where we have used $\langle \mathbf{u}_{\mathbf{k}} \rangle = \mathbf{0}$ and rapid decay of W_u in the $|\mathbf{u}| \to \infty$ limit.

To calculate the second term of Eq. (J.14), we use the following relation easily checked

$$\int_{\mathbf{u}} \mathscr{L}(1 - \mathscr{P}) = \int_{\mathbf{u}} \mathscr{L}_{\text{int}}(1 - \mathscr{P}) \tag{J.17}$$

Hence

$$\langle \mathscr{L}(1 - \mathscr{P})(\hat{\delta} - \mathscr{L})^{-1}(1 - \mathscr{P})\mathscr{L} \rangle_{\mathbf{u}} = \langle \mathscr{L}_{\text{int}}(1 - \mathscr{P})(\hat{\delta} - \mathscr{L})^{-1}(1 - \mathscr{P})\mathscr{L}_{\text{int}} \rangle_{\mathbf{u}} \tag{J.18}$$

As will be justified later, $\mathscr{L}_{\text{int}}$ can be regarded of order $\varepsilon^{1/2}$, so that we may replace $\mathscr{L}$ in Eq. (J.18) by $\mathscr{L}_0$ to order ε. Thus

$$\mathscr{L}_{\text{eff}} = \langle \mathscr{L}_0 \rangle_{\mathbf{u}} + \langle \mathscr{L}_{\text{int}}(1 - \mathscr{P})(\hat{\delta} - \mathscr{L}_0)^{-1}(1 - \mathscr{P})\mathscr{L}_{\text{int}} \rangle_{\mathbf{u}} \tag{J.19}$$

To proceed further, we split $\mathscr{L}_{\text{int}}$ into two parts

$$\mathscr{L}_{\text{int}} = \mathscr{L}_1 + \mathscr{L}_2 \tag{J.20}$$

where

$$\mathscr{L}_1 = -\lambda_0 \int_{\mathbf{k}} \int d\tau \frac{\delta}{\delta \mathbf{c}(\tau)} \mathbf{u}_{\mathbf{k}} \, e^{-i\mathbf{c}(\tau) \cdot \mathbf{k}} \tag{J.21}$$

$$\mathscr{L}_2 = \lambda_0 \int_{\mathbf{k}} \int d\tau \frac{\delta}{\delta \mathbf{u}_{\mathbf{k}}} \left[\frac{\delta \mathscr{H}_E}{\delta \mathbf{c}(\tau)} \right]_{\perp} e^{i\mathbf{k}\cdot\mathbf{c}(\tau)} \tag{J.22}$$

Note that $\langle \mathscr{L}_2(1-\mathscr{P})[\delta - \mathscr{L}_0]^{-1}(1-\mathscr{P})\mathscr{L}_{\text{int}} \rangle_{\mathbf{u}} = 0$, this can be seen by integration by parts. Thus we have only to calculate $\mathscr{L}_1 - \mathscr{L}_1$ and $\mathscr{L}_1 - \mathscr{L}_2$ terms.

$$\begin{aligned}
&\langle \mathscr{L}_1(1-\mathscr{P})(\hat{\delta} - \mathscr{L}_0)^{-1}(1-\mathscr{P})\mathscr{L}_1 \rangle_{\mathbf{u}} \\
&\quad = \lambda_0^2 \int_0^\infty dt \int_{\mathbf{u}} \int_{\mathbf{k}} \int d\tau \frac{\delta}{\delta \mathbf{c}(\tau)} \mathbf{u}_{\mathbf{k}}(t)\, e^{-i\mathbf{k}\cdot\mathbf{c}(\tau)} \int_{\mathbf{k}'} \int d\tau' \frac{\delta}{\delta \mathbf{c}(\tau')} \mathbf{u}_{\mathbf{k}'}(0) \\
&\quad \times e^{-i\mathbf{k}'\cdot\mathbf{c}(\tau')} W_u = \lambda_0^2 \int d\tau \int d\tau' \int_{\mathbf{k}} \int_{\mathbf{k}'} \int_0^\infty dt\, \langle u_{\mathbf{k}}(t) u_{\mathbf{k}'}(0) \rangle \\
&\quad \times \frac{\delta}{\delta \mathbf{c}(\tau)} e^{-i\mathbf{k}\cdot\mathbf{c}(\tau)} \frac{\delta}{\delta \mathbf{c}(\tau')} e^{-i\mathbf{k}'\cdot\mathbf{c}(\tau')}
\end{aligned} \tag{J.23}$$

In deriving Eq. (J.23) we have used the fact that the correlation time of the velocity field is much smaller than that for the conformation;

$$(\hat{\delta} - \mathscr{L}_0)^{-1} = \int_0^\infty dt\, e^{t\mathscr{L}_0} \simeq \int_0^\infty dt\, e^{t\mathscr{L}_u} \tag{J.24}$$

where $\mathscr{L}_u$ is the second term of Eq. (J.4). Furthermore, we have used the relation

$$\langle \mathbf{u}_{\mathbf{k}}\, e^{t\mathscr{L}_0} \mathbf{u}_{\mathbf{k}'} \rangle_{\mathbf{u}} = \langle \mathbf{u}_{\mathbf{k}}(t) \mathbf{u}_{\mathbf{k}'}(0) \rangle \tag{J.25}$$

where $\mathbf{u}_{\mathbf{k}}(t)$ is the result of the free motion of the solvent (without polymers). We know from $\langle \mathbf{u}_{\mathbf{k}}(t)\mathbf{u}_{\mathbf{k}'}(0)\rangle = e^{-\eta_0 k^2 t}(1 - \mathbf{k}\mathbf{k}|k^2)\,\delta(\mathbf{k}+\mathbf{k}')$

$$\int_0^\infty dt \langle \mathbf{u}_k(t) \mathbf{u}_{k'}(0) \rangle = \delta(\mathbf{k} + \mathbf{k}') T_{\mathbf{k}} \tag{J.26}$$

where $T_{\mathbf{k}}$ is the Fourier transform of the Oseen tensor. Therefore Eq. (J.23) finally becomes

$$\lambda_0^2 \int d\tau \int d\tau' \frac{\delta}{\delta \mathbf{c}(\tau)} T(\tau, \tau') \frac{\delta}{\delta \mathbf{c}(\tau')} \tag{J.27}$$

Analogously we can transform the $\mathscr{L}_1 - \mathscr{L}_2$ term into

$$\lambda_0^2 \int d\tau \int d\tau' \frac{\delta}{\delta \mathbf{c}(\tau)} T(\tau, \tau') \frac{\delta \mathscr{H}_E}{\delta \mathbf{c}(\tau')} \tag{J.28}$$

Thus combining Eqs. (J.16), (J.27), and (J.28) gives the $\mathscr{L}_{\text{eff}}$, which is exactly the $\mathscr{L}_F^*$ operator [Eq. (315)]. We know from the discussion in Section V.A the second term of Eq. (315) is order ε smaller than its first term. This implies that λ_0 may be considered as the parameter of order $\varepsilon^{1/2}$.

We can systematically improve our calculation of $\mathscr{L}_{\text{eff}}$. To order ε^2 we realize that the Oseen tensor is *insufficient* to describe the effect of the solvent motion. Thus we conclude that the full-diffusion equation is justifiable only to order ε, if we require an internally consistent description of polymer solutions.

The calculation we have given above is essentially the same as is given in Appendix D of Onuki and Kawasaki.[132] The derivation of the Oseen tensor through the reduction of kinetic models was done by Kawasaki[171] for the first time (see also Ref. 172).

References

1. H. Yamakawa, *Modern Theory of Polymer Solutions*, Harper & Row, New York, 1971.
2. S. G. Whittington, "Statistical mechanics of polymer solutions and polymer adsorption," in *Advances in Chemical Physics*, Vol. LII, I. Prigogine and S. A. Rice, eds., Wiley, New York, 1983.
3. D. S. McKenzie, *Phys. Rep.* **27C**, 35 (1976).
4. P. G. de Gennes, *Scaling Concepts in Polymer Physics*, Cornell University Press, Ithaca, 1980.
5. P. G. de Gennes, *Phys. Lett.* **A38**, 339 (1972).
6. J. des Cloizeaux, *J. Phys.* (France) **36**, 281 (1975), see Ref. 177 also.
7. K. G. Wilson and J. Kogut, *Phys. Rep.* **12C**, 75 (1974).
8. S. K. Ma, *Modern Theory of Critical Phenomena*, Benjamin, New York, 1976.
9. G. Toulouse and P. Pfeuty, *Introduction to Renormalization Groups and Critical Phenomena*, Wiley, New York, 1977.
10. M. E. Fisher, "Scaling, Universality and Renormalization Group Theory", in *Lecture Notes in Physics*, Vol. 186, F. J. W. Hahne, ed., Springer, New York, 1983.
11. B. Shapiro, *J. Phys.* **C11**, 2829 (1978).
12. M. Napiorkowski, E. M. Hauge, and P. C. Hemmer, *Phys. Lett.* **72A**, 193 (1979).
13. F. Family, *J. Phys.* **A13**, L325 (1980).
14. J. Malakis, *Physica* **104A**, 427 (1980).
15. S. Redner and P. J. Reynolds, *J. Phys.* **A14**, 2679 (1981).
16. S. L. A. de Queiroz and C. M. Chavez, *Z. Physik* **B40**, 99 (1980).
17. B. Derrida, *J. Phys.* **A14**, 15 (1981).

18. A. Baumgärtner, *J. Phys.* **A13**, L39 (1980).

19a. H. E. Stanley, P. J. Reynolds, S. Redner, and F. Family, "Position-space renormalization group for models of linear polymers, branched-polymers, and gels," in *Real-Space Renormalization*, T. W. Burkhardt and J. M. J. van Leeuwen, eds., Springer, New York, 1982.

19b. F. Family, "Polymer Statistics and Universality: Principles and Applications of Cluster Renormalization," in *Random Walks and Their Applications in the Physical and Biological Sciences*, M. F. Schlessinger and B. J. West, eds., AIP Proceedings 109 (1984).

20. P. J. Flory, *Principles of Polymer Chemistry*, Cornell University Press, Ithaca, 1953.

21. D. Brydges and T. Spencer, "Self-avoiding walk in 5 or more dimensions," preprint (1984).

22. See, for example, J. O. Hirschfelder, C. F. Curtiss, and R. B. Bird, *Molecular Theory of Gases and Liquids*, Wiley, New York, 1954; T. Kihara, *Molecular Forces*, Wiley, New York, 1978.

23. S. F. Edwards, *Proc. Phys. Soc.* **85**, 613 (1965).

24. R. P. Feynman, *Statistical Mechanics: a Set of Lectures*, Benjamin, Reading, MA, 1972, Chapter 3.

25. I. M. Gel'fand and A. M. Yaglom, *J. Math. Phys.* **1**, 48 (1960).

26. K. Itô, *Proc. Fourth Berkeley Symposium on Mathematical Statistics and Probability*, vol. II 227 (1965).

27. R. P. Feynman, *Rev. Mod. Phys.* **20**, 367 (1948).

28. M. Kac, *Trans. Am. Math. Soc.* **65**, 1 (1949).

29. See, for example, J. Glimm and A. Jaffe, *Quantum Physics, a Functional Integral Point of View*, Springer, New York, 1981.

30. E. Teramoto, *Busseiron-Kenkyu* **39**, 1 (1951).

31. R. J. Rubin, *J. Chem. Phys.* **20**, 1940 (1952).

32. F. Bueche, *J. Chem. Phys.* **21**, 205 (1953).

33. B. H. Zimm, W. H. Stockmayer, and M. Fixman, *J. Chem. Phys.* **21**, 1716 (1953).

34. T. B. Grimley, *J. Chem. Phys.* **21**, 185 (1953); H. M. James, *J. Chem. Phys.* **21**, 1628 (1953); N. Saitô, *J. Phys. Soc. Jpn.* **9**, 780 (1954).

35. M. Muthukumar and B. Nickel, *J. Chem. Phys.* **80**, 5839 (1984).

36. H. Yamakawa and G. Tanaka, *J. Chem. Phys.* **47**, 3991 (1967).

37. C. Domb, *Adv. Phys.* **19**, 339 (1970).

38. Y. Chikahisa, *J. Chem. Phys.* **52**, 206 (1970).

39. M. Gordon, S. Ross-Murphy, and H. Suzuki, *Europ. Polym. J.* **12**, 733 (1976).

40. S. F. Edwards, *J. Phys.* **A8**, 1171 (1975).

41. Y. Oono, *J. Phys. Soc. Jpn.* **39**, 25 (1975).

42. Y. Oono, *J. Phys. Soc. Jpn.* **41**, 787 (1976).

43. G. Tanaka, *Macromolecules* **13**, 1513 (1980).

44. K. Symanzik, "Euclidean Quantum Field Theory," in *Local Quantum Theory*, R. Jost, ed., Academic Press, New York, 1969.

45. M. J. Westwater, *Commun. Math. Phys.* **72**, 131 (1980).

46. A. Bovier, G. Felder, and J. Frölich, *Nuclear Phys.* **B230**, 119 (1984).

47. J. Dimock, *Commun. Math. Phys.* **35**, 347 (1974).

48. J. Magnen and R. Sénéor, *Commun. Math. Phys.* **56**, 237 (1977).

49. See, for example, J. Zinn-Justin, *Phys. Rep.* **70**, 102 (1981); M. J. Westwater, *Commun. Math. Phys.* **84**, 459 (1982).
50. See, for example, P. J. Flory, *Statistical Mechanics of Chain Molecules*, Wiley, New York, 1969.
51. P. J. Flory, *J. Chem. Phys.* **17**, 303 (1949).
52. M. Fixman, *J. Chem. Phys.* **23**, 1656 (1955); *J. Chem. Phys.* **36**, 306, 3123 (1962).
53. M. Kurata, W. H. Stockmayer, and A. Roig, *J. Chem. Phys.* **33**, 151 (1960).
54. O. B. Ptitsyn, *Vysokomol. Soed.* **3**, 1673 (1961).
55. H. Reiss, *J. Chem. Phys.* **47**, 186 (1967).
56. P. G. de Gennes, *Rep. Progr. Phys.* **32**, 187 (1969).
57. K. F. Freed, *Adv. Chem. Phys.* **22**, 1 (1972).
58. A. Kyselka, *J. Phys.* **A7**, 315 (1974).
59. S. F. Edwards, *Proc. Phys. Soc.* **88**, 265 (1966).
60. Y. Oono, *J. Phys. Soc. Jpn.* **41**, 228 (1976).
61. M. A. Moore, *J. Phys.* **A10**, 305 (1977).
62. S. Redner and K. Kang, *Phys. Rev. Lett.* **51**, 1729 (1983).
63. B. Nienhuis, *Phys. Rev. Lett.* **49**, 1062 (1982).
64. A. Malakis, *J. Phys.* **A9**, 1283 (1976).
65. Y. Oono, *Kobunshi* **28**, 781 (1979) (relevant part is translated into English in Y. Oono and K. F. Freed, *J. Chem. Phys.* **75**, 993 (1981).)
66. T. Oyama and K. Shiokawa, preliminary report given at the polymer workshop at Kyoto (1983).
67. Y. Shapir and Y. Oono, *J. Phys.* **A17**, L39 (1984).
68. Y. Oono, *J. Phys. Soc. Jpn.* **40**, 917 (1975); *J. Phys. Soc. Jpn.* **41**, 2095 (1976). It is conceivable that if we use properly renormalized interaction parameters the mean-field or interdimensional scaling argument can give a very accurate exponent ν. Indeed $\nu = 2/(1 + d)$ is very close to the tricritical exponent (see A. Kholodenko and K. F. Freed, *J. Phys.* **A17**, L191 (1984)). See also J. Issacson and T. C. Lubensky, *J. Phys. (France) Lett.* **41**, L469 (1980); T. C. Lubensky and J. Vannimenus, *J. Phys. (France) Lett.* **43**, L377 (1982); S. Redner and A. Coniglio, *J. Phys.* **A15**, 2273 (1982).
69. Y. Imry, *Phys. Lett.* **41A**, 381 (1972).
70. P. G. de Gennes, *J. Phys. Lett.* **36**, L55 (1975).
71. Y. Oono and T. Oyama, *J. Phys. Soc. Jpn.* **44**, 301 (1978).
72. A. R. Khokhlov, *Polymer* **19**, 1387 (1978).
73. B. Duplantier, *J. Phys. (France) Lett.* **41**, L409 (1980); B. Duplantier, *J. Phys. (France)* **43**, 991 (1982).
74. A. Kholodenko and K. F. Freed, *J. Chem. Phys.* **80**, 900 (1984).
75. See, for example, J. A. Barker and D. Henderson, *Rev. Mod. Phys.* **48**, 587 (1976).
76. L. P. Kadanoff, *Physics* **2**, 263 (1966).
77. B. Mandelbrot, *The Fractal Geometry of Nature*, Freeman, San Francisco (1982).
78. K. G. Wilson, *Phys. Rev.* **B4**, 3174 and 3184 (1971).
79. F. Spitzer, *Principles of Random Walk*, Van Nostrand, Princeton, 1964.
80. See, for example, W. Feller, *An Introduction to Probability Theory and Its Applications*, Vol. II, Wiley, New York, 1971. Chapters VIII and IX.

81. S. Chandrasekhar, *Rev. Mod. Phys.* **15**, 1 (1945).

82. F. Knight, *Trans. Am. Math. Soc.* **103**, 218 (1962).

83. See, for example, Feller,[80] p. 253.

84. K. Itô and H. P. McKean, Jr., *Diffusion Processes and their Sample Paths*, Springer, New York, 1974.

85. See, for example, E. W. Montroll and G. H. Weiss, *J. Math. Phys.* **6**, 167 (1965).

86a. A. Dvoretsky, P. Erdös, S. Katutani, and S. J. Taylor, *Proc. Camb. Phil. Soc.* **53**, 856 (1957).

86b. A. Dvoretsky, P. Erdös, and S. Kakutani, *Acta Scientarum Matematicarum* (Szeged) **12**, 75 (1950).

87. See, for example, N. N. Bogoliubov and D. V. Shirkov, *Introduction to the Theory of Quantized Fields* (Interscience, NY 1959); D. J. Amit, *Field Theory, the Renormalization Group, and Critical Phenomena*, McGraw-Hill, New York, 1978.

88. K. G. Wilson and M. E. Fisher, *Phys. Rev. Lett.* **38**, 240 (1972).

89. See, for example, F. John, *Partial Differential Equations*, Springer, New York, 1982.

90. E. Brezin, J. C. Le Guillon, and J. Zinn-Justin, in *Phase Transitions and Critical Phenomena*, Vol. 6, C. Domb and M. S. Green, eds., Academic Press, New York, 1976.

91a. Y. Oono, T. Ohta, and K. F. Freed, *J. Chem. Phys.* **74**, 6458 (1981).

91b. A. Kholodenko and K. F. Freed, *J. Chem. Phys.* **78**, 7390 (1983).

92. P. Baldwin, private communication.

93. M. E. Fisher, *J. Chem. Phys.* **44**, 616 (1966).

94. J. des Cloizeaux, *Phys. Rev.* **A10**, 1665 (1974).

95. T. A. Witten, Jr., and L. Schäfer, *J. Phys.* **A11**, 1843 (1978).

96. T. A. Witten, Jr., and L. Schäfer, *J. Chem. Phys.* **74**, 2582 (1981).

97. C. Domb, J. Gillis, and G. Wilmers, *Proc. Phys. Soc.* **85**, 625 (1965).

98. Y. Oono and K. F. Freed, *J. Phys.* **A15**, 1931 (1982); see for a bit different approaches I. D. Lawrie, *J. Phys.* **A9**, 961 (1976); D. J. Elderfield, *J. Phys.* **A11**, 2483; *J. Phys.* **C13**, 5883.

99. B. Bagchi and Y. Oono, *J. Chem. Phys.* **78**, 2044 (1983).

100. C. Domb and F. T. Hioe, *J. Phys.* **C3**, 2223 (1970).

101. F. T. Wall, *J. Chem. Phys.* **63**, 3713 (1975).

102. F. T. Wall and W. A. Seitz, *J. Chem. Phys.* **70**, 1860 (1979).

103. M. Daoud, J. P. Cotton, B. Farnoux, G. Jannink, G. Sarma, H. Benoit, R. Duplessix, C. Picot, and P. G. de Gennes, Macromolecules **8**, 804 (1975).

104. T. Ohta, private communication (1980); T. Ohta and Y. Oono, *Phys. Lett.* **A89**, 460 (1982).

105. For another nonfield theoretic approach, see J. des Cloizeaux, *J. Phys.* (*France*) **41**, 749, 769 (1980).

106. M. Kac, in *Applied Probability*, L. A. MacColl, ed., McGraw-Hill, New York, 1971.

107. See, for example, C. Itzykson and J. B. Zuber, *Quantum Field Theory*, McGraw-Hill, New York, 1980, Chapters 4 and 9.

108. P. Wiltzius, H. R. Haller, D. S. Cannell, and D. W. Schaefer, *Phys. Rev. Lett.* **51**, 1183 (1983).

109. M. D. Donsker, "On function space integrals," in *Analysis in Function Space*, W. T. Martin and I. E. Segal, eds., MIT Press, Boston, 1963.

110. H.-H. Kuo, "Donsker's Delta Function as a Generalized Brownian Functional and Its

Application," in *Theory and Application of Random Fields*, G. Kallianpur, ed. (Lecture Notes in Control and Information Sciences 49) Springer, New York, 1983.

111. I. Kubo, "Ito Formula for Generalized Brownian Functionals," in *Theory and Application of Random Fields*, G. Kallianpur, ed., (Lecture Notes on Control and Information Sciences 49) Springer, New York, 1983.
112. T. Ohta and A. Nakanishi, *J. Phys.* **A16**, 4155 (1983).
113. Y. Oono, *Phys. Rev.*, **A. 30**, 986 (1984).
114. Y. Oono and M. Kohmoto, *J. Chem. Phys.* **78**, 520 (1983).
115. Y. Oono, *J. Chem. Phys.* **79**, 4629 (1983).
116. J. G. Kirkwood and J. Riseman, *J. Chem. Phys.* **16**, 565 (1948).

117a. R. B. Bird, O. Hassager, R. C. Armstrong, and C. F. Curtiss, *Dynamics of Polymeric Liquids*, Vol. 2, Wiley, New York, 1976.

117b. M. Bixon, *Ann. Rev. Phys. Chem.* **27**, 65 (1976).

118. H. A. Kramers, *J. Chem. Phys.* **14**, 415 (1946).

119a. K. F. Freed, "Polymer Dynamics and the Hydrodynamics of Polymer Solutions," in *Progress in Liquid Physics*, C. A. Croxton, ed., Wiley, New York, 1978.

119b. M. Muthukumar, *J. Phys.* **A14**, 2129 (1981).

120. A. Wilemski and G. Tanaka, *Macromolecules* **14**, 1531 (1981).
121. C. M. Pyun and M. Fixman, *J. Chem. Phys.* **42**, 3838 (1965); *J. Chem. Phys.* **44**, 2107 (1965).
122. M. Fixman, *J. Chem. Phys.* **42**, 3831 (1975).
123. M. Bixon and R. Zwanzig, *J. Chem. Phys.* **68**, 1890 (1978).
124. D. Jasnow and M. A. Moore, *J. Phys. (France)* **38**, L467 (1978).
125. G. F. Al-Noaimi, G. C. Martinez-Mekler, and C. A. Wilson, *J. Phys. (France)* **39**, L373 (1978).
126. Y. Oono and T. Ohta, *Phys. Lett.* **85A**, 480 (1981).
127. Y. Oono and M. Kohmoto, *Phys. Rev. Lett.* **49**, 1397 (1982).
128. S. F. Edwards and K. F. Freed, *J. Chem. Phys.* **61**, 1189 (1974).
129. J. G. Kirkwood, *J. Polym. Sci.* **12**, 1 (1954).
130. Y. Oono and K. F. Freed, *J. Chem. Phys.* **75**, 1009 (1981).
131. K. Kawasaki and J. Gunton, in *Progress in Liquid Physics*, C. A. Croxton, ed., Wiley, New York, 1978.
132. A. Onuki and K. Kawasaki, *Ann. Phys.* **121**, 456 (1979).
133. Y. Shiwa, private communication (1982); *Phys. Lett.* **A103**, 211 (1984).
134. Y. Shiwa and K. Kawasaki, *J. Phys.* **C15**, 5345 (1982).
135. A. Z. Akcasu and M. Gurol, *J. Polym. Sci.* **14**, 1 (1977); M. Benmouna and A. Z. Akcasu, *Macromolecules* **11**, 1187 (1978); see also, G. Tanaka and W. H. Stockmayer, *Proc. NAS* **79**, 6401 (1982).
136. A. Lee, P. W. Baldwin, and Y. Oono, *Phys. Rev.* **A 30**, 968 (1984).
137. L. Mandelkern and P. J. Flory, *J. Chem. Phys.* **20**, 212 (1952); H. A. Scheraga and L. Mandelkern, *J. Am. Chem. Soc.* **75**, 179 (1953).
138. G. Weill and J. des Cloizeaux, *J. Phys. (France)* **40**, 99 (1979).
139. A. Z. Akcasu, M. Benmouna, and S. Alkahafaji, *Macromolecules* **14**, 147 (1981).
140. W. H. Stockmayer and A. C. Albrecht, *J. Polym. Sci.* **32**, 215 (1958).

141. M. J. Pritchard and D. Carolin, *Macromolecules* **14**, 424 (1981); P. Vidakovic and F. Rondelez, *Macromolecules* **16**, 253 (1983).

142. C. C. Han and A. Z. Akcasu, *Macromolecules* **14**, 1080 (1981).

143. J. François, T. Schwartz, and G. Weill, *Macromolecules* **13**, 564 (1980); H. Fujita and T. Norisuye, *Macromolecules* **14**, 1130 (1981).

144. T. Witten, Jr., *J. Chem. Phys.* **76**, 3300 (1982).

145. T. Ohta, Y. Oono, and K. F. Freed, *Phys. Rev.* **A25**, 2801 (1982).

146. I. Noda, M. Imai, T. Kitano, and M. Nagasawa, *Macromolecules* **16**, 427 (1983).

147. A. Nakanishi and T. Ohta, *J. Phys.* **A18**, 217 (1985); T. Ohta, preprint.

148. L. Schäfer, *Macromolecules* **15**, 652 (1982); L. Schäfer, *Macromolecules* **17**, 1357 (1984).

149. P. G. de Gennes, *J. Polym. Sci. Polym. Symposium* **61**, 313 (1977).

150. M. Daoud and F. Family, *J. Phys. (France)* **45**, 151 (1984).

151. M. Lipkin, Y. Oono, and K. F. Freed, *Macromolecules* **14**, 1270 (1981).

152. J. J. Prentice, *J. Chem. Phys.* **76**, 1574 (1982).

153. A. Miyake and K. F. Freed, *Macromolecules* **16**, 1228 (1983).

154. K. F. Freed, *J. Chem. Phys.* **79**, 3121 (1983); E. S. Nikomarov, *Physica* **120A**, 647 (1983).

155. See, for example, P. S. Virk, *AIChE Journal* **21**, 625 (1975).

156. K. Yamazaki and T. Ohta, *J. Phys.* **A15**, 287 (1982).

157. A. Baumgärtner, *Polymer* **22**, 1308 (1981).

158. T. Norisuye, K. Kawahara, A. Teramoto, and H. Fujita, *J. Chem. Phys.* **49**, 4330 (1968); K. Kawahara, T. Norisuye, and H. Fujita, *J. Chem. Phys.* **49**, 4339 (1968).

159. G. Tanaka, S. Imai, and H. Yamakawa, *J. Chem. Phys.* **52**, 2639 (1970).

160. Y. Miyaki, Y. Finaga, H. Fujita, and M. Fukuda, *Macromolecules* **13**, 588 (1980).

161. T. Nose and B. Chu, *Macromolecules* **12**, 1122 (1979).

162. Y. Miyaki, Y. Einaga, T. Hirosye, and H. Fujita, *Macromolecules* **10**, 1356 (1977).

163. T. Hirosye, Y. Einaga, and H. Fujita, *Polymer J.* **11**, 819 (1979).

164. T. Kato, K. Miyaso, I. Noda, T. Fujimoto, and M. Nagasawa, *Macromolecules* **3**, 777 (1970); I. Noda, K. Mizutani, T. Kato, T. Fujimoto, and M. Nagasawa, *Macromolecules* **3**, 787 (1970).

165. Y. Miyaki, Y. Einaga, and M. Fujita, *Macromolecules* **11**, 1180 (1978).

166. H. U. ter Meer, W. Burchard, and W. Wunderlich, *Colloid. Poly. Sci.* **258**, 675 (1980).

167. B. Zimm, *Macromolecules* **13**, 592 (1980).

168. J. G. de la Torre, A. Jimenéz, and J. J. Freire, *Macromolecules* **15**, 148 (1982).

169. M. Schmidt and W. Burchard, *Macromolecules* **14**, 210 (1981).

170. C. M. Guttman, F. I. McCrackin, and C. C. Han, *Macromolecules* **15**, 1205 (1982).

171. K. Kawasaki, in *Synergetics*, H. Haken, ed., Teubner, Stuttgart, 1973.

172. R. Kapral, D. Ng, and S. G. Whittington, *J. Chem. Phys.* **64**, 538 (1976); R. S. Adler, *J. Chem. Phys.* **69**, 2849 (1978).

173. M. Fixman, preprint; *J. Chem. Phys.* **78**, 1594 (1983); *Macromolecules* **14**, 1710 (1981).

174. I. Majid, Z. V. Djordjevic, and H. E. Stanley, *Phys. Rev. Lett.* **51**, 1282 (1983); J. C. Le Guillou and J. Zinn-Justin, *Phys. Rev. Lett.* **39**, 95 (1977).

175a. Y. Tsunashima, N. Nemoto, and M. Kurata, *Macromolecules* **16**, 1184 (1983).

175b. N. Nemoto, Y. Makita, Y. Tsunashima, and M. Kurata, *Macromolecules* **17**, 425 (1984).

176. I. Noda, N. Kato, T. Kitano, and H. Nagasawa, *Macromolecules* **14**, 668 (1981).

177. A very important progress is being made by P. J. Gujrati, *Phys. Rev. Lett.* **53**, 2453 (1984); *Phys. Rev. B* (April 1, 1985) and preprint (1985).

Glossary of Symbols

A, α	A_2	Osmotic second virial coefficient
	a	Cutoff contour length
	α	$R^2/2N_0$ [Eq. (73)]
	α_R	Expansion factor for the mean-square end-to-end distance
	α_G	Expansion factor for the mean-square radius of gyration
B, β	B	Suffix B denotes bare quantities
	β, β_ξ	"Beta" functions [e.g., Eqs. (45), (98) and (291)]
C	C	Dimension of the contour variable
	C	Conformation
	c	Polymer number density
	$\mathbf{c}(\tau)$	Position vector of the τth monomer unit
	$\tilde{\mathbf{c}}(\tau)$	Position vector of the τth monomer unit w.r.t. the center of mass of the chain
D, δ	$\mathscr{D}$	Functional measure [Appendix B]
	d	Spatial dimensionality
	Δ	Laplacian
	δ	Scaling variable describing the draining effect [Eq. (311)]
	$\delta/\delta\cdot$	Functional derivative
E, ε	$\mathbf{e}_x, \mathbf{e}_y, \ldots$	Unit vector in the $x, y, \ldots$ direction
	ε	$4 - d$
F	$\mathbf{F}(\tau)$	Force exerted on solvent by the τth monomer unit [Section V.A.1]
	f	Translational friction constant [Eq. (254)]
G, γ	G_0	Gaussian distribution [Eq. (70)]
	g	Shear strength [Eqs. (252) and (253)]
	γ	Exponent for the total number of conformations [Eq. (111)]
	$\hat{\gamma}$	Euler's constant ($\simeq 0.577$)
	γ_Q	Coefficients in the renormalization-group equation [e.g., Eqs. (46) and (47)]
	γ_Q^*	γ_Q at the fixed point [Eqs. (60) and (103)]
H, η	$\mathscr{H}_E$	The Edwards Hamiltonian [Eq. (11)]
	η_0	Viscosity of the solvent
	$[\eta]$	Intrinsic viscosity
K, ξ	k_B	Boltzmann constant
	ξ, ξ_0	Reduced friction coefficients [Eqs. (261) and (283)]
	ξ^*	ξ at fixed points [Section V.A.5]
L, λ	L	Phenomenological length scale [Section III.B]

	$\mathscr{L}, \mathscr{L}.,$	Fokker-Planck operators [Sections V.B and V.C]
	λ	Eigenvalues of the Fokker-Planck operator [Section V.C]
M, μ	M	Molecular weight
	M_n	Number-averaged molecular weight
	M_w	Weight-averaged molecular weight
	m, m_0	Order of FSAW [Eq. (170)]
	μ	Polydispersity parameter = M_n/M_w [Eq. (209)]
N, ν	N, N_0	Total contour length
	N_A	Avogadro's constant
	$\mathscr{N}_0$	Normalization constant [Eq. (66)]
	n	Total number of polymers
	ν	Exponent for the radius
	ν_{SAW}	The value of ν in the good-solvent limit
	ν_G	Effective exponent for the static radius [Eq. (379)]
	ν_H	Effective exponent for the hydrodynamic radius [Eq. (380)]
0	0	Suffix 0 denotes bare quantities, or unperturbed quantities
P, π, ψ	P, P_0	Molecular weight distribution function in Section IV, and (time-dependent) distribution functions in Section V
	π	Osmotic pressure
	ψ	Kac's random field [Section V.C]
Q	Q	Representative quantity in Section III
R, ρ	$\mathbf{R}$	End-to-end vector
	$\langle R^2 \rangle$	Mean-square end-to-end distance
	$\langle R_G^2 \rangle$	Mean-square radius of gyration
	ρ	Monomer number density (field) [Eqs. (9) and (232)]
S, σ	S_0	Static scattering function for the Gaussian chain [Eq. (199)]
	$S(\mathbf{k})$	Scattering function [Eq. (237)]
	σ	Contour variable
T, τ	T	Oseen tensor
	t	Dilation parameter [Eq. (55)]
	τ	Contour variable
	Θ	Suffix Θ denotes the theta state
	θ	Relative order m/N [Eq. (172)]
	θ	Homotopy parameter in Section IV.D
U	$U_{X/Y}$	Universal ratios [Section VI.A]

	u, u_0	Dimensionless excluded-volume parameter [Eqs. (34) and (35)]
	u^*	u at the fixed point
	$\mathbf{u}(\tau)$	Velocity of the τth monomer unit
V	V	Volume of the system
	v, v_0	Excluded-volume parameter
	$\mathbf{v}, \tilde{\mathbf{v}},$	Solvent velocity field
W, Ω	w_0	Three-body interaction parameter
	w	$u/(u^* - u)$ [Eq. (156)]
	Ω	Initial decay rate [Section V.C.2]
X	x	Scaling variable [Eqs. (115) and (133)]
	$X, \hat{X}$	Scaling variable for semidilute solutions [Eqs. (212) and (227)]
Z, ζ	Z	Partition function
	Z	Scaling variable for dilute solutions [Eq. (167)]
	Z_Q	Renormalization constants [Section III.B] for Q
	z	Conventional z-parameter in the two parameter theory
	$\hat{z}$	Intermediate scaling variable [Eq. (307)]
	ζ	True scaling variable for dilute solutions [Eq. (147)]
	ζ_0	Bare translational friction constant for a chain unit
	$\hat{\zeta}$	Renormalized ζ_0 [Eq. (282)]

AUTHOR INDEX

Numbers in parentheses are reference numbers and indicate that the author's work is referred to although his name is not mentioned in the text. Numbers in *italic* show the pages on which the complete references are listed.

SUBJECT INDEX